*The Social Ecology of Infectious Diseases*

# The Social Ecology of Infectious Diseases

*Edited by*
Kenneth H. Mayer and H.F. Pizer

AMSTERDAM • BOSTON • HEIDELBERG • LONDON • NEW YORK • OXFORD
PARIS • SAN DIEGO • SAN FRANCISCO • SINGAPORE • SYDNEY • TOKYO
Academic Press is an imprint of Elsevier

Academic Press is an imprint of Elsevier
84 Theobald's Road, London WC1X 8RR, UK
30 Corporate Drive, Suite 400, Burlington, MA 01803, USA
525 B Street, Suite 1900, San Diago, California 92101-4495, USA

First edition 2008



**Library of Congress Cataloging-in-Publication Data**
A catalog record for this book is available from the Library of Congress

**British Library Cataloguing-in-Publication Data**
A catalogue record for this book is available from the British Library

ISBN: 978-0-12-370466-5

# Dedications

Kenneth Mayer dedicates this book to the memory of Betty Mayer, who passed away on 20 December 2005, and of Paul Mayer, who passed away on 8 August 1995. My parents taught me to be curious about the world around me, to enjoy the many wonderful gifts that living provides, and to be aware of the suffering of others. I am also very appreciative of my sister, Arlene Shainker, who acts as my best advisor and closest friend, her wonderful husband, Stuart, and my delightful niece and nephew, Haley and Danny. I am grateful for close family support that has enabled me to travel far and wide to learn more about the universe, without feeling adrift.

H.F. Pizer dedicates this book to Chris, Katie and Bruce.

# Dedications

Kenneth Mayer dedicates this book to the memory of Betty Mayer, who passed away on 20 December 2003, and of Paul Mayer, who passed away on 8 August 1994. My parents taught me to be curious about the world around me, to enjoy the many wonderful gifts that living provides, and to be aware of the suffering of others. I am also very appreciative of my sister, Arlene Shanker, who acts as my best advisor and closest friend, her wonderful husband, Stuart, and my delightful niece and nephew, Haley and Sammy. I am grateful for close family support that has enabled me to travel far and wide to learn more about the universe, without feeling adrift.

H.F. Pizer dedicates this book to Chris, Katie and Drew.

# Contents

# About the Editors

**Kenneth H. Mayer, MD**, is Professor of Medicine and Community Health at Brown University, Director of the Brown University AIDS Program, and Attending Physician in the Infectious Disease Division of the Miriam Hospital in Providence, Rhode Island. In addition, he is an Adjunct Professor at Harvard University's School of Public Health, and Medical Research Director at Boston's Fenway Community Health Center, where since 1983 he has conducted studies of the natural history and transmission of HIV. In the early 1980s, as a research fellow studying infectious diseases at Harvard Medical School and working at Fenway Community Health Center, Dr Mayer was one of the first clinical researchers in New England to care for patients living with AIDS.

In 1983, Dr Mayer co-authored (with H.F. Pizer) *The AIDS Fact Book*, the first book about AIDS written for the general public. In 1984 he began one of the first studies of the natural history of HIV infection, and was subsequently funded by the NIH and CDC to study the dynamics of the sexual transmission of HIV, the natural history of HIV in women, and new approaches to HIV prevention – ranging from vaccines (HIVNET, HVTN) to microbicides, behavioral and other interventions (HPTN). In the late 1980s he initiated the first community-based clinical trials for people living with HIV/AIDS in New England, and helped amFAR to develop its national Community-Based Clinical Trials Network (CBCTN). He was subsequently elected to the Board of Directors of amFAR and is currently a member of amFAR's Program Board.

Dr Mayer is the Director of the Brown and Tufts Universities' Fogarty (NIH) AIDS International Research and Training Program, which has trained almost 100 laboratory and clinical investigators from East Asia. He has worked increasingly in India and has participated in many regional conferences on biological and behavioral approaches to prevention research, and the development of community-based clinical research activities in Asia. He is currently a consultant with the Clinton Foundation's HIV/AIDS Initiative, designed to increase the capacity to provide comprehensive care for people living with HIV across the globe. Dr Mayer co-edited (with H.F. Pizer) *The Emergence of AIDS: Impact on Immunology, Microbiology, and Public Health*, published in 2000 by the American Public Health Association Press, and *The AIDS Pandemic: Impact on Science and Society*, published in 2005 by the Academic Press (Elsevier).

Dr Mayer has served on the Data Safety and Monitoring Board of the NIH's AIDS Clinical Trials Group and sits on several editorial boards of scientific publications, including *Clinical Infectious Disease* and *AIDS Patient Care and STDs*. He has co-authored more than 350 articles, chapters, and other publications on AIDS and related infectious disease topics, and is a frequent lecturer and presenter at national and international conferences and symposia. He is a former national board member of the HIV Medicine Association and the Gay and Lesbian Medical Association. He has received awards of recognition from the Governor of Massachusetts, the Rhode Island Department of Health, the AIDS Action Committee of Massachusetts, AIDS Project Rhode Island, amFAR, and the Greater Boston Business Council. In 2001, he and Dr Judith Bradford were named Co-Directors of The Fenway Institute, which is designed to conduct population-based research, develop professional and community educational programs, and disseminate information related to best practices and model clinical programs relevant to the global health needs of lesbian, gay, bisexual, and transgendered (LGBT) individuals and communities. In collaboration with Drs Harvey Makadon and Jenny Potter, he is editing the first comprehensive medical textbook of LGBT health for primary care medical providers, to be published by the American College of Physicians in 2007.

Dr Mayer received his BA in Psychology from the University of Pennsylvania, and his MD from Northwestern University Medical School. He completed his residency and internship in Internal Medicine at Boston's Beth Israel Hospital, while also completing a clinical fellowship in Medicine at Harvard Medical School. From 1980 to 1983 he completed an Infectious Diseases fellowship at Brigham and Women's Hospital and Harvard Medical School.

**H.F. Pizer, BA, PA** is a medical writer, health-care consultant, and physician assistant. He has written and edited 14 books and numerous articles about health and medicine. With Kenneth Mayer he co-authored the first book about AIDS for the general public, *The AIDS Fact Book* (Bantam Books, 1983), and co-edited *The Emergence of AIDS: Impact on Immunology, Microbiology, and Public Health* (American Public Health Association Press, 2000) and *The AIDS Pandemic: Impact on Science and Society* (Academic Press, 2005). With Chris Beyrer, he recently co-edited *Public Health and Human Rights: Evidence-Based Approaches*, to be published in 2007 by Johns Hopkins University Press. His other works cover a variety of subjects in health and medicine, including the first books for the general public on organ transplants (*Organ Transplants: A Patient's Guide* with the Massachusetts General Organ Transplant Teams; Harvard University Press, 1991) and stroke (*The Stroke Fact Book*, with Conn Foley, Bantam Books, 1985; Courage Press and the American Heart Association) and, in women's health, on family planning (*The New Birth Control Program*, with Christine Garfink, RN, Bolder Books, New York, 1977; Bantam Books, New York, 1979), parenting (*The Post Partum Book*, with Christine Garfink, RN, Grove Press, New York, 1979), miscarriage (*Coping With A Miscarriage*, with Christine O'Brien Palinski, The Dial Press, 1980), and artificial insemination (*Having a Baby Without A Man*, with Susan Robinson, MD, Simon & Schuster, 1985). He also co-authored *Confronting Breast Cancer* (with Sigmund Weitzman and Irene Kuter, Random House, 1987). From 1984 to 1994 he was founder and President of New England Medical Claims Analysts, a health-care consultancy, and during that time wrote and lectured on health-care cost containment. Presently he is Co-founder and Principal of Health Care Strategies, a consulting firm in Cambridge, Massachusetts, that provides program evaluation and management consulting services to community health-care providers, health-care systems, and social service organizations. Health Care Strategies specializes in working with clients to help them design and implement low-cost, practical systems for program evaluation. The goal is for providers and community-based organizations to be able to document program performance and conduct regular internal monitoring and quality improvement He is former President of the Massachusetts Association of Physician Assistants. His books have been published in English and in translation for overseas distribution by trade, mass market and academic publishers.

# Notes on Contributors

**Joseph J. Amon PhD MSPH** is the director of the HIV/AIDS Program at Human Rights Watch based in New York. Before joining Human Rights Watch in 2005, Dr Amon worked for more than 15 years conducting research, designing programs, and evaluating interventions related to AIDS in Africa, the Caribbean region, and Eastern Europe. He has worked for the US Centers for Disease Control and Prevention, the Walter Reed Army Institute of Research, and a wide range of non-governmental organizations working on AIDS prevention and treatment. In addition to his work on HIV/AIDS and human rights, Dr Amon has conducted research on the molecular biology and epidemiology of malaria, hepatitis and AIDS-related opportunistic infections.

**Andrew W. Artenstein MD FACP FIDSA** is the Chief, Department of Medicine, Memorial Hospital of Rhode Island; the founder and Director of the Center for Biodefense and Emerging Pathogens; and an Associate Professor of Medicine and Community Health at Brown Medical School. Prior to coming to Brown, he served as a Principal Scientist and Infectious Disease Officer and Head of the Section of Protective Immunity in the Division of Retrovirology at the Walter Reed Army Institute of Research, dividing his time between his lab in Washington DC and field sites in Thailand. His primary research efforts involve anthrax toxin pathogenesis and inhibition, biodefense vaccines, biomedical engineering approaches to infectious disease diagnostics, and bringing biodefense to vulnerable populations within communities. He serves on the editorial advisory board of the *Journal of Infectious Diseases*, and as a consultant on numerous governmental and academic advisory panels regarding biodefense and emerging infectious diseases.

**Susan H. Baker MS Management** has worked in the area of overseas development and emergency relief for the past 25 years and specializes in community development and institution building, with an emphasis on practical management and operational processes. Ms Baker has lived and worked in numerous countries in sub-Saharan Africa, as well as in the former Soviet Union. Her titles and responsibilities in the field have included community development consultant, assistant country director, country director, and emergency coordinator. Ms Baker has been an integral part of the international response to a number of emergencies, including those involving drought and famine, floods, genocide, and civil war. Ms Baker is Co-founder and Associate Director of the Center for Global Health at New York University School of Medicine, where she co-designed an integrated global health curriculum. The focus of her academic interests are the planning and coordination of outside aid interventions and the education and training of international and local field staff working in development and emergency relief. She holds a Master of Science degree in International Public Sector Management from the Robert F. Wagner School at New York University.

**Stefan David Baral MD MSc MPH** is a Community Medicine Resident at the University of British Columbia, and has just completed an MPH and MBA at the Johns Hopkins School of Public Health. Stefan completed his undergraduate degree specializing in immunology and microbiology from McGill University in Montreal, Quebec. He then went on to graduate school at McMaster University in Hamilton, Ontario, where he studied novel vaccination strategies and was part of the Canadian Network for Vaccines and Immunotherapeutics (CANVAC). Stefan completed his medical school at Queen's University in Kingston, Ontario, with a focus on international and public health issues. Dr Baral has worked in clinical settings ranging from inner-city methadone clinics in Winnipeg, Manitoba, to large tertiary care institutions in Kampala and Santiago.

**John G. Bartlett MD** is Professor of Medicine in the Division of Infectious Diseases at the Johns Hopkins University School of Medicine, Baltimore, Maryland. He served as Chief of the Infectious Disease Division at the School for 26 years, stepping down in June of 2006. Dr Bartlett received his undergraduate degree at Dartmouth College, Hanover, New Hampshire, and his medical degree at Upstate Medical Center, Syracuse, New York. He trained in internal medicine at the Peter Bent Brigham Hospital, Boston, Massachusetts, and the University of Alabama, Birmingham, and he completed his fellowship training in Infectious Diseases at the University of California, Los Angeles (UCLA). Before accepting his position at the Johns Hopkins University, Dr Bartlett served as a faculty member at UCLA and Tufts University School of Medicine in Boston, Massachusetts, and was Associate Chief of Staff for research at the Boston VA Hospital. Dr Bartlett has worked in several areas of research, all related to his specialty in infectious diseases. His major research interests have included anaerobic infections, pathogenic mechanisms of *Bacteroides fragilis*, anaerobic pulmonary infections, and *Clostridium difficile*-associated colitis. Since moving to Johns Hopkins, his major interests have been HIV/AIDS, managed care of patients with HIV infection, and bioterrorism. Dr Bartlett is a member of the Institute of Medicine, a master of the American College of Physicians, past president of the Infectious Diseases Society of America (IDSA), and a recipient of the Kass Award from the IDSA. In 2005, Dr Bartlett was awarded the

Alexander Fleming Award by the IDSA and the Finland Award from the National Foundation for Infectious Diseases (NFID). He has authored over 500 articles and reviews, more than 280 book chapters, and over 60 editions of 18 books, and has served on editorial boards for 19 medical journals.

**Chris Beyrer MD MPH** is Professor of Epidemiology and International Health at the Johns Hopkins Bloomberg School of Public Health. He serves as Director of Johns Hopkins Fogarty AIDS International Training and Research Program, and as Founder and Director of the Center for Public Health and Human Rights at Johns Hopkins. He also serves as Senior Scientific Liaison and Chair of the Injecting Drug Use Working Group of the HIV Vaccine Trials Network (HVTN). He has an undergraduate degree in History from Hobart & Wm. Smith Colleges, obtained his medical degree from the State University of New York, Downstate Medical Center in Brooklyn, NY, and did his public health and infectious diseases training at Johns Hopkins. Active in the international health arena since 1991, he has done research and public health work on substance use and HIV in Thailand, China, Burma, India, Laos, Tajikistan, and the Russia Federation. He is the author of the 1998 book *War in the Blood: Sex, Politics and AIDS in Southeast Asia* (Zed Books, London; St Martins Press, New York), and editor and contributor to *Public Health and Human Rights: Evidence-Based Approaches* with H.F. Pizer (Johns Hopkins University Press, Baltimore, MD, 2007). Dr Beyrer has published extensively on HIV/AIDS epidemiology and prevention research, HIV vaccine research, and public health and human rights, and is the author of numerous articles and scientific papers. He has served as a consultant to the World Bank Institute, the World Bank Thailand Office, the Office for AIDS Research of the US NIH, the Levi Strauss Foundation, the US Military HIV Research Program, the Henry M. Jackson Foundation for the Advancement of Military Medicine, the Open Society Institute, the Royal Thai Army, and numerous other organizations.

**Kenneth Bridbord MD MPH** devoted the first 12 years of his 35-year federal government career to environmental health, initially with the US Environmental Protection Agency and later with the National Institute for Occupational Safety and Health. For the past 12 years he has been responsible for the extramural programs of the Fogarty International Center; these programs are devoted to building research capacity in low- and middle-income countries to address global health threats from both communicable and non-communicable diseases.

**David D. Celentano ScD MHS** is Professor and Director of the Infectious Diseases Epidemiology Program and Deputy Chair of the Department of Epidemiology at the Johns Hopkins Bloomberg School of Public Health. He is Principal Investigator of the HIV Prevention Trials Unit for Thailand, for the NIMH Collaborative HIV/STD Prevention Trial in India, and for a series of investigator-initiated NIH grants in Thailand. He has been working on social and behavioral factors in HIV/AIDS since 1983, and has conducted a series of epidemiologic and behavioral intervention trials in the USA, Thailand, India, and Vietnam. He serves as Behavioral Sciences Work Group Chair for the HPTN, and has been a consultant to the Office of AIDS Research (NIH) and to various Institutes and Centers of the NIH. Dr Celentano received his ScD in Behavioral Sciences from the Johns Hopkins University.

**Lin H. Chen MD FACP** is the Director of the Travel Medicine Center at Mount Auburn Hospital in Cambridge, Massachusetts, and an Assistant Clinical Professor of Medicine at Harvard Medical School in Boston, Massachusetts. Her interests focus on travel medicine and tropical medicine, particularly emerging infectious diseases and their impact on travelers. Her additional interests include immigrant health, international adoption, and vaccines. She has served as an Associate Editor for *Travel Medicine Advisor* since 1997, and is a Fellow of the American College of Physicians. She serves on the Certificate Examination Committee of the American Society of Tropical Medicine and Hygiene, and the Professional Education Committee and the Research Committee of the International Society of Travel Medicine.

**Jonathan Cohen LLB MPhil** is the director of the Law and Health Initiative at the Open Society Institute. He oversees a range of legal assistance, litigation, and law reform efforts to advance public health goals worldwide. Mr Cohen was previously a researcher with the HIV/AIDS and Human Rights Program at Human Rights Watch, where he conducted numerous investigations of human rights violations linked to AIDS epidemics in sub-Saharan Africa, Southeast Asia, and North America. A Canadian lawyer, Mr Cohen served as a law clerk at the Supreme Court of Canada in 2001 and was co-editor-in-chief of the *University of Toronto Faculty of Law Review*. He holds degrees from Yale College, the University of Cambridge, and the University of Toronto Faculty of Law.

**Thomas J. Daniels PhD** is Associate Research Scientist in the Department of Biological Sciences at Fordham University. He is Co-director of the Vector Ecology Laboratory at the Louis Calder Center located in Westchester County, NY, a focus of emerging arthropod-borne diseases in the US. His research interests include population dynamics of arthropod vectors, disease emergence and changes in landscape, and host–parasite interactions.

**Wendy Davis EdM** is a research coordinator with the Department of Epidemiology at the Johns Hopkins Bloomberg School of Public Health. She received her Master's degree from Harvard University's School of Education, and worked for the American Psychiatric Association on the development of the fourth edition of the *Diagnostic and Statistical Manual for Mental Disorders*, the DSM-IV.

She was the DSM-IV Project Coordinator and later the DSM-IV Editorial Coordinator and project coordinator for the DSM-IV Primary Care version. While at the APA she authored and edited numerous papers on the development of the DSM-IV, and was the editor of a series of columns on specific issues in the development of DSM-IV for the journal *Hospital and Community Psychiatry*. At Johns Hopkins, she is involved with a series of projects considering HIV/AIDS and marginalized populations.

**Richard C. Falco PhD** is a medical entomologist and Associate Research Scientist at Fordham University, and is Co-Director of the Vector Ecology Laboratory, located at the Louis Calder Center, Fordham's biological field station in Armonk, New York. He also holds positions at New York Medical College, located in Valhalla, NY, where he is an Adjunct Assistant Professor in the Department of Medicine and a lecturer in the School of Public Health. Dr Falco received his PhD from Fordham University in 1987, and his dissertation was one of the first in the nation on the topic of deer-tick ecology. Dr Falco's research interests include the ecology and epidemiology of vector-borne diseases, risk assessment and development of surveillance strategies for tick and mosquito-borne diseases, biological control agents of vectors, the epidemiology of tick bites, and the ecology of invasive species.

**Pierce Gardner MD** is a consultant for the Fogarty Clinical Research Scholars Program at the Fogarty International Center, National Institutes of Health, and Professor of Medicine and Public Health at the Medical School at Stony Brook University, New York. For nine years Dr Gardner served as the liaison representative of the American College of Physicians to the Advisory Committee on Immunization Practices at the Centers for Disease Control (CDC). He also served in the Epidemic Intelligence Service at CDC, where he was the Chief of the Central Nervous System Viral Surveillance Unit. Dr Gardner has done extensive international work and has been a consultant for the World Health Organization and CDC. Dr Gardner has published more than 125 articles, reviews, and books, primarily dealing with immunization issues and health issues of international travel. He has a longstanding interest in adult immunization, and has served as editor of the most recent edition of the *Guide for Adult Immunization*, published by the American College of Physicians and Infectious Diseases Society of America. Dr Gardner graduated from Harvard Medical School and trained in Internal Medicine at the University of Washington and at Case Western Reserve. Dr Gardner did his fellowship training in Infectious Diseases at the Massachusetts General Hospital. His major academic appointments were at Harvard Medical School, the University of Chicago, and Stony Brook University.

**Vivian Go PhD MA** is Assistant Professor of Epidemiology at the Johns Hopkins Bloomberg School of Public Health. She received her MA in International

Economics from the Johns Hopkins School of Advanced International Studies and her PhD in Social and Behavioral Sciences from the Johns Hopkins Bloomberg School of Public Health. She has worked extensively in Asia, including Vietnam, India, and Thailand. Her research involves using qualitative and quantitative methods to examine barriers to HIV prevention among vulnerable populations. She is currently the Principal Investigator of a randomized controlled trial of a network-oriented HIV/STD behavioral prevention intervention among injection drug users in northern Vietnam, and co-investigator of an HIV/AIDS prevention trial in Chennai, India.

**Duane J. Gubler ScD MS** is Professor and Chair, Department of Tropical Medicine, Medical Microbiology and Pharmacology, John A. Burns School of Medicine, University of Hawaii. He serves as Director, Asia-Pacific Institute of Tropical Medicine and Infectious Diseases. Dr Gubler is a graduate of the Johns Hopkins University School of Hygiene and Public Health, and has spent his entire career working on tropical infectious diseases, with extensive field experience in Asia, the Pacific, tropical America, and Africa. He has published extensively in the area of vector-borne infectious diseases, and has served as Director of the Division of Vector-Borne Infectious Diseases, the National Center for Infectious Diseases, and the Centers for Disease Control and Prevention for 15 years. He is a Fellow of the Infectious Disease Society of America and of the American Association for the Advancement of Science (AAAS), and is past president of the American Society of Tropical Medicine and Hygiene.

**Ronald Jay Lubelchek MD** is Attending Physician in the Division of Infectious Diseases at the John H. Stroger, Jr Hospital of Cook County, Chicago, Illinois, and Assistant Professor of Medicine at Rush University Medical Center in Chicago, Illinois. He is a graduate of Wesleyan University and the University of Illinois School of Medicine, and a Member of the Infectious Diseases Society of America.

**Heather J. Lynch MD MPH** is a general pediatrician with a strong interest in health promotion and disease prevention. She is a graduate of the Yale School of Medicine and the University of Washington Pediatric Residency Program. She was a Robert Wood Johnson Clinical Scholar at the University of Washington, where she received her Master of Public Health in Health Services with a research focus in pediatric oral health.

**Edgar K. Marcuse MD MPH** is Professor of Pediatrics and adjunct Professor of Epidemiology at the University of Washington (UW) Schools of Medicine and Public Health and Community Medicine, and Associate Medical Director at the Children's Hospital and Regional Medical Center in Seattle, WA. He graduated from Oberlin College (AB), Stanford University School of Medicine (MD), and the University of Washington School of Public Health and Community Medicine (MPH); trained in pediatrics at Boston's Children's Hospital and Seattle's Children's Hospital, and was a CDC Epidemic Intelligence Service Officer assigned to Washington State. He has served as a member and Chair of the National Vaccine Advisory Committee, a member of the AAP Committee on Infectious Disease, Associate Editor of the *Red Book*, and a member of the USPHS Advisory Committee on Immunization Practice (ACIP). He has been involved in immunization, pediatrics, public health, and related education and research for more than 35 years. He has numerous publications relating to immunization, general pediatrics, and public health. He is co-editor of the American Academy of Pediatrics' *AAP Grand Rounds*, a monthly publication critiquing new studies relevant to pediatric practice.

**Troy Martin MD** received his medical degree from the University of Washington in 1999. He attended residency training in internal medicine, and served a fellowship in Infectious Diseases at Brown University. In 2004 he joined the Brown University faculty, based at the Memorial Hospital of Rhode Island, and is currently an Assistant Professor in Medicine. While there, he was a faculty member of the Center for Biodefense and Emerging Pathogens. In June 2006 he moved to Hanoi, Vietnam, where he is currently working for the William Jefferson Clinton HIV/AIDS Initiative.

**Anthony J. (Tony) McMichael MBBS PhD** is Director of the National Centre for Epidemiology and Population Health at the Australian National University, Canberra. He was previously Professor of Epidemiology at the London School of Hygiene and Tropical Medicine, UK. His research interests have spanned occupational–environmental risks to health; food, nutrition, and disease; and, more recently, environmental changes and their impacts on health. Since 1993 he has participated in coordinating and reviewing the scientific assessment of health impacts for the UN's Intergovernmental Panel on Climate Change (IPCC), and has played a corresponding role in the international Millennium Ecosystem Assessment Project (2001–2005). He has been an advisor on environment and health to WHO, UN Environment Program and World Bank. Within the Earth System Science Partnership (International Council of Science), he co-chairs the new international research network on Global Environmental Change and Health. In addition to his many peer-reviewed published papers he has authored several books, including *Human Frontiers, Environments and Disease: Past Patterns, Uncertain Futures* (Cambridge University Press, 2001), and (as senior editor) *Climate Change and Human Health: Risks and Responses* (WHO/UNEP/WMO, 2003).

**Shruti Mehta PhD MPH** is an Assistant Professor in the Department of Epidemiology at the Johns Hopkins Bloomberg School of Public Health. She has an undergraduate degree in Art Theory and Practice from Northwestern University, a Master's degree in Public Health from the Johns Hopkins School of Public Health, and a PhD in infectious disease epidemiology again from the Johns Hopkins School of Public Health. She is the principal investigator of a longitudinal study on the incidence of HIV infection among injection drug users in Baltimore, Maryland. She has published in the areas of HCV epidemiology and prevention, HIV and HCV co-infection, and access to care and treatment for HIV and HCV among injection drug users.

**Marguerite A. Neill MD** is Associate Professor of Medicine in the Department of Medicine of the Brown Medical School. She obtained her MD degree from the George Washington University School of Medicine, and was an Epidemic Intelligence Service officer at CDC. In research conducted during an Infectious Diseases fellowship at the University of Washington, she was the first to demonstrate that *E. coli* O157:H7 is the predominant cause of the hemolytic uremic syndrome in the United States. She served on the National Committee for Microbiological Criteria in Foods, and was a member of the US Delegation of the Codex Alimentarius Commission of the FAO. She has served as an advisor and expert consultant to the USDA, FDA, and WHO. She is Associate Director of the Center for Biodefense and Emerging Pathogens at Memorial Hospital of Rhode Island, and is the Chair of the Bio-emergency Work Group of the Infectious Disease Society of America. Dr Neill is involved in bio-emergency education and preparedness locally and nationally. Her efforts have focused on improving clinical preparedness for bioterrorism and breaking public health events.

**André-Jacques Neusy MD DTM&H** is an Associate Professor of Medicine and the founding Director of the Center for Global Health at the New York University School of Medicine. He has served as coordinator of health programs for various international organizations, both in development and disaster settings, in Africa and the Balkans. Dr Neusy has first-hand experience in developing and implementing health strategies and programs in disaster and development settings, as well as creating cross-disciplinary education and training of health professionals involved in such efforts. An advisor to the Ministry of Health of Rwanda and the National University of Rwanda, his current research centers on global health workforce development. Dr Neusy, an active member of the Global Health Education Consortium, is also its immediate past president. He served as a consultant to and committee member of the Institute of Medicine's Board of Global Health, and is actively involved with various organizations working in poverty reduction and global health.

**Peter L. Page MD** is Vice President of the Atlantic Division of the American Red Cross Blood Services. After being on the staff at Boston's Beth Israel Hospital, he has had 28 years experience with the Red Cross blood program, initially in a medical role, but assuming more operational responsibilities for blood collection, processing, testing, and distribution to hospitals in the Northeastern and Western United States. He has also been medical officer at the Red Cross headquarters in Washington DC. Trained in hematology and medical oncology at Harvard, he is also board-certified in blood banking (Pathology). He has served on the board of the National Marrow Donor Program, and serves on a number of committees of the American Association of Blood Banks.

**Bjorg Palsdottir MPA** is an organizational development consultant for global health, and Co-founder and Associate Director of the Center for Global Health at New York University School of Medicine. Prior to working for the Center Ms Palsdottir worked for the International Rescue Committee, an emergency relief and development organization, first at their headquarters in New York and then as a regional information coordinator for East and Central Africa. Her research interests focus on issues related to health systems and organizational development in the health sector, particularly in low-income countries. Most recently, Ms Palsdottir's work has included recommending models for human capacity development for the Institute's Committee on Options for Overseas Placement of US Health Professionals, used to advise the US Government as part of the President's $15 billion Emergency Plan for AIDS Relief; the development of a management training program for field staff of development and relief agencies; and an evaluation of the American International Health Alliance's emergency medical training programs in the Ukraine, Russia, and Uzbekistan. Prior to working for humanitarian organizations she worked as a journalist for *The Economist* Intelligence Unit in New York, as well as for the *Palestine-Israel Journal* in Jerusalem. She holds a BA in economic journalism, a Master's degree in Public Administration, and a certificate in Management Training and Organizational Development from New York University.

**Robert F. Pass MD** is a Professor of Pediatrics and Microbiology at the University of Alabama at Birmingham School of Medicine. He received his medical degree from the University of Alabama, completed pediatric residency at Stanford University Medical Center, and trained in virology and infectious diseases at UAB. His research has been focused on congenital cytomegalovirus (CMV) infection, moving from studies of natural history and pathogenesis to epidemiological studies aimed at identifying sources of maternal infection, to vaccine clinical trials. He is currently conducting a NIAID (DMID) sponsored clinical trial of a CMV glycoprotein B subunit vaccine aimed at prevention of maternal and congenital CMV infection.

**Aron Primack MD MA** is a program officer at the Fogarty International Center of the National Institutes of Health, which provides for training for medical professionals and scientists from developing countries. He is a graduate of the Northwestern University School of Medicine, and completed a fellowship in

Oncology at the National Cancer Institute. For almost two decades he was a clinical medical oncologist and taught on the faculties of both Georgetown Medical School and George Washington Medical School. Dr Primack also has a Master's degree in Cultural Anthropology from the Catholic University, and served as Area Peace Corps Medical Officer for Niger, Chad, Mali, and Mauritania. He served as Medical Director in Program Integrity and Medical Director for the Center for Health Plans and Providers of the US Health Care Financing Administration in Washington, DC. Dr Primack also served as Associate Professor in the Department of Preventive Medicine and Biometrics at the Uniformed Services University of the Health Sciences, where he was responsible for the International Health courses and developed the curriculum for and taught medical anthropology.

**Joshua P. Rosenthal PhD** is an ecologist with a longstanding interest in the integration of environment, public health, and international development. He manages two interagency research and capacity-building programs on behalf of the Fogarty International Center. The International Cooperative Biodiversity Groups combine biodiversity-based drug discovery, bioinventory, intellectual property management, and research capacity building in 15 countries around the world. The Ecology of Infectious Diseases program integrates field and mathematical analysis of disease dynamics with measures of population and environmental change to yield predictive tools. Dr Rosenthal is also Deputy Director of the Division of International Training and Research of the Fogarty International Center. He has authored a variety of technical, policy, and popular publications, including research reports, research topic reviews, magazine articles, opinion pieces, and one edited book on *Biodiversity and Human Health*. Dr Rosenthal serves on advisory panels for various US Government, United Nations, and World Health Organization programs on conservation of biodiversity, bioinformatics, genetic resources and biomedicine.

**Frangiscos Sifakis PhD MPH** is faculty Assistant Scientist in the Department of Epidemiology at the Johns Hopkins Bloomberg School of Public Health. Dr Sifakis received his MPH in Epidemiology from the University of Alabama at Birmingham School of Public Health, and his PhD in Infectious Disease Epidemiology from the Johns Hopkins Bloomberg School of Public Health. He directs the CDC-funded National HIV Behavioral Surveillance effort in Baltimore, Maryland. Dr Sifakis has worked extensively with hard-to-reach populations at the greatest risk of contracting HIV – such as men who have sex with men, and injection drug users. Current research concentrations of Dr Sifakis include developing and implementing surveillance methods for evolving environments of social and sexual interaction, including the Internet.

**Ronald Waldman MD MPH** has been working in humanitarian emergencies since the late 1970s. With colleagues at the Centers for Disease Control and

Prevention, he co-authored a series of papers that helped establish the epidemiology of refugee health. He is the founding Director of the Program on Forced Migration and Health at the Mailman School of Public Health of Columbia University, where he is currently Professor of Clinical Population and Family Health. His recent emergency and post-conflict work has been in Afghanistan, Liberia, the Democratic Republic of Congo, and South Sudan.

**Robert A. Weinstein MD** is Chief Operating Officer of the outpatient Ruth M. Rothstein CORE Center for the Prevention, Care, and Research of Infectious Diseases. He is also Chair of the Division of Infectious Diseases at the John Stroger (formerly Cook County) Hospital, Director of Infectious Disease Services for the Cook County Bureau of Health Services, and Professor of Medicine at Rush University Medical College. He directs the Cook County component of the Rush/Cook County Infectious Diseases fellowship program. Dr Weinstein's clinical and research interests focus on hospital-acquired infections (particularly the epidemiology and control of antimicrobial resistance and infections in intensive care units), rapid HIV testing, and health-care costs and outcomes for patients with HIV/AIDS. Dr Weinstein is a past president of the Society for Healthcare Epidemiology of America (SHEA), and immediate past-chair of the US Center for Disease Control's Federal Healthcare Infection Control Practices Advisory Committee (HICPAC). He was the first recipient of the National Association of Public Hospitals and Health Systems Clinical Research Award, was the 2005 recipient of the SHEA Lectureship Award, and has published over 250 scientific articles and book chapters, 2 books, 21 CDs, and Internet educational materials.

**Bruce A. Wilcox MS PhD** is Professor of Tropical Medicine, Director of the Asia-Pacific Center for Infectious Disease Ecology, Chair of the Division of Ecology and Health, and on the Graduate Faculty of the Program in Ecology, Evolution, and Conservation Biology at the University of Hawaii at Manoa. As a graduate student at UCSD he co-founded the field of conservation biology and co-published its first text. He subsequently held research positions at Stanford University and in the private sector, where he continued to work at the interface of the environmental sciences, biodiversity conservation, and health. He joined the University of Hawaii's School of Medicine in 2001 to establish the first ecology and health unit in a US medical school, and led the creation of the peer review journal *EcoHealth*, serving as its Editor-in-Chief. He also established and directs a university-wide collaborative graduate research and training program on the ecology of infectious diseases of the tropical Asia-Pacific region.

**Mary E. Wilson MD FACP** is Associate Clinical Professor of Medicine at Harvard Medical School and Associate Professor of Population and International Health at the Harvard School of Public Health. Dr Wilson's main academic interests include tuberculosis, the ecology of infections and emergence of microbial threats, travel medicine, and vaccines. She served as Chief of Infectious Diseases at Mount Auburn Hospital in Cambridge for more than 20 years. She is a Fellow in the Infectious Diseases Society of America. She served on the Advisory Committee for Immunization Practices (ACIP) of the Centers for Disease Control from 1988 to 1992, and has been a member of the Academic Advisory Committee for the National Institute of Public Health in Mexico. She has also served on four committees for the Institute of Medicine of the National Academies, including the Committee on Emerging Microbial Threats to Health in the 21st Century. She is author of *A World Guide to Infections: Diseases, Distribution, Diagnosis* (Oxford University Press, New York, 1991) and senior editor, with Richard Levins and Andrew Spielman, of *Disease in Evolution: Global Changes and Emergence of Infectious Diseases* (New York Academy of Sciences, 1994).

**Rosalie E. Woodruff BAComm MPH PhD** is a research fellow at the National Centre for Epidemiology and Population Health, The Australian National University. Dr Woodruff is the convenor of the Environmental Health research group. Her area of expertise is climatic influences on health, in particular on patterns of mosquito-borne disease transmission, and the seasonality of respiratory and cardiovascular infections. She has undertaken a number of climate-change impact assessments, including a major assessment for the Australian Department of Health and Ageing in 2003, and for the Australian Medical Association and the Australian Conservation Foundation in 2005. She is a contributor and reviewer for the *Intergovernmental Panel on Climate Change Fourth Assessment Report* (Chapters on Australia and New Zealand, and Human Health). She is contributing to the development of global and national climate-change burden of disease guidelines for the World Health Organization.

**Gary P. Wormser MD** is Professor of Medicine and Pharmacology, Chief of the Division of Infectious Diseases, and Vice Chairman of the Department of Medicine at New York Medical College. He is Chief of the Section of Infectious Diseases at Westchester Medical Center, and Director and Founder of the Lyme Disease Practice, a well-respected walk-in clinic for the care and study of patients with tick-borne infections. Dr Wormser is a graduate of the Johns Hopkins University School of Medicine, and completed his Infectious Diseases fellowship at The Mount Sinai Hospital in New York City. His principal research interests include Lyme disease and human granulocytic anaplasmosis, as well as HIV infection, infection control, and investigational antimicrobial agents and vaccine preparations. He is on the editorial board of *Clinical Infectious Diseases, Vector-Borne and Zoonotic Diseases* and *Wiener Klinische Wochenschrift*. Dr Wormser is the author of more than 350 published papers, and editor of a leading textbook on AIDS, *AIDS and Other Manifestations of HIV Infection* (Academic Press, 2004), now in its fourth edition.

**Stephen H. Zinner MD**, a board-certified specialist in Internal Medicine and Infectious Diseases, is the Charles S. Davidson Professor of Medicine at the Harvard Medical School, and Chair of the Department of Medicine at Mount Auburn Hospital in Cambridge, Massachusetts. Formerly he was Professor of Medicine at Brown University School of Medicine and Director of the Division of Infectious Diseases at Rhode Island and Roger Williams Hospitals, both in Providence, Rhode Island. Dr Zinner received his BA from Northwestern University in Evanston, Illinois, and his medical degree from the University of Pennsylvania School of Medicine in Philadelphia. He completed a residency in Internal Medicine at the University of Chicago Hospitals in Chicago, Illinois, followed by research fellowships in Medicine and in Bacteriology and Immunology at Harvard Medical School and Boston City Hospital. He served an Infectious Diseases fellowship at the Channing Laboratory and Thorndike Memorial Laboratory, Harvard Medical Service at the Boston City Hospital. Dr Zinner's research is focused on antimicrobial pharmacodynamics and infections in cancer patients.

# Preface

In 1971, I appeared before my state medical licensing board for a required routine interview before being able to practice medicine in Rhode Island. I was asked what my specialty was, and when I replied "Infectious Diseases," the group of wizened physicians on the Board looked at me rather incredulously and asked, "Are you going to have enough to do?" They were not being sarcastic or even naïve. In fact, these experienced physicians were convinced that infectious diseases were solved, cured by antibiotics or eliminated by vaccines.

Over the past four decades enormous changes in clinical infectious diseases have been recognized. We have seen a myriad of infecting agents "emerge," and with them new clinical syndromes and diseases. It is not that the microbes themselves are necessarily new, as the microbial world has existed for millions and millions of years, but rather that new clinical illnesses have been described and ultimately discovered to be caused by micro-organisms hitherto either unrecognized or clearly known to cause disease. Human behavior and social ecology surely have had major influences on these phenomena. In addition, we have recently encountered growing problems with antimicrobial resistance among bacteria that for the past 50 years had been treatable with available antibiotics, but which now threaten our ability to cure common infections.

Social changes have had enormous impact on infectious diseases. For example, the death rate from tuberculosis in the United States declined considerably as housing conditions improved, well in advance of any active treatments for the tubercle bacillus. Similarly, the incidence of rheumatic fever and acute glomerulonephritis, sequellae of infections with *Streptococcus pyogenes*, also known as "Group A beta-hemolytic streptococci," declined with decreased crowding in households. It is clear that penicillin was and is indeed a wonder drug in terms of eradicating streptococcal skin and pharyngeal infections, but social changes impacted these diseases greatly as well.

Over the past half century, several "new" infections have been identified or have increased in number as a direct or indirect result of social changes. For example, it is highly likely that Lyme disease "existed" for centuries before it was identified in the last half of the twentieth century, and its specific pathogenic bacteria were identified and characterized so beautifully by modern epidemiology and molecular biology. It is highly likely that suburbanization and exurbanization was responsible for bringing more and more people in direct contact with the Ixodes tick that transmits the causative organism of Lyme disease, *Borrelia burgdorferi*.

Legionnaire's disease was first recognized in the 1970s after veterans attending a Legionnaires' convention in Philadelphia became ill with a respiratory illness. The causative organism, *Legionella pneumophila*, surely existed for eons before these first cases came to light. Modern epidemiology and bacteriology relatively quickly associated the dissemination of these water-borne bacteria with air-conditioning cooling systems – an example of modern technology making possible the convening of large groups of people in relative comfort. It has also been shown that this organism lives comfortably in the scaly sediment on the inner surface of water pipes in many areas. Outbreaks of Legionnaire's disease have been associated with building construction, as the "jackhammering" of ground near water pipes helps these organisms to leave their comfortable environment in the pipe scale and enter the water supply itself.

As illustrated in this book, there is no more obvious association of changing human behavior and infectious diseases than that linking social change and sexually transmitted infections (STI). Fifty years ago, any textbook on sexually transmitted diseases would have included chapters on diseases caused by two or three microbes known to be transmitted sexually. The enormous changes in sexual mores, possibly initiated by the introduction of oral contraceptives, have led to a dramatic expansion in the number of pathogenic microbes that can be spread sexually. Well over 35 micro-organisms are known to cause human infection in association with sexual behavior. For example, a formerly rarely seen infection of the uterus with *Actinomyces* has been well described in association with the use of certain indwelling contraceptive devices or coils. Oral contraceptives were introduced in the 1960s, and this was followed by dramatic increases in syphilis and gonorrhea as well as the description of "new" genital infections caused by *Chlamydia trachomatis*, *Herpes simplex* virus, and, of course, Human Immunodeficiency Virus (HIV) and AIDS.

The introduction of the Internet has led to new outlets for "instantaneous" sexual mating, and recent outbreaks of syphilis and other STIs are well described. The gay liberation movement in the past century made it possible for more and more homosexuals to live openly, and in the years prior to the HIV/AIDS epidemic sexual contacts were facilitated and transmission of many organisms occurred. The introduction of the Human Immunodeficiency Virus in the population of men who have sex with men is well described in this book. HIV existed long before its first recognized clinical manifestations 25 years ago, but the importance of social changes in sexual mores of gay men in the developed world as well as heterosexual Africans and Asians, where this infection has spread rampantly, are clear examples of the interaction of social behavior, social ecology, and the microbial world. Women's movements have also impacted STIs, first liberating them to have more sexual encounters (the so-called "swingers" of the early 1970s) but now empowering women to take more proactive roles in the prevention of STI transmission.

Over the past five or six decades, intravenous drug use was well known to be a risk factor for staphylococcal, pseudomonas, and fungal infections, and physicians became familiar with the occurrence of serious bloodstream, cardiac, and bone infections caused by these organisms in IV drug users. HIV, hepatitis B and C, and other blood-borne pathogens are increasingly being recognized in patients who use intravenous drugs. It is also not uncommon for physicians to overlook the possibility of drug use in their patients. The frequency of drug use across all strata of society is not well understood, but an important chapter in this book on this topic brings this clearly into focus.

As the world has been "shrunk" by thousands of daily flights from everywhere to anywhere, the influence of travel on infectious diseases has been magnified. Years ago international travel was available only to the very rich, who rarely came in contact with sick people during their trips. The recent epidemics of SARS and West Nile Virus (WNV) in North America call attention to the ease with which infectious agents can be transported around the world by people incubating infectious diseases. While most people do not travel when they are very ill, many people can and do travel while they are in the "incubation phase" of an infection. Neither the West Nile Virus nor the coronavirus associated with SARS evolved overnight, but the ability to fly quickly from continent to continent clearly allowed these agents to enter new areas of the world and spread rapidly to close contacts. The spread of WNV from the eastern to the western part of the United States over a few summers indicates how susceptible the population is to infection with a newly introduced virus. (Infection with WNV itself is not necessarily associated with symptomatic illness.) Outbreaks of tuberculosis and food- and water-borne infections are well described aboard airplanes and cruise liners. Travel-associated infections are also being increasingly recognized as an outcome of "adventure travel" and other trips that bring tourists and business people into areas not previously frequented by these groups. The new discipline of Travel Medicine, championed by Dr Mary Wilson and others, has arisen in response to imported infections in returning travelers.

The association of pestilence and war is well established in medical history. In early wars, death from infected wounds was the major fate of those who survived the initial trauma of battle. Today, soldiers billeted in parts of the world where vector-borne parasitic diseases are common are obviously exposed to agents they had never encountered in basic training. War wrenchingly disrupts society, and often leads to the displacement of large numbers of people from their homes. Prisons and refugee camps are well known to allow the rampant spread of infections, and these are all too common in these socially displaced people. Prisons have a special ecology, and have been associated with the transmission of drug resistant tuberculosis, hepatitis, HIV and other STIs and, more recently, staphylococcal skin infections caused by antibiotic-resistant bacteria. Society is now being challenged by the specter of bioterrorism-related spread of infectious agents. The still unsolved distribution of anthrax via the mail a few years ago reminds us how fragile this ecology is, and how aberrant behavior armed with biological weapons could affect populations around the globe. The tragic juxtaposition of terror and pathogens is a warning for all concerned people of the world, and this arena is brought into clear if dangerous focus in this book.

Since their introduction in the middle of the twentieth century, antibiotics have been seen as wonder drugs. The early drugs, penicillins and cephalosporins, were incredibly effective in the treatment of bacterial diseases caused by streptococci, staphylococci, and common gram-negative bacteria such as *E. coli* – the principal cause of urinary tract infections. Morbidity and mortality associated with common bacterial illnesses such as pneumonia, bacteremia and endocarditis, urosepsis, osteomyelitis, peritonitis, syphilis, gonorrhea, and many, many others dramatically decreased as a direct result of these life-saving wonder drugs. The early antibiotics were extraordinarily safe and not associated with serious toxicity (other than anaphylaxis in patients with severe antibiotic allergies). Their wide availability, safety, and relative affordability led to rampant social demand, subsequent use and surely eventual misuse. In the mid-1970s, enough antibiotic was marketed to treat every man, woman, and child in the United States for two bacterial infections per year – far more than the actual number of bacterial infections that occurred. It has been estimated that about 50 percent of antibiotic use now is for the treatment of respiratory infections in children, the vast majority of which are due to viruses that do not respond in any way to antibiotics.

As bacteria are able to rapidly "evolve" systems to overcome the actions of antibiotics, it is no wonder that many antibiotics in common use only 10 years ago are losing their punch. Bacteria have "learned" how to destroy the penicillins and cephalosporins, as well as aminoglycosides, with enzymes that they manufacture, and they can modify their structures with slight genetic changes that result in alterations in the targets of the heavily-used and successful fluoroquinolone and macrolide classes of antibiotics. They have evolved "efflux pumps" that can eliminate the antibiotic from the bacterial cell and make it difficult or impossible for the drug to reach its target. The clinical result of these manipulations is increasing failure of antibiotics to cure more and more infections, the causative organisms of which have been called "killer bacteria," such as methicillin-resistant *Staphylococcus aureus* (MRSA), currently causing large numbers of skin infections in many otherwise healthy people around the world. These infections are still susceptible to some old and new antibiotics, but how long this will continue is arguable. Another "social" change is the recent decline in the number of pharmaceutical companies willing to invest in the discovery and development of new antibiotics in the face of increasing antibiotic resistance – surely an ominous situation. It could be argued that antibiotics have made possible so many advances in modern medicine – including bone-marrow and solid organ transplantation, and the treatment of cancer with drugs that temporarily ablate the immune system, rendering people very susceptible to bacterial (and other) infections – that it is hard to imagine continued success in these important advances without antibiotics to treat life-threatening bacterial infections in these seriously ill patients.

It seems obvious to me after almost four decades of work in the field of infectious diseases that there is still much to be done. The brilliant text that follows gives testimony to the critical interaction of germs and people, of societal upheavals and infectious diseases. The world can learn much from the past, and hopefully our leaders can apply these lessons to achieving a healthier future for us all.

*Stephen H. Zinner MD*

# Acknowledgments

The authors wish to recognize the assistance of Ms Lola Wright in manuscript preparation, and KHM wishes to express appreciation to colleagues at Brown University, Harvard University, Northwestern University, and the University of Pennsylvania, as well as the Miriam Hospital and Fenway Community Health, who have helped him increase his understanding of the dynamic and complex interactions between microbial organisms and humans.

# Introduction: What constitutes the social ecology of infectious diseases?

**Kenneth H. Mayer and H.F. Pizer**

*Infectious disease is one of the great tragedies of living things – the struggle for existence between different forms of life. Man sees it from his own prejudiced point of view; but clams, oysters, insects, fish, flowers, tobacco, potatoes, tomatoes, fruit, shrubs, trees, have their own varieties of smallpox, measles, cancer, or tuberculosis. Incessantly, the pitiless war goes on, without quarter or armistice – a nationalism of species against species ... The important point is that infectious disease is merely a disagreeable instance of a widely prevalent tendency of all living creatures to save themselves the bother of building, by their own efforts, the things they require. Whenever they find it possible to take advantage of the constructive labors of others, this is the direction of the least resistance ... About the only genuine sporting proposition that remains unimpaired by the relentless domestication of a once free-living human species is the war against these ferocious little fellow creatures, which lurk in the dark corners and stalk us in the bodies of rats, mice, and all kinds of domestic animals; which fly and crawl with the insects, and waylay us in our food and drink and even in our love.*

Hans Zinsser, 1935, in *Rats, Lice and History*

*Most people associate ecology with efforts to preserve nature, to recycle waste products and conserve resources and heal or prevent the wounds people inflict upon the world's ecosystems. But although these are important results of ecological thinking, ecology itself is much more. In its original sense, it is the science of connections. It seeks to describe the vast web of interrelationships that tie living things to their environments, its fundamental premise being that a change in any part of one of the*

> *"tangled banks" of life we call ecosystems can have broad and often unexpected implications for any living thing seeking to survive within them.*
>
> Gabriel Rotello, 1997, in *Sexual Ecology AIDS and the Destiny of Gay Men*

This book was conceived to expand upon previous efforts to explore the social ecology of infectious diseases, which we define as the scientific study of the ways by which human activities enable microbes to disseminate and evolve, creating favorable conditions for the diverse manifestations of communicable diseases. Despite advances in living standards, public health, and medical technologies (including antimicrobial drugs and vaccines), infectious and parasitic diseases cause about a third of deaths worldwide and are the second leading cause of mortality and disability (Fauci, 2001). The rapidity by which changes in human behavior can result in severe epidemics is well illustrated by HIV, the cause of AIDS, which is spread primarily by sexual activity and injecting drugs using unsterile equipment. But AIDS is only one of multiple microbial threats, and other human activities can cause devastating infectious epidemics. The goal of this text is to analyze the wide range of activities and behaviors that influence the evolution and dissemination of infectious disease epidemics. They include the migration of people and the transport of goods; the production and distribution of food; human-influenced changes in demography and the environment; advances in medical technologies; and patterns of governance, conflict, natural disaster and the dislocation of populations. Taken together, we believe these diverse activities are essential determinants of the prevalence and incidence of the current array of infectious diseases that continue to be major causes of morbidity and mortality across the planet.

In the electronic media era, on almost any given day, it seems the news is highlighting a new infectious disease pathogen as the next great threat to humanity. In 2003 it was SARS. In 2006 it was avian influenza. Before that there were Legionnaires' disease and Toxic Shock Syndrome, and earlier tuberculosis was (as coined by Rene and Jean Dubos) known as "the white plague" (Dubos and Dubos, 1952). While the battle between humans and constantly evolving microbes is continuous, it would be wrong to be unduly pessimistic about the future. The microbes adapt to their hosts and humans respond. Each year, more people have access to modern medicines. Health programs expand to more communities, and more young professionals are trained in clinical care and public health. In response to the AIDS epidemic, the international health workforce is expanding and improving at an unprecedented pace. While more always needs to be done, there now is a historic commitment to improving health, especially in the developing world. What is different now from the unwarranted optimism of the 1960s is the realization that medical technology by itself will not prevent or fully contain infectious diseases. Vaccines have been effective for some diseases, but many microbial pathogens are highly adaptable and human activities are often amenable to creating niches and opportunities for the spread of infectious diseases. This is demonstrated by the re-emergence of old diseases like dengue fever, yellow fever, measles, cholera, leishmaniasis, and malaria, and the appearance of new ones like AIDS, the hemorrhagic fevers, SARS, and avian influenza. Although new antimicrobial agents continue to be developed, as they become increasingly used by humans and in veterinary medicine, pathogens often develop resistance. Some vectors have become resistant to insecticides, complicating the control of several common diseases, such as malaria.

Advances in medical technology create new opportunities for nosocomial pathogens to proliferate in immunocompromised hosts and prosthetic tissues. Although secondary spread of these opportunistic infections is uncommon, nosocomial outbreaks of multi-resistant bacteria and fungi have plagued many intensive care units and cancer treatment centers. Advances in surgery and life support create new niches for microbes to proliferate, ranging from in-dwelling catheters to surgical wounds. As life-saving advances such as organ transplants, immunosuppressive therapy of autoimmune diseases like rheumatoid arthritis, and chemotherapy for malignancies become more accessible in under-resourced communities, the ability of a diverse array of microbes to create new clinical challenges will grow.

Moreover, the continued existence of millions of impoverished and malnourished individuals throughout the world creates a vast reservoir for the development and dissemination of new plagues. The elimination of poverty through strategies to promote human development would not eradicate all infectious diseases, but could certainly decrease global morbidity and mortality substantially. We believe the contributors to this text make forceful arguments that protection against new infectious disease epidemics necessitates the use of traditional, as well as innovative, public health strategies that include:

- conducting continuous, concerted epidemiological surveillance
- ensuring that communities have clean and safe water supplies, proper sanitation, housing and vector control
- effective oversight over travel and the transport of goods
- government regulation of food and blood supplies, as well as pharmaceuticals
- vaccination of children and regulation of health-care and day-care facilities
- identification and treatment of sexually transmitted infections and, where appropriate, partner notification.

There is a need to emphasize public health in medical and other clinical education, to expand the international health workforce, and to improve global responsiveness to natural and man-made disasters, as well as to take advantage of newer technologies – such as computerized surveillance, web links to report new outbreaks, and videoconferencing – to share relevant outbreak information.

These strategies can provide valuable new tools to address the intractable, as well as emerging, epidemics of the twenty-first century.

We begin the text with a discussion of the role modern travel plays in infectious diseases. Population migrations and trade have always spread epidemics, but the speed, scope, and volume of modern travel is qualitatively different from any period before in history. Over the last half-century, international travel grew 32-fold. Two million people cross international borders every day, and many more travel within their country's borders but still over significant distances. In the last 200 years, the average daily distance a Frenchman now travels has increased 1000-fold. The world is covered and linked by an efficient network of roads, and rail, water, and air routes. In March and April 2003 a businessman flew from Hong Kong to Frankfurt, and then took seven flights within Europe, before returning to Hong Kong, where he was diagnosed with SARS. Travelers invariably and inadvertently expose themselves and others to new microbes, and today's fast and efficient travel network has created a global gene pool that links even the remote regions of the world.

In the United States and other developed nations, a sexual revolution began in the 1960s. An unprecedented number of young people started to leave home for college. Effective hormonal contraception freed them from the fear of unplanned pregnancy. Young people delayed marriage. Women gained new independence through increased workforce participation. A new literature on lovemaking became popular, encouraging heterosexual men to become more competent lovers and women to be orgasmic and assertive about their sexual desires. Meanwhile, the divorce rate increased, and with it the number of sexually experienced unmarried adults available for intimate relationships. From 1971 to 1988, the number of never-married women in the United States between the ages of 15 and 19 that had had sexual intercourse increased from 28 percent to over 50 percent. The gay pride and sexual liberation movements created vastly expanded opportunities for homosexually active men to meet and have sex. All of these social changes helped set the stage for epidemics of AIDS and hepatitis A and B, as well as the resurgence of other sexually transmitted infections, such as syphilis. The Internet now offers a new, efficient, inexpensive, and anonymous tool for sexually active individuals to communicate and then meet for sex, abetted by the ease of rapid travel.

International drug epidemics have contributed to the widespread transmission of HIV, hepatitis B and C, malaria, tetanus, and syphilis and other sexually transmitted infections. Drug trafficking and drug injecting flourish where people encounter adverse economic and social conditions, and where there is political unrest and societal disruption. While different patterns of recreational drug use may be associated with a range of health risks (for example, non-parenteral drug use is associated with behavioral disinhibition leading to sexually transmitted infections), drug injecting and needle sharing may still account for 10 percent of all new AIDS cases worldwide. The editors of this book and the authors of the chapter on substance use believe that prohibitionist approaches exacerbate drug problems and their health consequences. Limiting drug supplies makes drugs more expensive and encourages injecting, which is the most efficient route of pathogen administration. Harsh legislation and law enforcement drive injecting drug users away from testing, education, and treatment programs. Another option is to combat the adverse downstream effects of drug use with education, drug treatment, and drug substitution therapy, access to clean needles and syringes, and routine screening for, and treatment of, sexually transmitted infections in drug users.

Over the past half-century, humankind has gone through a revolutionary historic shift from rural to urban living. Two hundred years ago, Beijing was the only city in the world with a population of more than a million; now there are at least 400. Most of the growth has been in tropical zones and in congested, sprawling urban environments rife with neighborhoods that lack adequate electricity, drinking water, sewage systems, and waste management. Public health and infectious disease control generally are under-funded or absent. At some point a critical threshold is reached where there are enough susceptible individuals who have not been previously exposed to a particular infectious agent, and an outbreak occurs. Uncontrolled urbanization across the developing world is now a major factor in the eruption of epidemics of old and new infectious diseases, and provides hubs for their spread to nearby communities and to distant countries and continents. Our contributors focus on dengue fever to discuss this phenomenon, but their observations also apply to other well-characterized diseases like yellow fever, as well as newly recognized and emerging infectious diseases. The social ecology of dengue fever involves the interaction of humans and the mosquito vector *Aedes aegytpi*, which over time evolved from its origins in the tropical rainforest to thriving on humans in sprawling tropical cities. In overcrowded, impoverished urban centers, every water container and domestic animal can be a potential source for maintaining and propagating *A. aegypti*, and each mosquito bite has the potential for spreading dengue within urban neighbourhoods, which then radiates out from the city by the continuous flow of inter-city commuters, urban–rural migrants, animal reservoirs, and vectors. Between 2.5 and 3 billion people, half the world's population, now live in areas at risk for dengue, and each year an estimated 50–100 million people are infected with dengue flaviruses. At least four dengue serotypes that cause human disease have been isolated, and infection with one subtype does not confer immunity to the other subtypes. While it is not possible to kill every mosquito, prior success at controlling dengue provides lessons for the future – better laboratory-based epidemiological surveillance and improved training for frontline health practitioners; community-based public strategies such as targeting open-air water reservoirs and installing air conditioning and screens on windows; targeted spraying; and educating the public to avoiding peak exposure hours, especially when surveillance has provided early warning of an outbreak.

For more than half a century the suburbs of affluent nations have been replacing farms and forest with lawns and parks – so, for example, over the past 20 years the greater Chicago area has sprawled out by 46 percent, even though the population has grown by only 4 percent. While the net amount of forest in the country may not have changed significantly in the last two decades, expanding suburbanization has created a huge patchwork of forest interspersed with residential communities where more than half the US population now lives. This is the ideal set-up for putting humans, wildlife, and ticks in close proximity. Conservation efforts, including the protection of tick-infected deer, have expanded the pool of potential vectors for several infectious pathogens. The result has been a rapid increase in the number of humans who have developed Lyme disease, caused by a spirochete, and now the most common tick-borne disease in the word. It came to public attention in the late 1970s, in large part through the efforts of two suburban mothers in Connecticut. Almost 17,000 cases are reported each year in the United States alone by individuals who may suffer serious joint, nervous system, and cardiac involvement. However, seroepidemiological studies suggest even greater prevalence of Lyme spirochetal infection. Areas of high exposure risk include parks and the mowed back lawns of single-family suburban homes that abut woodlands. Meanwhile, general public opposition to widespread insecticide application and to measures to reduce the deer population have helped to expand the pool of infected ticks. Thus the responsibility for preventing exposure to Lyme disease largely falls on the individual through behavior modification, insect repellent use, and prompt removal of attached ticks. However, these measures have had limited effectiveness, so the incidence of Lyme disease continues to rise.

Changes in child-rearing practices, ranging from decreased breastfeeding to leaving children in day care while mothers work, can alter the natural history of communicable diseases. For example, in affluent societies it is now common for women of childbearing age to work, and for young children to be in group day care where they are exposed to common viruses and bacteria. There have been many infectious agents associated with group day care, such as *Hemophilus influenzae* b (prior to the advent of an effective vaccine), cytomegalovirus, *Streptococcus pneumoniae*, *Neisseria meningitides*, rotavirus, Salmonella, Shigella, *Giardia lamblia*, *E. coli*, respiratory viruses, varicella, hepatitis A, and parvovirus B 19. Children are often separated by age and cared for in a single room, and each year a new group of children with limited prior exposure to infectious diseases enters the day-care center. Until they acquire immunity to common infectious agents, children in group day care experience a higher incidence of diarrhea, otitis media, and respiratory tract infections than children who stay home. Studies of hepatitis A outbreaks have illustrated how an infectious agent spreads in group day-care settings and then from the day-care center to the home and even to the wider community. Another well-studied example is CMV infection, which is heavily influenced by both breastfeeding and the group care of children. Once infected with CMV, a child can shed virus for years in saliva and urine. Because young children cannot control their body fluids, day-care centers have become ideal settings for the transmission of CMV between children, and from children to adult caregivers and to family members. The prevalence of CMV infection among day-care workers ranges from 8 percent to 20 percent, while the average is about 2 percent among adults in general. Until recently, relatively few infants were breastfed or attended group day care, so many of today's mothers have no immunity to CMV. If a pregnant woman without immunity sends her child to day care and that child becomes infected with CMV, the mother is then at risk for having her next newborn develop congenital CMV, which can cause hearing loss, mental retardation, or cerebral palsy.

Modern medical care would not be possible without a safe and available blood supply. With the population aging and more medical uses for blood each year, the demand for safe blood and blood products continues to grow. Even though millions of blood units are collected each year, more seems always to be needed in developed nations, and the need is even greater in developing nations, which typically do not have well-regulated and well-organized national voluntary donor systems. Voluntary donation is the proven key to a safe blood supply. For example, from 1970 to 2000 the United States instituted a national voluntary system and the overall risk of acquiring a transfusion-transmitted infection dropped from 1 in 70,000 to 1 in 11.15 million units transfused. For hepatitis B, the risk dropped from one in 855 to one in 138,700 units transfused. (Compare this with the risk of dying from motorcycling, which is one in 50 per person per year, and from playing soccer, which is one in 5000 per person per year.) Regardless of whether a blood unit is contaminated with a well-characterized infectious agent such as malaria or a newly recognized one like HIV or *Babesia microti*, the likely causes of the human infection and entry into the blood supply include intravenous drug use with needle sharing, high-risk multiple-partner sexual activity, exposure to insect vectors, travel, population migration, and medical negligence. Each of these factors may be influenced by economic, political, and social conditions. While laboratory testing to uncover infectious agents in the blood supply is constantly improving, there probably never will be a fully effective laboratory test for every infectious disease that can be transmitted by blood transfusion, particularly if the infections are very recent with a low micro-organism burden. Moreover, the next infectious threat is likely to be present in asymptomatic blood donors before an effective diagnostic test is developed for it. The blood system's safety therefore is based on relatively low-cost, low-technology public health strategies: altruistic donation, public education, rigorous professional training, continuous epidemiological surveillance, and consistent frontline screening of prospective donors before they give blood.

The mass production and distribution of food is one of the great triumphs of the twentieth century. A hundred years ago, about 40 percent of the population in the United States lived and worked on family farms where food animals were raised in small herds, slaughtered locally, and, without refrigeration, could be transported only a short distance. Now food production and distribution are highly industrialized, and an unprecedented variety and quantity of products are transported over large distances. While safer and less expensive than ever before, the mass production, co-mingling and shipment of food products over large distances and across international borders has the potential for producing large-scale outbreaks of infectious diseases. For example, a single contaminated batch of mozzarella cheese from one small producer shredded into products by four other manufacturers created a multi-state outbreak of salmonellosis. One hamburger patty can have meat from hundreds of different cows. Even though agricultural products require extensive handling, at present the United States does not have microbiological regulatory criteria for fresh produce. Between 1977 and 1997, outbreaks due to fresh produce increased 10-fold. While food contamination is most common and serious in developing nations, it should be no surprise that even in a developed one like the United States about 20 percent of the population contracts acute gastroenteritis each year. While most cases of diarrhea are self-limited, there also are serious cases of hepatitis, sepsis, meningitis, and even paralysis. The spread of well-known food-borne diseases like typhoid fever by cooks and food handlers (for example, Typhoid Mary) have largely been eliminated in the developed world, but serious new pathogens have emerged such as non-typhoidal, drug-resistant strains of *S. typhimurium*. New and emerging food-borne pathogens continue to be recognized, and almost surely more will arise in the future.

The discovery and widespread use of antibiotics is one of the most important advances in humankind's history, but it also has had the unavoidable downside of creating an environment for drug-resistant bacteria developing a selective advantage. Antibiotic resistance and nosocomial infection are now significant threats to global health. Hospitals are a major source of the problem because they must use antibiotics regularly to care for increasingly ill and elderly patients. Another factor is the inappropriate outpatient use of antibiotics to treat self-limited illnesses such as viral upper respiratory infections. The overuse and misuse of antibiotics is also common where they are sold without a prescription. Drug pricing is another factor – for example, France has some of the lowest drug prices in Europe but also one of the highest *per capita* rates of use of antibiotics. Some believe that the most significant problem is in agriculture, where millions of pounds of antibiotics are used each year to treat sick animals, prevent infection in herds raised in close proximity, and promote growth. Animal husbandry has found that sub-therapeutic use of oral antibiotics is associated with rapid animal growth, but these doses are ideal for selecting resistant bacterial mutants.

There is no alternative to using antibiomicrobial drugs, and no single strategy for stopping the ability of micro-organisms to adapt to them. Hospitals can probably do a better job with infection control, surveillance, professional education, encouraging the use of less expensive antibiotics that are targeted to microbiologically-diagnosed infections, using computers, and requiring infectious disease consultant's approval before more expensive antibiotics are prescribed. Despite understandable professional concern about prescribing restrictions, studies have shown that sensible limits on provider prescribing have not produced adverse patient outcomes. Similar efforts are needed to improve outpatient prescribing practices, and the public needs to be more aware that taking antibiotics is not always in an individual's interest or in the interest of public health. Legal and economic forces that encourage unnecessary antibiotic use must be addressed – especially in developing countries, where antibiotics often can be purchased without a prescription, there is a ready black market for them; and pharmacists with limited medical training regularly prescribe antibiotics or refill prescriptions without physician approval. With increasing numbers of immunocompromised patients living after chemotherapy or organ transplants, wider use of antibiotics and the subsequent evolution of resistance will be inevitable, even if judicious policies are put in place.

The development of safe, effective, and cost-effective vaccines is one of the great medical achievements in history. It has had enormous impact on society by dramatically increasing childhood survival, and decreasing morbidity and disability. Smallpox was eradicated, and previously common childhood diseases like diphtheria, whooping cough, measles, and tetanus are so rare that many physicians have never seen a case. Since the late 1980s, the global incidence of polio has been reduced by 99 percent. There are now vaccines against 15 infectious diseases in the US childhood immunization schedule, and national immunization programs currently reach 80 percent of the world's children. Perversely, some parents in the developed world now refuse to immunize their children, refusing to tolerate even a very small risk of side effects. Moreover, urban myths about the association of vaccinations with the development of childhood autism and other chronic diseases have been abetted by the proliferation of disinformation on the Internet. Mass immunization programs raise thorny issues of individual rights versus public good, the risks and rewards of inaction, and social justice. New vaccines continue to be developed, like those against human papillomavirus (which is thought to account for about 70 percent of cervical cancer cases) and against rotavirus (which causes an estimated 600,000 deaths and 2 million hospitalizations, primarily in children due to diarrheal disease, mainly in poor countries). Two main challenges lie ahead: one is to discover new vaccines against the most important infectious agents, especially HIV; and the other is to make vaccines available to the poor in developing nations, which is a complex matter of cost, access to health-care, education, culture, and infrastructure. There is a worldwide consensus that the problems of both new vaccine development and expanding access must be tackled and solved through international public–private partnerships. Historically, vaccines have been developed through collaboration between academia, government, and industry, with non-profit foundations such as the Rockefeller Foundation playing an important role. Unfortunately, the cost of developing new vaccines has been rising faster than the ability of individuals and governments (especially in developing nations) to pay for them, which is why the increasing involvement of newer groups, such as the Bill and Melinda Gates Foundation, offers great promise. Even in a wealthy country like the United States, the Federal Government purchases more than half of all childhood vaccines. The primary body in Latin America is the Pan American Health Organization, and UNICEF covers much of the rest of the developing world. A bright spot is that local pharmaceutical companies are now manufacturing vaccines in developing nations. Meanwhile, there is an unprecedented commitment to support vaccine research, manufacture, and distribution. For example, in November 2005 the Bill and Melinda Gates Foundation and the country of Norway announced a 10-year $1-billion commitment to support childhood vaccinations globally. The WHO and UNICEF have their ambitious Global Immunization Vision and Strategy (GIVS) to immunize more people, especially children in low-income countries; to introduce new vaccines and technologies; and to provide basic health-care services in tandem with immunization. The GIVS goal is for each country to be immunizing at least 80 percent of its population by 2010.

Bioterrorism is the deliberate and malicious use of microbial agents or their toxins as weapons of mass destruction. Bioweapons are relatively inexpensive, simple, available, easy to conceal and produce, and easy to disseminate in a stealth attack. In 1972, under the auspices of the United Nations, more than 100 nations agreed to prohibit the development, production, and stockpiling of biological WMD, but at least 11 nations are known to have programs to develop them. Open societies are vulnerable, and a bioterrorism attack would most likely come without specific warning. Biological agents have an incubation period of days or weeks before symptoms arise, so infected persons might travel over long distances before their status is known. Initial symptoms also might not be diagnostic, and it is likely there would be additional delay in recognition because most medical professionals have never seen a patient infected with a biological WMD. The technology to produce these weapons is "dual use," which means it can serve legitimate scientific and medical purposes, or malicious ones. This serves to insulate rogue states from international scrutiny. While advances in molecular biology and biotechnology provide the possibilities to develop protective vaccines, medications, and diagnostics, they are also the tools for producing ever more lethal weapons of hybrid organisms and drug-resistant and/or particularly virulent mutants. A WHO study modeled the release of 50 kg of aerosolized anthrax spores in a city of 500,000. Depending on whether this were to occur in a developed or developing setting, the agent would likely spread at least 20 km downwind and kill or injure between 84,000 and 210,000 people. In an age of global terrorism, bioterrorism and its state sponsorship must be taken seriously.

Climate sets geographic and seasonal limits on the expression of infectious agents. The current consensus projection among climate scientists is that Earth will warm by 2–3°C this century in response to human activities, in particular the burning of fossil fuels that produce greenhouse gases. Among the numerous possible impacts of climate change on the environment are alterations in the natural systems that affect microbes, vectors, and hosts. The reproduction rate of many common pathogens and arthropod vectors, like mosquitoes and ticks, increases with rising temperatures. Demography, land use, human migration, industry, technology, and other factors also interact with climate, so that the relationship of temperature change and infectious disease transmission is not likely to be linear. While it is not possible to predict with certainty where and when a disease outbreak will occur, it is possible to model probabilities based on what is known about the effect of temperature on specific pathogens, vectors, and hosts. For example, if the "freeze line" moves north in places where schistosomiasis (a worm parasite of humans in tropical climates) is endemic, it is reasonable to predict that more people would be at risk. Malaria, which is carried by mosquitoes, is likely to spread more rapidly in warmer temperatures. Cholera thrives in warmer waters. The risk of food poisoning from salmonella and other common agents also increases as temperature and humidity rise. Because climate is affected by a wide array of economic and social activities, interdisciplinary effort will be needed to understand how climate change may affect infectious diseases. One concern is that the more rapid the pace of climate change, the more difficult it may be for humans to adapt with new technologies and public health strategies.

In the aftermath of the atrocities committed by Nazi scientists during World War II, the United Nations adopted the Universal Declaration of Human Rights, which proclaims each individual's right to the highest attainable level of health. Subsequent proclamations obligate governments to provide high-quality health services and prevent rights abuses that would have negative impacts on public health. Measures of health status – like life expectancy, maternal and infant mortality, immunization, and the prevalence of preventable and treatable infectious diseases – can be used to assess the health of populations. They may also be reasonably used to assess how well governments are fulfilling their responsibilities to populations. There is an increasing awareness that there is a link between personal economic and health status, including infectious diseases, and how governments behave with regard to public accountability, adherence to the rule of law, tolerance of an independent civil society, social stability, and respect for human rights. While more solid research in this area needs to be done, organizations like Freedom House and the World Bank have made efforts to document these links. It is clear that infectious disease risks are not spread evenly across populations and countries. Poverty leading to crowding and malnutrition, the subordination of women and minorities, as well as the mistreatment of or denial of services to other socially vulnerable populations (like injecting drug users and sex workers), are factors known to increase the risk for infectious disease transmission. For example, in Russia today there is large and growing disparity in income and employment between Moscow and a few other large cities, and the vast geographic area of the Russian federation. With the demise of the former Soviet Union and widening income disparities, there has been a sizeable increase in the number of women migrating to Moscow to perform sex work at truck stops, in hotel lobbies, and through escort services. While prostitution is legal, it takes a special permit to live in Moscow, and these women live illegally in the city. In effect, they are stateless in their own country, and thus unable to access basic health and social services. The police, their pimps, and anyone else who realizes they are not legal residents can "shake them down". This is a set-up for the transmission of HIV and other sexually transmitted infections, as well as the continued exploitation of the women. While more research is needed to document the impact of governance and poverty elimination on infectious diseases, it seems likely that these links are real.

Throughout history, poverty, war, civil unrest, political violence, social dislocation, and natural catastrophe have been responsible for establishing the underlying conditions that promote infectious epidemics. These events produce hunger and malnutrition, exposure to extremes of heat and cold, injuries and wounds, the disruption of sanitation and clean water supplies, and the abrupt dislocation of peoples leading to refugee populations. In the fifth century BCE, a plague killed one-quarter to one-third of the population of Athens because residents were crowded into and largely confined to the city as a consequence of the Peloponnesian War. Spanish conquistadors brought smallpox and measles from Europe to the New World and devastated the Incas and Aztecs. In 1741, Prague surrendered to the French, in large part because 30,000 defending Austrian soldiers were stricken with typhus. Between 1917 and 1923, millions in Europe were infected with typhus carried by soldiers from the battlefront. Conditions during World War I almost certainly facilitated the great influenza pandemic of 1918. War, civil strife, social dislocation, and natural disasters have cause millions of deaths since World War II as a direct or indirect consequence of infectious diseases such as HIV/AIDS, hepatitis A, cholera, shigella, leptospirosis, trypanosomiasis, malaria, tuberculosis, influenza, dengue, typhoid, pneumocystis carinii, pseudomonas species, and other pathogens. The fact that major infectious epidemics did not occur in the wake of the enormous undersea earthquake and resulting massive Asian tsunami that struck Indonesia, Thailand, India, and Sri Lanka in December 2004 provides important lessons for dealing with future man-made and natural disasters. Loss of life due to infectious disease can be prevented with prompt and massive international response based on providing clean water, sanitation, and shelter; limiting the size and population density of refugee camps; conducting extensive immunization campaigns; and treating diarrheal and other infectious illnesses promptly and on an empirical basis.

For both humanitarian reasons and self-preservation in an age where pathogens can so readily be transported across the globe, there is an immediate need to close the enormous health-care training gap between wealthy and poor nations. While many of the health-worker shortages in Africa are because of the advent of AIDS, other diseases and settings also urgently need highly trained health professionals. Compounding the lack of training opportunities in their home country, skilled health-care professionals in developing nations are often motivated to migrate to developed ones in order to enjoy expanded career opportunities, avoid civil strife, and enjoy an improved lifestyle for themselves and their families. Training the next generation of health professionals for practice in resource-constrained settings requires awareness of, and a multidisciplinary approach towards, the social, ecological, and political factors that impact infectious diseases. Fortunately, new cooperative training ventures are being developed to serve both poor and rich nations, so the need and rewards are mutual. Collaborative training programs build bridges across nations and cultures, students from affluent countries are enriched in their medical knowledge and understanding of the world, and linked programs provide essential venues for conducting research. There are bilateral training programs supported through US governmental organizations such as the NIH, CDC, USAID, and PEPFAR, as well as multilateral programs supported by the United Nations, the Pan American Health Organization, the Global Fund to Fight AIDS, TB and Malaria, and non-profit foundations such as the Gates and Clinton Foundations. Two examples of the effort to expand a trained international health workforce are the NIH's Fogarty International Center's Clinical Scholars Program and the US Centers for Disease Control's Epidemiology Intelligence Service, which now collaborates with 60 countries. Graduates form a core of committed individuals with clinical skills, and training in public health, public policy, government, and education. One of the first programs to recognize the need for expanding international training is the Epidemic Intelligence Service (EIS) of the Centers for Disease Control. Its mission is to train future frontline responders to assist local and state health departments in the United States and overseas to conduct epidemiological investigations, research, and public health surveillance; and to publish and present their work to the academic community, the media, and the public. There are educational opportunities in surveillance, communication, coordination, and the prompt application of public health principles to curtail remote epidemics. EIS emphasizes practical skills in areas such as computers, biostatistics, and epidemiology through day-to-day access to experienced epidemiologists who act as mentors and primary supervisors. Since the program began in the early 1950s, EIS has worked on more than 10,000 projects and expanded its geographic scope and areas of investigation. Recent infectious disease projects include polio eradication in Africa and Asia, investigation of a Hantavirus outbreak in the south-western United States, and training in bioterrorism. There are EIS field programs on every continent except Antarctica.

The Fogarty International Center has used its resources to train a large cadre of both domestic and international clinical investigators, many of whose insights have advanced global health, often with very practical public health implications. Observations that short courses of antiretroviral therapy can dramatically decrease mother-to-child HIV transmission have created moral imperatives that have facilitated treatment scale-up programs in sub Saharan Africa and other heavily affected, low-resource countries. The training of international medical and public health professionals in the conduct of clinical trials and related research will have beneficial effects for many years, based on the assumption that those trained will train other local colleagues. At the same time, the reality of "brain drain" and the limitation in dedicated resources to support clinical care and public health infrastructure in developing countries means that the implementation of new research insights may not be able to be immediately translated into appreciable benefits for those who are most in need.

The response to an infectious disease outbreak almost always begins at the local level. In a poor nation, the first responder might be a community health worker in a refugee camp or a health-care professional in the emergency room of a hospital in a large city. If the situation is grave, help will be needed from the outside. WHO regulations require nations to report outbreaks promptly and take measures to prevent their spread, while minimizing the negative impacts on trade and travel. Compliance depends largely on the action of individual governments. While the World Health Organization and the US CDC have made great strides in improving coordination, the international organizational response network remains largely a loose collection of multinational organizations, national, regional and local governments, private health-care institutions, health and education ministries, local and international NGOs, religious groups, research centers and universities, and UN programs, agencies, and funders. And there are more programs and partnerships for dealing with long-term health problems. Capacity on paper does not always translate into effective action, even in an affluent country like the United States – as evidenced when Hurricane Katrina hit the Gulf Coast in August 2005. The WHO currently is the first and leading multilateral organization for dealing with infectious disease threats. It has separate divisions for dealing with AIDS, TB, malaria, and childhood immunization, while all other infectious diseases come under Communicable Diseases. There is a global emergency response division. WHO relies on its member nations to conduct research. With at least 160 staff in more than 43 countries, the US CDC is the global swat team; after receiving an invitation from a government, they come in and help with surveillance, epidemic investigations, laboratory and epidemiological research, and training and public education to prevent and control communicable diseases. As fast travel and widening trade make the world smaller, there is an increasing need to improve the international response to infectious diseases. Hurricane Katrina showed that even in a wealthy nation the capacity on paper to respond to a disaster does not always easily translate to effective action in real life. Also, the system needs to move past emphasizing current disease threats to improving local primary-care medical and public health systems.

Descriptions of the social determinants of infectious diseases date back to antiquity. Ovid wrote: "What timid man does not avoid contact with the sick, fearing lest he contract a disease so near?" (*Pontic Epistles*, III.ii.13). Humankind has been perpetually aware of the presence of invisible agents that could devastate communities with great rapidity. However, until the development of the germ theory of microbiology, the understanding of the transmission of infectious agents was based on empirical observations – such as John Snow's recognition that people who obtained their water from the High Street pump in London were more likely to become ill with cholera. Advances in microbiology enabled nineteenth-century investigators to identify etiological agents of communicable disease outbreaks and to develop more rational approaches to disease prevention and treatment. Ironically, during this recent era of impressive advances in the understanding of the pathogenesis and transmission of infectious diseases, which have led to evidence-based approaches to treatment and prevention, new epidemics have emerged that have taken advantage of diverse technological breakthroughs, ranging from rapid global travel to blood transfusions and immunosuppression after organ transplants. Ultimately, humans must recognize that we live in a global gene pool, with any microbe within days of potential contact with any population. As human activities change, microbial evolution will continue to occur to enable organisms to take advantage of newly available niches and opportunities for replication. It is neither feasible nor desirable to live in a sterile world. Although microbes can cause devastating diseases, and plagues have altered the course of human history, some, like lactobacilli in the female genital tract, are commensual and assist in keeping pathogens at bay. Others have been domesticated to allow humans to enjoy cheese, beer, and wine. The ultimate purpose of this text is to enable the reader to become more knowledgeable about some of the most recent ways that micro-organisms have evolved in the modern era, and to stimulate thinking about how the clinical and public health communities can use the rapidly evolving tools of enhanced communication technologies, such as the Internet, to disseminate information as new epidemics emerge, and to share insights that may mitigate the harm that infectious disease pathogens may cause. Some of the best ways to avoid epidemics do not rely on new technologies, but derive from insights as old as humankind: people who are well-fed and not crowded are less likely to be sick, and destructive wars provide a milieu for contagious disease to spread rapidly. Thus, if we can learn from the past and utilize some of the amazing panoply of new technologies for information dissemination, infectious diseases will not disappear but their dissemination may be most efficiently curtailed.

## References

Dubos, R. and Dubos, J. (1952). *The White Plague: Tuberculosis, Man, and Society*. Boston: Little, Brown and Company.

Fauci, A.S. (2001). Infectious diseases: considerations for the 21st century. *Clinical Infectious Diseases* **32**, 675–685.

Rotello, G. (1997). *Sexual Ecology AIDS and the Destiny of Gay Men*. New York: Dutton Books.

Zinsser, H. (1935) *Rats, Lice and History*. New York: Bantam Books.

# Travel 1

**Mary E. Wilson and Lin H. Chen**

We live in a world of moving parts, fragmented, altered landscapes, and vast networks of connections. Human travel today is unprecedented in volume, reach, and speed (Wilson, 1995a). Spatial mobility for the average person increased more than 1000-fold between 1800 and 2000 (Cliff and Haggett, 2004; see Figure 1.1). Two million people cross international borders each day (WTO, 2006a). Movement of humans may be the result of planned travel – for business, tourism, education or research, missionary or volunteer work, visiting friends or family, or military purposes – but abrupt displacement of populations can also be a consequence of war, natural disasters, economic, political, or environmental events (Myers, 2001). Human activities have led to unprecedented movement or displacement of other species through intent (e.g. trade) or inadvertence (Smolinski *et al.*, 2003). The conveyances of travel, such as ships and airplanes, can become places for transmission of infectious diseases. The roads and railroad tracks built to carry people and products from one location to another can also fragment habitats and can lead to decreased biodiversity. Table 1.1 provides examples of infectious disease dissemination associated with human travel.

Infectious diseases in people, plants, and animals are dynamic and are influenced by biological, environmental, socioeconomic, political, demographic, and genetic factors (Wilson *et al.*, 1994; Wilson, 1995b). The abundance and variety of microbial life and its ongoing evolution mean that we will continue to encounter new infections or altered expressions of old ones in the foreseeable future. Among the characteristics of the global population today that favor the appearance and spread of infections – size, density, location, vulnerability, inequalities, and mobility – the latter will be the focus of this chapter.

Why does travel matter in the epidemiology of infectious diseases? Although many microbes that cause infectious diseases are found throughout the world, many others have a focal distribution because of the need for specific geoclimatic conditions or a particular intermediate or reservoir host, poor sanitation or control efforts, or other factors. Even for microbes that are globally distributed, risk of

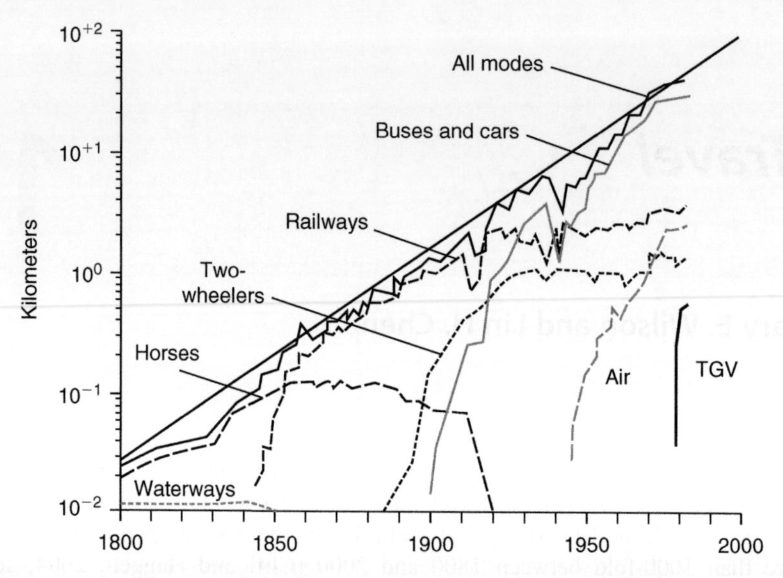


**Figure 1.1** Spatial mobility: increased spatial mobility of the population of France, changes over a 200-year period, 1800–2000. The vertical scale is logarithmic, showing that growth in average travel distance has increased exponentially over time. Based on Cliff and Haggett (2004).

exposure may vary greatly from one geographic area to another (Freedman *et al.*, 2006). Humans are interactive biological units; when they travel they can pick up pathogenic microbes that may make them sick (Wilson, 2003a). Whether they get sick or not, they may also carry pathogens or microbial genetic material (including resistance genes and virulence factors) in or on their bodies to a new location. Depending on the type of organism, its transmission mechanisms, and immunity of contacts, travelers may be able to introduce a pathogen into a new population.

Humans are social creatures who carry with them their traditions, customs, practices, dress codes, and values that may be observed and imitated or scorned and rejected by the people they visit. Travel typically is not just an origin and destination, but also includes stops and interactions along the way. Travelers have contact not only with local populations in areas they visit, but also with other travelers – who may reflect a wide range of geographic origins and carry their own microbiological baggage and customs that influence spread.

When people travel, they may engage in activities that they might not undertake at home. The experience in a new environment may lead them to have sexual contacts with new partners, try new foods or types of preparations, engage in risky water and other sports activities, receive injections or tattoos at local facilities, or handle or come into contact with animals. Many of these activities can put them at risk for infections that did not exist in their home environment.

**Table 1.1 Examples of infectious disease dissemination associated with human travel**

| Pathogen | Site of origin | Mode of spread | Consequences | Reference |
|---|---|---|---|---|
| HIV | Sub-Saharan Africa | Truck routes in Africa; air travel | 14.5 million infected outside site of origin; approximately 40 million people lived with HIV worldwide at the end of 2005, 4 million became newly infected, and 2.8 million died. | UNAIDS, 2006 |
| Measles | Asia (China, India), Europe | Air travel | Outbreaks associated with international adoption have occurred. An 11-year-old North Carolina resident traveled home from the United Kingdom via New York and Connecticut and transmitted measles to an infant contact; transmission to multiple states and countries was possible due to infectious period during flights. A 17-year-old adolescent who contracted measles in Bucharest, Romania, initiated an outbreak in Indiana, with 34 confirmed cases. An outbreak occurred in the Greater Boston area since May 2006, following an index case that had arrived recently from India. | CDC, 2004a, 2004b, 2005a, 2005b; Parker *et al.*, 2006; Massachusetts DPH, 2006 |
| Poliomyelitis | Six endemic countries (Afghanistan, Egypt, India, Niger, Nigeria, Pakistan) | Migration; air travel | Between 2002 and 2005, wild poliovirus spread to 21 previously polio-free countries; introduction led to sustained transmission in 13 countries. | CDC, 2005c, 2006a, 2006b |
| SARS | China, Hong Kong | Air travel following exposure in infected countries | From 1 November 2002 to 31 July 2003, SARS spread to > 25 countries, causing 8096 reported infections and 774 deaths. | WHO, 2003 |

(*Continued*)

**Table 1.1 (Continued)**

| Pathogen | Site of origin | Mode of spread | Consequences | Reference |
|---|---|---|---|---|
| Dengue | Tahiti | Air travel; association with a returning traveler with dengue-like symptoms | The first autochthonous outbreak in Hawaii since 1944 occurred in 2001–2002 and caused 122 laboratory-confirmed cases. | Effler *et al.*, 2005 |
| West Nile Virus | Israel | Unknown; possibly air travel via an infected person, bird, or mosquito | Since its initial detection in New York, 16,706 cases were reported to the CDC between 1999 and 2004. WNV has also spread to Canada, the Caribbean, and Latin America. | Hayes *et al.*, 2005 |
| Influenza | Worldwide | Cruise ships | Outbreaks occurred among cruise-ship passengers between New York and Montreal, Tahiti and Hawaii, and Alaska and the Yukon Territory. | CDC, 1997; Uyeki *et al.*, 2003 |
| Norovirus | Worldwide | Cruise ships, air travel | The Vessel Sanitation Program at the CDC identified > 12 outbreaks on cruise ships in 2002. An outbreak occurred among the crew of a flight, with limited transmission to passengers. | Widdowson *et al.*, 2004, 2005 |
| Tuberculosis, including multi-drug resistant tuberculosis | Worldwide, Saudi Arabia | Air travel | The 2005 TB rate in foreign-born persons in the US was 8.7 times that in US-born persons; the incidence of MDR TB is higher in low- and middle-income countries. Comparison of TB tests using a whole-blood assay (Quanti-FERON TB assay) prior to and after return from the Hajj showed 10% conversion consistent with exposure during the pilgrimage. | CDC, 2006c, 2006d; Gushulak and MacPherson, 2004; Wilder-Smith *et al* 2005 |

| | | | | |
|---|---|---|---|---|
| Mumps | UK | Air travel | Summer-camp outbreak in New York involved 31 cases and was associated with a counselor from the UK; attack rate was 5.7%. Outbreak began in Iowa in December 2005, and 2597 cases were reported from 11 states between 1 January and 2 May 2006; some cases were potentially infectious during air travel. | CDC, 2006e, 2006f, 2006g |
| Meningo-coccal meningitis | Saudi Arabia | Air travel | Pilgrims to the Hajj became infected and spread strains into other regions. | Moore *et al.*, 1989; CDC 2000, 2001; Wilder-Smith *et al.*, 2002; Dull *et al.*, 2005 |

The development of the travel industry has also altered the landscape in many countries, where luxury hotels occupy prime real estate and new infrastructure is created to serve travelers. Although much travel tends to follow well-trodden paths to well-known places, increasingly travelers are seeking remote locations – and today's transportation technology makes it increasingly easier than ever before to reach these areas.

All of this massive movement of the human population and trade is occurring as an overlay on the background of the natural migration of animals. Although birds, land animals, and marine animals can move thousands of kilometers in seasonal migrations, humans today have access to more parts of the Earth than any other species. Humans have also altered the planet (land, sea, and atmosphere) more than any other species.

## A brief history of travel and the movement of microbes

Throughout history, travelers have carried microbes to new geographic areas and susceptible populations and provided the spark that could ignite epidemics. Recurrent bubonic plague occurred along the Silk Road as early as the 600s. Plague followed the routes of trade caravans in medieval Europe over the years 1347–52, killing an estimated 20 million people (Cliff *et al.*, 2004). It reached Europe from Central Asia, arriving at Kaffa on the Black Sea in about 1347; from there it was carried by ships to the major ports of Europe and North Africa, and then spread over land routes. Explorers introduced smallpox, measles, and other infectious diseases into the New World, causing an estimated 56 million deaths (Black, 1992) and contributing to the collapse of the Aztec and Inca civilizations (Crosby, 1972).

Although Columbus is described as discovering the New World, the historian Carmichael describes the event as the creation of one new world from two old worlds. Cultural as well as biological transformation of the world followed the expanded movement of European overseas trade and exploration (Carmichael, 2006). Global human disease patterns also changed after 1500.

Aided by the modern steamship, which carried infected rats and fleas, a third bubonic plague pandemic circled the globe between 1884 and the early 1900s (Echenberg, 2002). Carried in grain wagons and other forms of transport, it moved inland from the port cities across southern China, reaching Hong Kong in 1894, Singapore and Bombay in 1896, Alexandria, Oporto, and Honolulu in 1899, Sydney, Buenos Aires, Rio de Janeiro, and San Francisco in 1900, and Cape Town in 1901. Many port cities had overcrowded urban tenements with impoverished populations, often immigrants, who were most often infected and were subjected to isolation, rejection, and stigmatization (Echenberg, 2002).

In 1787, the time to travel from England to Australia by sailing vessel was about a year (Cliff and Haggett, 2004). Many infections transmitted from person to person, like measles, were no longer a threat by the time of arrival because passengers were dead or immune. After 1860 the faster steamship quickly replaced the sailing vessel, and with that technological advance came shorter boat trips that increased the risk of transmitting infectious agents across oceans. The time to travel from England to Australia had dropped from 100 days (by clipper in 1840) to 50 days by the early 1900s. The development of relatively inexpensive air travel further increased the risk of spreading infectious agents across vast distances. Now a traveler can reach almost any major city on Earth within 24 hours – less time than the incubation period of most infectious diseases.

Last century, the Spanish flu of 1918–19 spread around the world in three waves, carried by humans, and killed as many as 50 million people (Barry, 2004). In the latter decades of the twentieth century, the virus that causes AIDS was carried throughout the world with human travelers as the primary transporters, who disseminated it to all countries. The best evidence suggests that the virus emerged after multiple introductions of related simian viruses in African monkeys and apes into the human population and the subsequent evolution to the human immunodeficiency virus (HIV) (Hahn *et al.*, 2000). The connectedness and ease of travel in the late twentieth century allowed it to spread widely before the magnitude of the threat was recognized. Infections that can be transmitted from person to person can be carried by travelers to any part of the Earth (Wilson, 2003b).

The number of plant and animal species is higher in tropical areas and decreases as the distance from the equator increases; this is known as the latitudinal species diversity gradient. A recent analysis of species that cause infectious diseases (including parasitic species) shows a link between latitude and the spatial pattern of human pathogens, and suggests that climatic factors play a primary role in this pattern (Guernier *et al.*, 2004).

Even today, the spectrum of disease from infections varies in relation to place. A recent analysis of sentinel surveillance data on >17,000 ill, returned travelers seen at GeoSentinel sites (staffed by clinicians knowledgeable about clinical tropical medicine) found significant regional differences in proportionate morbidity for 16 broad syndromic categories (Freedman *et al.*, 2006). Systemic febrile illness occurred disproportionately among travelers returning from sub-Saharan Africa and Southeast Asia.

## Modern global travel

The global population increased from 1.6 billion to 6.1 billion in the twentieth century, and reached 6.5 billion in 2006; everyone born before 1960, when the population was 3 billion, has lived through a doubling of the global population. Global travel has increased even more rapidly. Between 1950 and 2005, international tourist arrivals increased 32-fold, reaching 808 million in 2005 (Table 1.2; WTO, 2006a).

People travel for many reasons – tourism, work, research, study, humanitarian aid, religious purposes or missionary work, visits to friends and relatives, displacement due to catastrophic events, for economic incentives, and due to environmental disasters or sociopolitical upheaval. According to the data from the World Tourism Organization (WTO), international trips to visit friends and relatives and for health or religious purposes rose from 19.7 percent of 441 million in 1990 to 24.2 percent of 763 million in 2004; business and professional travel also increased, from 13.7 percent to 15.7 percent. In contrast, the proportion of visits for leisure and holiday travel declined from 55.4 percent to 51.8 percent in the same period of time (see Table 1.3).

The volume of travel has grown exponentially with an average annual growth of international tourist arrivals of 6.5 percent (WTO, 2006a). In the US, 18 airports receive more than 500,000 international arrivals by air annually (16 have a total of >25 million air travelers per year) and 14 ports each receive more than 150,000 maritime passengers (Sivitz *et al.*, 2006: 127). Figure 1.2 displays the civilian global aviation network and the extensive interconnections. In total, the US has 19,500 airports, and at peak times 5000 airplanes are aloft in US airspace.

**Table 1.2 Growth in world population and international tourist arrivals**

| Year | World population (millions) | International tourist arrivals (millions) |
|---|---|---|
| 1950 | 2557 | 25.3 |
| 2005 | 6451 | 808 |
| Change | ×2.5 | ×32 |

Data from the US Census Bureau (http://www.census.gov/ipc/www/worldpop.html) and World Tourism Organization (http://www.world-tourism.org/facts/menu.html).

**Table 1.3 Arrivals by purpose of visit**

| Reason for travel | 1990 (international tourist arrivals = 441 million) % of total international arrivals | 2004 (international tourist arrivals = 763 million) % of total international arrivals |
|---|---|---|
| Leisure, holiday | 55.4 | 51.8 |
| Business, professional | 13.7 | 15.7 |
| VFR, health, religion | 19.7 | 24.2 |
| Not specified | 11.2 | 8.3 |

Data from World Tourism Organization: Tourism Indicators (available at http://www.world-tourism.org/facts/menu.html).

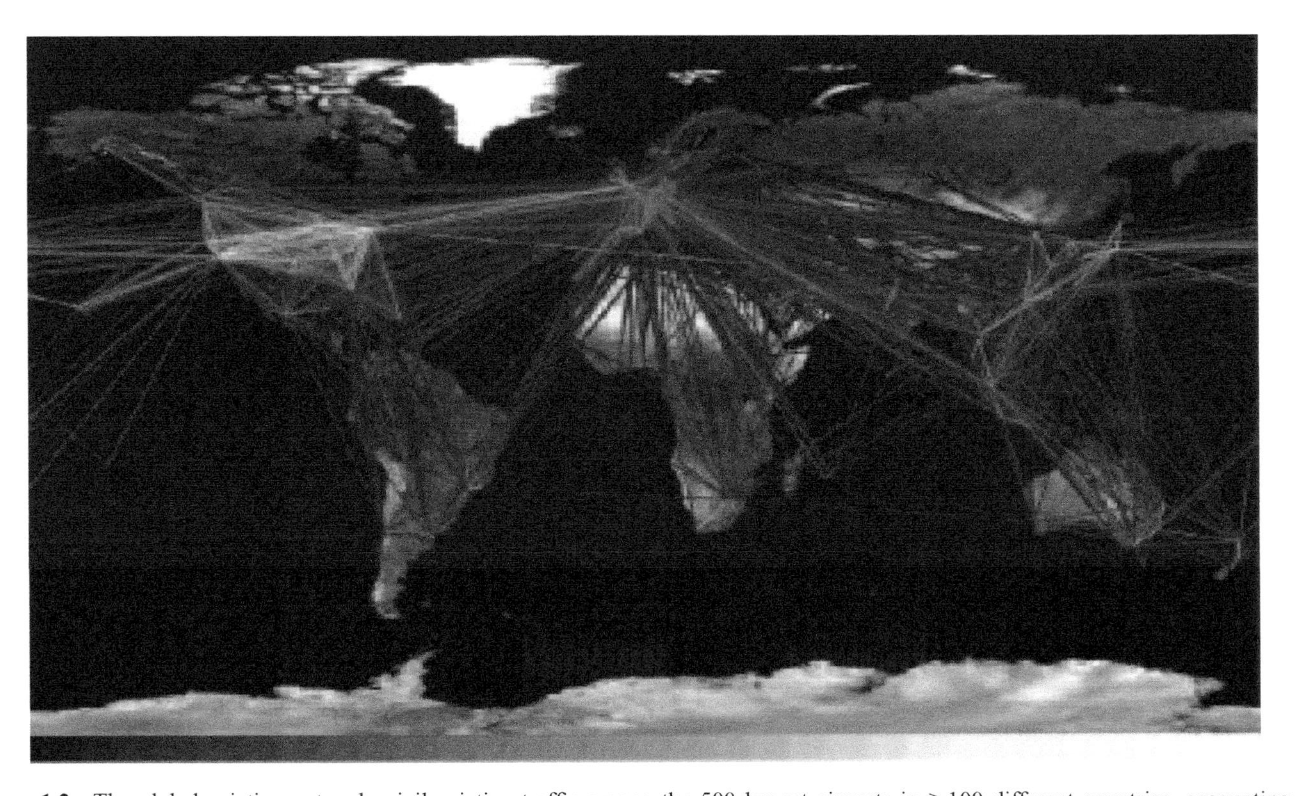

**Figure 1.2** The global aviation network: civil aviation traffic among the 500 largest airports in >100 different countries, accounting for >95 percent of international civil aviation traffic. Each line represents a direct connection between airports, and its shade encodes the number of passengers per day (see shaded bar at the bottom) traveling between two airports. From Figure 1 in Hufnagel *et al.* (2004).

Travel by cruise ship is also growing rapidly. In 2003, 184 cruise ships served the US cruise market and 7.4 million cruise passengers went through US ports (Sivitz *et al.*, 2006).

Today, approximately 2 percent of the world's population, or >200 million people, reside outside their country of birth (Gushulak and MacPherson, 2004). These include immigrants, migrant workers, refugees, asylum seekers, and international students. Those who return to their home country to visit friends and relatives (VFR) are more likely to visit rural areas or to stay in accommodations that lack good sanitary facilities and safe water. Because they are "going home," they may be unaware of risks to themselves and their children (who may have been born in a developed country) and thus have inadequate preparation (e.g. vaccines, chemoprophylaxis). VFRs are at higher risk for malaria, typhoid fever, and other infectious diseases than are people traveling for other purposes (Bacaner *et al.*, 2004; Leder *et al.*, 2007) – for example, 43 percent of imported malaria cases in Europe during the period January 1991 to September 2001 occurred in VFRs and other migrant populations (Schlagenhauf *et al.*, 2003). The population of foreign-born in the US is estimated to be 34.2 million, with more than half from Latin America and about a quarter from Asia.

In 2004, international tourist arrivals traveled mainly by air or land, with sea travel contributing only 7.4 percent of the total (WTO, 2006b; Table.1.4). For Africa, America, and the Asia-Pacific regions, air travel surpassed land travel. This indicates a greater proportion of long-haul travel, likely associated with the use of large aircraft, and possibly greater potential for dispersal of organisms into and out of these regions.

## Patterns of world travel today

A person with SARS in 2003 illustrates the potential consequences of today's travel patterns (Breugelmans *et al.*, 2004). A 48-year-old businessman flew from

**Table 1.4 Arrivals by mode of transport, 2004**

| | Air (%) | Land (%) | Water (%) | Unspecified route (%) |
|---|---|---|---|---|
| World | 43 | 49.3 | 7.4 | 0.3 |
| Africa | 48 | 43.9 | 7.8 | 0.3 |
| America | 53.1 | 41.3 | 5.5 | 0.1 |
| Asia and Pacific | 46.2 | 31.9 | 10.7 | 1.2 |
| Europe | 38.1 | 54.8 | 7.0 | 0.1 |
| Middle East | 45.3 | 49.5 | 5.3 | 0.0 |

Data from World Tourism Organization: Tourism Indicators (at http://www.world-tourism.org/facts/menu.html).

Hong Kong to Frankfurt, Germany, on 30 March 2003. He traveled on seven flights throughout Europe during a five-day period from 31 March to 4 April 2003, with stops in Barcelona, London, Munich, and Hong Kong. He was admitted to a hospital in Hong Kong on 8 April with possible SARS, which was confirmed on April 10. His potential contacts spanned many countries.

The "sphere of travel" has enlarged over the years. Bradley described the spatial range of travel in sequential generations of his own family, demonstrating a 10-fold increase with each generation (Bradley, 1989). As travel has become faster, cheaper, and safer, people take more trips and travel longer distances. The average daily distance that an individual in France travels has increased over 1000-fold over the past 200 years, and presumably is similar in other populations in industrialized countries (Cliff and Haggett, 2004). Social and demographic changes have affected travel patterns, and global immigration, migration, and changes in family structure today lead to frequent long-distance travel, often by air, to visit family.

Aircraft and cruise ships are also increasing in size, which expands the population of fellow passengers, often from multiple geographic regions, who come into close contact for a period of hours to days. Assuming homogeneous mixing of passengers, the risk of being exposed to a person with a communicable disease increases four-fold when an aircraft doubles in size – for example, from 200 to 400 passengers (Bradley, 1989). The WTO forecast of growth in long-haul travel between regions over intraregional travel also illustrates the concept of an enlarging sphere of travel (WTO, 2006b). In addition, regional shares for international tourist arrivals are shifting. The WTO predicts an average global growth of 4.1 percent in international tourist arrivals per year, but greater than average growth for Africa, and even greater growth of >6 percent for East Asia and the Pacific, South Asia, and the Middle East (see Figure 1.3). Although Europe is projected to remain the most popular destination, its overall share in the market is predicted to decline.

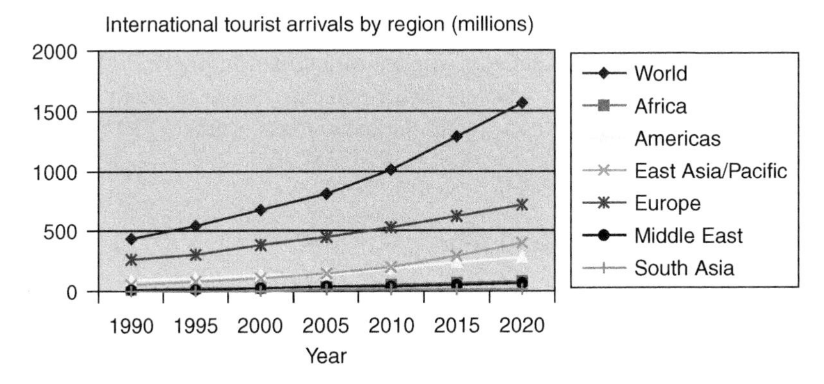


**Figure 1.3** International tourist arrivals by region (millions) with forecast. Data from WTO (2006a).

While Europe's and America's combined share in world tourist arrivals was >95 percent in 1950, it declined to 82 percent in 1990 and 76 percent in 2000. The shift of international travel to developing regions in tropical and subtropical regions also increases potential exposure to microbes and vectors endemic in those regions.

Big international airports are typically situated near large metropolitan areas. In many of the developing, low-latitude regions (most of them tropical or subtropical), large periurban slums surround large cities, often populated by people with families in rural areas. Regularly scheduled international flights and their passengers bring these populations into potential contact, and link urban and rural biota of the world.

## Travelers' risk behavior

During travel and exploration of regions far from home, individuals may engage in risky activities that can lead to potential exposure to pathogens in blood and body fluids, including HIV, hepatitis C, hepatitis B, CMV, HTLV-1, and other sexually transmitted infections. Some travelers choose their destinations for sex tourism, and many engage in casual sex with new partners (Marrazzo, 2005). A survey of >9000 European travelers regarding their potential exposure to hepatitis B through sex or other contacts found 6.6–11.2 percent to be at high risk (with 24.4 percent vaccinated), 60.8–75.8 percent had potential risk (with 19.2 percent vaccinated), and only 33.4 percent had no identifiable risk of exposure (Zuckerman and Steffen, 2000).

Among Canadian travelers surveyed, 15 percent had potential exposure to blood and body fluids; 9 percent had sexual intercourse with a new partner, 5 percent shared implements such as a razor or toothbrush, 3.2 percent had an injection for medical treatment, 1 percent had acupuncture or other percutaneous non-traditional treatment, 0.5 percent received tattoos or body piercing, and 0.5 percent experienced abrasive injury (Correia *et al.*, 2001).

In a study of tourists departing from Cuzco, Peru, 5.6 percent indicated that they had engaged in sexual activity with a new partner during their stay there, most commonly with other travelers (54.3 percent). Some reported sex with local partners (40.7 percent) and with commercial sex workers (2.15 percent) (Cabada *et al.*, 2003).

## The impact of rumors and fear

Fear of disease can deter travel even if risk is minimal or absent. Individuals, organizations, and countries may not base decisions on sound science. An infectious disease may have a severe economic impact on a region, making governments reluctant to acknowledge its presence. Diseases are not neutral in reputation. Infections such as plague (Wilson, 1995c) and cholera carry stigma, and may lead to irrational decisions about travel and trade. For example, after reports of a plague outbreak in India in 1994, many countries stopped importing foodstuffs and textiles from India even though the World Health Organization requested that no travel or trade restrictions be imposed on the country. Travel to India dropped, with the loss of at least 2.2 million tourists during one season. The estimated losses secondary to the reported outbreak were more than US$2 billion (Cash and Narasimhan, 2000).

The number of international tourist arrivals declined in 2003 in response to outbreaks of SARS and the associated travel advisories. The WTO reported that arrivals at some affected countries in Asia plunged to below 50 percent of their usual levels in April and May (WTO, 2005). Although the region rebounded quickly, SARS was responsible for a 9 percent overall loss in travel volume for Asia for the year 2003. Estimates of the global cost of SARS associated with lost economic activity have been estimated at about US$40 billion, and perhaps as high as US$54 billion (Lee and McKibbin, 2004).

Rumors and fear can also affect the control of infectious diseases. Rumors that polio vaccine contained the AIDS virus and hormones that could sterilize girls in the largely Muslim population led Nigeria's northern states to halt polio vaccination in mid-2003 (Heymann and Aylward, 2004; Samba *et al.*, 2004). Political and religious factors contributed to the decision. By the time additional testing had been completed and the vaccine declared safe, polio outbreaks had spread in Nigeria and fueled outbreaks in other countries (see Figure 1.4).

## The human transport of microbes and microbial genetic material

Travelers carry microbes and microbial genetic material, a large portion of it the trillions of bacteria that form the commensal flora (Wilson, 2003a; Sears, 2005). They may carry and transmit pathogenic microbes and microbial genetic material, including resistance and virulence genes, sometimes in the absence of symptoms. Carriage may be transient or long term. Chronic infections, such as HIV, hepatitis C, and hepatitis B, can persist and be transmitted even if the individual is unaware that infection is present. Infections can be latent (e.g. tuberculosis) and can potentially reactivate at a time and place remote from acquisition. Potentially pathogenic microbes can be carried on the skin, in the respiratory tract, in the genital tract, in the gastrointestinal tract, or in blood and body fluids. There are thus multiple potential ways that pathogens can be transmitted.

Even when an infection is already found in a geographic region, travelers may introduce a new serotype or strain that is more virulent or resistant to antimicrobials. Wide use of antimicrobial agents puts pressure on microbial populations and contributes to the emergence of resistant microbes, but travelers may

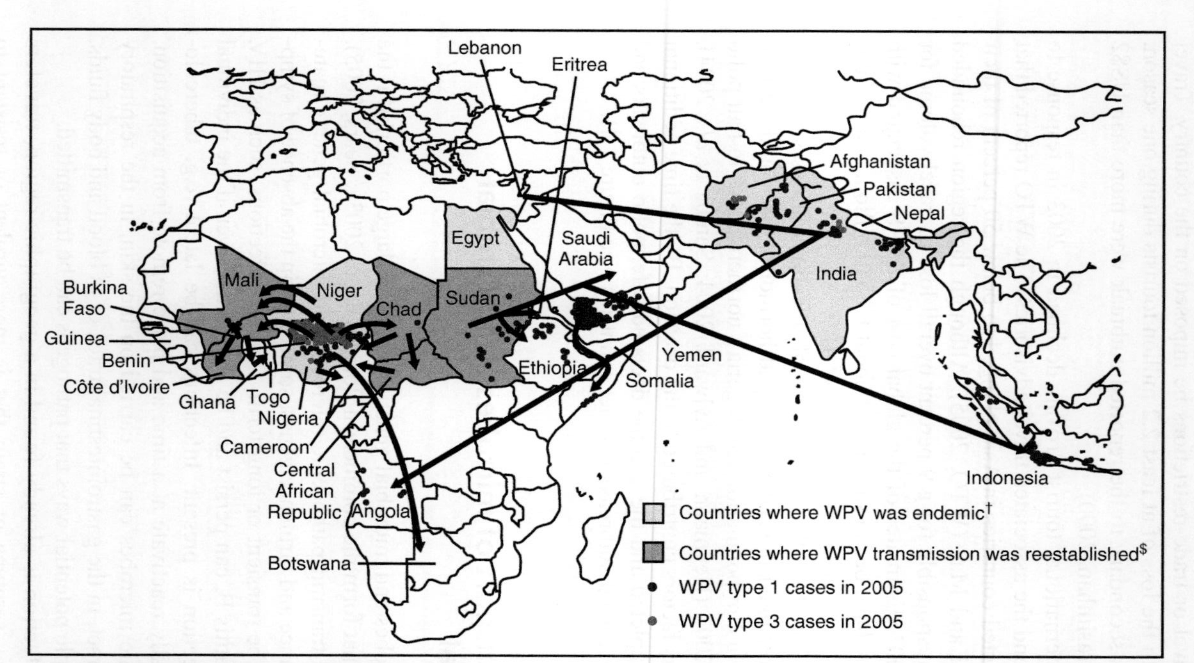


*Routes (not all importation events) indicated by arrows.
†As of February 1, 2006, Niger and Egypt were considered no longer endemic for WPV because neither country had indigenous transmission during the preceding 12 months.
$Countries were considered to have reestablished transmission if WPV was detected for >1 year after importation. The majority of these countries have not experienced WPV type 1 transmission since July 2005.

**Figure 1.4** Wild poliovirus (WPV) cases in 2005 and WPV importation routes during the period 2002–2005 worldwide. From CDC (2006b).

be significant in their dissemination. For example, only 10 clones of penicillin-resistant *Streptococcus pneumoniae* were responsible for 85 percent of invasive disease caused by this organism in the US in 1998 (Corso *et al.*, 1998).

## Transmission mechanisms

Which means of transmission are most important in serious infectious diseases? A breakdown of the infections causing death globally in 1995 found that infections that are transmitted from person to person (e.g. tuberculosis, measles, HIV) caused 65 percent of deaths; infections acquired from contaminated food, water, or soil (e.g. cholera and many other infections causing diarrhea, hepatitis A) caused 22 percent; vector-borne infections (primarily malaria) caused 13 percent; and infections from animals (e.g. rabies) caused 0.3 percent (WHO, 1996). These numbers are useful as general approximations, but vastly underestimate the role that animals play as a source of human pathogens. Many major human pathogens evolved in the recent or remote past from related pathogens in animals (e.g. HIV, measles, tuberculosis, malaria) (Weiss, 2001), and the majority of recently emerged infectious diseases are zoonoses, many of them viruses (Woolhouse, 2002; Smolinski *et al.*, 2003; see also Chapter 4).

Ease of transmission varies considerably by pathogen and by site of infection. Some microbes are able to spread in multiple ways. For example, the seasonal influenza viruses can be spread by direct contact (e.g. by contaminated objects or from hands of infected person), by large or medium droplets, or by droplet nuclei (tiny-droplet aerosol) (Musher, 2003). Tuberculosis, on the other hand, is spread by droplet nuclei that are small enough to bypass the trapping mechanisms of the upper airway and can reach the lungs. *Neisseria meningitidis* (the cause of meningococcal meningitis and sepsis) is spread by direct contact and large or medium droplets, but not by tiny-droplet aerosol. The mechanisms of transmission and transmissibility are key characteristics of microbes that influence ease of spread and determine what interventions will be effective. Timing of onset of infectiousness relative to onset of symptoms and duration of infectiousness are important factors in transmission dynamics.

Environmental conditions may also influence transmissibility. Influenza virus survives better in cool temperatures and at low humidity, which may partially explain the seasonality of influenza in temperate areas. Several of the summertime outbreaks of influenza have involved transmission in air-conditioned indoor spaces, where air was cool and dried, and people were in close contact (Uyeki *et al.*, 2003).

## Basic reproductive rate

Whether infection can be sustained and will spread in a new population depends on the basic reproductive rate ($R_0$), which is the average number of secondary infections produced by an infected individual in a susceptible population (Anderson and May, 1991). For an infection to be sustained in a new population, the basic reproductive rate must exceed one (May *et al.*, 2001). The $R_0$ for a specific infection may vary depending on time, place, and population. For example, in the pre-vaccine era the $R_0$ for measles was 16–18 in England and Wales in 1950–68, and 5–6 in Kansas, USA, in 1918–21. A plausible explanation for the difference is differences in population density and mixing of populations. In the early 1980s, the $R_0$ for HIV was 11–12 in Nairobi, Kenya, in a prostitute population and 2–5 in male homosexuals in England and Wales (Anderson and May, 1991). Based on detailed epidemiologic data from Singapore and other locations, Lipsitch and colleagues estimated that the reproductive number for SARS was about 3 (Lipsitch *et al.*, 2003). SARS was barely containable using isolation and quarantine measures. One biological characteristic of SARS that favored its control was that virtually all transmission occurred after patients had developed symptoms. This is not the case for seasonal influenza, HIV, and many other infections that can be transmitted in the absence of symptoms (Fraser *et al.*, 2004).

Multiple factors can influence the $R_0$, including population density (noted above), behavior, genetics of the host population, and evolution of the pathogen. If a population includes a large number of hosts who are immunocompromised and remain infected and infectious for a prolonged period, this could also influence $R_0$. A pathogen with a high mutation rate that allows it to adapt more readily to new hosts is more likely to be successful in emergence (Antia *et al.*, 2003).

## Crossing the species barrier

Crossing the species barrier requires several steps, the first being contact with the microbe – and travel today provides more potential exposure events with a wide range of other species or microbes in the environment than ever before. To cross successfully from one species to another, a virus must be able to find a cell type that it can infect (Webby *et al.*, 2004). Even if it can replicate, it must be able to avoid or overcome host immune response, exit from the cell, and infect additional cells. Finally, it must be able to exit the host in such a way that it can be transmitted to other humans. For many animal viruses, humans are always or usually a dead-end host (e.g. hantaviruses causing hantavirus pulmonary syndrome), and are irrelevant in maintenance of the microbe in nature. The avian influenza A virus (H5N1) has not transmitted efficiently from person to person as of mid-2006, and a potential explanation is that the cells in the upper respiratory tract of humans do not have receptors to which the virus can bind. The virus can replicate efficiently only in the lower respiratory tract (alveolar cells) of humans, from which it is not transmitted easily by sneezing and coughing (Shinya *et al.*, 2006).

RNA viruses have been prominent among microbes causing emerging infections. They are extremely mutable; many mutations can occur during replication, and

RNA viruses have few or no proofreading mechanisms (Webby *et al.*, 2004). These characteristics generate viral diversity. Variants with any advantage in a particular host or under specific environmental conditions are selected and can amplify. These viruses can also generate diversity through recombination and reassortment.

## Superspreading and heterogeneity of transmission

In SARS and some other infections, a few individuals seem to account for a disproportionate number of the secondary cases. In Singapore, for example, 162 (81 percent) of probable SARS patients showed no evidence of secondary transmission, whereas 5 persons each transmitted infection to 10 or more others (see Figure 1.5; CDC 2003). Lloyd-Smith and colleagues showed that heterogeneity in transmission, instead of being an anomalous event, was a common feature of disease spread (Lloyd-Smith *et al.*, 2005). In several examples, 20 percent of the cases caused 80 percent of the transmission. This is of more than theoretical interest, as it can affect the course of an epidemic and can influence the choice of interventions to try to curb the spread of infection (Galvani and May, 2005). If highly contagious (or highly connected, for sexually transmitted infections) individuals can be identified, it may be possible to contain spread with fewer resources.

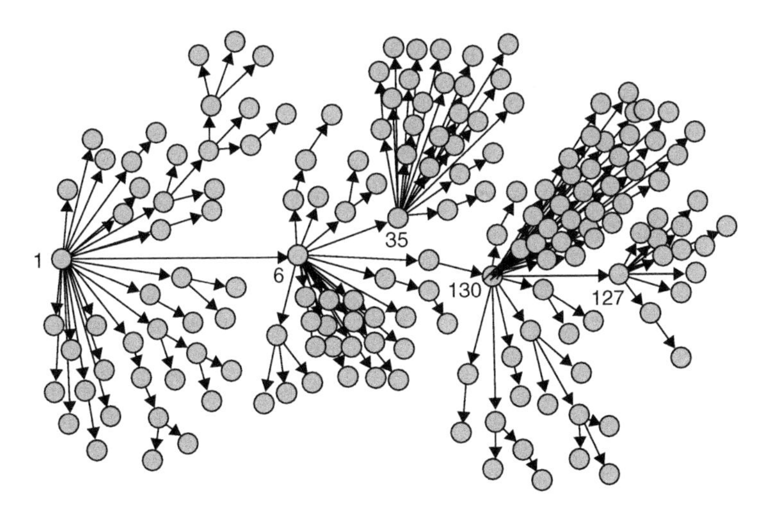


* Patient 1 represents Case 1; Patient 6, Case 2; Patient 35, Case 3; Patient 130, Case 4; and Patient 127, Case 5. Excludes 22 cases with either no or poorly defined direct contacts or who were cases translocated to Singapore and the seven contacts of one of these cases. *Reference*: Bogatti SP. Netdraw 1.0 Network Visualization Software. Harvard, Massachusetts: Analytic Technologies, 2002.

**Figure 1.5** Superspreaders. Probable SARS cases by reported source of infection, Singapore (25 Feb–30 Apr 2003). From CDC (2003).

A small study that may be relevant to the topic of heterogeneity in transmission of respiratory pathogens assessed exhaled aerosols. Eleven healthy adult volunteers (18–65 years old) were connected to a device that measured the number and size of particles exhaled during quiet breathing. The number of exhaled particles varied dramatically among the subjects, ranging from 1 per liter to >10,000 per liter. Over a 3-month period, the high producers remained high producers on repeat assessment (Edwards *et al.*, 2004).

## Receptivity of places and populations

Receptivity of places and populations to the introduction of a microbe by a traveler varies greatly. Many infections require the presence of a specific arthropod vector, such as a mosquito, to support the replication and development of the microbe and to carry it to a susceptible host. Others require a specific animal, such as a rodent, as a reservoir host. Animals such as birds, pigs, and others may serve as an amplifying host – i.e. an animal in which a microbe can replicate to such a level that it can be transmitted, often via an arthropod vector. Environmental conditions (e.g. temperature, humidity, and rainfall, among others) have a profound effect on the survival and abundance of many vectors and reservoir hosts.

West Nile Virus (WNV) became established for the first time in the Western Hemisphere in 1999 in New York, and has since spread across the United States, into Canada, Mexico, Central America, and the Caribbean. Whether the virus initially arrived in an infected person, bird, or mosquito, the virus found a congenial environment in which to persist and expand its geographic reach. Already in place were abundant mosquitoes of multiple species that were competent to support its growth and had frequent contact with humans. Dozens of species of birds were susceptible, and hundreds of thousands of birds died from WNV infection. Crows seemed especially vulnerable. In fact, dead-crow density was used as a marker for the arrival of the virus into new regions, as deaths in birds were often observed before human cases were diagnosed (Eidson *et al.*, 2001).

The physicochemical environment may influence the presence or absence, variety, and abundance of potential vectors and reservoir hosts. Additional factors also affect the receptivity of an area or population to the introduction of pathogens (Wilson, 1995b). These include housing, sanitation, general living conditions, and education. To become infected, a person must be exposed. Crowded living conditions, absence of sanitation, lack of clean water, poor housing (e.g. absence of screens and doors, lack of good ventilation), lack of knowledge regarding preventive measures, and close contact with animals (including rodents, pets, and food animals) all increase the probability of exposures to microbes that are transmitted from person to person, from contaminated food and water, from vectors, and from animals.

In Laredo (Texas) and Nuevo Laredo (Mexico), two urban areas separated only by the Rio Grande River, researchers found that residents of Nuevo Laredo were significantly more likely to have antibodies indicating recent or past infection with dengue virus, a mosquito-borne infection, than were those of Laredo, Texas (Reiter *et al.*, 2003). Environmental sampling showed that the mosquito vector (*Aedes aegypti*) was present in both cities, and infested containers were more common on the Texas side. In a multivariate analysis, residences without air-conditioning were 2.6-fold more likely to have a dengue-positive occupant, suggesting that socioeconomic factors were key determinants in exposure to infection (see also Chapter 4).

When a human pathogen is introduced into a new geographic region, socioeconomic, political, demographic, and other factors potentially influence the geographic extent of spread and the burden of disease in morbidity and mortality. In general, the poorest populations are the most likely to suffer and die from infectious diseases. In the poorest 20 percent of the world population, communicable diseases led to 59 percent of the deaths and 64 percent of the DALYs (disability adjusted life years) lost, whereas in the richest 20 percent communicable diseases accounted for only 8 percent of the deaths and 44 percent of DALYs lost (Gwatkin *et al.*, 1999).

In some instances, however, individuals who are extensively connected because they have resources that allow them to travel may be more likely to come into contact with unusual pathogens. In fact, in some countries, such as Tanzania and Kenya, household wealth is positively associated with HIV prevalence (Shelton *et al.*, 2005). Wealth may be associated with mobility and resources to maintain social interactions and concurrent sexual partnerships. In any country, availability of a strong health-care system and wide access to medical resources make it more likely that a disease will be recognized early and diagnosed. Access to treatment may limit morbidity and mortality, as well as spread of infection. Transparency about reporting infectious diseases and the capacity to organize and support appropriate public health investigation and response may limit the impact of an introduced infection. Social support may also be important in the successful completion of treatment for infections that require prolonged therapy, such as tuberculosis.

Potential receptor populations differ in size, density, location, mobility, vulnerability, and demographics (Wilson, 2003b). The human population today is larger than ever in history, and more urbanized, with about half of the world's population now living in urban areas. The populations that are growing the most rapidly are urban populations in developing countries – typically in low latitude areas. Large periurban slums surround many of these megacities that are linked to the rest of the world by air travel. Two types of vulnerability that characterize populations today (in addition to undernutrition, mentioned below) are AIDS and aging. Both make populations more vulnerable to a variety of infections.

## Immunity

In addition to the mechanisms noted above, populations can be completely or relatively resistant to the introduction and establishment of a microbe because of innate or acquired immunity. Levels of immunity may be high because of past infection or immunization. Levels of immunity to hepatitis A are high in many developing countries where most people are infected at a young age because the virus circulates widely. Paradoxically, as socioeconomic status and sanitation in a country improve, outbreaks in older children and young adults begin to occur because exposure has not occurred at a young age (Jacobsen and Koopman, 2004). Hepatitis A as a cause of outbreaks becomes more visible because of the shifting upward of the age at which infection occurs.

High levels of immunity from immunization may protect a population from introduction of infection. In the US today, circulation of endemic measles has been interrupted, yet infections continue to occur sporadically in travelers from endemic regions or their contacts. Measles has a high $R_0$, and spread can occur even in populations with high levels of immunity. Figure 1.6 shows measles importations into the Americas in 1997, with source countries in Europe and Asia (de Quadros *et al.*, 2004). Fourteen measles outbreaks were identified in the US during 2001–2004 (CDC, 2005a), seven of which originated with a US

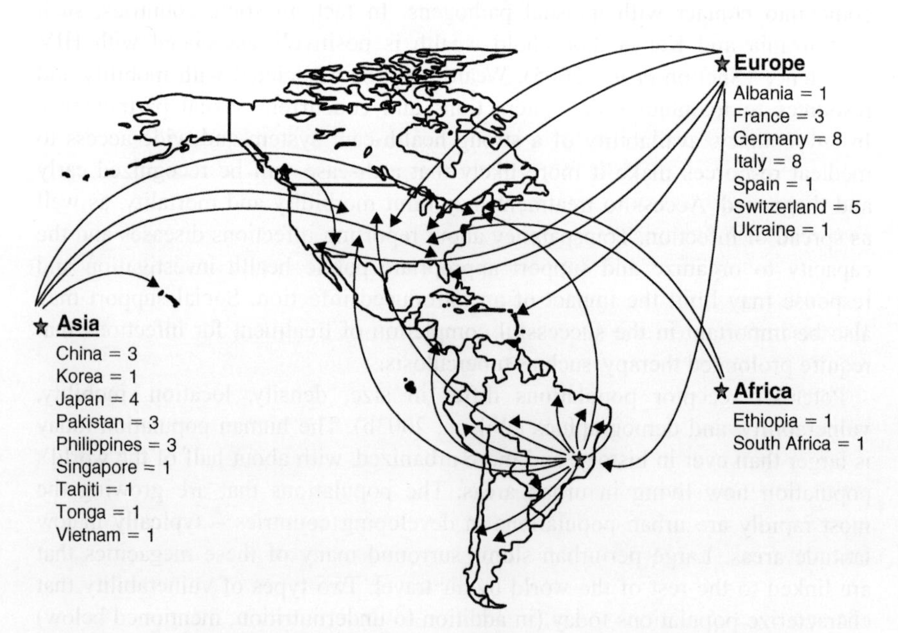


**Figure 1.6** Measles importations into the Americas, 1997. From de Quadros *et al.* (2004).

resident traveler. Genetic factors can influence susceptibility to infection or its expression, though for most infections this is not well defined. Nutritional status can influence outcome of infection and may determine the burden from infection in a population. Globally, undernutrition and micronutrient deficiencies are a leading cause of health loss and contribute to deaths from many infections, including malaria, measles, diarrhea, and pneumonia (Ezzati *et al.*, 2002).

The immune deficiency that accompanies HIV infection has led to recognition of infections previously undefined. As a sentinel population, their infections have also helped to chart the presence and geographic distribution of infections that may have previously gone unrecognized. Because of the potential for late reactivation of latent infections, such as visceral leishmaniasis, Chagas disease, and tuberculosis, it is important to review the history of travel and past residence of HIV-infected individuals.

Some types of immune deficiency can predispose to prolonged excretion of live vaccine-associated poliovirus (MacLennan *et al.*, 2004). Virulent vaccine-derived poliovirus has caused cases of paralytic polio in several countries (China, Egypt, Haiti, Madagascar, and the Philippines) (Kew *et al.*, 2002). In March 2005, a 22-year-old woman from Arizona contracted paralytic polio caused by vaccine-derived poliovirus (CDC, 2006a) after travel for a study-abroad program. In Costa Rica, she had had contact with an infant recently vaccinated with live oral poliovirus. She was unvaccinated against polio because of a religious exemption. Cases of vaccine-associated paralytic poliomyelitis (VAPP) had occurred in the past in the US, but no cases had been reported since 1999. The ongoing movement of poliovirus is further highlighted by events in the US. In September 2005, vaccine-derived poliovirus was found in an unvaccinated, immunocompromised infant in Minnesota (CDC, 2005c). Another four other children in the community also had asymptomatic infection. The virus differed about 2.3 percent from the parent Sabin vaccine strain, suggesting that it had been replicating for about two years and most likely originated from a visitor to the US who had received OPV elsewhere. Neither the infant nor the family had traveled internationally.

Dengue fever is an example that illustrates the interplay of multiple factors in the emergence and spread of an infection (see also Chapter 4). Dengue, a flavivirus transmitted by *Aedes* mosquitoes in tropical and subtropical regions, is increasing in incidence and in geographic range. An estimated 100 million cases of dengue fever and 25,000 cases of dengue hemorrhagic fever (DHF), and 25,000 deaths, occur annually (Gubler, 2002). Four serologically distinct serotypes exist. Because infection confers immunity only to the infecting serotype, a person can be infected up to four times. Having had prior infection with a dengue virus increases the likelihood of severe dengue fever (hemorrhagic fever or shock syndrome) if there is then infection with a different serotype. Viremic travelers are essential for the movement of the virus from place to place. The incubation period (usually 2–7 days) allows a person to move to another location before infection manifests. If mosquitoes competent to transmit dengue are present and one or more bite the infected person at the time the virus is in the bloodstream (median of 5 days but range of 1–12 days) (Gubler *et al.*, 1981) and survive to transmit the virus to another human, infection can be established in a new area. The main mosquito vector, *Aedes aegypti,* is well adapted to the contemporary urban landscape. It breeds well in discarded plastic cups, flower pots, used tires, and other urban litter, it prefers human blood, and it enters homes. Among the multiple reasons for the increase in dengue globally are: increased regional and global travel (including especially travel to tropical and subtropical areas) and trade; poor vector control and pesticide resistance, which has allowed expansion of the areas infested with *Aedes aegypti*; poor housing (absence of screens, doors) and water supply (water stored in containers that are good breeding sites for mosquitoes); growth of urban populations in tropical areas globally; and increase in number of urban areas that have reached a population size sufficient to sustain the ongoing circulation of dengue viruses. Over the past two centuries, the number of dengue lineages has been increasing roughly in parallel with the size of the human population.

In addition, another competent vector, *Aedes albopictus*, the Asian tiger mosquito, has also been introduced into a number of countries. In the US it was introduced in 1985, probably via used tires shipped from Asia, and has subsequently spread (Moore and Mitchell, 1997). Its dispersal followed interstate highways. It was the mosquito vector responsible for an outbreak in Hawaii in 2001 (Effler *et al.*, 2005). Phylogenic analysis found that the Hawaiian virus was similar to contemporaneous dengue isolates from Tahiti, suggesting that viremic travelers introduced the virus from the South Pacific.

In some dengue-endemic countries, such as Thailand and Vietnam, epidemics tend to occur in cycles of every three to five years, postulated to be because of the introduction of new genetic variants with greater epidemic potential and virulence. In an analysis that modeled data from 850,000 dengue infections from 72 provinces in Thailand during the period 1983–97, Cummings and colleagues found a spatial–temporal traveling wave in the incidence of DHF (Cummings *et al.*, 2004; Gubler, 2004). Furthermore, they found that the wave started in Bangkok, the largest city, and moved radially at about 148 km per month. Although all four serotypes are continuously present in Bangkok, smaller communities may experience periods without infection from specific serotypes. In epidemic years, a lag of 10 months may occur before outbreaks develop in some areas. It suggests that dense urban centers are the genesis of viruses that cause epidemics that then migrate (in humans) to surrounding regions.

## Movement of animals (vectors and intermediate hosts)

Migrating birds can move potential human pathogens into new regions, as has been shown with West Nile virus and H5N1 influenza A (avian) virus (Rappole *et al.*, 2000; Chen *et al.*, 2005; Liu *et al.*, 2005). Much movement of animals (live, dead, parts) is orchestrated by humans, sometimes with disastrous consequences. The importation of wild animals from Ghana that were then housed with wild-caught prairie dogs from the United States led to an outbreak of monkeypox (a disease previously known to exist only in Africa) in the Midwest (Reed *et al.*, 2004). Hunting of bushmeat, as local populations seek a source of protein and as logging roads make trade easier, threatens some wild animal species. Killing and butchering wild animals also puts people at risk for becoming infected with pathogens in animals (Wolfe *et al.*, 2005). The global market in bushmeat has grown. A report from the UK Department of Environment, Food and Rural Affairs estimated that 11,600 tons of illegal bushmeat were smuggled into the UK in 2003 (*Daily Telegraph*, 5 September 2004; http://www.telegraph.co.uk/news/main.jhtml). Shipments included monkey, rat, bat, gorilla, camel, and elephant. Consumption is on the increase outside of Asia, Africa, and South America as immigrant populations expand in Europe, North America, and other areas.

Illegal trade may also be a source of imported pathogens. Two Crested Hawk-Eagles were confiscated in Brussels after being carried from Thailand via Vienna in hand luggage placed in an overhead storage bin on airplanes. Although the birds did not appear ill, both were found to be infected with highly pathogenic H5N1 virus (van Borm *et al.*, 2005).

Mosquitoes are regularly carried on commercial flights and ships, and can be introduced into new regions where they may be able to survive and become established locally (Sutherst, 2004). Depending on the local environment and opportunities, they may be able to transmit tropical infections to humans. During the period 1969 through August 1999, 12 countries reported 89 cases of confirmed or probable cases of "airport malaria" – instances when an infected mosquito was transported by plane from a malaria-endemic region to an area without malaria and survived long enough to transmit malaria to a local resident. Most countries were in Europe, but the US, Australia, and Israel also reported cases (Muentener *et al.*, 1999; Gratz *et al.*, 2000).

## Vehicles of travel

Outbreaks in travelers can be traced to infection acquired in the vehicle of travel – *en route* transmission. Food-borne outbreaks (e.g. salmonellosis, shigellosis, cholera, staphylococcal food poisoning, and others) have been linked to food or beverages served on flights (Mangili and Gendreau, 2005). Infections transmitted directly from person to person, and those that are airborne, also pose a risk to airplane travelers. Published reports describe transmission of tuberculosis, SARS, measles, and influenza on aircraft. In one outbreak, 72 percent of 54 passengers on a plane that had been grounded for Three hours because of a failed air-circulation system developed influenza-like illness; influenza A was documented (Moser *et al.*, 1979). Norovirus, a cause of acute gastroenteritis, has probably also been transmitted on an airplane (Widdowson *et al.*, 2005). Following a single 3-hour flight between Hong Kong and Beijing that included an ill passenger who later died of SARS, confirmed or probable SARS developed in 18 of 120 passengers. Those seated within three rows in front of the index case were most likely to become infected (Olsen *et al.*, 2003).

People tend to think of cruises as an outdoor activity, but on large cruise ships passengers spend much time in indoor shared spaces. Gastrointestinal infections have long been associated with cruise ships, but more recently outbreaks of other infections have been reported (Minooee and Rickman, 1999). These include legionellosis (CDC, 2005d; Kura *et al.*, 2006) and infections, like influenza (Brotherton *et al.*, 2003) and rubella, with person-to-person spread. Noroviruses have emerged as a major problem for cruise ships. Not only can these viruses be spread from person to person by fecal–oral transmission and airborne transmission, they can also be spread via contaminated food and water and by contact with contaminated surfaces in the environment (Bull *et al.*, 2006). They can persist in the environment despite disinfection efforts (Widdowson *et al.*, 2005). New virus variants have emerged and have been associated with an increase in outbreaks on cruise ships (Lopman *et al.*, 2004; Widdowson *et al.*, 2004). The dispersal of passengers after cruise-ship travel may allow seeding of multiple communities with infections.

Persons who are older and have chronic medical problems make up a high percentage of travelers on some cruises. On one cruise ship with an outbreak of influenza, 77.4 percent of 1448 passengers were 65 years of age or older, and 26.2 percent had chronic medical problems (CDC, 1997).

The arrival of travelers changes places and affects the host population. This can occur through building hotels, providing food and other services for travelers, building new roads, and changing the habitat or landscape to make it more desirable to visitors. These activities can bring economic benefits to a region and provide new jobs for local residents. There are also negative consequences – destruction of pristine habitats, alteration of ecosystems and fragmentation of habitats; increased traffic and pollution; increased inequalities; disruption or corruption of local activities and customs; sexual tourism; and increased crime as wealthy tourists have close contact with impoverished local residents. Much of the wealth from tourism to a region may flow out of the country in cases where hotels and other concessions are owned by foreign investors.

## Tools

Powerful tools now aid in answering epidemiologic questions. Advances in laboratory methods enable surveillance at the microbial level. Computer technology can estimate disease transmission and control. Clinicians also participate in the surveillance of diseases, especially specialists in travel and tropical medicine.

## Laboratory techniques

Molecular markers can determine the similarity or divergence between two isolates of an organism, and assess whether a patient has a new infection or a relapse of an old one. When multiple infections occur in a population, it can be determined whether infections were related to the same source. It is possible to pinpoint an animal, arthropod, food or environmental source. It is also often possible to characterize the patterns of spread and the mode of transmission. Other questions that can be addressed are whether some strains are more transmissible, more virulent, or associated with different clinical characteristics (e.g. longer carriage). Devising sensible and effective interventions is more likely if the routes of transmission and epidemiological patterns are known.

## Use of predictive models

Models have been developed to simulate the estimated impact of infectious diseases, and some have focused on air travel or have integrated data about air travel into the models (Brockmann *et al.*, 2005, 2006; Colizza *et al.*, 2006). The rapid spread of SARS showed that estimates on the spread of modern epidemics need to consider the pattern of travel, the global aviation network, the number of flights departing from and arriving at airports, the number of passengers carried, and the size of aircraft (Hufnagel *et al.*, 2004). Simulations can be used to model the potential impact of different interventions on the control of epidemics – for example through vaccination (where reducing the susceptible population leads to fewer "reproducers") as well as travel restrictions, and especially by isolation of largest cities.

The present pattern of air travel is expected to alter the dynamics of an influenza pandemic compared to past pandemics. The pandemic influenza of 1918–19 spread by ships, over land, and reached some remote areas by dog sled. In 1968–69, the time of the Hong Kong pandemic, 160 million persons traveled internationally on commercial flights. The Hong Kong influenza strain diffused through the network of cities globally by air travel, first to northern and then to southern latitudes (Rvachev and Longini, 1985). Simulation of the epidemic using 2000 air-transportation data for 52 global cities showed the virus spread concurrently to cities in both northern and southern hemispheres, resulting in little seasonal swing and a very short time for public health intervention (Grais *et al.*, 2003). The simulation showed that disease would spread first to nearby cities, but also to distant cities with high air-travel volumes, such as Sydney, Singapore, Johannesburg, Melbourne, Perth, and Wellington. The estimated time for the pandemic to reach northern-hemisphere cities, using the 2000 travel data, was 111 days shorter than in 1968 (Grais *et al.*, 2003).

## Travel medicine: origin and evolution of travel medicine as a specialty

Over the past 25 years a new specialty has evolved to address the needs of travelers, especially those who visit resource-poor regions (Hill, 2006). Travel medicine has been defined as the "discipline devoted to the maintenance of the health of international travelers through health promotion and disease prevention" (Kozarsky and Keystone, 2004). It is an interdisciplinary specialty with a global focus, by integrating an understanding of global health issues into the care of travelers. Pioneers in the field advocated analyzing epidemiological data to define risks associated with travel, and ways to reduce them (Steffen, 1991).

Travel medicine focuses on prevention involving such specialties as clinical tropical medicine, infectious diseases, public health, occupational health, high-altitude and diving medicine, wilderness medicine, travel-related women's health and pediatric medicine, psychiatry, migration medicine, environmental health, and military medicine. The International Society of Travel Medicine (ISTM) was established in 1991 from this multi-disciplinary interest. In the globalized economy, the business community is also drawing upon the expertise of travel-medicine specialists as companies send workers and their families to other countries for extended periods.

## GeoSentinel and other networks

Travel and tropical medicine clinics are in a position to detect disease outbreaks that may be heralded by returning travelers. Surveillance networks of travel and tropical medicine clinics have been established to systematically collect data on travel-associated diseases and be on the alert for disease outbreaks or changes in patterns of infections. GeoSentinel is such a network, established in 1995 through collaborative agreement between the International Society of Travel Medicine and the Centers for Disease Control and Prevention (Freedman *et al.*, 1999). TropNetEurop is another network that primarily has sites in Europe. The GeoSentinel network now has 31 sites located in six continents. Studies published by members of the GeoSentinel network are also contributing to the database that will allow clinicians to make recommendations based on evidence.

GeoSentinel analysis of morbidity in travelers can provide guidance on identifying common illnesses in travelers and risk of travel to health (Freedman *et al.*, 2006). Moreover, GeoSentinel has disseminated alerts regarding a number of diagnoses in returning travelers – leptospirosis in participants returning from Eco-Challenge in Borneo, Malaysia; SARS in travelers returning to Canada from Asia; and malaria in travelers to resorts in Punta Cana, Dominican Republic. Subsequent to these notifications, clinicians identified additional cases in travelers, and public health responses were initiated.

ProMED, a program of the International Society of Infectious Diseases, has also accelerated the dissemination of outbreak reports through the use of the Internet. Networks such as GeoSentinel and ProMED have enabled rapid information-sharing between public health authorities and clinicians.

## Conclusions

Travel changes people and places. The traveler can be the target and sufferer from microbial threats; the traveler can also provide the microbial transport system. Traveling humans alter infectious diseases patterns by introducing pathogenic microbes or resistance or virulence factors into new populations. The movement of pathogenic microbes through history has been intimately linked to the capacity of humans to travel and to migrate to new locations. Today, humans have the capacity to reach virtually any city in the world within a day or two. Travel to tropical and developing countries is increasing more rapidly than travel to developed countries, and all projections suggest that human travel will continue to increase in the foreseeable future. Infectious diseases influence decisions about travel and trade; and travelers change the epidemiology of infectious diseases.

Travel has played an essential role in the movement of HIV/AIDS throughout the world. Rapid international air travel moved the virus that causes SARS to multiple countries within weeks. Travel has been and will continue to be important in the movement of human influenza.

Recent experiences with SARS and avian influenza underscore several observations:

- travel and trade are major forces in the global economy
- movement of humans and animals influences the biogeography of infectious diseases
- many important human pathogens originate from related microbes in animals
- interactions between humans and animals are common and widespread
- animal populations (for food) are growing even faster than human populations
- animals and birds can migrate
- humans have some control over animal trade (though illegal movement and marketing is huge)
- humans cannot control movement of microbes through bird migration
- surveillance is critical and must be global and linked to response capability
- communication among scientists must be frequent and open
- isolates of microbes must be made available to scientists
- international collaboration is essential and must be multidisciplinary
- communication about risks must be timely and accurate
- poverty and ignorance are risk factors for infectious diseases
- poverty limits choices.

Travel consists of sequential shared environments, often with people from diverse regions of the world. It should be considered a loop, and not just an origin and destination. Travel has created one, extensively interconnected world in terms of microbial threats. Approaches must consider the global community.

## References

Anderson, R.M. and May, R.M. (1991). *Infectious Diseases of Humans*. Oxford: Oxford University Press, p. 70.

Antia, R., Regoes, R.R., Koella, J.C. and Bergstrom, C.T. (2003). The role of evolution in the emergence of infectious diseases. *Nature* **426**, 658–661.

Bacaner, N., Stauffer, B., Boulware, D.R. *et al.* (2004). Travel medicine considerations for North American immigrants visiting friends and relatives. *Journal of the American Medical Association* **291**(23), 2856–2864.

Barry, J.M. (2004). *The Great Influenza. The Epic Story of the Deadliest Plague in History*. New York: Viking.

Black, F.L. (1992). Why did they die? *Science* **258**, 1939–1940.

Bradley, D.J. (1989). The scope of travel medicine. In: R. Steffen (ed.), *Travel Medicine: Proceedings of the First Conference on International Travel Medicine*. Berlin: Springer Verlag, pp. 1–9.

Breugelmans, J.G., Zucs, P., Porten, K. *et al.* (2004). SARS transmission and commercial aircraft. *Emerging Infectious Diseases* **10**(8), 1502–1503.

Brockmann, D., Hufnagel, L. and Giesel, T. (2005). Dynamics of modern epidemics. In: A. McLean, R.M. May, J. Pattison and R.A. Weiss (eds), *SARS. A Case Study in Emerging Infections*. Oxford: Oxford University Press, pp. 81–91.

Brockmann, D., Hufnagel, L. and Geisel, T. (2006). The scaling laws of human travel. *Nature* **439**, 462–465.

Brotherton, J.M.L., Delpech, V.C., Gilbert, G.L. *et al.* (2003). A large outbreak of influenza A and B on a cruise ship causing widespread morbidity. *Epidemiology and Infection* **130**, 263–271.

Bull, R.A., Tu, E.T.V., McIver, C.J. *et al.* (2006). Emergence of a new norovirus genotype II.4 variant associated with global outbreaks of gastroenteritis. *Journal of Clinical Microbiology* **44**(2), 327–333.

Cabada, M.M., Montoya, M., Echevarria, J.I. *et al.* (2003). Sexual behavior in travelers visiting Cuzco. *Journal of Travel Medicine* **10**(4), 214–218.

Carmichael, A.G. (2006). Infectious disease and human agency: an historical overview. In: M.O. Andreae, U. Confalonieri and A.J. McMichael (eds), *Global Environmental Change and Human Health*. Vatican City: The Pontifical Academy of Sciences, Vol. 106, pp. 3–46.

Cash, R.A. and Narasimhan, V. (2000). Impediments to global surveillance of infectious diseases: consequences of open reporting in a global economy. *Bulletin of the World Health Organization* **78**(11), 1358–1367.

Centers for Disease Control and Prevention (CDC) (1997). Update: influenza activity – United States, 1997–1998 season. *Morbidity and Mortality Weekly Report* **46**, 1094–1098.

Centers for Disease Control and Prevention (CDC) (2000). Serogroup meningococcal disease among travelers returning from Saudi Arabia, United States, 2000. *Morbidity and Mortality Weekly Report* **49**, 345–346.

Centers for Disease Control and Prevention (CDC) (2001). Public health dispatch: update, assessment of risk for meningococcal disease associated with the Hajj 2001. *Morbidity and Mortality Weekly Report* **50**(12), 221–222.

Centers for Disease Control and Prevention (CDC) (2003). Severe acute respiratory syndrome – Singapore, 2003. *Morbidity and Mortality Weekly Report* **52**(18), 405–411.
Centers for Disease Control and Prevention (CDC) (2004a). Update, multistate investigation of measles among adoptees from China – April 16, 2004. *Morbidity and Mortality Weekly Report* **53**, 323–324.
Centers for Disease Control and Prevention (CDC) (2004b). Measles outbreak associated with an imported case in an infant – Alabama, 2002. *Morbidity and Mortality Weekly Report* **53**(2), 30–33.
Centers for Disease Control and Prevention (CDC) (2005a). Preventable measles among US residents, 2001–2004. *Morbidity and Mortality Weekly Report* **54**(33), 817–820.
Centers for Disease Control and Prevention (CDC) (2005b). Import-associated measles outbreak – Indiana, May–June 2005. *Morbidity and Mortality Weekly Report* **54**, 1073–1075.
Centers for Disease Control and Prevention (CDC) (2005c). Poliovirus infections in four unvaccinated children – Minnesota, August–October 2005. *Morbidity and Mortality Weekly Report* **54**, 1053–1055.
Centers for Disease Control and Prevention (CDC) (2005d). Cruise-ship associated Legionnaires' disease, November 2003–May 2004. *Morbidity and Mortality Weekly Report* **54**, 1153–1155.
Centers for Disease Control and Prevention (CDC) (2006a). Imported vaccine-associated paralytic poliomyelitis – United States, 2005. *Morbidity and Mortality Weekly Report* **55**(4), 97–99.
Centers for Disease Control and Prevention (CDC) (2006b). Resurgence of wild poliovirus type 1 transmission and consequences of importation – 21 countries, 2002–2005. *Morbidity and Mortality Weekly Report* **55**(6), 145–150.
Centers for Disease Control and Prevention (CDC) (2006c). Trends in tuberculosis – United States, 2005. *Morbidity and Mortality Weekly Report* **55**(11), 305–308.
Centers for Disease Control and Prevention (CDC) (2006d). Emergence of Mycobacterium tuberculosis with extensive resistance to second-line drugs – worldwide, 2000–2004. *Morbidity and Mortality Weekly Report* **55**(11), 301–305.
Centers for Disease Control and Prevention (CDC) (2006e). Mumps outbreak at a summer camp in New York, 2005. *Morbidity and Mortality Weekly Report* **55**(7), 175–177.
Centers for Disease Control and Prevention (CDC) (2006f). Exposure to mumps during air travel – United States, April 2006. *Morbidity and Mortality Weekly Report* **55**(14), 401–402.
Centers for Disease Control and Prevention (CDC) (2006g). Update: multistate outbreak of mumps – United States, January 1–May 2, 2006. *Morbidity and Mortality Weekly Report* **55**(20), 559–563.
Chen, H., Smith, G.J.D., Zhang, S.Y. *et al.* (2005). H5N1 virus outbreak in migratory waterfowl. *Nature* **436**, 191–192.
Cliff, A. and Haggett, P. (2004). Time, travel and infection. *British Medical Bulletin* **69**, 87–99.
Cliff, A., Haggett, P. and Smallman-Raynor, M. (2004). *World Atlas of Epidemic Diseases*. New York: Oxford University Press.
Colizza, V., Barrat, A., Berthelemy, M. and Vespignani, A. (2006). The role of the airline transportation network in the prediction and predictability of global epidemics. *Proceedings of the National Academy of Sciences* **103**(7), 2015–2020.
Correia, J.D., Shafer, R.T., Patel, V. *et al.* (2001). Blood and body fluid exposure as a health risk for international travel. *Journal of Travel Medicine* **8**, 263–266.
Corso, A., Severina, E.P., Petruk, V.F. *et al.* (1998). Molecular characterization of penicillin-resistant *Streptococcus pneumoniae* isolates causing respiratory disease in the United States. *Microbiology and Drug Resistance* **4**, 325–337.

Crosby, A.W. (1972). *The Columbian Exchange. Biological and Cultural Consequences of 1492*. Westport: Greenwood Press.
Cummings, D.A.T., Irizarry, R.A., Huang, N.E. *et al.* (2004). Travelling waves in the occurrence of dengue haemorrhagic fever in Thailand. *Nature* **427**(22 Jan), 344–347.
de Quadros, C., Izurieta, H., Venczel, L. and Carrasco, P. (2004). Measles eradication in the Americas: progress to date. *Journal of Infectious Diseases* **189**(Suppl. 1), S227–S235.
Dull, P.M., Abdelwahab, J. and Sacchi, C.T. (2005). *Neisseria meningitidis* serogroup W-135 carriage among US travelers to the 2001 Hajj. *Journal of Infectious Diseases* **191**, 33–39.
Echenberg, M. (2002). Pestis redux, the initial years of the third bubonic plague pandemic, 1894–1901. *Journal of World History* **13**(2), 429–449.
Edwards, D.A., Man, J.C., Brand, P. *et al.* (2004). Inhaling to mitigate exhaled bioaerosols. *Proceedings of the National Academy of Sciences* **101**, 17,383–17,388.
Effler, P.V., Pang, L., Kitsutani, P. *et al.* (2005). Dengue fever, Hawaii, 2001–2002. *Emerging Infectious Diseases* **11**(5), 742–749.
Eidson, M., Kramer, L., Stone, W. *et al.* (2001). Dead crow densities and human cases of West Nile virus, New York State, 2000. *Emerging Infectious Diseases* **7**(4), 631–635.
Ezzati, M.K., Lopez, A.D., Rodgers, A. *et al.* (2002). Selected major risk factors and global and regional burden of disease. *Lancet* **360**, 1347–1360.
Fraser, C., Riley, S., Anderson, R.M. and Ferguson, N.M. (2004). Factors that make an infectious disease outbreak controllable. *Proceedings of the National Academy of Sciences* **101**, 6146–6151.
Freedman, D.O., Kozarsky, P.E., Weld, L.H. and Cetron, M.S. (1999). GeoSentinel, the global emerging infections sentinel network of the International Society of Travel Medicine. *Journal of Travel Medicine* **6**(2), 94–98.
Freedman, D.O., Weld, L.H., Kozarsky, P.E. *et al.* (2006). Spectrum of disease and relation to place of exposure among ill returned travelers. *New England Journal of Medicine* **354**(2), 119–130.
Galvani, A.P. and May, R.M. (2005). Dimensions of superspreading. *Nature* **438**(17), 293–295.
Grais, R.F., Ellis, J.H. and Glass, G.E. (2003). Assessing the impact of airline travel on the geographic spread of pandemic influenza. *European Journal of Epidemiology* **18**, 1065–1072.
Gratz, N.G., Steffen, R. and Cocksedge, W. (2000). Why aircraft disinfection? *Bulletin of the World Health Organization*. **78**(8), 995–1004.
Gubler, D.J. (2002). Epidemic dengue/dengue hemorrhagic fever as a public health, social and economic problem in the 21st century. *Trends in Microbiology* **10**(2), 100–102.
Gubler, D.J. (2004). Cities spawn epidemic dengue viruses. *Nature Medicine* **10**, 129–130.
Gubler, D.M., Suharyono, W., Tan, R. *et al.* (1981). Viremia in patients with naturally acquired dengue infection. *Bulletin of the World Health Organization* **59**(4), 623–630.
Guernier, V., Hockberg, M.E. and Guegan, J.-F. (2004). Ecology drives the worldwide distribution of human diseases. *Public Library of Science Biology* **2**(6), 740–746.
Gushulak, B.D. and MacPherson, D.W. (2004). Globalization of infectious diseases, the impact of migration. *Clinical Infectious Diseases* **38**, 1742–1748.
Gwatkin, D.R., Guillot, M. and Heuveline, P. (1999). The burden of disease among the global poor. *Lancet* **354**, 586–589.
Hahn, B.H., Shaw, G.M., DeCock, K.M. and Sharp, P.M. (2000). AIDS as a zoonosis, scientific and public health implications. *Science* **287**, 607–614.
Hayes, E.B., Komar, N., Nasci, R.S. *et al.* (2005). Epidemiology and transmission dynamics of West Nile virus disease. *Emerging Infectious Diseases* **11**(8), 1167–1173.

Heymann, D.L. and Aylward, R.B. (2004). Eradicating polio. *New England Journal of Medicine* **351**(13), 1275–1277.
Hill, D.R. (2006). The burden of disease in international travelers. *New England Journal of Medicine* **354**(2), 115–117.
Hufnagel, L., Brockmann, D. and Geisel, T. (2004). Forecast and control of epidemics in a globalized world. *Proceedings of the National Academy of Sciences* **101**(42), 15,124–15,129.
Jacobsen, K.H. and Koopman, J.S. (2004). Declining hepatitis A weroprevalence: a global review and analysis. *Epidemiology and Infection* **132**, 1005–1022.
Kew, O., Morris-Glasgow, V., Landaverde, M. *et al.* (2002). Outbreak of poliomyelitis in Hispaniola associated with circulating type1 vaccine-derived poliovirus. *Science* **296**, 356–359.
Kozarsky, P.E. and Keystone, J.S. (2004). Introduction to travel medicine. In: J.E. Keystone, P.E. Kozarsky, D.O. Freedman *et al.* (eds), *Travel Medicine*. Edinburgh: Mosby, pp. 1–3.
Kura, F., Amemura-Maekawa, J., Yagita, K. *et al.* (2006). Outbreak of Legionnaires' disease on a cruise ship linked to spa-bath filter stones contaminated with *Legionella pneumophila* serogroup 5. *Epidemiology and Infection* **134**, 385–391.
Leder, K., Tong, S., Weld, L. *et al.* (2007). Illness in travelers visiting friends and relatives, a review of the GeoSentinel Network. *Clinical Infectious Diseases* **43**(9), 1185–1193.
Lee, J.-W. and McKibbin, W.J. (2004). Estimating the global economic costs of SARS. In: S. Knobler, A. Mahmoud, S. Lemon *et al.* (eds), *Learning from SARS: Preparing for the Next Disease Outbreak*. Institute of Medicine Workshop Summary. Washington, DC: The National Academies Press, pp. 92–109.
Lipsitch, M., Cohen, T., Cooper, B. *et al.* (2003). Transmission dynamics and control of severe acute respiratory syndrome. *Science* **300**, 1966–1970.
Liu, J., Xiao, H., Lei, F. *et al.* (2005). Highly pathogenic H5N1 influenza virus infection in migratory birds. *Science* **309**, 1206–1207.
Lloyd-Smith, J.O., Schreiber, S.J., Kopp, P.E. and Getz, W.M. (2005). Superspreading and the effect of individual variation on disease emergence. *Nature* **438**, 355–359.
Lopman, B., Vennema, H., Kohli, E. *et al.* (2004). Increase in viral gastroenteritis outbreaks in Europe and epidemic spread of new norovirus variant. *Lancet* **363**, 682–688.
MacLennan, C., Dunn, G., Huissoon, A.P. *et al.* (2004). Failure to clear persistent vaccine-derived neurovirulent poliovirus infection in an immunodeficient man. *Lancet* **363**, 1509–1513.
Mangili, A. and Gendreau, M.A. (2005). Transmission of infectious diseases during commercial air travel. *Lancet* **365**, 989–996.
Marrazzo, J.M. (2005). Sexual tourism, implications for travelers and the destination culture. *Infectious Disease Clinics of North America* **19**, 103–120.
Massachusetts Department of Public Health. (2006). *Measles Alert June 7, 2006*. Available at: http://www.mass.gov/dph/cdc/epii/imm/alerts/measles_alert_20060607.pdf (accessed 14 June 2006).
May, R.M., Gupta, S. and McLean, A.R. (2001). Infectious disease dynamics: what characterizes a successful invader? *Philosophical Transactions of the Royal Society, London B* **356**, 901–910.
Minooee, A. and Rickman, L.S. (1999). Infectious diseases on cruise ships. *Clinical Infectious Diseases* **29**, 737–744.
Moore, C.G. and Mitchell, D.J. (1997). Aedes albopictus in the United States, ten-year presence and public health implications. *Emerging Infectious Diseases* **3**(3), 329–334.
Moore, P.S., Harrison, L.H., Telzak, E.E. *et al.* (1989). Group A meningococcal carriage in travelers returning from Saudi Arabia. *Journal of the American Medical Association* **260**(18), 2686–2689.

Moser, M.R., Bender, T.R., Margolis, H.S. *et al.* (1979). An outbreak of influenza aboard a commercial airliner. *American Journal of Epidemiology* **110**, 1–6.

Muentener, P., Schlagenhauf, P. and Steffen, R. (1999). Imported malaria (1985–95): trends and perspectives. *Bulletin of the World Health Organization* **77**, 560–566.

Musher, D.M. (2003). How contagious are common respiratory tract infections? *New England Journal of Medicine* **348**, 1256–1266.

Myers, N. (2001). Environmental refugees. *Philosophical Transactions of the Royal Society London, Biology* **357**, 609–613.

Olsen, S.J., Chang, H.-L., Cheung, T.Y.-Y. *et al.* (2003). Transmission of the severe acute respiratory syndrome on aircraft. *New England Journal of Medicine* **349**, 2416–2422.

Parker, A.A., Staggs, W., Dayan, G.H. *et al.* (2006). Implications of a 2005 measles outbreak in Indiana for sustained elimination of measles in the United States. *New England Journal of Medicine* **355**(5), 447–455.

Rappole, J.H., Derrickson, S.R. and Hubalek, Z. (2000). Migratory birds and spread of West Nile virus in the Western Hemisphere. *Emerging Infectious Disease* **6**(4), 319–328.

Reed, K.D., Melski, J.W., Braham, M.B. *et al.* (2004). The detection of monkeypox in humans in the Western Hemisphere. *New England Journal of Medicine* **350**, 342–350.

Reiter, P., Lathrop, S., Bunning, M. *et al.* (2003). Texas lifestyle limits transmission of dengue virus. *Emerging Infectious Diseases* **9**(1), 86–89.

Rvachev, L. and Longini, I. (1985). A mathematical model for the global spread of influenza. *Mathematics and Bioscience* **75**, 3–22.

Samba, E., Nkrumah, F. and Leke, R. (2004). Getting polio eradication back on track in Nigeria. *New England Journal of Medicine* **350**(7), 645–646.

Schlagenhauf, P., Steffen, R. and Loutan, L. (2003). Migrants as a major risk group for imported malaria in European countries. *Journal of Travel Medicine* **10**, 106–107.

Sears, C.L. (2005). A dynamic partnership, celebrating our gut flora. *Anaerobe* **11**(5), 247–251.

Shelton, J.D., Cassell, M.M. and Adetunji, J. (2005). Is poverty or wealth at the root of HIV? *Lancet* **366**, 1057–1058.

Shinya, K., Ebina, M., Yamada, S. *et al.* (2006). Influenza virus receptors in the human airway. *Nature* **440**, 435–436.

Sivitz, L.B., Stratton, K. and Benjamin, G.C. (eds) (2006). *Quarantine Stations at Ports of Entry: Protecting the Public's Health.* Committee on Measures to Enhance the Effectiveness of the CDC Quarantine Station Expansion Plan for US Ports of Entry. Institute of Medicine of the National Academies. Washington DC: The National Academies Press.

Smolinski, M.S., Hamburg, M.A. and Lederberg, J. (eds) (2003). *Microbial Threats to Health: Emergence, Detection, and Response.* Institute of Medicine of the National Academies. Washington, DC: The National Academy Press.

Steffen, R. (1991). Travel medicine – prevention based on epidemiological data. *Transactions of the Royal Society of Tropical Medicine and Hygiene* **85**, 156–162.

Sutherst, R.W. (2004). Global change and human vulnerability to vector-borne diseases. *Clinical Microbiology Review* **17**(1), 136–173.

UNAIDS (2006). *2006 Report on the Global AIDS Epidemic. Joint United Nations Programme on HIV/AIDS.* Available at: http://data.unaids.org/pub/GlobalReport/2006/2006_GR-ExecutiveSummary_en.pdf (accessed 14 June 2006).

Uyeki, T.M., Zane, S.B., Bodnar, U.R. *et al.* (2003). Large summertime influenza A outbreak among tourists in Alaska and the Yukon territory. *Clinical Infectious Diseases* **36**(9), 1095–1102.

van Borm, S., Thomas, I., Hanquet, G. *et al.* (2005). Highly pathogenic H1N1 influenza virus in smuggled Thai eagles, Belgium. *Emerging Infectious Diseases* **11**(5), 702–705.
Webby, R., Hoffmann, E. and Webster, R. (2004). Molecular constraints to interspecies transmission of viral pathogens. *Nature Medicine Supplement* **10**(12), S77–81.
Weiss, R.A. (2001). The Leeuwenhoek Lecture 2001. Animal origins of human infectious diseases. *Philosophical Transactions of the Royal Society London, Biology* **356**, 957–977.
Widdowson, M.-A., Cramer, E.H., Hadley, L. *et al.* (2004). Outbreaks of acute gastroenteritis on cruise ships and on land: identification of a predominant circulating strain of norovirus – United States 2002. *Journal of Infectious Diseases* **190**, 27–36.
Widdowson, M.-A., Glass, R., Monroe, S. *et al.* (2005). Probable transmission of norovirus on an airplane. *Journal of the American Medical Association* **293**, 1859–1860.
Wilder-Smith, A., Barkham, T.M.S., Earnest, A. and Paton, N.I. (2002). Acquisition of W135 meningococcal carriage in Hajj pilgrims and transmission to household contacts: prospective study. *British Medical Journal* **325**, 365–366.
Wilder-Smith, A., Foo, W., Earnest, A. and Paton, N.I. (2005). High risk of Mycobacterium tuberculosis infection during the Hajj pilgrimage. *Tropical Medicine and International Health* **10**(4), 336–339.
Wilson, M.E. (1995a). Travel and the emergence of infectious diseases. *Emerging Infectious Diseases* **1**, 39–46.
Wilson, M.E. (1995b). Infectious diseases: an ecological perspective. *British Medical Journal* **311**, 1681–1684.
Wilson, M.E. (1995c). The power of plague. *Epidemiology* **6**(4), 458–460.
Wilson, M.E. (2003a). The traveller and emerging infections, sentinel, courier, transmitter. *Journal of Applied Microbiology* **94**, S1–11.
Wilson, M.E. (2003b). Health and security, globalization of infectious diseases. In: L. Chen, J. Leaning and V. Narasimhan (eds), *Global Health Challenges for Human Security*. Cambridge: Harvard University Press, pp. 86–104.
Wilson, M.E., Levins, R. and Spielman, A. (eds) (1994). *Disease in Evolution: Global Changes and Emergence of Infectious Diseases*, Vol. 740. New York: New York Academy of Sciences.
Wolfe, N.D., Daszak, P., Kilpatrick, A.M. and Burke, D.S. (2005). Bushmeat hunting and deforestation in prediction of zoonoses emergence. *Emerging Infectious Diseases* **11**(12), 1822–1827.
Woolhouse, M.E.J. (2002). Population biology of emerging and re-emerging pathogens. *Trends in Microbiology* **10**(10), S3–7.
World Health Organization (1996). *The World Health Report 1996: Fighting Disease, Fostering Development*. Geneva, World Health Organization, pp. 23–24.
World Health Organization (2003) *Summary of Probable SARS Cases with Onset of Illness from 1 November 2002 to 31 July 2003*. Available at:http://www.who.int/csr/sars/country/table2004_04_21/en/ (accessed 1 March 2006).
World Tourism Organization (2005). *Tourism Highlights, Edition 2004*. Available at: http://www.world-tourism.org (accessed 27 October 2005).
World Tourism Organization (2006a). *UNWTO World Tourism Barometer 4(1)*. Available at: http://www.world-tourism.org/facts/menu.html (accessed 5 March 2006).
World Tourism Organization (2006b). *Tourism Indicators*. Available at: http://www.world-tourism.org/facts/menu.html (accessed 5 March 2006).
Zuckerman, J.N. and Steffen, R. (2000). Risks of hepatitis B in travelers as compared to immunization status. *Journal of Travel Medicine* **7**, 170–174.

# Changing sexual mores and disease transmission

# 2

**David D. Celentano, Frangiscos Sifakis, Vivian Go and Wendy Davis**

> *I won't stand in your bedroom and shake my finger under your nose and say, "Now, don't do that, you're going to catch something," but I'm telling you here, now, that's how we think you're catching it. It doesn't go through the air. You can't cough it into somebody's face. You can't get it from a telephone or from shaking hands. But you can get it sexually, and if you can't stop this extreme sexual activity, at least cut it down so your Russian roulette gun will have two bullets instead of six bullets in the chamber. Because right now, the way you're going, you've got six bullets in the chamber.*
>
> (Dr Selma Dritz, epidemiologist with the San Francisco Department of Public Health, speaking to a packed recreation hall in the Castro District, San Francisco's gay neighborhood, in the fall of 1981; Dritz, 1997).

An understanding of the transmission dynamics of infectious diseases, particularly those that are spread sexually, can explain epidemiologic trends that we see, and can aid in setting priorities for data collection and prevention programs (Anderson, 1999). Specific behaviors can foster the transmission of sexually transmitted infections (STI), determine differences in risk, and define rates of disease in selected groups. Variations in the impact of the HIV/AIDS epidemic, for example, can be tied to different patterns of specific sexual behaviors. Responding to emerging epidemics within subpopulations with circumscribed behavior change can have a major impact on ensuing epidemiology. Accounting for the varying burden of disease within an epidemic can be reduced to recognizing the interplay of human behavior and infectious disease biology. For example, researchers have long recognized that partner selection varies considerably within and between racial/ethnic groups, and this has led to a higher burden of STI in the Black community.

To illuminate the dynamics of the social ecology of STI, including HIV/AIDS, this chapter reviews three fundamental sociocultural movements which have been linked to how and when individuals find sexual partners. First, we look at a series of social changes in the second half of the last century, and evaluate the impact they had on sexual behaviors and STI in the US. These include the growth in numbers of young Americans attending college and delaying marriage, the introduction of hormonal contraception, the movement of women into the workforce, the sexual liberation movement of the 1960s, and changes in the population distribution of Blacks in post-World War II America. The connection between these changes and prevalence rates of common STI is reviewed.

The emergence of the gay pride movement in the US was a central social movement of the second half of the century – one that played a tremendously powerful role in escalating transmission rates of STI in the 1970s, and the establishment of the HIV/AIDS epidemic in the 1980s. A sexual revolution among gay men in the late 1960s facilitated the rapid transmission of the AIDS virus by inspiring widespread participation in sexual risk-taking. High HIV infection rates among men who have sex with men (MSM) and a mounting death toll throughout the 1980s devastated the gay community and led, in turn, to significant cultural and behavioral changes, with initial substantive declines in STI and HIV incidence. Medical breakthroughs, including the development of highly active antiretroviral therapy (HAART) in 1996, encouraged yet another shift in behavior and disease transmission rates, with therapeutic optimism making HIV seem manageable, resulting in new increases in sexual risk-taking behavior.

A third cultural trend, the technological innovations that dominated the end of the twentieth century and the beginning of the twenty-first century, have influenced all aspects of our society, including sexual behavior. The adoption of the Internet as a major source of communication has influenced, among other things, the practice of sexual partner selection. The Internet provides a unique source of sexual partners and allows individuals anonymously to seek specific sexual activities at a specific time and place. Use of the Internet to find sexual partners is now being linked to elevated rates of unsafe sexual practices and, once again, a significant rise in human STI.

These three sociocultural trends suggest that patterns in the human behavior of seeking out companionship and establishing a sexual relationship are intimately associated with patterns in the transmission of STI, including HIV/AIDS. We explore these three themes here.

## Postwar social and sexual trends: evolution and revolution of sexual mores

Throughout most of the twentieth century, sexual attitudes and behaviors became steadily more permissive in the United States (Gillmore *et al.*, 1999). Several societal trends and innovations in the latter half of the twentieth century were associated with this shift in sexual attitudes and behaviors. The impact of these trends and innovations, which included a steep rise in the number of young Americans going to college, a growing number of women in the labor force, an increase in divorce rates, the invention of the contraceptive pill, and the Women's Liberation Movement, were magnified by the fact that the surge of individuals born after the end of World War II, the baby boomers, began their reproductive years in the 1960s and 1970s. These trends and innovations, and the changes in sexual behavior they encouraged, were in turn associated with new patterns of transmission rates of STI.

## Social trends

In the 1960s, young men and women increasingly delayed marriage in order to pursue higher education. The median age of marriage among women rose from 20 in 1955 to 22 in 1979 (Bureau of the Census, 1999), while the number of young adults attending college more than quadrupled in that same time period (Bureau of the Census, 1998). The rising age at first marriage, combined with a growing number of students living away from parents' direct supervision, provided the time and freedom for young, single adults to explore their sexuality.

Women were not only delaying marriage and attending college; they were also entering the labor force at increasing rates. In the period between 1948 and 1987, the labor force participation of mothers with preschoolers increased five-fold (Bureau of Labor Statistics, 1988). A combination of factors, including women's rising wages, consumerism, and changing attitudes, facilitated women's workforce participation in the post-World War II period (Cain, 1966; Bowen and Finegan, 1969; Gordon and Kammeyer, 1980). In the 1970s, high inflation and economic uncertainty encouraged families to adopt new work–family arrangements and further accelerated women's entry into the labor force (Edwards, 2001). The transition from homemaking to paid labor-force participation increased women's autonomy and provided opportunities for sexual liaisons (Gillmore *et al.*, 1999).

The movement of women into the workforce has also been linked to increases in the rate of divorce (Heer and Grossbard-Schectman, 1981; Corley and Woods, 1991; Davanzo and Rahman, 1993). Indeed, the incidence of divorce per 1000 married women aged 14 and over increased from 9.2 in 1960 to 20.3 in 1975 (Bureau of the Census, 1976). As a result of the increasing divorce rate, the proportion of sexually experienced, unmarried adults also increased (Kinsey *et al.*, 1953). The increasing prevalence of divorce, the growing number of educated women, and the increasing accessibility of financial independence were all slowly altering the traditional landscape of sexual behavior and disease.

The introduction of the use of oral contraceptives (*aka* "the pill") in the early 1960s fueled these trends and the sexual revolution (Asbell, 1995; Watkins,

2001). Its effectiveness gave women unprecedented control over their fertility, offering them sexual freedom that had previously only belonged to men. The pill enabled couples to have intercourse as an expression of love or for physical pleasure separate from procreation. The social impact of the pill was profound and ultimately provided, in concert with other social trends, the momentum for the Women's Liberation Movement. The publication of a report from the Kennedy Administration's Commission on the Status of Women (United States President's Commission on the Status of Women, 1963) and Betty Friedan's *The Feminine Mystique* (Friedan, 1963), both of which documented discrimination against women in virtually every area of American life, were published in 1963 and marked the beginning of the Women's Liberation Movement. As it gained momentum, several legislative victories were celebrated by the movement, including the passage of Title VII of the Civil Rights Act of 1964, which made it illegal to practice employment discrimination on the basis of sex as well as race, religion, and national origin; of Title IX, in the Education Amendments of 1972, which forbade discrimination in the field of education; and of Roe vs. Wade in the US Supreme Court, which legalized abortion in all 50 states.

The Women's Liberation Movement challenged American ideology about female sexuality. A new literature emerged representing women's perspectives in sexual relationships; books such as the *Joy of Sex* (Comfort, 1972) focused on both men's and women's roles in the quest for better lovemaking, and became best sellers. As Gillmore and colleagues described, "The pressure on men to become competent lovers, for women to be orgasmic and assertive in their sexual desires changed the meaning and experience of sex for both men and women" (Gillmore *et al.*, 1999). This revised ideology was accompanied by increasingly permissive attitudes about sex (Reiss, 1960; Glenn and Weaver, 1979). The recognition of women as sexually independent actors represented a new phase in American sexuality.

The sexual and social freedom this series of developments fostered in the 1960s and 1970s was reflected in dramatic increases in the rates of STI such as syphilis (Nakashima *et al.*, 1996), gonorrhea (CDC, 2005a), chlamydia (Holmes, 1981), and genital herpes (Becker *et al.*, 1985). Individuals at the forefront of social and sexual change in this period showed some specific vulnerability to STI. Young people (Zaidi *et al.*, 1983), separated and divorced women (Manhart *et al.*, 2004), women using the pill as their method of contraception (Berger *et al.*, 1975; Richmond and Sparling, 1976; Arya *et al.*, 1981), and women in general (Zaidi *et al.*, 1983) were all more likely to be diagnosed with certain STI.

After two decades during which Americans experienced dramatic changes in marriage and divorce patterns, labor-force participation by women, and social attitudes towards sexuality, these social developments are continuing, but at a slower pace. The divorce rate has begun to stabilize, though it remains at the highest level ever recorded, with half of first marriages estimated to end in divorce (Glick, 1984; Norton and Moorman, 1987). The age at first marriage continues to increase as men and women seek additional education and work experience before marriage (Lugaila, 1992). In the US, the percentage of adults aged 15 and older who were married declined from 69.3 percent and 65.9 percent in 1960 to 57.1 percent and 54 percent in 2003 for men and women, respectively (US Census Bureau, 2003).

This less dramatic pace of social change has been reflected in trends of reported STI. Rates of syphilis and gonorrhea have declined nationwide in the past two decades. From a high of 467.7 cases per 100,000 population reported in 1975, the US gonorrhea rate declined by 76 percent to 113.5 (CDC, 2005b). While the number of reported cases of chlamydia has increased steadily among women since the 1980s, it is likely that this increase reflects the continued expansion of screening efforts and increased use of more sensitive diagnostic tests rather than an actual increase in new infections (CDC, 2005b).

## Population movements and STI segregation

Another broad social trend of the second half of the twentieth century had a unique influence on the transmission of STI in the Black population. During World War II, a need for labor for the production of war material encouraged a major migration of Blacks from the rural, agrarian South to factories in the North. In subsequent decades this migration intensified, with significant shifts in the demographics of large cities in the East and Mid-West – notably Chicago and Detroit. One consequence of this influx was an intensification of racial segregation. As the National Advisory Committee on Civil Disorders stated in their report to President Lyndon Johnson, "Our nation is moving toward two societies, one Black, one White – separate and unequal" (US National Advisory Commission on Civil Disorders, 1968). Differences in socioeconomic status, political power, access to medical care, and preventive health services (LaVeist, 2002), along with racial differences in sexual partner preferences, were associated with perpetually high rates of STI in the Black community. These differences remain today, with Blacks continuing to experience disproportionately high levels of chlamydia and gonorrhea (Adimora and Schoenbach, 2005).

## Trends in sexual behavior

In post-World War II America, social norms prohibited sex outside of marriage (Heer and Grossbard-Schectman, 1981). The Kinsey studies (Kinsey *et al.*, 1953, 2006) of the 1940s and 1950s, however, suggested that Americans were engaging in a wide range of sexual activities. Three of Kinsey's most surprising findings were the frequent use of prostitutes by married (10–20 percent) and unmarried (70 percent) men, the prevalence of homosexual activities (37 percent of males and 13 percent of females had had at least one homosexual experience in their lifetimes), and the substantial rate of extramarital affairs. Kinsey's studies hinted at an American public that was not necessarily abiding by strict social expectations. The extent to which basic expectations about sexual behavior were adjusted in the years following Kinsey's research is discussed next.

## Adolescents and premarital sex

In the late 1940s and early 1950s, Kinsey and colleagues found that 71 percent of males and 33 percent of females had experienced premarital intercourse by the age of 25 (Kinsey *et al.*, 1953, 2006). Beginning in the late 1960s, the proportion of teenage women who had ever had sexual intercourse increased substantially through 1988 (Hofferth *et al.*, 1987; Forrest and Singh, 1990; CDC, 1991). A national survey of never-married women aged 15–19 years in the US in 1971 found that 28 percent had experienced sexual intercourse (Kantner and Zelnick, 1972). A second national survey conducted in 1976 revealed that the proportion of individuals in this same group had increased to 34.9 percent (Zelnick and Kanter, 1977). By 1982, 47.1 percent of teenage women had ever had sex; by 1988, 53.2 percent of teenage women were reporting ever having had sex (Forrest and Singh, 1990). Data since 1988, however, suggest that this historical trend has stopped and perhaps reversed (Abma and Sonenstein, 2001; CDC, 2002). Between 1991 and 2001, the percentage of women in high school who had ever had sexual intercourse decreased from 51 percent to 43 percent (CDC, 2002) Sexual experience among young men has followed a similar pattern. In 1979, 66 percent of males aged 17 to 19 had ever had sex. By 1988, this statistic had reached 76 percent for this cohort, but by 1995 had dropped to 68 percent (Ku *et al.*, 1998). In 2001, 49 percent of male teens reported ever having had sex (CDC, 2006a).

Contraceptive use among teens has also changed significantly over time. In the early 1970s, Kantner and Zelnick reported that half of sexually active 15–19-year-old women failed to use contraception at the most recent sexual intercourse (Kantner and Zelnik, 1973). Between 1971 and 1976 the use of oral contraceptives nearly doubled, while other methods, including condoms, rhythm, and withdrawal methods, decreased (Zelnick and Kantner, 1977). By the mid-1970s, the birth-control pill was the most common contraceptive method, followed by condoms and withdrawal (Biddlecom, 2004). In the early 1980s, nearly half of 20–29-year-olds reported condom use at first intercourse, but only about 14 percent reported condom use at last intercourse compared to more than half reporting use of the pill at last intercourse (Tanfer and Horn, 1985). Partly in response to the AIDS epidemic and subsequent education programs, condom use among adolescents increased dramatically in the 1980s, while the use of birth-control pills declined (Biddlecom, 2004). In 1991, 46 percent of sexually active high-school students reported condom use at last intercourse, and by 2003 this proportion had increased to 63 percent (CDC, 2006a).

Some possible reasons for the recent delayed initiation of sexual intercourse and improved contraceptive practice are that teenagers are more fearful of acquiring a STI, have better access to effective hormonal contraception, are responding to the increased emphasis on abstinence, or that changes in welfare reform had an effect on delaying childbearing (Sonenstein, 2004).

The connection between age of initiation of first sexual intercourse, condom use and STI is not well documented, although some patterns in the prevalence of STI in adolescents have emerged. Between 1970 and the late 1980s, when age for the initiation of sexual intercourse was dropping among young women (CDC, 1991), the prevalence of genital herpes among adolescents increased (Fleming *et al.*, 1997). The prevalence of gonorrhea in adolescents also increased between 1981 and 1986, but then dropped between 1986 and 1996 (Fox *et al.*, 1998) as condom use increased (Biddlecom, 2004). Since 1990, the prevalence of genital chlamydia among a sample of young women entering job-training programs has also dropped (Mertz *et al.*, 2001). Diagnoses of HIV/AIDS among adolescents in the second half of the 1990s also decreased (Biddlecom, 2004). As Biddlecom (2004) observes, despite limitations in the available data on adolescents and STI, a connection can be made between decreases in the numbers of adolescents who have ever had sexual intercourse, increases in the use of condoms among adolescents, and decreases in several primary STI.

## Sexual activity among adults

A consistent, comparable series of surveys of adults similar to those collected among adolescents does not exist in the US, making it more difficult to assess whether sexual behaviors of older American adults have changed in the post-World War II era. Catania and colleagues analyzed three national-level behavioral data sets collected at different points and found that sexual behaviors among heterosexual adults aged 18 to 49 were relatively constant from 1988 to 1996, whereas condom use increased significantly in the same time period (Catania *et al.*, 2001). A study of sexual behaviors of a representative sample of Seattle residents comparing sexual behaviors in 1995 and 2004 also found little difference between time periods. At both measurement periods, the median number of lifetime sex partners and the proportion of participants having concurrent partnerships had changed little (Aral *et al.*, 2005). Similarly, however, this study population also reported an increase in use of condoms. Increases in condom use among women aged 18–44 years between 1988 and 1995 were also reported by the National Survey of Family Growth (Anderson *et al.*, 2003).

Recent evidence suggests that this pattern of increasing condom use has not continued. In 2000, adult condom use with primary partners was low even among those at highest risk of sexually transmitted disease (Anderson *et al.*, 2003). Data suggest, however, that adults may be practicing more protective behavior with non-primary partners. The General Social Survey (GSS) found, among adults sampled between 1996 and 2002, a statistically significant trend toward greater condom use with non-regular partners (Anderson *et al.*, 2003).

Trends in STI have generally corresponded to trends in sexual behaviors and condom use. As condom use has increased over the past several decades, rates of STI have generally decreased (CDC, 2005a, 2005b). The HIV/AIDS epidemic and the response of heterosexuals to the epidemic has been more complex. While the epidemic has generally stabilized within the United States, it continues to expand within certain segments of the heterosexual population. The proportion of AIDS cases among females increased from 18 percent in 1985 to 27 percent in 2004, and while the number of AIDS cases among men and women has dropped since its peak in the mid 1990s, the decrease has been less pronounced for women (13 percent for women compared to 35 percent for men) (CDC, 2006b). As with other STI, racial and ethnic minority populations have been disproportionately affected by the HIV epidemic. Among women, rates among Blacks are now 21 times higher than for Whites (CDC, 2006c). These statistics suggest that this evolving epidemic may demand an ever more comprehensive understanding of the interplay of social trends and sexual behaviors.

## Out of the closet – homosexuality in America

The progression of the AIDS epidemic among MSM in the United States over the past 25 years provides an important example of the integral role human behavior can play in the spread of an infectious agent. In this population, behavior and disease have performed a surprisingly coordinated dance where carefree behavior has led to wild surges in the prevalence of infection, and behavior modulated by fear has brought in turn a tempering of infection rates. The social trends that engendered these behaviors and their connections to behavior and disease are reviewed here.

### "Stonewall" and the emergence of the sexual revolution among gay men

Despite Kinsey's studies of human sexuality, which openly described relatively prevalent homosexual behavior, gay lifestyles remained largely hidden until the late 1960s and early 1970s. The loosening of sexual mores and the climate of rebellion (including the Women's Liberation Movement) which typified the late 1960s, however, encouraged a demand for gay rights and the freedom to engage in a gay lifestyle. Resistance to a police raid of a well-known gay bar in Greenwich Village, the Stonewall Inn, on the evening of 27 June 1969, was a public first step towards this goal. The raid of the Stonewall Inn precipitated several days of rioting, and within weeks the formation of the Gay Liberation Front (d'Emilio, 2002). Members of the Gay Liberation Front ultimately formed the Gay Activists Alliance, a gay rights group which organized "gay-ins" in which gays gathered for picnics and dancing in public areas, encouraged the public demonstration of affection in same-sex couples, and campaigned for the repeal of sodomy laws (Andriote, 1999). The fight for gay civil rights was energized over the next decade by events such as Anita Bryant's successful campaign to overturn a gay rights ordinance in Dade County, Florida, in 1977 and in 1979, the minimal sentence given to the murderer of San Francisco mayor George Moscone and his openly gay city supervisor Harvey Milk. Later that year, on 14 October 1979, over 100,000 gay men, lesbians, and their supporters participated in the first National March on Washington for Lesbian and Gay Rights.

Dramatic changes in gay lifestyles mirrored this dramatic progress in gay rights. As the locations where gay men had historically gathered, including gay bathhouses, gay bars, sex clubs, and adult bookshops, were no longer targeted by police, the number of these establishments grew. Gay bathhouses, in particular, encouraged a sense of gay community and supported an active gay lifestyle. In the 1970s, the proliferation of bathhouses gave gay men easy access to other gay men interested in sex. The bathhouse environment facilitated sexual activity with multiple partners, and made anonymous sex possible. Increases in the number of sexual partners and anonymous partners in the 1970s were linked to increases in reports of STI among gay men (Fichtner *et al.*, 1983; Ostrow and Altman, 1983). Higher rates of STI in gay men were reported as compared with heterosexual men (Judson *et al.*, 1980), and in gay men tested at bathhouses as compared with gay men tested in gay health clinics (Carlson *et al.*, 1980). The recognition that hepatitis B was epidemic among gay men (Rotello, 1997) reinforced growing concerns about the public health implications of the gay lifestyle of the 1970s.

## The emergence of HIV/AIDS

It was within the epicenters of gay life – San Francisco, New York, and Los Angeles – that omens of an epidemic were first recognized. In 1980 and 1981, physicians in these cities began diagnosing rare conditions, *Pneumocystis carinii* pneumonia and Kaposi's Sarcoma (a skin cancer), in young homosexual men. The fact that these diseases were very rare, were being seen in an unexpected context, and were clustering in homosexual men hinted at the possibility of an emerging epidemic. The first published reports of the sudden outbreak of these conditions came from clinicians in these three cities who simultaneously noted that what they were observing might be related to a "homosexual lifestyle" (CDC, 1981a, 1981b). That several of the individuals represented in these reports had already died underscored the seriousness of this fledgling epidemic. Other unusual opportunistic infections were also showing up in gay men, prompting conjecture that a common underlying immune suppressing disease was involved (Shilts, 1987). The presence of *Pneumocystis carinii* pneumonia in intravenous drug users (IDUs) (Masur *et al.*, 1981) and hemophiliacs (CDC, 1982a) suggested the likelihood of a blood-borne infectious agent that could also be transmitted sexually, like hepatitis B.

This newly recognized immune-suppressing disease was initially referred to as Gay-Related Immune Deficiency (GRID). By mid-1982, however, GRID had also been diagnosed in IDUs, hemophiliacs, and individuals from Haiti, and was renamed AIDS. That AIDS was an epidemic was clear by the end of 1982. By 15 September 1982, the Centers for Disease Control (CDC) had received reports of a total of 593 cases of AIDS and noted that the incidence of AIDS by date of diagnosis had essentially doubled every six months since the second half of 1979 (CDC, 1982b). In San Francisco, the number of AIDS cases diagnosed in the second half of 1982 equaled the number of cases that had been diagnosed since the epidemic was first recognized (Moss *et al.*, 1983). While the rate at which cases of AIDS were emerging was causing alarm among public health officials and gay activists, governmental and public response to the epidemic was minimal, especially when compared with other recent public health scares including toxic shock syndrome and Legionnaire's disease (Shilts, 1987). Many, including Larry Kramer, co-founder of the Gay Men's Health Crisis, the first AIDS advocacy and fund-raising organization, argued that public neglect of the burgeoning epidemic stemmed from the fact that the majority of the first victims were gay men (Shilts, 1987). This reality may also have hampered early prevention efforts.

In March of 1983, the CDC shared the results of their epidemiologic investigations into AIDS, observing that among homosexual men those with multiple sexual partners appeared to be at greater risk of contracting the disease, and that the period between exposure and the manifestation of illness could be as long as two years (CDC, 1983). These findings led the CDC to recommend that individuals avoid sexual contact with "persons known or suspected to have AIDS." A *New York Times* article published in February explained that the only protection against AIDS that clinicians could offer their homosexual patients was behavior change. In particular, it was suggested that homosexual men practice monogamy and, ideally, abstain from anal intercourse altogether (Henig, 1983). Many gay men balked at these recommendations, feeling that they undermined a sexual freedom so recently gained (Shilts, 1987).

Concern about the transmission of AIDS led inevitably to concerns about gay bathhouses. As early as March 1983, the closing of gay bathhouses in San Francisco was proposed (Shilts, 1987). In deference to AIDS and the environment of fear it was creating, some gay bathhouses made changes. They distributed and displayed safe-sex posters, brochures, and condoms; boarded up orgy rooms; and introduced "jack-off" nights (Berube, 2003). Despite these innovations, activists and officials in San Francisco wrestled with the issue of gay bathhouses and sex clubs, debating whether or not they should be closed, regulated, or encouraged as the ideal setting in which to foster AIDS awareness (Disman,

2003). Ultimately, in October 1984, Dr Mervyn Silverman, San Francisco's Public Health Director, ordered that gay bathhouses be closed (Disman, 2003). While the legality of this move continued to be debated, the decision itself reflected disenchantment with the bathhouses and the risk they symbolized in the midst of a deadly epidemic.

Awareness of AIDS increased as the number of AIDS cases escalated and AIDS became the "leading cause of premature mortality" in never-married men aged 25 to 44 years in New York and San Francisco (Jaffe *et al.*, 1985). By 1985, the effects of AIDS awareness and prevention campaigns were reflected in reductions of self-reported risk behaviors by homosexual men in San Francisco (CDC, 1985) and New York (Martin, 1987). Notable drops in rates of gonorrhea in homosexually active men New York City (CDC, 1984), Denver (Judson, 1983), and Seattle (Handsfield, 1985) provided further evidence of a reduction in sexual risk-taking behavior. Behavioral prevention efforts were strengthened by the formation of groups such as Stop AIDS, whose members pledged to eliminate AIDS by practicing safe sex (Andriote, 1999). By the beginning of 1986, the incidence of the virus which causes AIDS, Human Immunodeficiency Virus (HIV), among a sample of MSM in San Francisco had dropped to 4.2 percent from a high of 18.4 percent in mid-1982 (Winkelstein *et al.*, 1987). "Voluntary activist organizations" within the homosexual community were credited with achievements in behavior change and reduction in infection rates (Institute of Medicine, 1986).

Success in curbing the spread of HIV/AIDS was accompanied by increasingly vocal AIDS activism on the part of homosexual men. One of the most successful AIDS activist organizations, the AIDS Coalition to Unleash Power (ACT UP), was formed in 1987 in response to concerns about the FDA drug approval process and its impact on the testing and approval of drugs for the treatment of HIV/AIDS (Andriote, 1999). ACT UP chapters formed throughout the country and internationally (Andriote, 1999). Pressure from ACT UP improved the accessibility and pricing of AIDS drugs, and the speed with which AIDS drugs were developed and made available to HIV/AIDS infected individuals. This process undoubtedly altered the long-term impact of the AIDS epidemic for many, many individuals.

## Sexual behaviors in the gay '90s and beyond

By the late 1980s, the gay community's ability to institute widespread behavioral change was well documented (Stall *et al.*, 1988), and rates of AIDS infection among homosexual men were slowly dropping (CDC, 1990). In the midst of this success, however, data were beginning to surface that suggested that some MSM were struggling to maintain safe-sex behavior, and some had never conformed to the new norms (Stall *et al.*, 2000). Younger men and men who did not have a close friend or lover with AIDS were more likely to engage in sexual risk-taking behavior (Ekstrand and Coates, 1990), as were men of racial and ethnic minorities (Stall *et al.*, 2000). Men who used drugs in conjunction with sexual activity were also more likely to engage in sexual risk behavior (Ostrow *et al.*, 1990). Some studies also indicated that feeling a part of a gay community (Joseph *et al.*, 1991) and community expectations (Kelly *et al.*, 1992) were positively associated with risk-reduction behavior. Stop AIDS, the San Francisco advocacy group formed in 1985 to promote a community-wide commitment to safe-sex behavior, had shut down in 1987, confident that this commitment was well established (Andriote, 1999), but reopened their doors in 1990 amid concerns about rising levels of unprotected sex and new HIV infections (Andriote, 1999). Reports from other cities of increases in sexually transmitted diseases (Handsfield and Schwebke, 1990) and rates of HIV infection among homosexual men (Kingsley *et al.*, 1991) generated renewed concern.

In the mid-1990s, the promise of highly active antiretroviral therapy (HAART) furthered worry that risk reduction might not remain a priority. Several studies did show an association between lessened concern about infection with HIV/AIDS because of HAART, and sexual risk-taking (Stall *et al.*, 2000, Ostrow *et al.*, 2002). In young MSM, treatment optimism was higher among men who perceived themselves at greater risk of being HIV infected (Huebner *et al.*, 2004), while young MSM who were less concerned with the seriousness of infection were more likely to have more partners and to be men of color (Koblin *et al.*, 2003). Throughout the 1990s both young MSM (Katz *et al.*, 1998; Seage *et al.*, 1997) and ethnic- or racial-minority MSM (CDC, 2001a; Blair *et al.*, 2002) consistently demonstrated relatively higher HIV infection rates.

In 1998 the CDC launched the multi-site Young Men's Study with the goal of quantifying and better understanding the dynamics of HIV infection among young MSM. Data from this study showed that young MSM had high rates of HIV infection (Valleroy *et al.*, 2000), that young Black MSM had higher rates of HIV than young White MSM (Celentano *et al.*, 2005), and that recreational drug use was associated with HIV infection in young MSM (Celentano *et al.*, 2006a). These three findings highlight areas of concern in the HIV/AIDS epidemic among MSM today.

In summary, young MSM (CDC, 2001a, 2001b), Black MSM (Millett *et al.*, 2006), and MSM who use drugs or alcohol at the time of sexual intercourse (Colfax and Guzman, 2006; Koblin *et al.*, 2006; Mansergh *et al.*, 2006a) are at disproportionate risk of becoming HIV-infected. In each of these instances, the interplay of individual, environment, and disease is apparent. Young MSM coming to grips with their sexual identity and unaware of both historical and potential costs of infection with HIV/AIDS can be less likely to adopt the older gay community's norm of safe sex. Black MSM experience disproportionately high rates of STI, which can facilitate the transmission of HIV/AIDS. At the same time, they are less likely to be aware of their HIV-positive status than other

MSM (Millet *et al.*, 2006), which also impedes the reduction of disease. MSM who are participating in a social life that includes partying at clubs and circuit parties, and using drugs in conjunction with sexual behavior, are adhering to a completely different set of community norms. Within this community, "club" drugs are associated with social disinhibition and heightened sexual experiences (Romanelli *et al.*, 2003). Targeting these populations of MSM who are at particular risk is essential to limiting the HIV/AIDS epidemic in the United States.

As the United States marks the twenty-fifth anniversary of the HIV/AIDS epidemic, half a million Americans are living with HIV/AIDS, and 60 percent of these are MSM (CDC, 2005c). Since the very earliest days of the epidemic in the United States, MSM have been most affected by the disease. The response of this population to this epidemic has helped to chart its course, its successes, and failures. As the still single largest transmission category, the steps this group takes to alter and adopt risk reduction behaviors will have a critical impact on the ultimate course of this epidemic.

The experiences of American MSM may also provide insight into the progression of the HIV/AIDS epidemic among MSM in other societies. In the US, reductions in HIV/AIDS and other STI among gay men have been associated with a visible and active gay community. In many societies sexual activity between men remains a taboo, and in approximately 70 countries is even illegal (Timberlake, 2006). In such environments, successful surveillance and prevention activities are not possible. In some societies, a reluctance to acknowledge gay lifestyles can affect even the most basic prevention messages. In Thailand, for example, where the prevalence of HIV among MSM nearly doubled between 2003 and 2005 (van Griensven *et al.*, 2006), a third of participants in a recent surveillance study did not understand at a basic level how HIV is transmitted (Mansergh *et al.*, 2006b). The only hope for prevention of HIV/AIDS among MSM in such countries is a societal shift in the acceptance of certain sexual behaviors and lifestyles.

## The Internet and human sexual behavior: the evolution of a new technology

The advent of the Internet, which revolutionized the flow of information and streamlined communications, also permanently altered the American social landscape. More than 70 percent of American adults now use the Internet, and more and more feel that the Internet has "improved their ability" to do their jobs, and "improved the way" they "pursue their hobbies and interests" and the "way they get information about health care" (Madden and Lenhart, 2006). The Web's promise as a means of making social connections has also been recognized, and 74 percent of Internet users who are single and looking for romantic partners have used the Internet as an aid in this pursuit (Madden and Lenhart, 2006).

The public health implications of the fast and easy Internet in the world of individual social connections, as well its promise for population-based research and prevention, are reviewed in this section.

## A new medium for sexual partnerships

The Internet constitutes a forum where individuals from different geographic areas can meet and discuss a wide variety of issues, ranging from personal growth to politics and current events. Its use as a means of forging personal social relationships has also been realized. Almost half of those individuals who have visited online dating websites have actually gone on a date with someone they met through a website, and, of these, 17 percent (nearly 3 million adults) are in a long-term committed relationship with someone they met online (Madden and Lenhart, 2006). Recent studies have also demonstrated that individuals are using the Internet to actively solicit both heterosexual and homosexual partnerships (Cooper and Sportolari, 1997; Wysocki, 1998; Toomey and Rothenberg, 2000; Bull *et al.*, 2001).

The virtual anonymity of the Internet has enhanced the interactions of socially disenfranchised groups, such as gay and bisexual individuals, allowing individuals with similar personal interests to approach each other without social stigmatization. Indeed, those who seek sex partners online can peruse cyberspace for a partner of very specific demographic and behavioral characteristics (e.g. unprotected sex, drug use) (Bull and McFarlane, 2000), doing so anonymously and from a variety of private and public access points.

## MSM and the Internet

The MSM community has embraced the accessibility, anonymity, and reach of online environments. As many as two-thirds of MSM are reported to actively use the Internet (Weatherburn *et al.*, 2003), while up to one-third of MSM use the Internet to find male sex partners (Elford *et al.*, 2001; Kim *et al.*, 2001; Benotsch *et al.*, 2002). The discretion and accessibility inherent in online socializing may make it easier for younger men, and men who do not necessarily identify as gay, to seek other men for sex. The Internet, and the online search engines, chat rooms, message boards, and media-sharing it provides, allow MSM to circumvent the potentially daunting and public ambiance of traditional gay venues, such as clubs, bars, bathhouses, and public sex environments (Rietmeijer *et al.*, 2001; McFarlane *et al.*, 2002). It may be for these reasons that MSM are significantly more likely to sexually engage with an online partner when compared with heterosexual men and women (McFarlane *et al.*, 2000; Bull *et al.*, 2001; Kim *et al.*, 2001). MSM who seek sex partners online are also comparatively younger (Kim *et al.*, 2001; Benotsch *et al.*, 2002), are more likely to identify as bisexual or heterosexual, and are more likely to report sex with women (Ross *et al.*, 2000; Rhodes *et al.*, 2002; Weatherburn *et al.*, 2003).

Accessing online sex environments has been associated with a greater likelihood of engaging in sexual risk behaviors among MSM. MSM who seek sex partners online have been shown to have more casual sexual partners (Kim *et al.*, 2001; Benotsch *et al.*, 2002), report higher rates of unprotected sex (Kim *et al.*, 2001; Benotsch *et al.*, 2002; Hospers *et al.*, 2002), and report sex with an HIV-positive individual (Kim *et al.*, 2001) than their counterparts who seek sexual partners offline. Furthermore, use of recreational drugs, such as MDMA (ecstasy), nitrites (poppers), methamphetamines, and sexual-performance enhancing medications (e.g. Viagra) is reported at significantly higher proportions by MSM who seek sex partners online (Benotsch *et al.*, 2002; Mettey *et al.*, 2003; Hirshfeld *et al.*, 2004; Taylor *et al.*, 2004).

Meeting sexual partners online has also been associated with higher rates of STI among MSM. In 1999, an outbreak of syphilis among MSM in San Francisco was tracked to individuals meeting online (Klausner *et al.*, 2000). Similarly, an investigation into cases of rectal gonorrhea in San Francisco revealed that infected MSM were three times more likely to have met sexual partners online than were uninfected MSM (Kim *et al.*, 2000). Studies of STI clinic clients have also shown that a large proportion of MSM clients use the Internet to find sex partners (Klausner *et al.*, 2000; Hospers *et al.*, 2002; McFarlane *et al.*, 2002; CDC, 2003). At the same time, MSM who use the Internet to seek sex partners are also significantly more likely to report history of an STI than those who do not (Elford *et al.*, 2001; Lau *et al.*, 2003). MSM with a history of an STI are at higher risk for the transmission and acquisition of future STI, including HIV, by way of a documented record of high-risk behavior.

While the Internet has increased the risk of transmission of disease for some MSM, some HIV-infected MSM are employing the Internet as an agent for reducing risk. In this process, known as "serosorting," HIV-positive men seek sex partners online who are also HIV positive (Suarez and Miller, 2001; Hirshfeld *et al.*, 2004), and engage in unprotected anal intercourse (Elford *et al.*, 2001; Weatherburn *et al.*, 2003) with these seroconcordant partners. The virtual anonymity and safety of the Internet enables HIV-positive MSM to disclose their HIV serostatus without fear of stigma, and expect that their online partners will do the same (Elford *et al.*, 2001; Reitmeijer *et al.*, 2001). Although some may argue that serosorting may have an impact in the reduction of HIV transmission, there is evidence that it may also promote the transmission of other STI (Elford *et al.*, 2001), as well as acquisition of drug-resistant strains of HIV. A more troubling phenomenon is that some MSM who are HIV-uninfected report that they are more likely to have unprotected sex with HIV-infected partners (Elford *et al.*, 2001), and sex in exchange for money or drugs (Kim *et al.*, 2001) with men they have met online. Risk behavior among HIV-serodiscordant individuals who meet online increases the risk of HIV and STI transmission, facilitating the efficient propagation of infectious diseases.

## A new medium for prevention and research

The exact same elements of the Internet (namely its anonymity, reach, and accessibility) which make it a potential threat for the spread of infectious diseases may also make it an ideal tool for curbing the spread of disease. Never before has information been so widely available to so many people of such a wide variety of demographic and social backgrounds in so many geographic locations. The reach and accessibility of the Internet can be harnessed to inform and protect against disease acquisition; anonymity can be a convincing ally in the fight against disease transmission.

Although the extraordinary capacity of the Internet has yet to be fully tapped for this purpose, fledgling efforts have demonstrated its potential. It has been used for partner notification and the distribution of education, prevention, and health information. In 1999 in San Francisco and in 2003 in Los Angeles, public health officials notified users of gay chat rooms and sexual contacts of known syphilis cases of the potential for syphilis exposure, and urged them to seek medical care (Klausner *et al.*, 2000; CDC, 2004). In the Netherlands, MSM recruited in chat rooms indicated that they would utilize a website which posted HIV/STI-prevention information, and provided question-and-answer, e-mail and live-chat forum services devoted to safe-sex topics (Hospers *et al.*, 2002). MSM in the UK also reported positive attitudes toward receiving health information online and interacting live with public health experts (Chiasson *et al.*, 2005). Structured HIV-prevention interventions need to utilize the Internet to recruit affected and disenfranchised individuals for offline programs, and develop the ability to implement intervention curricula online; a few (Gaither, 2000; Davis *et al.*, 2004) are following in the footsteps of successful online efforts, such as online smoking cessation programs.

This generation's adolescents and young adults are already using the Web as a source of health information (Kaiser Family Foundation, 2001). The vast majority of 15- to 24-year-olds access the Internet, and two-thirds have accessed health information. Of these, four in ten have accessed information on pregnancy, HIV/AIDS, STI, and birth control (Kaiser Family Foundation, 2001). Minority youth and teenage youth seem to be the most receptive to such information, and report intent to change behavior (Kaiser Family Foundation, 2001). The fact that most remain skeptical as to the validity of the posted information, however, underscores the need for public health agencies to maintain current and reliable information.

Recent HIV/STI prevention, outreach, and education efforts in the US on both local (McFarlane *et al.*, 2005) and federal (Anderton and Valdiserri, 2005) levels, however, have encountered a number of barriers. Evaluation of online efforts, proper training of involved staff, and a close collaboration between online venue owners and public health workers are critically important (McFarlane *et al.*, 2005). The core differences that exist between the corporate and public health worlds also present a challenge. Internet service providers and website owners have legal liability and confidentiality issues, and must conform to a market-driven, business environment. Meanwhile, public health practitioners and researchers request access to these same clients with the goal of discussing sensitive issues (Anderton and Valdiserri, 2005). Only through mutual understanding and shared benefit can the two worlds collaborate successfully.

As Internet-based research and prevention grows, clear priorities are emerging. Methods for online behavioral and prevention research need to be further developed, and standardized methods for online outreach and educational activities need to be further refined. Sound research methodology will not only inform prevention programs, but also evaluate them. The strengths of the Internet – its ability to reach and recruit large numbers of geographically, demographically, and socially isolated or disenfranchised individuals and subgroups in a cost-effective way – must be balanced against concerns about biased recruitment methods, the validity and generalizability of data collected online, and the ethical implications of recruiting participants online (Chiasson *et al.*, 2006). Online surveys do eliminate interviewer and response bias, allowing participants to report their experiences in greater detail and in an anonymous fashion. Inconsistent results from online studies thus far, however, also suggest the possibility of sampling bias, and that sampling and recruitment methods need greater standardization. Furthermore, while the promise of anonymity may encourage individuals to participate, ethical questions remain about true anonymity (e.g. e-mail addresses, Internet Protocol (IP) addresses, etc.), as well as duplicate participation, and validity of informed consent. Regardless, use of the Internet as an additional tool for research in the area of HIV continues to grow. At the 2006 International AIDS Conference, the number of abstracts referencing HIV and the Internet had increased by 41 percent since the 2004 meeting.

The connection between online and offline behavioral research and infectious disease surveillance systems also needs to be considered. In today's social climate, assessments of the prevalence of infectious diseases and behavioral outcomes in various populations should take into consideration online populations and behaviors resulting from online interactions (Link and Mokdad, 2005). Research efforts that include both online and offline sample populations will also need to evaluate any qualitative differences between these participants.

The Internet is the latest in a series of social forces that offers both real opportunity and real risk. The degree to which these opportunities and risks are considered and managed will determine what role the Internet ultimately plays in infectious diseases that are transmitted through sexual contact. With thoughtful analysis and guidance, the Internet, with its broad reach and wide accessibility, has the potential to become a force of unprecedented power in the pursuit of widespread public health.

## Lessons learned for the journey ahead

The journey our society has traveled over the last 50 years in its expectations of social and sexual behaviors has been a dramatic one. It has been characterized by victory and loss, freedom and imprisonment, peril and promise. Some women and men have watched their worlds broaden to more than they could ever imagine, while others just as rapidly watched as every dream shattered. At times, these men and women were one and the same. Throughout the journey we have learned much about how the behaviors of individuals and communities can shape the destiny of whole populations.

Of the lessons learned, one of the most striking is the power of an intentional community. The response of the gay community to the crisis of AIDS is particularly notable. With urgency and creativity, gay men constructed not only a self-imposed behavioral response which interrupted the transmission of infectious disease, but also a coordinated assault on industry and the government to ensure that progress be made in slowing the progression of disease. We have also witnessed dramatic changes in condom use adherence in diverse groups of sex workers in Thailand (Hanenberg *et al.*, 1994) and the Dominican Republic (Kerrigan *et al.*, 2003), as well as drug users (Institute of Medicine, 2006), reducing their sexual risks after HIV voluntary counseling and testing. Another central lesson has been the disruptive power of societal change. A transformation in the fundamental roles of American women in the second half of this century engendered a permanent alteration in basic expectations of the social and sexual behavior of young women and young men. Finally, a social change of particular power has been the broad acceptance of computers and the Internet, as well as other technologies such as cell phones. Core elements of our society have been restructured, and this restructuring has again permanently altered some of our most basic behaviors.

If these and other lessons have been learned, how can we apply them to populations both here and abroad as they face epidemics of STI, particularly HIV/AIDS? One population of real concern in our own country is minorities. Black women and men, in particular, are disproportionately burdened by HIV/AIDS; 50 percent of all Americans diagnosed with HIV/AIDS in 2004 were African-Americans (CDC, 2006d). Some have suggested that the lesson of community may be usefully applied in this context. Advocates have turned to Black churches to help them spread information and understanding about the spread of AIDS in the Black community (Avery and Bashir, 2003). Organizations such as The Balm in Gilead, which sponsors The Black Church Week of Prayer for the Healing of AIDS (Avery and Bashir, 2003), have become an effective community voice in the fight against HIV/AIDS.

Other lessons may be usefully applied to international communities and their burgeoning HIV/AIDS epidemics. The social and technological changes that have characterized our country's development over the last several decades are now occurring in other countries. In Thailand, for example, dramatically shifting sexual roles are starting to be reflected in alarming rates of STI among young Thai women (Celentano *et al.*, 2006b). The spread of technology in Thailand, where almost half the population has cell phones and more than a tenth have Internet access (CIA, 2006), is also encouraging greater connections between young people. Preparing for the consequences of these developments, by implementing educational programs which are aimed at this segment of the population and which make use of available technologies, may be one way of applying lessons we have learned from epidemics in this country to potential epidemics in other countries.

Clearly, understanding the relationships between social developments and their impact on social structures and human behaviors, specifically sexual behaviors, is an essential component in understanding the sexual transmission of disease. We have learned much in this country about the interplay of these processes. Applying this knowledge prospectively both in our country and in other countries remains one of our greatest challenges in the field of STI.

## References

Abma, J.C. and Sonenstein, F. (2001). Sexual activity and contraceptive practices among teenagers in the United States, 1988 and 1995. *Vital Health Statistics* **23**, 1–79.

Adimora, A.A. and Schoenbach, V.J. (2005). Social context, sexual networks, and racial disparities in rates of sexually transmitted infections. *Journal of Infectious Disease* **191**, S115–122.

Anderson, J.E., Santelli, J. and Mugalla, C. (2003). Changes in HIV-related preventive behavior in the US population: data from national surveys, 1987–2002. *Journal of Acquired Immune Deficiency Syndrome* **34**, 195–202.

Anderson, R.M. (1999). Transmission dynamics of sexually transmitted infections. In: K.K. Holmes, P.D. Sparling, P.-A. Mardh *et al.* (eds), *Sexually Transmitted Diseases*, 3rd edn. New York: McGraw-Hill, pp. 25–37.

Anderton, J.P. and Valdiserri, R.O. (2005).Combating syphilis and HIV among users of internet chatrooms. *Journal of Health Communication* **10**, 665–671.

Andriote, J. (1999). *Victory Deferred*. Chicago: The University of Chicago Press.

Aral, S.O., Patel, D.A., Holmes, K.K. and Foxman, B. (2005). Temporal trends in sexual behaviors and sexually transmitted disease history among 18- to 39-year-old Seattle, Washington, residents: results of random digit-dial surveys. *Sexually Transmitted Disease* **32**, 710–717.

Arya, O.P., Mallinson, H. and Goddard, A.D. (1981). Epidemiological and clinical correlates of chlamydial infection of the cervix. *British Journal of Venereal Disease* **57**, 118–124.

Asbell, B. (1995) *The Pill: A Biography of the Drug that Changed the World*. New York: Random House.

Avery, B. and Bashir, S. (2003). The road to advocacy – searching for the rainbow. *American Journal of Public Health* **93**, 1207–1210.

Becker, T.M., Blount, J.H. and Guinan, M.E. (1985). Genital herpes infections in private practice in the United States, 1966 to 1981. *Journal of the American Medical Association* **253**, 1601–1603.
Benotsch, E.G., Kalichman, S., and Cage, M. (2002). Men who have met sex partners via the Internet: prevalence, predictors, and implications for HIV prevention. *Archives of Sexual Behavior* **31**, 177–183.
Berger, G.S., Keith, L. and Moss, W. (1975). Prevalence of gonorrhoea among women using various methods of contraception. *British Journal of Venereal Disease* **51**, 307–309.
Berube, A. (2003). The history of gay bathhouses. In: W.J. Woods and D. Binson (eds), *Gay Bathhouses and Public Health Policy*. Binghamton: Harrington Park Press, pp. 33–53.
Biddlecom, A.E. (2004). Trends in sexual behaviours and infections among young people in the United States. *Sexually Transmitted Infections* **80**(Suppl. 2), 74–79.
Blair, J.M., Fleming, P.I. and Karon, J.M. (2002). Trends in AIDS incidence and survival among racial/ethnic minority men who have sex with men, United States, 1990–1999. *Journal of Acquired Immune Deficiency Syndrome* **31**, 339–347.
Bowen, W.G. and Finegan, T.A. (1969). *The Economics of Labor Force Participation*. Princeton: Princeton University Press.
Bull, S.S. and McFarlane, M. (2000). Soliciting sex on the Internet: what are the risks for sexually transmitted diseases and HIV? *Sexually Transmitted Disorders* **27**, 545–550.
Bull, S., McFarlane, M. and Rietmeijer, C. (2001). HIV and sexually transmitted infection risk behaviors among men seeking sex with men on-line. *American Journal of Public Health* **91**, 988–989.
Bureau of Labor Statistics (1988). *Labor Force Statistics Derived from the Current Population Survey, 1948–87*. Washington, DC: US Government Printing Office.
Bureau of the Census (1976). *1976d Statistical Abstract of the United States*. Washington, DC: US Government Printing Office.
Bureau of the Census (1998). *School Enrollment – Social and Economic Characteristics of Students: October 1996 (update)*. Atlanta: US Department of Commerce, Bureau of the Census.
Bureau of the Census (1999). *Table MS-2. Estimated Median Age at First Marriage & Sex: 1890 to the Present*. Atlanta: US Department of Commerce, Bureau of the Census.
Cain, G. (1966). *Married Women in the Labor Force: An Economic Analysis*. Chicago: University of Chicago Press.
Carlson, B.L., Fiumara, N.J., Kelly, R. and McCormack, W.M. (1980). Isolation of Neisseria meningitis from anogenital specimens from homosexual men. *Sexually Transmitted Diseases* **7**, 71–73.
Catania, J.A., Canchola, J., Binson, D. *et al.* (2001). National trends in condom use among at-risk heterosexuals in the United States. *Journal of Acquired Immune Deficiency Syndrome* **27**, 176–182.
Celentano, D.D., Sifakis, F., Hylton, J. *et al.* (2005). Race/ethnic differences in HIV prevalence and risks among adolescent and young adult men who have sex with men. *Journal of Urban Health* **82**, 610–621.
Celentano, D.D., Valleroy, L.A., Sifakis, F. *et al.* (2006a). For the Young Men's Survey Study Group F. Associations between substance use and sexual risk among very young men who have sex with men. *Sexually Transmitted Diseases* **33**, 265–271.
Celentano, D.D., Sirirojn, B., Sherman, S. *et al.* (2006b). STI prevalence among adolescent methamphetamine smokers requires new HIV/AIDS prevention strategies.

*Abstract TUPE0287*. XVI International Conference on AIDS. Toronto, Canada, 13–18 August 2006.
Centers for Disease Control and Prevention (1981a). *Pneumocystis* pneumonia – Los Angeles. *Morbidity and Mortality Weekly Report* **30**, 250–252.
Centers for Disease Control and Prevention (1981b). Kaposi's Sarcoma and *Pneumocystis* pneumonia among homosexual men – New York City and California. *Morbidity and Mortality Weekly Report* **30**, 305–308.
Centers for Disease Control and Prevention (1982a). Epidemiologic notes and reports *Pneumocystis carinii* pneumonia among persons with hemophilia A. *Morbidity amd Mortality Weekly Report* **31**, 365–367.
Centers for Disease Control and Prevention (1982b). Current trends update on acquired immune deficiency syndrome (AIDS) – United States. *Morbidity and Mortality Weekly Report* **31**, 513–514.
Centers for Disease Control and Prevention (1983). Current trends prevention of acquired immune deficiency syndrome (AIDS): report of inter-agency recommendations. *Morbidity and Mortality Weekly Report* **32**, 101–103.
Centers for Disease Control and Prevention (1984). Declining rates of rectal and pharyngeal gonorrhea among males – New York City. *Morbidity and Mortality Weekly Report* **33**, 295–297.
Centers for Disease Control and Prevention (1985). Epidemiologic notes and reports: self-reported behavioral change among gay and bisexual men – San Francisco. *Morbidity and Mortality Weekly Report* **34**, 613–615.
Centers for Disease Control and Prevention (1990). Current trends update: Acquired Immunodeficiency Syndrome – United States, 1989. *Morbidity and Mortality Weekly Report* **39**, 81–86.
Centers for Disease Control and Prevention (1991). Premarital sexual experience among adolescent women – United States, 1970–1988. *Morbidity and Mortality Weekly Report* **39**, 929–932.
Centers for Disease Control and Prevention (2001a). *HIV Prevalence Trends in Selected Populations in the United States: Results from National Serosurveillance, 1993–1997.* Atlanta: Centers for Disease Control and Prevention.
Centers for Disease Control and Prevention (2001b). HIV incidence among young men who have sex with men – seven US cities, 1994–2000. *Morbidity and Mortality Weekly Report* **50**, 440–444.
Centers for Disease Control and Prevention (2002). Trends in sexual risk behaviors among high school students – United States, 1991–2001. *Morbidity and Mortality Weekly Report* **51**, 856–859.
Centers for Disease Control and Prevention (2003). Internet use and early syphilis infection among men who have sex with men – San Francisco, California, 1999–2003. *Morbidity and Mortality Weekly Report* **52**, 1229–1232.
Centers for Disease Control and Prevention (2004). Using the Internet for partner notification of sexually transmitted diseases – Los Angeles County, California, 2003. *Morbidity and Mortality Weekly Report* **53**, 129–131.
Centers for Disease Control and Prevention (2005a). *Sexually Transmitted Disease Surveillance 2004 Supplement: Gonococcal Isolate Surveillance Project (GISP) Annual Report – 2004*. Atlanta: US Department of Health and Human Services.
Centers for Disease Control and Prevention (2005b). *Sexually Transmitted Disease Surveillance, 2004*. Atlanta: US Department of Health and Human Services.
Centers for Disease Control and Prevention (2005c). *HIV/AIDS Surveillance Report, 2004*, Vol. 16. Atlanta: US Department of Health and Human Services, Centers for Disease Control and Prevention.

Centers for Disease Control and Prevention (2006a). *National Youth Risk Behavior Survey: 1991–2005, Trends in the Prevalence of Sexual Behaviors*. Atlanta: US Department of Health and Human Services.

Centers for Disease Control and Prevention (2006b). Epidemiology of HIV/AIDS – United States, 1981–2005. *Morbidity and Mortality Weekly Report* **55**, 589–592.

Centers for Disease Control and Prevention (2006c). Racial/ethnic disparities in diagnoses of HIV/AIDS – 33 states, 2001–2004. *Morbidity and Mortality Weekly Report* **55**, 121–125.

Centers for Disease Control and Prevention (2006d). *HIV/AIDS Among African Americans*. Atlanta: US Department of Health and Human Services.

Chiasson, M., Hirshfield, S., Humberstone, M. *et al.* (2005). A comparison of online and offline risk in MSM. Paper presented at the Twelfth Conference on Retroviruses and Opportunistic Infections, 25 February 2005, Boston, MA.

Chiasson, M.A., Parsons, J.T., Tesoriero, J.M. *et al.* (2006). HIV behavioral research online. *Journal of Urban Health* **83**, 73–85.

CIA (2006). *The World Fact Book: Thailand*. Available at: https://www.cia.gov/cia/publications/factbook/geos/th.html.

Colfax, G. and Guzman, R. (2006). Club drugs and HIV infection: a review. *Clinical Infectious Diseases* **42**, 1463–1469.

Comfort, A. (1972). *The Joy of Sex: A Cordon Bleu Guide to Lovemaking*. New York: Crown.

Cooper, A. and Sportolari, L. (1997). Romance in cyberspace: understanding online attraction. *CyberPsychology and Behavior* **22**, 7–14.

Corley, C.J. and Woods, A.Y. (1991). Socioeconomic, sociodemographic and attitudinal correlates of the tempo of divorce. *Journal of Divorce and Remarriage* **16**, 47–68.

Davanzo, J. and Rahman, M.O. (1993). American families: trends and correlates. *Population Index* **59**, 350–386.

Davis, M., Bolding, G., Hart, G. *et al.* (2004). Reflecting on the experience of interviewing online: perspectives from the Internet and HIV study in London. *AIDS Care* **16**, 944–952.

d'Emilio, J. (2002). *The World Turned*. Durham: Duke University Press.

Disman, C. (2003). The San Francisco bathhouse battles of 1984: civil liberties, AIDS risk, and shifts in health policy. In: W.J. Woods and D. Binson (eds), *Gay Bathhouses and Public Health Policy*. Binghamton: Harrington Park Press, pp. 71–129.

Dritz, S.E. (1997). Charting the epidemiological course of AIDS, 1981–1984, an oral history conducted in 1992 and 1993 by Sally Smith Hughes, PhD. *The AIDS Epidemic in San Francisco: The Medical Response, 1981–1984* **III**, 73.

Edwards, M.E. (2001). Uncertainty and the rise of the work–family dilemma. *Journal of Marriage and Family* **63**, 183–196.

Elford, J., Bolding, G. and Sherr, L. (2001). Seeking sex on the Internet and sexual risk behaviour among gay men using London gyms. *AIDS* **15**, 1409–1415.

Ekstrand, M.L. and Coates, T.J. (1990). Maintenance of safer sexual behaviors and predictors of risky sex: the San Francisco Men's Health Study. *American Journal of Public Health* **80**, 973–977.

Fichtner, R.R., Aral, S.O., Blount, J.H. *et al.* (1983). Syphilis in the United States: 1967–1979. *Sexually Transmitted Diseases* **10**, 77–80.

Fleming, D.T., McQuillan, G.M., Johnson, R.E. *et al.* (1997). Herpes simplex virus type 2 in the United States, 1976–1994. *New England Journal of Medicine* **337**, 1105–1111.

Forrest, J.D. and Singh, S. (1990). The sexual and reproductive behavior of American women, 1982–1988. *Family Planning Perspective* **22**, 206–214.

Fox, K.K., Whittington, W.L., Levine, W.C. *et al.* (1998). Gonorrhea in the United States, 1981–1996: demographic and geographic trends. *Sexually Transmitted Diseases* **25**, 386–393.

Friedan, B. (1963). *The Feminine Mystique*. New York: Norton.

Gaither, C. (2000) Group roams chat rooms to talk to gay men about AIDS. *New York Times*, 9 November.

Gillmore, M.R., Schwartz, P. and Civic, D. (1999). The social context of sexuality: the case of the United States. In: K.K. Holmes, P.D. Sparling, P.-A. Mardh *et al.* (eds), *Sexually Transmitted Diseases*, 3rd edn. New York: McGraw-Hill, pp. 95–105.

Glenn, N.D. and Weaver, C.N. (1979). Attitutes toward premarital, extramarital, and homosexual relations in the US in the 1970s. *Journal of Sexual Research* **15**, 108.

Glick, P.C. (1984). Marriage, divorce and living arrangements: prospective changes. *Journal of Family Issues* **5**, 7.

Gordon, H.A. and Kammeyer, K.C.W. (1980). The gainful employment of women with small children. *Journal of Marriage and the Family* **42**, 327–336.

Handsfield, H.H. (1985) Decreasing incidence of gonorrhea in homosexually active men – minimal effect on risk of AIDS. *Western Journal of Medicine* **143**, 469–470.

Handsfield, H.H. and Schwebke, J. (1990). Trends in sexually transmitted diseases in homosexually active men in King County, Washington, 1980–1990. *Sexually Transmitted Diseases* **17**, 211–215.

Hanenberg R.S., Rojanapithayakorn W., Kunasol, P. and Sokal, D.C. (1994). Impact of Thailand's HIV-control programme as indicated by the decline of sexually transmitted diseases. *Lancet* **344**, 243–245.

Heer, D.M. and Grossbard-Schectman, A. (1981). The impact of the female marriage squeeze and the contraceptive revolution on sex roles and the women's liberation movement in the United States, 1960 to 1975. *Journal of Marriage and the Family* **43**, 49–65.

Henig, R. (1983). AIDS: a new disease's deadly odyssey. *New York Times*, 6 February.

Hirshfield, S., Remien, R., Humberstone, M. *et al.* (2004). Substance use and high-risk sex among men who have sex with men: a national online study in the USA. *AIDS Care* **16**, 1036–1047.

Hofferth, S.L., Kahn, R.J. and Baldwin, W. (1987). Premarital sexual activity among US teenage women over the past three decades. *Family Planning Perspective* **19**, 46–53.

Holmes, K.K. (1981). The chlamydia epidemic. *Journal of the American Medical Association* **245**, 1718–1723.

Hospers, H.J., Harterink, P., van Den, H.K. and Veenstra, J. (2002). Chatters on the Internet: a special target group for HIV prevention. *AIDS Care* **14**, 539–544.

Huebner, D.M., Rebchook, G.M. and Kegeles, S.M. (2004). A longitudinal study of the association between treatment optimism and sexual risk behavior in young adult gay and bisexual men. *Journal of Acquired Immune Deficiency Syndrome* **37**, 1514–1519.

Institute of Medicine (1986). *Confronting AIDS: Directions for Public Health, Health Care and Research*. Washington, DC: National Academy Press.

Institute of Medicine (2006). *Preventing HIV Infection Among Injecting Drug Users in High Risk Countries: An Assessment of the Evidence*. Washington, DC: National Academy Press.

Jaffe, H.W., Hardy, A.M., Morgan, W.M. and Darrow, W.W. (1985). The acquired immunodeficiency syndrome in gay men. *Annals of Internal Medicine* **103**, 662–664.

Joseph, K.M., Adib, S.M., Joseph, J.G. and Tal, M. (1991). Gay identity and risky sexual behavior related to the AIDS threat. *Journal of Community Health* **16**, 287–297.

Judson, F.N. (1983). Fear of AIDS and gonorrhoea rates in homosexual men. *Lancet* **2**(8342), 159–160.

Judson, F.N., Penley, K.A., Robinson, M.E. and Smith, J.K. (1980). Comparative prevalence rates of sexually transmitted diseases in heterosexual and homosexual men. *American Journal of Epidemiology* **112**, 836–843.
Kaiser Family Foundation (2001). *Generation Rx.com: How Young People Use the Internet for Health Information* (Pub. No. 3202). Kaiser Family Foundation, 11 December 2001.
Kantner, J.F. and Zelnick, M. (1972). Sexual experience of young unmarried women in the United States. *Family Planning Perspectives* **4**, 9–18.
Kantner, J.F. and Zelnik, M. (1973). Contraception and pregnancy: experience of young unmarried women in the United States. *Family Planning Perspective* **5**, 21–35.
Katz, M.H., McFarland, W., Guillin, V. *et al.* (1998). Continuing high prevalence of HIV and risk behaviors among young men who have sex with men: The Young Men's Survey in the San Francisco Bay Area in 1992 to 1993 and in 1994 to 1995. *Journal of Acquired Immune Deficiency Syndrome* **19**, 178–181.
Kelly, J.A., Murphy, D.A., Roffman, R.A. *et al.* (1992). Acquired Immunodeficiency Syndrome/Human Immunodeficiency Virus risk behavior among gay men in small cities: findings of a 16-city national sample. *Archives of Internal Medicine* **152**, 2293–2297.
Kerrigan, D., Ellen, J., Moreno, L. *et al.* (2003). Environmental-structural factors significantly associated with consistent condom use among female sex workers in the Dominican Republic. *AIDS* **17**, 415–423.
Kim, A., Kent, C. and Klausner, J. (2000). An investigation of rectal gonorrhea among men who have sex with men in San Francisco, 2000 {abstract P92}. Presented at the 2000 National STD Prevention Conference, Milwaukee, Wisconsin, 4–7 December 2000.
Kim, A.A., Kent, C., McFarland, W. and Klausner, J.D. (2001). Cruising on the Internet highway. *Journal of Acquired Immune Deficiency Syndrome* **28**, 89–93.
Kingsley, L.A., Zhou, S.Y., Bacellar, H. *et al.* (1991). Temporal trends in human immunodeficiency virus type 1 seroconversion 1984–1989. A report from the Multicenter AIDS Cohort Study (MACS). *American Journal of Epidemiology* **134**, 331–339.
Kinsey, A.C., Pomeroy, W.B. and Martin, C.E. (1953). *Sexual Behavior in the Human Female*. Philadelphia: Indiana University Press.
Kinsey, A.C., Pomeroy, W.B., Martin, C.E. and Gebhard, P.H. (2006). *Sexual Behavior in the Human Male*. Philadelphia: Indiana University Press.
Klausner, J., Wolf, W., Fischer-Ponce, L. *et al.* (2000). Tracing a syphilis outbreak through cyberspace. *Journal of the American Medical Association* **284**, 447–449.
Koblin, B.A., Perdue, T., Ren, L. *et al.* (2003). Attitudes about combination HIV therapies: the next generation of gay men at risk. *Journal of Urban Health* **80**, 510–519.
Koblin, B.A., Husnik, M.J., Colfax, G. *et al.* (2006). Risk factors for HIV infection among men who have sex with men. *AIDS* **20**, 731–739.
Ku, L., Sonenstein, F.L., Lindberg, L.D. *et al.* (1998). Understanding changes in sexual activity among young metropolitan men: 1979–1995. *Family Planning Perspectives* **30**, 256–262.
Lau, J.T., Kim, J.H., Lau, M. and Tsui, H.Y. (2003). Prevalence and risk behaviors of Chinese men who seek same-sex partners via the Internet in Hong Kong. *AIDS Education and Prevention* **15**, 516–528.
LaVeist, T.A. (2002). Segregation, poverty and empowerment: health consequences for African Americans. In: T.A. LaVeist (ed.), *Race, Ethnicity, and Health*. San Francisco: Jossey-Bass, pp. 76–96.
Link, M.W. and Mokdad, A.H. (2005). Alternative modes for health surveillance surveys: an experiment with web, mail, and telephone. *Epidemiology* **16**, 701–704.

Lugaila, T. (1992). Households, families and children: a 30-year perspective. *Current Population Reports*, 23–181.

Madden, M. and Lenhart, A. (2006). *Online Dating*. Washington, DC: Pew Internet and American Life Project.

Manhart, L.E., Aral, S.O., Holmes, K.K. *et al.* (2004). Influence of study population on the identification of risk factors for sexually transmitted diseases using a case-control design: the example of gonorrhea. *American Journal of Epidemiology* **160**, 393–402.

Mansergh, G., Shouse, R.L., Marks, G. *et al.* (2006a). Methamphetamine and sildenafil (Viagra) use are linked to unprotected receptive and insertive anal sex, respectively, in a sample of men who have sex with men. *Sexually Transmitted Infections* **82**, 131–134.

Mansergh, G., Naorat, S., Jommaroeng, R. *et al.* (2006b). Inconsistent condom use with steady and casual partners and associated factors among sexually-active men who have sex with men in Bangkok, Thailand. *AIDS and Behavior*, 20 May (epub.).

Martin, J.L. (1987). The impact of AIDS on gay male sexual behavior patterns in New York City. *American Journal of Public Health* **77**, 578–581.

Masur, H., Michelis, M.A., Greene, J.B. *et al.* (1981). An outbreak of community-acquired *Pneumocystis carinii* pneumonia: initial manifestation of cellular immune dysfunction. *New England Journal of Medicine*, **305**, 1431–1438.

McFarlane, M., Bull, S.S. and Rietmeijer, C.A. (2000). The Internet as a newly emerging risk environment for sexually transmitted diseases. *Journal of the American Medical Association* **284**, 443–446.

McFarlane, M., Bull, S.S. and Rietmeijer, C.A. (2002). Young adults on the Internet: risk behaviors for sexually transmitted diseases and HIV. *Journal of Adolescent Health* **31**, 11–16.

McFarlane, M., Kachur, R., Klausner, J.D. *et al.* (2005). Internet-based health promotion and disease control in the 8 cities: successes, barriers, and future plans. *Sexually Transmitted Disease* **32**, S60–64.

Mertz, K.J., Ransom, R.L., Groseclose, S.L. *et al.* (2001). Prevalence of genital chlamydial infection in young women entering a national job training program, 1990–1997. *American Journal of Public Health* **91**, 1287–1290.

Mettey, A., Crosby, R., DiClemente, R.J. and Holtgrave, D.R. (2003). Associations between internet sex seeking and STI associated risk behaviours among men who have sex with men. *Sexually Transmitted Infections* **79**, 466–468.

Millett, G.A., Peterson, J.L., Wolitski, R.J. and Stall, R. (2006). Greater risk for HIV infection of black men who have sex with men: a critical literature review. *American Journal of Public Health* **96**, 1007–1019.

Moss, A.R., Bacchetti, P., Gorman, M. *et al.* (1983). AIDS in the "gay" areas of San Francisco. *The Lancet* **1**(8330), 923–924.

Nakashima, A.K., Rolfs, R.T., Flock, M.L. *et al.* (1996). Epidemiology of syphilis in the United States, 1941–1993. *Sexually Transmitted Diseases* **23**, 16–23.

Norton, A.J. and Moorman, J.E. (1987). Current trends in marriage and divorce among American women. *Journal of Marriage and Family* **49**, 3.

Ostrow, D.G. and Altman, N.L. (1983). Sexually transmitted diseases and homosexuality. *Sexually Transmitted Diseases* **10**, 208–215.

Ostrow, D.G., VanRaden, M.J., Fox, R. *et al.* (1990). Recreational drug use and sexual behavior change in a cohort of homosexual men. The Multicenter AIDS Cohort Study (MACS). *AIDS* **4**, 759–765.

Ostrow, D.G., Fox, K.J., Chmiel, J.S. *et al.* (2002). Attitudes towards highly active anti-retroviral therapy are associated with sexual risk taking among HIV-infected and uninfected homosexual men. *AIDS* **16**, 775–780.

Reiss, I.L. (1960). *Premarital Sexual Standards in America*. New York: Free Press.
Rhodes, S.D., DiClemente, R.J., Cecil, H. *et al.* (2002). Risk among men who have sex with men in the United States: a comparison of an Internet sample and a conventional outreach sample. *AIDS Education and Prevention* **14**, 41–50.
Richmond, S.J. and Sparling, P.F. (1976). Genital chlamydial infections. *American Journal of Epidemiology* **103**, 428–435.
Rietmeijer, C.A., Bull, S.S. and McFarlane, M. (2001). Sex and the Internet. *AIDS* **15**, 1433–1434.
Romanelli, F., Smith, K.M. and Pomeroy, C. (2003). Use of club drugs by HIV-seropositive and HIV-seronegative gay and bisexual men. *Topics in HIV Medicine* **11**, 25–32.
Ross, M.W., Tikkanen, R. and Mansson, S.A. (2000). Differences between Internet samples and conventional samples of men who have sex with men: implications for research and HIV interventions. *Social Science and Medicine* **51**, 749–758.
Rotello, G. (1997). *Sexual Ecology: AIDS and the Destiny of Gay Men*. New York: Penguin Books.
Seage, G.R., Mayer, K.H., Lenderking, W.R. *et al.* (1997). HIV and hepatitis B infection and risk behavior in young gay and bisexual men. *Public Health Report* **112**, 158–167.
Shilts, R. (1987). *And the Band Played On*. New York: St Martin's Press.
Sonenstein, F.L. (2004). What teenagers are doing right: changes in sexual behavior over the past decade. *Journal of Adolescent Health* **35**, 77–78.
Stall, R.D., Coates, T.J. and Hoff, C. (1988). Behavioral risk reduction for HIV infection among gay and bisexual men: a review of results from the United States. *American Psychologist* **43**, 878–885.
Stall, R.D., Hays, R.B., Waldo, C.R. *et al.* (2000). The gay '90s: a review of research in the 1990s on sexual behavior and HIV risk among men who have sex with men. *AIDS* **14**(Suppl. 3), S101–114.
Suarez, T. and Miller, J. (2001). Negotiating risks in context: a perspective on unprotected anal intercourse and barebacking among men who have sex with men – where do we go from here? *Archives of Sexual Behavior* **30**, 287–300.
Tanfer, K. and Horn, M.C. (1985). Contraceptive use, pregnancy and fertility patterns among single American women in their 20s. *Family Planning Perspectives* **17**, 10–19.
Taylor, M., Aynalem, G., Smith, L. *et al.* (2004). Correlates of Internet use to meet sex partners among men who have sex with men diagnosed with early syphilis in Los Angeles County. *Sexually Transitted Disease* **31**, 552–556.
Timberlake, S. (2006). Men having sex with men and human rights: the UNAIDS perspective. ILGA World Conference, Pre-Conference: MSM and Gay Men's Health. Geneva, Switzerland, 29 March.
Toomey, K. and Rothenberg, R. (2000). Sex and cyberspace – virtual networks leading to high-risk sex. *Journal of the American Medical Association* **284**, 485–487.
United States President's Commission on the Status of Women (1963). *The Presidential Report on American Women*. Washington, DC: US Government Printing Office.
US Census Bureau (2003). America's families and living arrangements. (Annual Social and Economic Supplement: 2003 Current Population Survey.) *Current Population Reports*, 20–553.
US National Advisory Commission on Civil Disorders (1968). *Report of the National Commission on Civil Disorders*. Washington, DC: US Government Printing Office.
Valleroy, L., MacKellar, D.A., Karon, J.M. *et al.* (2000). HIV prevalence and associated risks in young men who have sex with men. Young Men's Survey Study Group. *Journal of the American Medical Association* **284**, 198–204.
van Griensven, F., Varangrat, A., Naorat, S. *et al.* (2006). Surveillance of HIV prevalence among populations of men who have sex with men in Thailand, 2003–2005.

*Abstract MOAC0101*. XVI International Conference on AIDS. Toronto, Canada, 13–18 August.

Watkins, E.S. (2001). *On the Pill: A Social History of Oral Contraceptives, 1950–1970*. Baltimore: Johns Hopkins University Press.

Weatherburn, P., Hickson, F. and Reid, D. (2003). *Gay Men's Use of the Internet and Other Settings Where HIV Prevention Occurs*. London: Sigma Research.

Winkelstein, W., Samuel, M., Padian, N.S. *et al.* (1987). The San Francisco Men's Health Study: III. Reduction in Human Immunodeficiency virus transmission among homosexual/bisexual men, 1982–86. *American Journal of Public Health* **76**, 685–689.

Wysocki, D.K. (1998). Let your fingers do the talking: sex on an adult chat-line. *Sexualities* **1**, 425–452.

Zaidi, A.A., Aral, S.O., Reynolds, G.H. *et al.* (1983). Gonorrhea in the United States: 1967–1979. *Sexually Transmitted Diseases* **10**, 72–76.

Zelnick, M. and Kantner, J.F. (1977). Sexual and contraceptive experience of young unmarried women in the United States, 1971 and 1976. *Family Planning Perspectives* **9**, 55–71.

# The international drug epidemic

# 3

**Chris Beyrer, Shruti Mehta and Stefan David Baral**

*How opium came to humans: A myth of the Akha people in northern Laos.*

*Seven men court a local beauty. She dies after making love to all of them – rather than having to choose only one suitor thereby avoiding bitterness and conflict. She promised a beautiful flower would emerge from her grave and those tasting its fruit would be compelled to do so again and again. She warned it would bring both good and evil...*

Cohen and Lyttleton, 2002

This story from one of the ethnic minority peoples who grow opium in the Golden Triangle region of Southeast Asia captures some core truths of the ambivalent nature of opiates – their seductive power, attraction to men, addictive capacity, and potential for both good and evil. As with opiates, still the most important class of drugs for severe pain across medicine, so too with alcohol. Wine has been called the nectar of the Gods, and is indeed essential to religious rituals, from the Catholic mass to shamanic rituals, among traditional peoples across Southeast Asia. But it is also a major source of morbidity and mortality, and is implicated as a lethal factor in threats as varied as motor vehicle accident fatalities, domestic violence, liver disease, and birth defects. These ambivalent features of psychoactive substances are as old as mankind, if not older, and they have generated a wide array of cultural, social, legal, political, and medical responses. Responses to the use of psychoactive agents range from attempts to control their use, tax them, restrict them to adults or to special populations (like some opiates reserved for the terminally ill), or ban them outright. Social responses have included tolerance to incarceration, legalization to aggressive promotion. There is perhaps no more compelling an example of the paradoxes in social responses to opiates than the First and Second Opium Wars of the nineteenth century. In these conflicts the British Empire, under the ostensibly conservative Queen Victoria, attempted to redress a grave trade imbalance by forcing opium sales on the Qing Dynasty of China. Opium grown in British-controlled India was forced on China at gunpoint, and the losses of the Qing in both wars led to the British right to sell opium throughout China, and to a colonial concession only recently returned to the Chinese – the opium port of Hong Kong (Waley, 1958). These complex responses to psychoactive substances have continued to have diverse downstream effects – some of profound social and public health import. Thus the now infamous "Rockefeller" era drug laws, harsh and mandatory sentences for possession and sale of cocaine, have been implicated in the disastrously high incarceration rates among the African-American poor in the United States (Correctional Association of New York, 2001). One global downstream effect of social ambiguity regarding substance use has been the widespread limits on safe and effective technologies for the prevention of HIV and HCV infections among injecting drug users. Restrictions such as the still active US Federal ban on funding for needle and syringe exchange have helped drive epidemics of life-threatening viral infections around the world (Beyrer, 2005).

Most striking, perhaps, of the varied social responses and subsequent health outcomes are the divergences in the cultivation of licit and illicit opium poppies – the beautiful flowers of the Akha. Licit poppy cultivation occurs primarily in controlled agricultural areas in Australia (on the island of Tasmania), India, and Turkey (Senlis Council, 2006). Australia supplies the raw material for most of the world's pharmaceutical-grade codeine, that balm for dental patients. Illicit poppy cultivation, principally in Afghanistan, Burma, and Laos, is at the center of an opiate addiction epidemic which stretches from Western Europe to the South China Sea and from Siberia to Indonesia, and which has had enormous health and social impacts, from burgeoning AIDS epidemics to the undermining of civil society in the production states (Beyrer, 2002). While the agent in question (opium in this case) is the same, the outcomes could hardly differ more sharply. And so, in trying to understand how these agents affect health and disease, we must explore not only the agents themselves but also our responses to them at community and societal levels. This approach is essential if we are to understand the interaction of one key mode of substance use, injection drug use (IDU), and the diseases with which it is associated.

HIV/AIDS is now the most severe infectious-disease threat to mankind. In 2005, a record 3.1 million people died of AIDS, and over 5 million became newly infected (UNAIDS, 2005). While sexual spread predominates in the most affected region in the world, sub-Saharan Africa, spread attributed to injection drug use accounted for an estimated 10 percent of all new infections in 2005, and 30 percent of all infections excluding those in sub-Saharan Africa (Beyrer, 2005). In the fastest-growing epidemic regions of Central Asia, the Russian Federation, and the former Soviet Union, spread by IDU was the predominate mode across this vast region (Aceijas *et al.*, 2004). And further east, in China, Indonesia, Vietnam, Malaysia and Burma, IDU spread was also the predominant mode of

| Emergent Epidemics | Established Epidemics |
|---|---|
| • **Belarus**<br>• **Estonia**<br>• **Kazakhstan**<br>• **Russia**<br>• **Ukraine**<br>• **Serbia & Montenegro** | • Burma<br>• China<br>• India<br>• Malaysia<br>• Thailand<br>• Vietnam |
| • **Iran** | • Italy<br>• Netherlands<br>• Portugal<br>• Spain |
| • **Nepal**<br>• **Indonesia** | • Argentina<br>• Brazil<br>• Uruguay |
| • **Libya**<br>• **Mauritius** | • Canada<br>• Puerto Rico<br>• USA |
| • **Kenya, Tanzania, Ghana, Nigeria**** | |

**African States with at least one published report of IDU risks

**Figure 3.1** States with at least one site with HIV prevalence >20 percent in IDU in 2006. Adapted from Aceijas *et al.* (2004).

transmission. IDU epidemics continue to emerge, with some of the most recently reported outbreaks in settings previously spared HIV spread – such as Iran and the Balkans. States with at least one outbreak with greater than 20 percent of injection drug users infected in 2004 include Iran, Libya, Serbia and Montenegro, Nepal, Tajikistan, Kazakhstan, and Belarus (Aceijas *et al.*, 2004; see also Figure 3.1). What so many of these states have in common, in addition to surges in narcotic availability and epidemics of injection use of drugs and of HIV infection among those injectors, are markedly limited HIV-prevention efforts to address emerging epidemics. The availability of narcotics and the addictive nature of opiates, and these limited prevention efforts, have led us in country after country to the same difficult reality – human behaviors are driving HIV epidemics at the individual level, and government responses are aiding and abetting these epidemics at community and population levels. We will explore this difficult reality, and attempt to propose ways forward that might both mitigate the devastating impacts of addiction and help control HIV spread.

## Injection drug use and disease spread

The injection of medications and the development of intravenous infusions have saved countless lives since these tools of modern medicine were developed. It is hard to imagine medicine without them, yet they are of relatively recent origin. The first syringes were devised as tools by the ancient Greeks to drain abscesses rather than for injection of materials into the body. It is thought that a barber in Alexandria around 280 BC first used what was called *pyulcos* (pus-puller) in Greek. The earliest reference to the use of this technique and tool is in the first-century AD treatise *Pneumatics*, by Hero of Alexandria (Duffin, 1999). Alexander Wood, a Scottish physician, was first given credit for inventing the hypodermic syringe. Remarkably, his syringe, developed in 1853, was used to inject morphine for neuralgic pain control (Bellis, 2005). The association of injection with opiates thus dates back to the very start of injecting itself. The first mass-produced syringe was developed by the Becton-Dickinson company for the administration of the newly developed injectable polio vaccine, given to over one million American children, by Jonas Salk in 1954 (Bellis, 2005).

Injection as a route of drug administration provides some obvious advantages to medicine: doses can be measured precisely; the absorptive limitations of the skin, gut, or respiratory mucosa can be bypassed; and agents can be introduced directly into the bloodstream (in intravenous injection) for rapid distribution to target tissues. Indeed, "keeping veins open" (KVO) lines are used for precisely this reason – to allow rapid access to the circulatory system. But these tremendous advantages can be misused and can lead to grave complications. As the epidermal barrier of the immune system is effectively bypassed, unclean needles and syringes can result in direct inoculation of pathogens into the system. As such, risk of bacterial endocarditis in drug users, along with acquisition and transmission of classic blood-borne pathogens, including HIV, HCV, HBV, malaria, tetanus, and syphilis, is exacerbated in IDU. Further, the direct introduction of agents into the bloodstream means rapid intake and distribution of psychoactive agents to the brain – markedly increasing the speed and intensity of the "high," but also increasing the likelihood of overdose (with heroin) and dependence and/or addiction (with heroin, cocaine, methamphetamine, and other injectables). The greater efficiency of drug action with injected doses is one of the key drivers of transitions to injection from other less efficient means, like snorting, sniffing, or smoking drugs, and has been reported in multiple settings (Griffiths *et al.*, 1994). As users become addicts and spend down their resources on drugs, the need to get the greatest effect from the drug used drives injection behavior. It should be clear that it is not the injection *per se* that leads to the most infectious disease risk for IDU; rather, it is the sharing of injection equipment with others.

## Patterns of IDU epidemics

Worldwide, the most commonly injected drugs are heroin, cocaine, and amphetamine (Deany, 2000). Which particular drug is injected where depends on a number of factors, including cost and availability, location relative to production areas or trafficking routes, and social norms. For example, heroin is the most commonly injected drug in Asian countries, whereas cocaine is the most frequently injected drug in South and Latin America (Mann *et al.*, 1992; Stimson and Choopanya, 1998). Injection of amphetamines has been reported in parts of Asia, including South Korea, Japan, Thailand, Laos, Indonesia, the Philippines, and China, while injection of pharmaceuticals has been commonly reported in India, Pakistan, and Bangladesh (Reid and Costigan, 2002). In the states of the former Soviet Union, where heroin is the most commonly injected drug, a large number of injectors also inject "home produced" drugs including "jeff," "vint" (ephedrine-based stimulants extracted from cough syrups or tablets) and "hanka" (a liquid derivative of the opium poppy) (Rhodes *et al.*, 2002).

In the past several years, because of increased law enforcement, processing of drugs has migrated closer to the areas where drugs are actually produced, resulting in an increase of transport and subsequent trafficking of illicit drugs through developing countries (Deany and Crofts, 2000). These changes in distribution have exposed new populations to injectable drugs, and in fact 10 million of the world's estimated injection drug users live in developing countries (Aceijas *et al.*, 2004; see also Figure 3.2). It is often the migrant and border populations that become first exposed to drug use and are thus at highest risk for HIV and hepatitis. In particular, the most affected tend to be the poorest and most marginalized populations, including children living in city slums, and persons who are dependent upon cash crops for their livelihood and subsequent survival. In many areas, including Asia and Eastern Europe, epidemics of drug use are tightly linked with disintegrating economies and social conditions, political unrest, lack of development and infrastructure, and high rates of unemployment (Deany, 2000). These conditions tend to create more demand for drugs and fuel economies of drug trafficking. In these environments, desperate, poor, and unemployed persons become dealers and users to survive (Ahmed, 1998).

## Transition to injection

Patterns of drug use can change over time, and tend to differ in places depending on which drug is most frequently injected – e.g. cocaine versus heroin. For example, cocaine users inject more frequently than heroin users, and thus may be at higher risk for disease transmission. Moreover, whereas some individuals begin drug-use careers with drugs of injection, the majority transition to them from non-injection drugs. Studies from the US have shown that as many as 85 percent of injection drug users transitioned from snorted heroin or cocaine or crack to injecting (Fuller *et al.*, 2001). In Asian countries, shifts have been noted from opium smoking to heroin smoking (commonly referred to as "chasing the dragon") to heroin injection (Stimson and Choopanya, 1998). The reasons for transition are both complicated and variable. In many countries, the custom of self-injection for

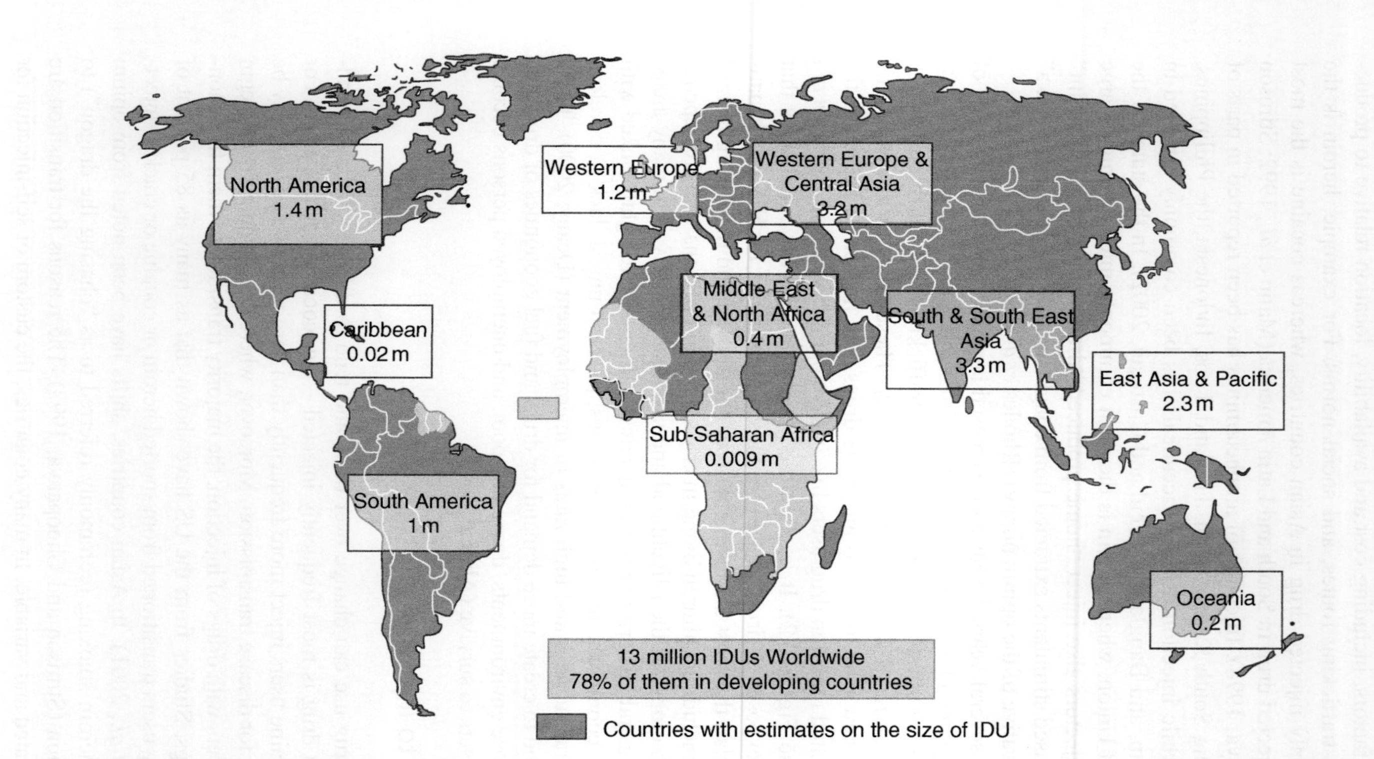


**Figure 3.2** Estimates on global prevalence of injection drug use, 1998–2003. Source: Reference Groups on the Prevention and Care of HIV/AIDS among IDUs (2002).

medication makes the practice of injecting very acceptable (Rhodes *et al.*, 1999a; UNDCP, 1998). Increased migration and communication also facilitate transitions toward injection, because new techniques for drug administration become easily transferable (Deany, 2000). However, most often transitions to drug injection are associated with law enforcement restricting drug supply and production, changes in drug production and distribution technologies, and the globalization of drug markets resulting in altered drug-trafficking routes and distribution networks (Crofts *et al.*, 1998; Rhodes *et al.*, 1999b; Stimson and Choopanya, 1998). Injection is a more effective method of drug delivery than smoking or sniffing, so people are especially likely to turn to injecting when the purity of the drug drops or the cost of the drug increases. In particular, when drugs become scarce as a result of law enforcement and drug-control efforts, it may become inefficient for a user to smoke or inhale the drug because much of the drug is lost in the smoke (des Jarlais *et al.*, 1992). Importantly, injectable forms of drugs are also more easily concealed (Deany and Crofts, 2000).

## Needle sharing and IDU networks

It is important to emphasize that it is not injection itself that puts users at increased risk for HIV and hepatitis, but rather the sharing of injection devices. Reasons for sharing injection equipment include individual-level factors, such as types and frequency of drugs used, longer duration of drug use, limited awareness, and perceived risk of HIV and hepatitis infection. Furthermore, there are social- or contextual-level factors involved, including cultural norms or practices; gender, ethnic or health inequalities; the political and social economy; scarcity or cost of needles and syringes; imprisonment; paraphernalia laws; and policing or law enforcement (Chaisson *et al.*, 1989; Muller *et al.*, 1995; Deany and Crofts, 2000; Bruneau *et al.*, 2001; Strathdee *et al.*, 2001; Rhodes *et al.*, 2002). For example, injection drug users may fear that while carrying needles and syringes they will be stopped by the police and arrested for a drug-related crime (Deany, 2000). There also exist additional means for disease transmission beyond direct needle sharing. Many injectors report practices where a drug solution is passed from a donor syringe into another, either by removing the plunger ("backloading") or the needle ("front loading") from the receiving syringe (Rhodes *et al.*, 2002). In addition, drug users often share the equipment (including cookers, water cups, filters, spoons, swabs, and other items) used for the preparation, storage and transport of drugs (Rhodes *et al.*, 2002). While sharing of contaminated injection equipment is not as efficient for HIV transmission, it highly facilitates HCV transmission (Hagan *et al.*, 2001).

Sharing often occurs in networks of drug users, in which drug use is not so much an individual act as a group activity, and sharing is part of the culture of injection. These networks can either increase or reduce the likelihood of transmission, depending on prevalence of disease within each network and also on whether the drug users are members of multiple networks (Lovell, 2002). If an injection drug user has only a limited number of partners and does not mix with other networks, then the likelihood of transmission may be reduced (des Jarlais *et al.*, 1995a). However, it is not uncommon for a female user to inject only with her male partner while her male partner may belong to multiple networks, thus putting women at higher risk for disease transmission. In some situations drug users may inject in shooting galleries, where a dealer provides needles and syringes and multiple networks mix. In these locations, drugs may be drawn from a common pot and injections are given in succession, usually without sterilization (Ball *et al.*, 1998). For some individuals, especially in developing countries (e.g. Pakistan, India, Bangladesh, Burma, and Vietnam), professional injectors will administer drugs to clients, and often these injectors will use the same needles for many clients (Ball *et al.*, 1998; Reid and Costigan, 2002).

## Sex and drug interactions

Female drug users are found in almost all countries, and while the number of female users appears to be increasing, the majority are male, young, and sexually active (Deany, 2000; Reid and Costigan, 2002). Thus men especially serve as a vehicle for HIV spread, not only through unsafe injection practices but also by being a key bridge population for sexual transmission. High-risk sexual behaviors (e.g. multiple sexual partners and sex without a condom) tend to go hand-in-hand with more frequent and risky drug use (UNDP, 1999). In many countries, HIV epidemics among injection drug users have been followed by epidemics via sexual transmission among non-injecting populations. Furthermore, in some cases sexual transmission from users to non-injecting partners has become the dominant mode of heterosexual transmission (des Jarlais *et al.*, 1992; Panda *et al.*, 1998). In this regard, drug use is like any other gendered phenomenon. Women are doubly vulnerable, through their own drug use but more frequently through a partner's drug use even when the women themselves are not using. Women may be introduced to drug use by their sexual partners, who actually inject the drugs for them. Some women may fear rejection from their partners if they do not use drugs and, what's more, some use drugs because they believe it will increase their pleasure during sex. Drug use and sexual transmission are especially linked in the countries of Latin American, where cocaine injection predominates. Depending on drug market patterns, crack cocaine (which is not injected) can replace cocaine. Crack has been strongly associated with high-risk sexual behavior and, though not associated with reduced injection-related HIV risk, the overall HIV risk remains high (Calleja *et al.*, 2002; Rossi *et al.*, 2003).

It is also not uncommon for men and women who use drugs to enter into commercial sex work to finance their drug dependence, thus further increasing their risk for HIV transmission (Reid and Costigan, 2002). In a study from North America, 70 percent of female injection drug users and 56 percent of male injectors had exchanged sex for either money or drugs (Rothenberg *et al.*, 2000). There are especially high estimates in places like Russia and the Ukraine, where increasing poverty and unemployment among geographically displaced populations have fueled the spread of sex work (Linglof, 1995; Karapetyan *et al.*, 2002). More recently, female users in a number of Asian countries, including China, Vietnam, the Philippines, Nepal, India, Bangladesh, Pakistan, Indonesia, and Sri Lanka, have become increasingly involved with sex work (Reid and Costigan, 2002). The overlap between injection drug use and commercial sex work serves to increase risk not only at the individual level but also at the population level as different networks interact.

## Global drug use and disease interactions

It is now a well-documented pattern worldwide that the use of parenteral drugs drives epidemics of HIV, HCV, and other infectious diseases, and that the social and political responses to substance use often exacerbate disease spread rather than control it.

### HIV

The highest rates of HIV infection in populations have been seen in sub-Saharan Africa, where approximately 25 million people living with the disease can be found, and where heterosexual transmission is the main form of spread (UNAIDS, 2005). However, with the exception of this region, there is mounting evidence that a principal mode of transmission of HIV in many regions is through the sharing of needles in IDU (Aceijas *et al.*, 2004). Globally, there are over 13 million injection drug users, with 25 countries having documented an HIV seroprevalence of over 20 percent among this group. The striking causal association between increasing IDU and the spread of HIV has been documented in many developed nations, such as the US, Canada, and Western Europe. Now, though, this route of exposure has taken the virus into regions spared in the early decades of AIDS – places like Central Asia and the Balkans, and remote regions far from capital cities, such as the far western province of Xinjiang in China, and Irkutsk in Russian Siberia, which now have some of the highest HIV rates among injection drug users reported for these large countries.

In the former Soviet Union, estimates range from 50 to 90 percent of HIV incidence being attributable to injection drug use (Rhodes *et al.*, 1999c). Before 1990, many states of the former Soviet Union had been largely spared from the HIV/AIDS pandemic, due in part to tight restrictions on population movement and contact with foreigners (Field, 2004). There was little evidence of spread of

HIV among injection drug users in the former Soviet Union in studies completed in 1995 (Rhodes *et al.*, 1999a). Since then, many countries have demonstrated rates of spread of the disease that are best described as explosive (Rhodes *et al.*, 1999a). A study completed with injection drug users in Togliatti City in Russia demonstrated a seroprevalence of 56 percent, with three-quarters of seropositive users being unaware of their status. This is in stark contrast to the situation in Togliatti City a few years earlier, when the prevalence of HIV was essentially zero (Rhodes *et al.*, 2002). This study demonstrates well a common theme of how quickly HIV can spread when the virus is introduced into a previously isolated population with established high-risk behaviors. Similar patterns have been seen in multiple states of the former Soviet Union, with the highest seroprevalence rates found in the Ukraine, Russia, and Belarus.

Explosive spread of HIV was also demonstrated in injection drug users in Bangkok, Thailand, when HIV seroprevalence increased from 1 percent to 40 percent within the single year of 1988 (Vanichseni *et al.*, 2004; see also Figure 3.3). Northern Thailand borders the Golden Triangle and, due to decreased availability of opium secondary to poppy eradication efforts, larger populations have turned to parenteral use of heroin – with its associated harmful health consequences – as a substitute (Gray, 1998). A cross-sectional study looking at associations between risk factors and HIV seroprevalence was completed in this region of Thailand on 1865 patients admitted for drug detoxification in one of the region's drug treatment centers. There, HIV prevalence in injection drug users was 30 percent (154/513), compared with 2.8 percent (33/1180) in non-users – an increased

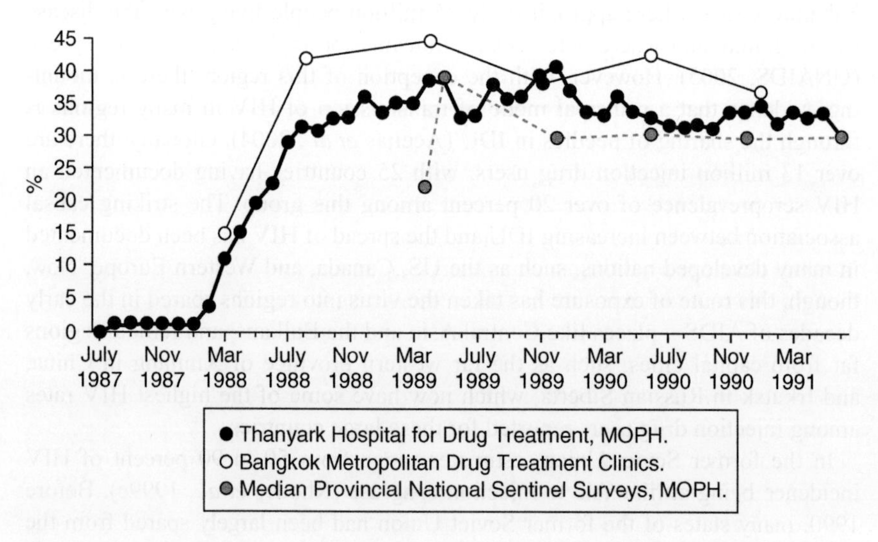


**Figure 3.3** HIV infection rates among IV drug users in Bangkok, and national median rates, 1985–1991.

risk of 14.8 times (95 percent CI, 10.2–21.6) (Razak *et al.*, 2003). In Thailand, control measures have largely focused on aggressive HIV/STD-risk education and enhancing provision of services, with the emphasis on sexual transmission. While measures including education and the "100 percent condom program" have mitigated the epidemic among the general population, there has been little effect on the epidemic among injection drug users (Beyrer *et al.*, 2003). The data from Thailand provide an excellent representation of the concept that HIV cannot be eradicated simply by directing prevention strategies at sexual transmission, as a large reservoir of disease remains among injection drug users.

Historians have speculated that sexually transmitted infections were first introduced to China by Portuguese traders in the sixteenth century, and the diseases were well established when Mao Zedong came to power (Cohen *et al.*, 1996). Maoist policies included very restrictive and heavily enforced prevention strategies and treatment, resulting in virtual eradication of the then prevalent STDs within a decade. However, with China's entrance to global markets in the 1980s there was a sharp increase in the prevalence of STD, associated primarily with commercial sex work (Zhang and Ma, 2002). Sporadic cases of HIV were also reported, but there was little evidence of epidemic spread of HIV in China until 1989. That year, there were reports of HIV infection among 146 drug users in the Yunnan province (Zhang *et al.*, 2002). Yunnan province is situated along the borders of Burma and Laos, collectively called the "Golden Triangle" and known to be responsible for approximately 60 percent of worldwide heroin production in 1990 (Beyrer *et al.*, 2000). Since that time, HIV infection has spread to all 31 provinces, with drug users making up 60–70 percent of the incidence (Portsmouth *et al.*, 2003). A recent cross-sectional study was completed in Chengdu, a city of 11 million in the Southwest, which is in close proximity to major heroin supply and trafficking areas. Earlier reports had demonstrated high-risk behavior among drug users including a high prevalence of injection behavior, extensive sharing of syringes, and an already high HIV seroprevalence among injection drug users (Wu *et al.*, 2004). In the current study of 266 users, sharing of injection materials was recorded in 39 percent (103/266), and such behavior was associated with being an ethnic minority or "internal migrant" ($P < 0.05$) (Chamla *et al.*, 2006). Another recent study evaluated indicators of high-risk behaviors among 1153 users in China, and found that a higher prevalence of sharing needles and unprotected sex was also most associated with being a temporary migrant ($P < 0.01$) or an ethnic minority ($P < 0.05$) (Yang *et al.*, 2005). These studies demonstrate that even within the population of injection drug users, there are sub-populations exhibiting higher-risk behaviors that further propagate the virus. What's more, the data reveal a common association of IDU and high-risk sexual behaviors, which functions to bridge this epidemic with the population as a whole.

Iran, a predominantly Muslim country situated in the Middle East, is experiencing an HIV epidemic predominantly within its injection drug user population (UNAIDS, 2005). Although there have been only 6700 reported cases of HIV/AIDS in Iran, of which about 85 percent are injection drug users and 10 percent secondary to sexual contact – it is estimated that there are closer to 30,000 people living with the disease (Razzaghi *et al.*, 2006). Iran is bordered by Afghanistan, which is known to produce a large proportion of the world's opium supply (see Figure 3.4). Opium is extensively trafficked from Afghanistan through Iran and, similar to the regions of China proximate to the Golden Triangle, there are exceptionally high numbers of injection drug users in this region (UNODC, 2005). A recent cross-sectional prevalence study was completed with 611 imprisoned drug users; both injecting and non-injecting. Among non-injectors, the HIV seroprevalence was 5.4 percent (21/390); the only statistically significant risk factor for HIV was having had sex but not used a condom (aOR 3.42; 95 percent CI, 1.25–9.36). The HIV seroprevalence was 15.2 percent (25/165) in injectors; the main risk factor was having shared injection materials while incarcerated (aOR 12.37; 95 percent CI, 2.94–51.97) (Zamani *et al.*, 2005). These results indicate an increased risk of HIV in drug users from injection. Among the non-injectors, increased risk was secondary to high-risk sexual behavior rather than drug use. The effects of imprisonment on high-risk drug use and spread of infectious disease will be explored in a later section of this chapter, as these study results introduce a well-established causal link.

## HCV

HCV is an RNA virus of the *Flaviviridae* family, first cloned in 1989 by Choo and colleagues, and soon after determined to be responsible for the vast majority of non-A, non-B viral hepatitis infections (Choo *et al.*, 1989; National Institutes of Health Consensus Development Conference Panel Statement, 2002). HCV epidemics associated with IDU have been remarkably consistent across wide geographic and social contexts. With worldwide average prevalence rates of around 80–90 percent, HCV is hyperendemic among injection drug users (Hagan *et al.*, 2005). A recent review from the European Union reinforced this, showing that the prevalence of HCV among injection drug users from all EU countries, with the exception of Luxembourg, was between 30 and 98 percent, with incidence rates ranging from 6.2 to 39.3 per 100 person years (Roy *et al.*, 2002). Even countries with traditionally low levels of injection drug use, such as Japan and Korea, are now experiencing epidemics of HCV among injection drug users (Satoh *et al.*, 2004; Suh and Jeong, 2006).

Multiple studies evaluating knowledge of hepatitis transmission and prevention have been completed in at-risk populations, and show similar results: little is known about viral transmission and prevention, and HCV infection is considered non-desirable but much less threatening than HIV (Carey *et al.*, 2005; Hagan *et al.*, 2005; Southgate *et al.*, 2005). Although HCV has an indolent natural history, with few symptoms for many years, nearly 80 percent will develop chronic

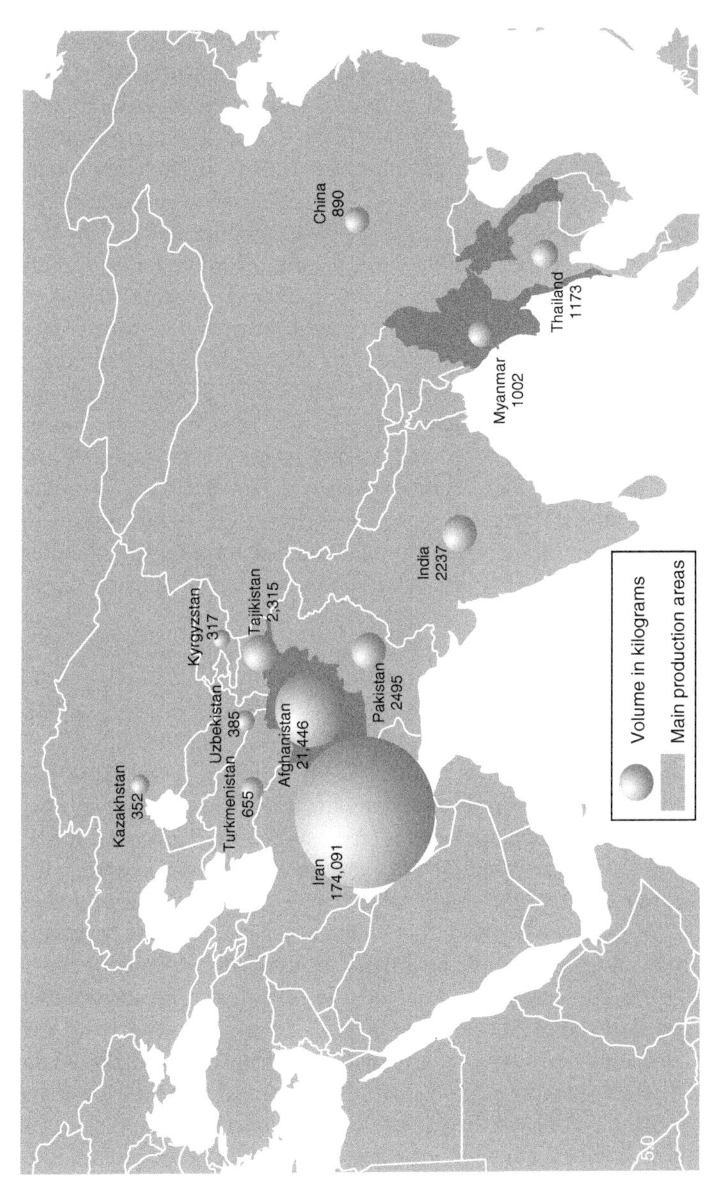


**Figure 3.4** Opiate seizures in Asia, 2004. Source: UNODC (2006).

infections, of which a quarter will manifest as complications including liver cirrhosis and hepatocellular carcinoma (des Jarlais *et al.*, 2005).

Inarguably, HCV is a pandemic. That said, it has received relatively little attention or funding in comparison with HIV. With few exceptions, the epidemiology of HCV, especially among injection drug users, has been estimated using either cross-sectional or retrospective study designs measuring prevalence represented by the seroprevalence of HCV antibodies (Cox *et al.*, 2005). It is from these prevalence studies that worldwide prevalence estimates of HCV are calculated. This extrapolation of data is responsible for the potentially inaccurate worldwide estimate (Hagan *et al.*, 2004). Several studies have demonstrated a delay between HCV exposure and infection, and diagnosis by the commonly used ELISA test for serum HCV antibodies (Beld *et al.*, 1999; Stramer *et al.*, 2004). These results imply that there is potential for the extremely high worldwide estimates of HCV seroprevalence rates in injection drug users to actually be considered conservative due to the delay of seroconversion. However, in 15 percent of documented HCV-infected people there is spontaneous HCV viral clearance in the absence of seroreversion (Thomas *et al.*, 2000), and results such as these imply, in contrast, that prevalence estimates using seroprevalence of anti-HCV overestimate the true global prevalence. The main lesson to be extracted from this discrepancy is a need for further prospective studies in injection drug users, using the presence of both HCV RNA and anti-HCV, which will better elucidate the true magnitude of the HCV pandemic.

## HBV

There are estimates that HBV has infected nearly 2 billion people worldwide, of whom 350 million have evidence of chronic infection which carries similar risks to those of HCV of complications such as cirrhosis and liver cancer (Kane, 1995). HBV is unique among the group of infectious diseases associated with IDU, as there has been an effective vaccine available for over two decades (Hutchinson *et al.*, 2004). However, studies have shown that vaccine uptake has been very low among injection drug users, ranging from 4 percent to 29 percent vaccination (Seal *et al.*, 2000). A recent study completed in New York City showed that factors associated with HBV vaccine completion included never having injected drugs ($P < 0.02$), and currently receiving public assistance (aOR 2.38; 95 percent CI, 1.24–4.56) (Ompad *et al.*, 2004). A study completed in New Haven verified the aforementioned factors associated with success of vaccine completion. What's more, this study showed that using syringe exchange sites as opportunities to provide vaccination was associated with completion of administration of two doses of vaccine in 77 percent (103/134) and the full three doses in 66 percent (89/134) of vaccine-eligible users (Altice *et al.*, 2005). As the WHO Expanded Program on Immunization added HBV in 1992, epidemiologists expect that incidence and prevalence will decrease in the next generation as a result of increased immunity (Ompad *et al.*, 2004).

## Efficacy of viral transmission by parenteral exposure

Injecting a substance contaminated with HCV, HBV, or HIV into the bloodstream is a particularly efficient means for transmitting infection – even more so than sexual transmission. Thus, drug use plays a key role in how epidemics first develop in regions. In the case of HIV, once it enters the drug-using population, epidemics have accelerated rapidly – starting in injection drug users and eventually spreading to the general population (Crofts *et al.*, 1998). The prevalence of HIV among people who inject drugs was found to have risen from 0 to 30 percent within 12 months in many cities, including New York, Edinburgh, Bangkok, Ho Chi Minh City, Manipur, Riuli, Kaliningrad, Svetigorsk, Odessa, and Nikolayav (Stimson *et al.*, 1998).

The trajectory of HCV and HBV infections in drug-injecting populations is even more dramatic. It is estimated that by the percutaneous route, HCV is 10 times and HBV is over 30 times more efficiently transmitted than HIV (Kiyosawa *et al.*, 1991; Mitsui *et al.*, 1992). This is evidenced by patterns in incidence and prevalence of these infections among new injectors. Data consistently suggest that HBV and HCV transmission increase rapidly within the first year of injection, whereas HIV transmission lags and increases at a slower rate (Garfein *et al.*, 1996; see also Figure 3.5). Rates of HCV and HBV incidence are generally between 10 and 35 per 100 person years in the first few years after injection

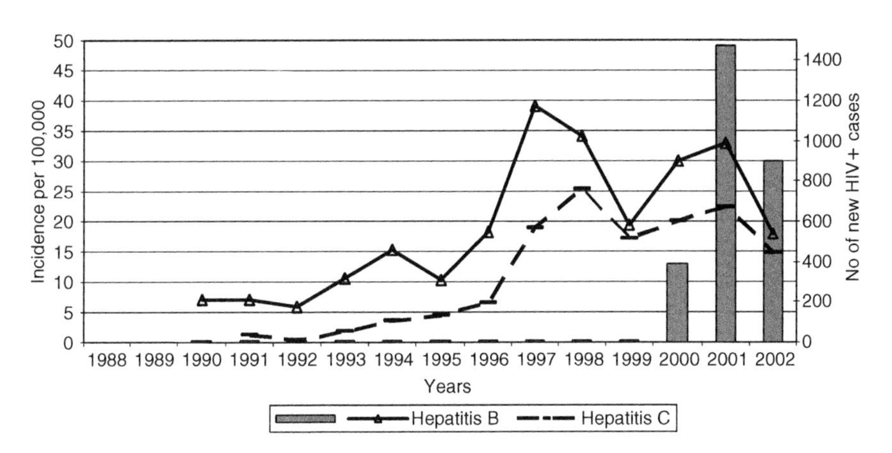


**Figure 3.5** Estonia HIV epidemic was preceded by… Incidence of reported HCV/HBV infections and absolute number of HIV-positive cases in Estonia, 1988–2002. Source: Uusküla *et al.* (2002).

initiation (Hagan *et al.*, 1995; Miller *et al.*, 2002). A study in Baltimore showed that the prevalence of HCV for injection drug users who had been injecting for 1 year or less was 65 percent (Garfein *et al.*, 1996). Among young injection drug users between the ages of 15 and 30 across the US, a recent multicenter study reported the incidences of HIV, HBV, and HCV to be 1.3, 11.3, and 15.8 per 100 person-years respectively (Garfein *et al.*, 2006). In Pakistan, the HIV infection rate among injection drug users in Lahore has been estimated to be 12 percent, but the prevalence of hepatitis C has been estimated to be as high as 89 percent. There is further evidence that transmission of hepatitis C (and hepatitis B) occurs not only through direct needle sharing, but also through sharing of drug-use paraphernalia – including cottons, cookers, and even drug tourniquets and environmental contamination (Hagan *et al.*, 2001; Thorpe *et al.*, 2002). Evidence consistently suggests that unless safe injecting practices are adopted throughout an injector's lifespan, he or she will most likely acquire HCV. Moreover, even low levels of needle sharing are probably sufficient to sustain an HCV epidemic. The same is not true for HIV (des Jarlais *et al.*, 1995b).

## Syphilis, tetanus, and Malaria

The main focus of research connecting IDU with infectious disease has been on the aforementioned viruses: HIV, HCV, and HBV. However, multiple recent reports indicate that injection drug users are at increased risk of a whole host of other blood-borne infections, including: syphilis, tetanus, wound botulism, and malaria (Passaro *et al.*, 1998; Chau *et al.*, 2002; Beeching and Crowcroft, 2005; Bradshaw *et al.*, 2005).

Incidence rates of syphilis infections, caused by the spirochete *Treponema pallidum*, have been steadily rising for the past two decades in at-risk populations in multiple areas throughout the world (Muga *et al.*, 1997), though it is not yet clear whether the association is related to transmission via sharing of syringes, or because injection drug users exhibit higher-risk sexual behaviors of which syphilis infection is representative (Karapetyan *et al.*, 2002). Studies evaluating syphilis prevalence among drug users have consistently shown that injectors practice higher-risk sexual behaviors (Lopez-Zetina *et al.*, 2000). Furthermore, there are significant racial, gender, and socioeconomic disparities in syphilis prevalence. In the US, syphilis prevalence rates among African-Americans have been estimated to be 60 times higher than those in non-Hispanic Whites (Lopez-Zetina *et al.*, 2000). In studies completed in both the former Soviet Union and the USA, there are much higher syphilis prevalence rates among women when compared with men, thought to be associated with high-risk commercial sex workers (CSW) (Muga *et al.*, 1997; Karapetyan *et al.*, 2002; Abdala *et al.*, 2003). While syphilis is a treatable condition, by virtue of the high prevalence rates among injection drug users it shares many risk factors with HIV (Fleming and

Wasserheit, 1999). Furthermore, syphilis-caused open genital lesions greatly enhance transmission of HIV, resulting in the increased spread of this non-treatable virus both within the injection drug user population and beyond, due to complex sexual networks.

The association between tetanus and IDU was first recorded in 1876 in the UK, and was further elucidated in the 1950s in Chicago (Levinson *et al.*, 1955; Beeching and Crowcroft, 2005). Tetanus is caused by infection with *Clostridium tetani*, which belongs to the family of obligate anaerobes *Clostridium*. Infections by this family of bacteria are traditionally associated with ingestion or with high-force crush injuries, but when injection drug users employ contaminated needles to inject areas of devitalized tissue, they also are at high risk of self-inoculation (Passaro *et al.*, 1998). Tetanus is the only member of this family for which there is a vaccine. Although its use does provide lifelong immunity, booster vaccines are generally recommended every 10 years. *C. Tetani* is not the only member of this family to cause morbidity and mortality in injection drug users; in 2000 there was an outbreak of an unexplained illness with extremely high mortality in England determined to be caused by *Clostridium novyi*. Over 90 percent of the cases were in injection drug users, and the main determinants of infection were being a current injector ($P < 0.05$) and sharing syringes ($P < 0.05$) (Bellis *et al.*, 2001). The toxin from *Clostridium botulinum* manifests in wound botulism, a condition associated with very high mortality rates. Since 1988, areas of California have experienced a significant increase in incidence associated almost exclusively with the injection of black-tar heroin – a black, gummy form of heroin produced in Mexico in makeshift factories proximate to opium poppy fields (Passaro *et al.*, 1998).

The transmission of malaria, a disease caused by the parasite *Plasmodium*, is commonly associated with the Anopheles mosquito in endemic regions of the world. However, a malaria epidemic among injection drug users was first described as early as 1928 in Egypt (Chau *et al.*, 2002). Researchers at that time theorized that sharing of injection materials was responsible for the epidemic, as the affected group had not come from an area where malaria was endemic. The occurrence of malaria epidemics among injection drug users in non-endemic areas has continued over the last century in various regions throughout the world. A recent study completed in Vietnam demonstrated a similar natural history of malaria in non-HIV-infected injection drug users in comparison with the general population where malaria is transmitted by mosquitoes (Chau *et al.*, 2002). However, among people presenting with the immunosuppressive manifestations of HIV there is higher incidence of malaria, with each episode associated with higher morbidity and mortality (Whitworth and Hewitt, 2005). Furthermore, as malaria has increasing incidence in injecton drug users with higher HIV seroprevalence rates, scientists have expressed concern regarding the development of a more resistant strain of malaria secondary to co-infection with HIV (Bastos *et al.*, 1999).

## Prevention and harm-reduction strategies

The most common approach to drug use, with many underlying political implications, has been to reduce either the availability or the supply of drugs. This prohibitionist approach has all too often served to exacerbate drug problems and their health consequences. Restricting drug supplies can actually encourage injection drug users to inject alternate substances, and can provide impetus for non-injectors to become injectors because injecting requires a smaller dose for an equivalent effect. Prohibition can even drive the drug trade into areas where individuals have not previously been exposed to injection drugs. In India, when the government tried to restrict the heroin trade, the price of heroin rose and addicts switched to synthetic opiates, yet actual injection behavior remained unchanged. Similarly, efforts to control opium smoking in Bangkok and Calcutta were followed by subsequent increases in heroin injection (Stimson and Choopanya, 1998). Even more telling is that, as a result of efforts to halt drug trade in other regions, West Africa has now emerged as an important transit point for cocaine originating in South America, and also for heroin from Southeast Asia. Threats of imprisonment have had similar deleterious effects and, furthermore, actual imprisonment tends to stimulate injection behavior rather than curtail it. As will be discussed in more detail later, since clean syringes are not widely available within prisons, prisoners often resort to shared or improvised equipment, which is difficult to sterilize (The World Bank, 1999).

### Prevention approaches for IDU and non-IDU disease transmission

What then are effective HIV prevention strategies that do not focus on supply reduction or insist on abstinence? Effective strategies range from drug treatment and drug substitution therapy, such as methadone maintenance therapy (MMT) and buprenorphine, to harm-reduction measures, including education regarding safer injecting and sex practices. Equally important are access to clean needles and syringes (e.g. through needle and syringe exchanges), voluntary counseling and testing, and treatment of sexually transmitted infections. Finally, peer educational and behavioral interventions are especially relevant and efficacious in at-risk populations, including sex workers who inject drugs, and prisoners.

### Methadone and buprenorphine

The efficacy of drug treatment and, more specifically, opiate substitution therapies (e.g. MMT and buprenorphine) for HIV prevention has long been well established. The use of substitution therapies facilitates the transition from IDU to non-IDU, and is generally accompanied by a reduction in injection-related risk behavior (Drucker *et al.*, 1998). In some treatment centers, sexual risk reduction counseling is also provided; however this is not universal (Sorensen and Copeland, 2000). In 1993, a prospective study by Metzger and colleagues demonstrated that the HIV incidence among methadone-treatment injection drug users was 3.5 percent compared with 22 percent in out of treatment users (Metzger *et al.*, 1993). Further evidence has suggested that substitution therapies are most effective when initiated early in the course of an HIV epidemic (Ball *et al.*, 1998; Drucker *et al.*, 1998). In contrast, decreasing the availability of MMT has been associated with rapid increases in HIV, as evidenced by the Spanish epidemic (de la Fuente *et al.*, 2003). However, despite the widespread evidence of the efficacy and effectiveness of MMT and buprenorphine, governments have been slow to officially adopt and fund such programs, and worldwide the availability of methadone treatment slots remains variable. Even in the United States, with the longstanding history of success of MMT, less than 15 percent of opiate addicts have access to MMT slots. MMT availability is further limited in parts of Eastern Europe and Asia (Grund, 1997; Drucker *et al.*, 1998). Broader expansion of MMT is still plagued by the perception that it represents the substitution of one addiction for another and, in spite of the evidence of program success, MMT implementation continues to encounter resistance from abstinence-oriented practitioners.

## Needle-exchange programs

Prevention and harm-reduction strategies face many of the same obstacles as drug treatment strategies. In general, harm reduction refers to strategies for reducing the harm associated with drug use, with emphasis on HIV and hepatitis. Central to the principle of harm reduction is that this is possible without requiring abstinence or a reduction in drug use itself (Heather *et al.*, 1993). Harm-reduction programs have been increasingly seen in developing countries, where a range of differing activities to decrease needle sharing and other unsafe injecting behavior has been developed (Crofts, 1999). Harm-reduction programs have typically been accompanied by reductions in rates of HIV transmission, and are more cost-effective than interdiction or incarceration of addicts, yet these programs remain controversial and present challenges to communities and governments (Crofts, 1999; Deany and Crofts, 2000). For example, there is consistent evidence that provision of clean needles and injecting equipment can halt or even reverse HIV/AIDS epidemics among injection drug users (Hurley *et al.*, 1997; Drucker *et al.*, 1998). Examples of effective programs come from Sydney, Glasgow, and Toronto. Needle-exchange programs reduce disease transmission directly by lowering rates of needle sharing and decreasing the prevalence of HIV among needles available for sharing, and also indirectly through provision of services including bleach distribution, referrals to drug treatment, provision of condoms, and education about risk behavior. An ecological study of 81 cities worldwide comparing those with and without needle-exchange programs found that the average seroprevalence increased by 5.9 percent per year in 52 cities without such programs, and decreased by 5.8 percent per year in 29 cities with programs (Hurley *et al.*, 1997). Further evidence suggests that HIV epidemics have been less explosive in cities and countries where needle-exchange programs have been established earlier in the epidemic. The role of such programs in preventing both HCV and HBV transmission has been far more controversial. Since most injection drug users acquire HCV and HBV earlier than HIV, in many places needle-exchange programs were established too late to be effective against HBV and HCV (Hagan *et al.*, 1999). However, in places where programs were established early, some success against HBV and HCV has been observed (Taylor *et al.*, 2000; Hope *et al.*, 2001). This is evidenced by a study in Glasgow, where the prevalence of HCV was significantly lower among injectors who initiated injection after rather than before the establishment of the needle-exchange program (Taylor *et al.*, 2000).

Yet despite this overwhelming evidence, the establishment of needle-exchange programs has often occurred too late in the course of an epidemic – well after the explosive spread of HIV – when reversal of the epidemic may not be possible. This has been the case in many countries in Eastern Europe. In 2002 in Togliatti City, Russia, where a dramatic increase in HIV prevalence between 2000 and 2002 was observed, less than 8 percent of injection drug users in one study obtained needles from the needle-exchange program (Rhodes *et al.*, 2002).

Safe injection facilities represent yet another effective harm-reduction tool. While these programs were initiated in Europe as early as 1970, in Amsterdam, the first safe injection facility in North America was opened in Vancouver in 2003 (Kerr, 2003; Kimber *et al.*, 2003). Within this facility, injection drug users can access sterile injection equipment, inject pre-obtained illicit drugs under the supervision of nurses, and also access nursing care and addiction counseling. These facilities enable use of pre-obtained drugs in a safe and anxiety-free atmosphere under hygienic and low-risk conditions (Kerr, 2003). A study from Vancouver showed that the use of medically supervised safer injection facility was associated with reduced syringe sharing in a community-based sample of injection drug users who reported similar rates of needle sharing before the opening of the facility (Kerr *et al.*, 2005, 2006). These facilities exist in Canada and some parts of Europe and Australia, but to date have not been adopted elsewhere and are likely to face the same type of political resistance encountered by MMT and needle-exchange programs.

## Education

Interventions that incorporate education of injection drug users regarding safe injection practices and sexual risk reduction are critical to prevention. Information and education can be provided through a number of avenues, including general awareness campaigns, though it can also be more focused and directed dissemination of information through the health and social services frequented by injection drug users. Education campaigns often utilize peer and drug-user networks and outreach workers (Ball *et al.*, 1998). Of key importance is easy access HIV testing, which is typically accompanied by risk-reduction counseling involving face-to-face communication. Finally, secondary prevention for individuals already infected with HIV or hepatitis is also critical. Providing access to primary care and antiretroviral treatment programs may facilitate behavior change and prevent further transmission. Comprehensive strategies are the key to effective prevention. For example, it is imperative that safe injection messages be complemented with safe-sex messages and condom promotion. In the case of MMT programs, for example, the effect on reduction in injection-related risk behavior is consistent; however, the evidence of its effect on sexual risk behavior is controversial. Although MMT offers an opportunity for comprehensive HIV prevention, many programs do not offer sexual risk-reduction counseling (Sorensen and Copeland, 2000).

## Pathogen specific prevention strategies

Risk-reduction strategies that are successful for HIV may be insufficient for the prevention of HCV because of the differences in dynamics of transmission. As previously discussed, HCV and HBV tend to be acquired within the first few years of injection, during which injection drug users are less likely to come into contact with prevention and treatment services (Garfein *et al.*, 1996). Persons entering methadone maintenance and attending needle exchanges tend to have on average 10 years of injection experience, so they have most likely already acquired HCV and HBV (Schutz *et al.*, 1994). Thus, modified strategies need to be explored to provide effective prevention from these viruses.

For hepatitis B prevention, effective use and distribution of the HBV vaccine is critical. However, although a safe and effective vaccine has been available for more than 20 years, rates among injection drug users remain low (Seal *et al.*, 2000). There have been successful efforts to vaccinate injection drug users, suggesting that this is a possibility, but infrastructure is needed to identify individuals who are both uninfected and at highest risk (Seal *et al.*, 2003). This means targeting individuals early, before, or soon after injection initiation.

Prevention of HCV is further complicated by the lack of an effective vaccine. Often hepatitis C is viewed by injection drug users as either unavoidable or of less importance than HIV (Davis *et al.*, 2004; Rhodes *et al.*, 2004). Risk reduction and prevention of HCV needs to be distinguished from that for HIV, and there is a need to develop HCV-specific prevention and treatment strategies. As with HBV, the risk in the first year of injection is so high that efforts need to be made to reach drug users before they initiate injection. However, accessing these populations has proven to be difficult. A more comprehensive approach, including transition prevention messages, in addition to the current education regarding the consequences of non-injection drug use, may be more effective at reducing HCV transmission (Wodak and Lurie, 1997; Vlahov *et al.*, 2004). Finally, relevant for prevention of all blood-borne infections but most critical for HCV, prevention messages need to alert injection drug users to the harm associated with *all* equipment sharing, rather than only injection sharing. Perhaps needle-exchange programs should also consider proactive distribution of injection paraphernalia, in particular cookers and cotton filters. Some studies have shown that HCV prevalence among injectors in places with comprehensive harm-reduction programs is lower than in other industrialized countries, again reinforcing the need for programs to be established early (Hope *et al.*, 2001).

## Synergies of prevention with HAART and HCV therapy

While secondary prevention of HIV and hepatitis C infection can be accomplished through behavioral risk modification at the level of the individual, treatment of HIV and HCV can prevent further spread at the level of the population. Studies have shown that HIV transmission is much less likely among individuals who have lower levels of circulating virus (Gray *et al.*, 2001). Currently, the best-known way to ensure undetectable levels of virus is through highly active antiretroviral therapy (HAART). However, despite the availability of HAART since 1996 and evidence that injection drug users do adhere to these therapies, achieving both virologic and immunologic response, users consistently lag behind other groups in HAART access (Mocroft *et al.*, 1999; Roca *et al.*, 1999; Celentano *et al.*, 2001). Access to HCV treatment among injection drug users is far worse than even HAART access. Until 2002, US national guidelines for HCV treatment called for exclusion of injection drug users from treatment for HCV, primarily due to concerns related to low adherence and a high risk of re-infection (National Institutes of Health Consensus Development Conference Panel Statement, 1997). However, a number of studies have demonstrated that injection drug users could successfully take HCV treatment and re-infection could be avoided if risk-reduction counseling was provided simultaneously (Backmund *et al.*, 2001; Sylvestre, 2002). In response to these findings and advocacy, the guidelines were reversed in 2002 and injection drug use was no longer considered to be an exclusion criterion (National Institutes of Health Consensus Development Conference Panel Statement, 2002). Even with this change, few injection drug users have received HCV treatment, and though approximately 30 percent of IDUs are co-infected with HCV and HIV, even fewer co-infected patients have received treatment (Restrepo *et al.*, 2005; Stoove *et al.*, 2005). HCV treatment is complicated to take and is associated with a high rate of side effects, including depression and suicidal ideation, which can be particularly problematic for this population who have high rates of psychiatric co-morbidity. Efforts are under way to develop new comprehensive care strategies that incorporate HIV treatment, HCV treatment, psychiatric treatment, and other social programs.

## Government responses

Given the extent and gravity of the health consequences of the global injection drug use epidemic, public health approaches would argue for rapid and widespread implementation of preventive approaches with evidence of efficacy for both drug treatment and mitigation of the harms associated with drug use. That this kind of response has proven to be so difficult, and to have become a politicized and passionately debated issue in so many settings, is undoubtedly due (at least in part) to the social ambivalence toward substance use and attitudes toward users. Are addicts patients or criminals? Do they deserve compassion or punishment or disdain? Should we provide services to addicts to prevent spread of diseases to the "general population," however defined, or should we deny services and resources (including access to anti-viral therapies for HIV and HCV infection) to the "guilty," who have acquired their infections through high-risk behavior? There are governments and societies who have accepted the harm-reduction approach to substance use, and who have implemented widespread provision of services to drug users – examples include Australia, the Netherlands, the UK, Hong Kong, and Canada. And there are governments that continue to criminalize users, incarcerate addicts without treatment, and deny those with histories of injection drug use access to life-saving therapies. The latter list is unfortunately long, and includes states in the Americas, Europe, the former Soviet Union, Asia, and Africa. A critical question in attempting to understand the social ecology of this problem is the extent to which government responses can drive (and have driven) infectious disease spread among drug users. This takes us beyond the level of individual risks and into the more difficult realm of attribution of social or ecologic risks for transmission. And it is here, too, that we must ask whether human rights violations against drug users are understudied risk factors for disease acquisition and transmission.

### Narcotics policies and drug use risks

A now classic examination of the ecological risks associated with injection drug use was reported by Hurley and colleagues (Hurley *et al.*, 1997), who performed a large ecologic study of 81 cities globally with and without needle and syringe exchange programs (NSEPs). Those without NSEPs saw an HIV prevalence rate increase among injection drug users of roughly 6 percent per year; those with NSEPs saw an average decline in HIV prevalence in injection drug users of 5.8 percent per year. This study, now almost 10 years old, and a great volume of subsequent work have demonstrated that these ecological factors can play powerful roles in control of infectious diseases. However, they continue to be resisted aggressively. As an example, in 2005 the United States Congress Subcommittee on Criminal Justice, Drug Policy and Human Resources held hearings (at which one of the current authors, Chris Beyer, testified) entitled "Harm Reduction or Harm Maintenance: Is There Such a Thing as Safe Drug Abuse?" in which a range of Republican witnesses were brought forward to question the evidence for harm reduction and to call it an ideological rather than an evidence-based approach to prevention. The positions put forward in support of the evidence for harm reduction from the American Medical Association, the American Nurses Association, and the American Public Health Association were all discounted. Advocates were told by sympathetic staffers that the federal needle and syringe exchange ban had become a "sacred cow," and that advocacy efforts to change the policy had little or no chance of success. With the US still the single largest funder of HIV/AIDS activities globally, this inability to respond to the evidence for harm reduction has global implications.

## Licit and illicit opium cultivation

A striking dichotomy of the nearly global criminalization of illicit opiate use is that opiates remain widely used in modern medicine and surgery, widely prescribed by physicians and other providers, and legally cultivated for these purposes. Drugs used in modern medicine and derived from the opium poppy include morphine, Demerol, codeine, and methadone, to name only a few (*Physicians' Desk Reference, 2005*). Anyone who has had major surgery, or the kind of severe pain associated with bone fractures, cancer, and dental emergencies, has likely known the extraordinary power of these agents to control human pain. They remain, for the most severe pains we suffer, the ultimate remedy – few of us who die in hospital will die without their benefit. But in their power lies their danger – as the Akha have told us. Opiates are highly addictive for humans precisely because the human nervous system is so richly endowed with opiate-like receptors – the famous endorphins, which were actually named for endogenous opiates (Heiss and Herholz, 2006). We are hard-wired for opiates. Yet we seek to isolate and punish those who grow opiates, but who do so out of the control of the "licit" market.

Illicit cultivation of opiates is concentrated, according to the terminology of the UN Office for Drugs and Crime, in a few chaotic and/or closed states; Afghanistan, Burma, and Laos together accounted for more than 90 percent of all illicit opium poppy cultivation in 2004 (UNODC, 2005). Licit cultivation, as we have seen, is concentrated in Australia, India, and Turkey. Licit growers produce their valuable crop under international treaty law, and are not seen as a threat to global security or health. Indeed, few people, even those concerned with the HIV epidemic among drug users, know that licit poppy cultivation continues in 2006. In contrast, the illicit growers have faced sanctions (in the case of

Burma) and nearly constant conflict with the international community. Yet illicit opiate volumes have increased in the past decade (UNODC, 2005). Afghanistan, in particular, has seen steadily increasing poppy production over 25 years of war and conflict (see Figure 3.6). The one exception to this was the year 2001, when the then leader of the Taliban regime, Mullah Omar, issued an Islamic Edict (*Fatwa*) against poppy cultivation. In 2002 the crop was again enormous, and the 2005 crop was estimated at 6100 metric tonnes of opium base – by far the world's highest single yield (UNODC, 2006). While the position of the UN and its members has been characterized as "zero tolerance" for illicit drug cultivation, there has been at least one proposal put forward regarding a new approach to the problem of Afghan heroin. The Senlis Council, a European-based narcotics policy organization, has proposed a feasibility study of transitioning Afghan opium from the illicit to the licit side of poppy cultivation (Senlis Council, 2006). One of their primary justifications for suggesting this change is a reported global shortage of opiates for pain relief. They make the novel suggestion that if the first world is serious about treating AIDS in Africa, and about relieving the pain of terminal AIDS and other fatal diseases in African and other developing country populations, the worldwide supply of opiates must dramatically increase, and their cost be reduced (Senlis Council, 2004). There is certainly evidence that the shortage of affordable pain medications for the poor globally, at least, is all too real.

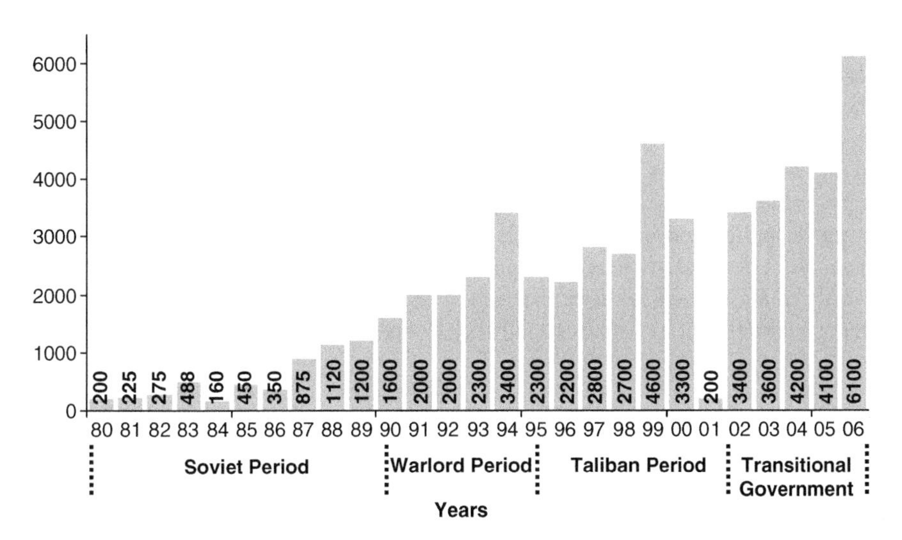


**Figure 3.6** Afghanistan opium production from 1980 to 2005, in metric tones. Modified from UNODC (2006).

## Punitive drug laws and disease spread

While producing narcotics has isolated countries like Burma, Afghanistan, and Colombia, laws and policies regarding use of these agents have operated at individual and community levels in perhaps even more profound ways. Harsh penalties for possession or use have been used by many governments in attempts to deter use and punish users. These approaches have often served to drive users underground, where they are harder to reach and more likely to engage in riskier behaviors in efforts to avoid arrest. Wodak and Lurie examined the contrasting approaches of Australia's relative tolerance and use of harm-reduction approaches and the policies of strict prohibition in the US, and found that the US policies were significant drivers of IDU risks (Wodak and Lurie, 1997). Related work has shown how policies toward illicit drug use, including police harassment and harsh sentencing laws for possession of small volumes, have driven HIV spread in Russia and Ukraine (Malinowska-Sempruch *et al.*, 2004).

These same forces are likely to be involved in the spread of HCV and HBV in marginalized and criminalized drug-using populations as well, although it is clearly in the HIV realm that more work on these unintended consequences of drug policies has been done. It is also clear that harsh policies on possession and use drive incarceration rates, as we have shown in Thailand, and that incarceration has played further roles in the social ecology of these diseases and behavior interactions (Beyrer *et al.*, 2003).

## Incarceration and spread of disease

The historical connection between infectious disease and incarceration dates back to 1666, when it was noted that while British prisoners were being tried for their crimes they often infected people in the courtroom (Stern, 2001). There are two main theories as to why there is such a high prevalence of infectious disease in jail: the first is that at-risk populations are concentrated in the prison environment, while the second focuses on the fact that the risk environment as a whole is much higher in prison. The truth is probably a combination of the two. Data from the US reveal that approximately 2.4 percent of the entire population – mostly minorities – is either incarcerated or awaiting trial, with almost a quarter being related to drug offenses, and these numbers continue to grow (Polonsky *et al.*, 1994; Fox *et al.*, 2005).

The burden of infectious disease in entrants to the penal system is significantly higher than in the general population, and is commonly related to a history of IDU (Horsburgh *et al.*, 1990; Simbulan *et al.*, 2001; Ruiz *et al.*, 2002; Alizadeh *et al.*, 2005; Drobniewski *et al.*, 2005; Kushel *et al.*, 2005). For example, the prevalence of HIV is over 10 times higher in incarcerated populations than in the general population (Vlahov *et al.*, 1991). Moreover, the risk environment (all risks external to the individual) becomes much more dangerous when a person is incarcerated (Watson *et al.*, 2004). The normally complex social networks of free injection drug users are inherently simplified while incarcerated, due to the limited population, thus causing exaggerated high-risk behaviors such as sharing injection devices (Rhodes *et al.*, 1999a). Secondary to the previously discussed strategies of zero-tolerance policing on drug use, most injection drug users will be incarcerated as a result of their drug use, and will continue to use drugs while in prison (Dufour *et al.*, 1996). Consequently, there has been incidence of infectious disease in prisons, including HIV, HCV, and HBV, secondary to drug use worldwide (Dolan and Wodak, 1999; Khan *et al.*, 2005; Taylor *et al.*, 1995, 2000; Thaisri *et al.*, 2003). These same conditions have resulted in multiple reports demonstrating that incarceration is an independent risk factor for the incidence of HIV, HCV, and HBV among injection drug users (Briggs *et al.*, 2001; Mishra *et al.*, 2003; Small *et al.*, 2005). Based on these reports, it is now time to implement the various harm-reduction strategies with proven efficacy, including education, opioid substitution therapy, and needle exchange, in prisons (Choopanya *et al.*, 2003; Dolan *et al.*, 2003, 2004; Nelles *et al.*, 1998).

## Alternatives

We have attempted to show how human behaviors related to parenteral substance use have driven epidemics of HIV, HCV, and other infectious diseases, and how in turn social and political responses to substance use have so often contributed to disease spread rather than control. Are there better ways? While attempts to reduce the absolute volume of illicit drugs worldwide will no doubt continue and should be supported, it is probably wise to consider that humans have used psychoactive drugs throughout our known history, and that addiction is a powerful force for continued use. A prudent and moderate approach would be to provide drug treatment to those who want it, to investigate new and better approaches to addiction therapy, and to attempt to minimize the health impacts to individuals and communities with the recognition that substance abuse is likely to continue for the foreseeable future. Simple first steps would be to take to scale the prevention measures we have with evidence for efficacy for reducing infectious disease spread among drug users. These include needle and syringe exchange programs, outreach and education, access to drug treatment (including substitution therapy), and access to effective clinical care for infectious diseases where such care is known to be beneficial. Anti-viral therapy for AIDS and HCV are examples. Taking this approach further, a human rights-based approach which recognizes that drug users are people deserving of compassion and care, and attempting to reduce the social isolation, official harassment, and criminalization of addiction, could go a long way to reducing the harm for individuals and communities. Human beings are the vectors for infections spread through injection drug use, and so this is one group of diseases for which humans can truly be said to be in control. That we have failed so miserably to control the global epidemics of infections among drug users says a great deal about our approach to these challenges. We have tried decades of prohibition and criminalization to little avail. It is well past time to consider other options.

## References

Abdala, N., Carney, J.M., Durante, A.J. *et al.* (2003). Estimating the prevalence of syringe-borne and sexually transmitted diseases among injection drug users in St Petersburg, Russia. *International Journal of STD & AIDS* **14**, 697–703.

Aceijas, C., Stimson, G.V., Hickman, M. and Rhodes, T. (2004). Global overview of injecting drug use and HIV infection among injecting drug users. *AIDS* **18**, 2295–2303.

Ahmed, A. (1998). UN General Assembly Special Session on the World Drug Problem. UN, New York.

Alizadeh, A.H., Alavian, S.M., Jafari, K. and Yazdi, N. (2005). Prevalence of hepatitis C virus infection and its related risk factors in drug abuser prisoners in Hamedan, Iran. *World Journal of Gastroenterology* **11**, 4085–4089.

Altice, F.L., Bruce, R.D., Walton, M.R. and Buitrago, M.I. (2005). Adherence to hepatitis B virus vaccination at syringe exchange sites. *Journal of Urban Health* **82**, 151–161.

Backmund, M., Meyer, K., von Zielonka, M. and Eichenlaub, D. (2001). Treatment of hepatitis C infection in injection drug users. *Hepatology* **34**, 188–193.

Ball, A.L., Rana, S. and Dehne, K.L. (1998). HIV prevention among injecting drug users: responses in developing and transitional countries. *Public Health Reports* **113**(Suppl. 1), 170–181.

Bastos, F.I., Barcellos, C., Lowndes, C.M. and Friedman, S.R. (1999). Co-infection with malaria and HIV in injecting drug users in Brazil: a new challenge to public health? *Addiction* **94**, 1165–1174.

Beeching, N.J. and Crowcroft, N.S. (2005). Tetanus in injecting drug users. *British Medical Journal* **330**, 208–209.

Beld, M., Penning, M., van Putten, M. *et al.* (1999). Low levels of hepatitis C virus RNA in serum, plasma, and peripheral blood mononuclear cells of injecting drug users during long antibody-undetectable periods before seroconversion. *Blood* **94**, 1183–1191.

Bellis, M.A. (2005). *History of the Hypodermic Needle or Syringe*. Available at: http://inventors.about.com/library/inventors/blsyringe.htm (accessed 9 March 2006).

Bellis, M.A., Beynon, C., Millar, T. *et al.* (2001). Unexplained illness and deaths among injecting drug users in England: a case control study using Regional Drug Misuse Databases. *Journal of Epidemiology and Community Health* **55**, 843–844.

Beyrer, C. (2002). Fearful symmetry: heroin trafficking and HIV spread. *Bulletin on Narcotics* **LIV**, 103.

Beyrer, C. (2005). *The Next Wave: Emerging Epidemics of HIV-1 infection in Eurasia*. Rio de Janeiro: International AIDS Society. Available at: http://www.kaisernetwork.org/health_cast/uploaded_files/Beyrer_slides.pdf#search=%22the%20next%20wave%20beyrer%22.

Beyrer, C., Razak, M.H., Lisam, K. *et al.* (2000). Overland heroin trafficking routes and HIV-1 spread in south and south-east Asia. *AIDS* **14**, 75–83.

Beyrer, C., Jittiwutikarn, J., Teokul, W. *et al.* (2003). Drug use, increasing incarceration rates, and prison-associated HIV risks in Thailand. *AIDS and Behavior* **7**, 153–161.

Bradshaw, C.S., Pierce, L.I., Tabrizi, S.N. *et al.* (2005). Screening injecting drug users for sexually transmitted infections and blood borne viruses using street outreach and self collected sampling. *Sexually Transmitted Infections* **81**, 53–58.
Briggs, M.E., Baker, C., Hall, R. *et al.* (2001). Prevalence and risk factors for hepatitis C virus infection at an urban Veterans Administration medical center. *Hepatology* **34**, 1200–1205.
Bruneau, J., Lamothe, F., Soto, J. *et al.* (2001). Sex-specific determinants of HIV infection among injection drug users in Montreal. *Canadian Medical Association Journal* **164**, 767–773.
Calleja, J.M., Walker, N., Cuchi, P. *et al.* (2002). Status of the HIV/AIDS epidemic and methods to monitor it in the Latin America and Caribbean region. *AIDS* **16**(Suppl. 3), S3–12.
Carey, J., Perlman, D.C., Friedmann, P. *et al.* (2005). Knowledge of hepatitis among active drug injectors at a syringe exchange program. *Journal of Substance Abuse Treatment* **29**, 47–53.
Celentano, D.D., Galai, N., Sethi, A.K. *et al.* (2001). Time to initiating highly active antiretroviral therapy among HIV-infected injection drug users. *AIDS* **15**, 1707–1715.
Chaisson, R.E., Bacchetti, P., Osmond, D. *et al.* (1989). Cocaine use and HIV infection in intravenous drug users in San Francisco. *Journal of the American Medical Association* **261**, 561–565.
Chamla, D., Chamla, J.H., Dabin, W. *et al.* (2006). Transition to injection and sharing of needles/syringes: potential for HIV transmission among heroin users in Chengdu, China. *Addictive Behaviors* **31**, 697–701.
Chau, T.T., Mai, N.T., Phu, N.H. *et al.* (2002). Malaria in injection drug abusers in Vietnam. *Clinical Infectious Diseases* **34**, 1317–1322.
Choo, Q.L., Kuo, G., Weiner, A.J. *et al.* (1989). Isolation of a cDNA clone derived from a blood-borne non-A, non-B viral hepatitis genome. *Science* **244**, 359–362.
Choopanya, K., Des, J., Vanichseni, S. *et al.* (2003). HIV risk reduction in a cohort of injecting drug users in Bangkok, Thailand. *Journal of Acquired Immune Deficiency Syndromes* **33**, 88–95.
Cohen, M.S., Henderson, G.E., Aiello, P. and Zheng, H. (1996). Successful eradication of sexually transmitted diseases in the People's Republic of China: implications for the 21st century. *Journal of Infectious Diseases* **174**(Suppl. 2), S223–229.
Cohen, P. and Lyttleton, C. (2002). Opium reduction programmes, discourses of addiction and gender in Northwest Laos. *Sojourn* **17**, 1–23.
Correctional Association of New York (2001). *Stupid and Irrational and Barbarous: New York Judges Speak Against the Rockefeller Drug Laws*. Public Policy Project of the Correctional Association of NY, New York. Available at: http://www.correctional association.org/general/about/contact.html.
Cox, A.L., Netski, D.M., Mosbruger, T. *et al.* (2005). Prospective evaluation of community-acquired acute-phase hepatitis C virus infection. *Clinical Infectious Diseases* **40**, 951–958.
Crofts, N. (1999). *The Manual for Reducing Drug Related Harm in Asia*. Thailand: The Center for Harm Reduction, Macfarlane Burnet Center for Medical Research, and the Asian Harm Reduction Network.
Crofts, N., Reid, G. and Deany, P. (1998). Injecting drug use and HIV infection in Asia. The Asian Harm Reduction Network. *AIDS* **12**(Suppl. B), S69–S78.
Davis, M., Rhodes, T. and Martin, A. (2004). Preventing hepatitis C: "common sense", "the bug" and other perspectives from the risk narratives of people who inject drugs. *Social Science & Medicine* **59**, 1807–1818.

Deany, P. (2000). *HIV and Injecting Drug Use: A New Challenge to Sustainable Human Development.* Geneva: UNDP Asia-Pacific Regional Programme on HIV and Development. Available at: http://www.undp.org/hiv/publications/deany.htm.

Deany, P. and Crofts, N. (2000). Harm reduction, HIV and development. *Development Bulletin* **52**, 45–48.

Dehovitz, J. and Uuskula, A. (2002). *International Journal of Infectious Disease* **6**(1), 23–27.

de la Fuente, L., Bravo, M.J., Barrio, G. *et al.* (2003). Lessons from the history of the human immunodeficiency virus/acquired immunodeficiency syndrome epidemic among Spanish drug injectors. *Clinical Infectious Diseases* **37**(Suppl. 5), S410–415.

des Jarlais, D.C., Friedman, S.R., Choopanya, K. *et al.* (1992). International epidemiology of HIV and AIDS among injecting drug users. *AIDS* **6**, 1053–1068.

des Jarlais, D.C., Hagan, H., Friedman, S.R. *et al.* (1995a). Maintaining low HIV seroprevalence in populations of injecting drug users. *Journal of the American Medical Association* **274**, 1226–1231.

des Jarlais, D.C., Friedman, S.R., Friedmann, P. *et al.* (1995b). HIV/AIDS-related behavior change among injecting drug users in different national settings. *AIDS* **9**, 611–617.

des Jarlais, D.C., Perlis, T., Arasteh, K. *et al.* (2005). Reductions in hepatitis C virus and HIV infections among injecting drug users in New York City, 1990–2001. *AIDS* **19**(Suppl. 3), S20–25.

Dolan, K.A. and Wodak, A. (1999). HIV transmission in a prison system in an Australian State. *Medical Journal of Australia* **171**, 14–17.

Dolan, K., Rutter, S. and Wodak, A.D. (2003). Prison-based syringe exchange programmes: a review of international research and development. *Addiction* **98**, 153–158.

Dolan, K.A., Bijl, M. and White, B. (2004). HIV education in a Siberian prison colony for drug dependent males. *International Journal for Equity in Health* **3**, 7.

Drobniewski, F.A., Balabanova, Y.M., Ruddy, M.C. *et al.* (2005). Tuberculosis, HIV seroprevalence and intravenous drug abuse in prisoners. *European Respiratory Journal* **26**, 298–304.

Drucker, E., Lurie, P., Wodak, A. and Alcabes, P. (1998). Measuring harm reduction: the effects of needle and syringe exchange programs and methadone maintenance on the ecology of HIV. *AIDS* **12** (Suppl. A), S217–230.

Duffin, J. (1999). *History of Medicine: A Scandalously Short Introduction.* Toronto: University of Toronto Press.

Dufour, A., Alary, M., Poulin, C. *et al.* (1996). Prevalence and risk behaviors for HIV infection among inmates of a provincial prison in Quebec City. *AIDS* **10**, 1009–1015.

Field, M.G. (2004). HIV and AIDS in the former Soviet Bloc. *New England Journal of Medicine* **351**, 117–120.

Fleming, D.T. and Wasserheit, J.N. (1999). From epidemiological synergy to public health policy and practice: the contribution of other sexually transmitted diseases to sexual transmission of HIV infection. *Sexually Transmitted Infections* **75**, 3–17.

Fox, R.K., Currie, S.L., Evans, J. *et al.* (2005). Hepatitis C virus infection among prisoners in the California state correctional system. *Clinical Infectious Diseases* **41**, 177–186.

Fuller, C.M., Vlahov, D., Arria, A.M. *et al.* (2001). Factors associated with adolescent initiation of injection drug use. *Public Health Reports* **116**(Suppl. 1), 136–145.

Garfein, R.S., Vlahov, D., Galai, N. *et al.* (1996). Viral infections in short-term injection drug users: the prevalence of the hepatitis C, hepatitis B, human immunodeficiency, and human T-lymphotropic viruses. *American Journal of Public Health* **86**, 655–661.

Garfein, R.S., Latka, M., Hagan, H. *et al.* (2006). *Opportunities for Reducing Sexual HIV Risk Among Young Injection Drug Users (IDUs) through Substance Abuse Treatment*. Philadelphia: CIDUS-III/DUIT.

Gray, J. (1998). Harm reduction in the hills of northern Thailand. *Substance Use & Misuse* **33**, 1075–1091.

Gray, R.H., Wawer, M.J., Brookmeyer, R. *et al.* (2001). Probability of HIV-1 transmission per coital act in monogamous, heterosexual, HIV-10-discordant couples in Rakai, Uganda. *Lancet* **357**, 1149–1153.

Griffiths, P., Gossop, M., Powis, B. and Strang, J. (1994). Transitions in patterns of heroin administration: a study of heroin chasers and heroin injectors. *Addiction* **89**, 301–309.

Grund, J.P. (1997). *Harm Reduction and Methadone Treatment in Eastern Europe*. New York: Open Society Institute.

Hagan, H., Jarlais, D.C., Friedman, S.R. *et al.* (1995). Reduced risk of hepatitis B and hepatitis C among injection drug users in the Tacoma syringe exchange program. *American Journal of Public Health* **85**, 1531–1537.

Hagan, H., McGough, J.P., Thiede, H. *et al.* (1999). Syringe exchange and risk of infection with hepatitis B and C viruses. *American Journal of Epidemiology* **149**, 203–213.

Hagan, H., Thiede, H., Weiss, N.S. *et al.* (2001). Sharing of drug preparation equipment as a risk factor for hepatitis C. *American Journal of Public Health* **91**, 42–46.

Hagan, H., Thiede, H. and des Jarlais, D.C. (2004). Hepatitis C virus infection among injection drug users: survival analysis of time to seroconversion. *Epidemiology* **15**, 543–549.

Hagan, H., Thiede, H. and des Jarlais D.C. (2005). HIV/hepatitis C virus co-infection in drug users: risk behavior and prevention. *AIDS* **19**(Suppl. 3), S199–207.

Heather, H., Wodak, A., Nadelmann, E. and O'Hare, P. (1993). *From Faith to Science: Psychoactive Drugs & Harm Reduction*. London: Whurr Publishers.

Heiss, W.D and Herholz, K. (2006). Brain receptor imaging. *Journal of Nuclear Medicine* **47**, 302–312.

Hope, V.D., Judd, A., Hickman, M. *et al.* (2001). Prevalence of hepatitis C among injection drug users in England and Wales: is harm reduction working? *American Journal of Public Health* **91**, 38–42.

Horsburgh, C.R. Jr, Jarvis, J.Q., McArther, T. *et al.* (1990). Seroconversion to human immunodeficiency virus in prison inmates. *American Journal of Public Health* **80**, 209–210.

Hurley, S.F., Jolley, D.J. and Kaldor, J.M. (1997). Effectiveness of needle-exchange programmes for prevention of HIV infection. *Lancet* **349**, 1797–1800.

Hutchinson, S.J., Wadd, S., Taylor, A. *et al.* (2004). Sudden rise in uptake of hepatitis B vaccination among injecting drug users associated with a universal vaccine programme in prisons. *Vaccine* **23**, 210–214.

Kane, M. (1995). Epidemiology of hepatitis B infection in North America. *Vaccine* **13**(Suppl. 1), S16–17.

Karapetyan, A.F., Sokolovsky, Y.V., Araviyskaya, E.R. *et al.* (2002). Syphilis among intravenous drug-using population: epidemiological situation in St Petersburg, Russia. *International Journal of STD & AIDS* **13**, 618–623.

Kerr, T. (2003). North America's first supervised injection site opens in Vancouver. *Canadian HIV AIDS Policy and Law Review* **8**, 24–25.

Kerr, T., Tyndall, M., Li, K. *et al.* (2005). Safer injection facility use and syringe sharing in injection drug users. *Lancet* **366**, 316–318.

Kerr, T., Stoltz, J.A., Tyndall, M. *et al.* (2006). Impact of a medically supervised safer injection facility on community drug use patterns: a before and after study. *British Medical Journal* **332**, 220–222.

Khan, A.J., Simard, E.P., Bower, W.A. *et al.* (2005). Ongoing transmission of hepatitis B virus infection among inmates at a state correctional facility. *American Journal of Public Health* **95**, 1793–1799.

Kimber, J., Dolan, K., van Beek, I. *et al.* (2003). Drug consumption facilities: an update since 2000. *Drug and Alcohol Review* **22**, 227–233.

Kiyosawa, K., Sodeyama, T., Tanaka, E. *et al.* (1991). Hepatitis C in hospital employees with needlestick injuries. *Annals of Internal Medicine* **115**, 367–369.

Kushel, M.B., Hahn, J.A., Evans, J.L. *et al.* (2005). Revolving doors: imprisonment among the homeless and marginally housed population. *American Journal of Public Health* **95**, 1747–1752.

Levinson, A., Marske, R.L. and Shein, M.K. (1955). Tetanus in heroin addicts. *Journal of the American Medical Association* **157**, 658–660.

Linglof, T. (1995). Rapid increase of syphilis and gonorrhea in parts of the former USSR. *Sexually Transmitted Diseases* **22**, 160–161.

Lopez-Zetina, J., Ford, W., Weber, M. *et al.* (2000). Predictors of syphilis seroreactivity and prevalence of HIV among street recruited injection drug users in Los Angeles County, 1994–6. *Sexually Transmitted Infections* **76**, 462–469.

Lovell, A.M. (2002). Risking risk: the influence of types of capital and social networks on the injection practices of drug users. *Social Science & Medicine* **55**, 803–821.

Malinowska-Sempruch, K., Hoover, J. and Alexandrova, A. (2004). Unintended consequences: drug policies fuel the HIV epidemic in Russia and Ukraine. In: K. Malinowska-Sempruch and S. Gallagher (eds), *War on Drugs, HIV/AIDS, and Human Rights*. New York: Idea Press.

Mann, J., Tarantola, D. and Netter, T. (1992). *AIDS in the World 1992*. Cambridge: Harvard University Press.

Metzger, D.S., Woody, G.E., McLellan, A.T. *et al.* (1993). Human immunodeficiency virus seroconversion among intravenous drug users in- and out-of-treatment: an 18-month prospective follow-up. *Journal of Acquired Immune Deficiency Syndromes* **6**, 1049–1056.

Miller, C.L., Johnston, C., Spittal, P.M. *et al.* (2002). Opportunities for prevention: hepatitis C prevalence and incidence in a cohort of young injection drug users. *Hepatology* **36**, 737–742.

Mishra, G., Sninsky, C., Roswell, R. *et al.* (2003). Risk factors for hepatitis C virus infection among patients receiving health care in a Department of Veterans Affairs hospital. *Digestive Diseases and Sciences* **48**, 815–820.

Mitsui, T., Iwano, K., Masuko, K. *et al.* (1992). Hepatitis C virus infection in medical personnel after needlestick accident. *Hepatology* **16**, 1109–1114.

Mocroft, A., Madge, S., Johnson, A.M. *et al.* (1999). A comparison of exposure groups in the EuroSIDA study: starting highly active antiretroviral therapy (HAART), response to HAART, and survival. *Journal of Acquired Immune Deficiency Syndromes* **22**, 369–378.

Muga, R., Roca, J., Tor, J. *et al.* (1997). Syphilis in injecting drug users: clues for high-risk sexual behavior in female IDUs. *International Journal of STD & AIDS* **8**, 225–228.

Muller, R., Stark, K., Guggenmoos-Holzmann, I. *et al.* (1995). Imprisonment: a risk factor for HIV infection counteracting education and prevention programmes for intravenous drug users. *AIDS* **9**, 183–190.

National Institutes of Health Consensus Development Conference Panel Statement (1997). *Management of Hepatitis C: 1997*. Bethesday: NIH.

National Institutes of Health Consensus Development Conference Panel Statement (2002). *Management of Hepatitis C: 2002*. Bethesda: NIH.

Nelles, J., Fuhrer, A., Hirsbrunner, H. and Harding, T. (1998). Provision of syringes: the cutting edge of harm reduction in prison? *British Medical Journal* **317**, 270–273.

Ompad, D.C., Galea, S., Wu, Y. *et al.* (2004). Acceptance and completion of hepatitis B vaccination among drug users in New York City. *Communicable Disease and Public Health* **7,** 294–300.

Panda, S., Sarkar, S., Bhattacharya, S.K. *et al.* (1998). HIV-1 in injecting-drug users and heterosexuals. *Lancet* **352**, 241.

Passaro, D.J., Werner, S.B., McGee, J. *et al.* (1998). Wound botulism associated with black tar heroin among injecting drug users. *Journal of the American Medical Association* **279**, 859–863.

*Physicians' Desk Reference* (2005). 59th edn. Bridgeport: Thomson Healthcare.

Polonsky, S., Kerr, S., Harris, B. *et al.* (1994). HIV prevention in prisons and jails: obstacles and opportunities. *Public Health Reports* **109**, 615–625.

Portsmouth, S., Stebbing, J., Keyi, X. *et al.* (2003). HIV and AIDS in the People's Republic of China: a collaborative review. *International Journal of STD & AIDS* **14**, 757–761.

Razak, M.H., Jittiwutikarn, J., Suriyanon, V. *et al.* (2003). HIV prevalence and risks among injection and noninjection drug users in northern Thailand: need for comprehensive HIV prevention programs. *Journal of Acquired Immune Deficiency Syndromes* **33**, 259–266.

Razzaghi, E., Nassirimanesh, B., Afshar, P. *et al.* (2006). HIV/AIDS harm reduction in Iran. *Lancet* **368**, 434–435.

Reid, G. and Costigan, G. (2002). *Revisiting the "Hidden Epidemic": A Situation Assessment of Drug Use in Asia in the Context of HIV/AIDS.* Melbourne: The Center for Harm Reduction, The Burnet Institute.

Restrepo, A., Johnson, T.C., Widjaja, D. *et al.* (2005). The rate of treatment of chronic hepatitis C in patients co-infected with HIV in an urban medical centre. *Journal of Viral Hepatitis* **12**, 86–90.

Rhodes, T., Stimson, G.V., Crofts, N. *et al.* (1999a). Drug injecting, rapid HIV spread, and the "risk environment": implications for assessment and response. *AIDS* **13**(Suppl. A): S259–269.

Rhodes, T., Stimson, G.V., Fitch, C. *et al.* (1999b). Rapid assessment, injecting drug use, and public health. *Lancet* **354**, 65–68.

Rhodes, T., Ball, A., Stimson, G.V. *et al.* (1999c). HIV infection associated with drug injecting in the newly independent states, Eastern Europe: the social and economic context of epidemics. *Addiction* **94**, 1323–1336.

Rhodes, T., Lowndes, C., Judd, A. *et al.* (2002). Explosive spread and high prevalence of HIV infection among injecting drug users in Togliatti City, Russia. *AIDS* **16**, F25–31.

Rhodes, T., Davis, M. and Judd, A. (2004). Hepatitis C and its risk management among drug injectors in London: renewing harm reduction in the context of uncertainty. *Addiction* **99**, 621–633.

Roca, B, Gomez, C.J. and Arnedo, A. (1999). Stavudine, lamivudine and indinavir in drug abusing and non-drug abusing HIV-infected patients: adherence, side effects and efficacy. *Journal of Infection* **39**, 141–145.

Rossi, D., Goltzman, P., Cymerman, P. *et al.* (2003). Human immunodeficiency virus/ acquired immunodeficiency syndrome prevention in injection drug users and their partners and children: lessons learned in Latin America – the Argentinean case. *Clinical Infectious Diseases* **37**(Suppl. 5), S362–365.

Rothenberg, R.B., Long, D.M., Sterk, C.E. *et al.* (2000). The Atlanta Urban Networks Study: a blueprint for endemic transmission. *AIDS* **14**, 2191–2200.

Roy, K., Hay, G., Andragetti, R. *et al.* (2002). Monitoring hepatitis C virus infection among injecting drug users in the European Union: a review of the literature. *Epidemiology and Infection* **129**, 577–585.
Ruiz, J.D., Molitor, F. and Plagenhoef, J.A. (2002). Trends in hepatitis C and HIV infection among inmates entering prisons in California, 1994 versus 1999. *AIDS* **16**, 2236–2238.
Satoh, Y., Hino, K., Kato, T. *et al.* (2004). Molecular epidemiologic analysis of hepatitis C virus infection in injecting drug users with acute hepatitis C in Japan. *Journal of Gastroenterology and Hepatology* **19**, 1305–1311.
Schutz, C.G., Rapiti, E., Vlahov, D. and Anthony, J.C. (1994). Suspected determinants of enrollment into detoxification and methadone maintenance treatment among injecting drug users. *Drug and Alcohol Dependence* **36**, 129–138.
Seal, K.H., Ochoa, K.C., Hahn, J.A. *et al.* (2000). Risk of hepatitis B infection among young injection drug users in San Francisco: opportunities for intervention. *West Journal of Medicine* **172**, 16–20.
Seal, K.H., Kral, A.H., Lorvick, J. *et al.* (2003). A randomized controlled trial of monetary incentives vs outreach to enhance adherence to the hepatitis B vaccine series among injection drug users. *Drug and Alcohol Dependence* **71**, 127–131.
Senlis Council (2004). *A Fourth International Convention for Drug Policy: Promoting Public Health Policies*, edited by D. Spivack.
Senlis Council (2006). *Drug Policy Advisory Forum*. London: Senlis Council.
Simbulan, N.P., Aguilar, A.S., Flanigan, T. and Cu-Uvin, S. (2001). High-risk behaviors and the prevalence of sexually transmitted diseases among women prisoners at the women's state penitentiary in Metro Manila. *Social Science & Medicine* **52**, 599–608.
Small, W., Kain, S., Laliberte, N. *et al.* (2005). Incarceration, addiction and harm reduction: inmates experience injecting drugs in prison. *Substance Use & Misuse* **40**, 831–843.
Sorensen, J.L. and Copeland, A.L. (2000). Drug abuse treatment as an HIV prevention strategy: a review. *Drug and Alcohol Dependence* **59**, 17–31.
Southgate, E., Weatherall, A.M., Day, C. and Dolan, K.A. (2005). What's in a virus? Folk understandings of hepatitis C infection and infectiousness among injecting drug users in Kings Cross, Sydney. *International Journal for Equity in Health* **4**, 5.
Stern, V. (2001). Problems in prisons worldwide, with a particular focus on Russia. *Annals of the New York Academy of Science* **953**, 113–119.
Stimson, G. and Choopanya, K. (1998). Global perspectives on drug injecting. In: G. Stimson, J.D. Des and A. Ball (eds), *Drug Injecting and HIV Infection: Global Dimensions and Local Responses*. London: UCL Press.
Stimson, G., Des, J.D. and Ball, A. (1998). *Drug Injecting and HIV Infection: Global Dimensions and Local Responses*. London: UCL Press.
Stoove, M.A., Gifford, S.M. and Dore, G.J. (2005). The impact of injecting drug use status on hepatitis C-related referral and treatment. *Drug and Alcohol Dependence* **77**, 81–86.
Stramer, S.L., Glynn, S.A., Kleinman, S.H. *et al.* (2004). Detection of HIV-1 and HCV infections among antibody-negative blood donors by nucleic acid-amplification testing. *New England Journal of Medicine* **351**, 760–768.
Strathdee, S.A., Galai, N., Safaiean, M. *et al.* (2001). Sex differences in risk factors for hIV seroconversion among injection drug users: a 10-year perspective. *Archives of Internal Medicine* **161**, 1281–1288.
Suh, D.J. and Jeong, S.H. (2006). Current status of hepatitis C virus infection in Korea. *Intervirology* **49**, 70–75.

Sylvestre, D.L. (2002). Treating hepatitis C in methadone maintenance patients: an interim analysis. *Drug and Alcohol Dependence* **67**, 117–123.
Taylor, A., Goldberg, D., Emslie, J. *et al.* (1995). Outbreak of HIV infection in a Scottish prison. *British Medical Journal* **310**, 289–292.
Taylor, A., Goldberg, D., Hutchinson, S. *et al.* (2000). Prevalence of hepatitis C virus infection among injecting drug users in Glasgow 1990–1996: are current harm reduction strategies working? *Journal of Infection* **40**, 176–183.
Thaisri, H., Lerwitworapong, J., Vongsheree, S. *et al.* (2003). HIV infection and risk factors among Bangkok prisoners, Thailand: a prospective cohort study. *BMC Infectious Diseases* **3**, 25.
The World Bank (1999). *Confronting AIDS, Public Health Priorities in a Global Epidemic*. Oxford: Oxford University Press.
Thomas, D.L., Astemborski, J., Rai, R.M. *et al.* (2000). The natural history of hepatitis C virus infection: host, viral, and environmental factors. *Journal of the American Medical Association* **284**, 450–456.
Thorpe, L.E., Ouellet, L.J., Hershow, R. *et al.* (2002). Risk of hepatitis C virus infection among young adult injection drug users who share injection equipment. *American Journal of Epidemiology* **155**, 645–653.
UNAIDS (2005). *Update on the Global HIV/AIDS Pandemic*. Geneva: UNAIDS.
UNDCP (1998). *Drug Control Brief, HIV/AIDS and Drug Abuse*. Bangkok: UNDCP Regional Office for East Asia.
UNDP (1999). *AIDS in South and Southwest Asia: A Development Challenge*. Delhi: UNDP HIV and Development Project South and South West Asia.
UNODC (2005). *World Drug Report*. Vienna: United Nations Publications.
UNODC (2006). *World Drug Report*. Geneva: United Nations Publications.
Uusküla, A., Kalikova, A., Zilmer, K. *et al.* (2002). The role of injection drug use in the emergence of Human Immunodeficiency Virus infection in Estonia. *International Journal of Infectious Diseases* **6**(1), 23–27.
Vanichseni, S., Des, J., Choopanya, K. *et al.* (2004). Sexual risk reduction in a cohort of injecting drug users in Bangkok, Thailand. *Journal of Acquired Immune Deficiency Syndromes* **37**, 1170–1179.
Vlahov, D., Brewer, T.F., Castro, K.G. *et al.* (1991). Prevalence of antibody to HIV-1 among entrants to US correctional facilities. *Journal of the American Medical Association* **265**, 1129–1132.
Vlahov, D., Fuller, C.M., Ompad, D.C. *et al.* (2004). Updating the infection risk reduction hierarchy: preventing transition into injection. *Journal of Urban Health* **81**, 14–19.
Waley, A. (1958). *The Opium War Through Chinese Eyes*. Stanford: Stanford University Press.
Watson, R., Stimpson, A. and Hostick, T. (2004). Prison health care: a review of the literature. *International Journal of Nursing Studies* **41**, 119–128.
Whitworth, J.A. and Hewitt, K.A. (2005). Effect of malaria on HIV-1 progression and transmission. *Lancet* **365**, 196–197.
Wodak, A. and Lurie, P. (1997). A tale of two countries: attempts to control HIV among injecting drug users in Australia and the United States. *Journal of Drug Issues* **27**, 117.
Wu, Z., Rou, K. and Cui, H. (2004). The HIV/AIDS epidemic in China: history, current strategies and future challenges. *AIDS Education and Prevention* **16**, 7–17.
Yang, H., Li, X., Stanton, B. *et al.* (2005). Heterosexual transmission of HIV in China: a systematic review of behavioral studies in the past two decades. *Sexually Transmitted Diseases* **32**, 270–280.

Zamani, S., Kihara, M., Gouya, M.M. *et al.* (2005). Prevalence of and factors associated with HIV-1 infection among drug users visiting treatment centers in Tehran, Iran. *AIDS* **19**, 709–716.
Zhang, K.L. and Ma, S.J. (2002). Epidemiology of HIV in China. *British Medical Journal* **324**, 803–804.
Zhang, C., Yang, R., Xia, X. *et al.* (2002). High prevalence of HIV-1 and hepatitis C virus coinfection among injection drug users in the southeastern region of Yunnan, China. *Journal of Acquired Immune Deficiency Syndromes* **29**, 191–196.

# Urbanization and the social ecology of emerging infectious diseases

# 4

**Bruce A. Wilcox, Duane J. Gubler and H.F. Pizer**

The twentieth century was a landmark in the history of mankind as a result of the widespread control and eradication of infectious diseases that historically had been the scourge of humans. The advent and effective use of new drugs, vaccines, insecticides, treatment and prevention strategies during and following World War II reinforced public health programs already in place, and provided the tools needed to bring many of the worst diseases under control. Smallpox was eradicated using a mass vaccination strategy. By the late 1960s, the "war on infectious diseases" was declared won by leading experts in the field and by the Surgeon General of the United States (Patlak, 1996).

Unfortunately, the major successes in controlling infectious diseases in the 1950s and 1960s was followed by two coincident global trends that would have an impact on the dramatic re-emergence of infectious diseases in the waning years of the twentieth century. The first was the redirection of the resources that were once used to control infectious diseases to other public health priorities, such as the "War on Cancer" in the early 1970s. The perception that infectious diseases were no longer a problem led to decreased resources, widespread deterioration of public health infrastructure to deal with infectious diseases, and complacency among government and public health officials as well as the public (Smolinski *et al.*, 2003). This trend included medical education with a de-emphasis on preventive medicine and a strong focus on curative medicine in medical schools. Today, training in preventive medicine is not included in the curriculum of most medical schools in the US.

The second trend was the sharply increasing and unprecedented rate of human population growth following World War II that has continued for 60 years. Increasing human numbers have been a principal factor leading to uncontrolled urbanization, changes in agriculture, land use and animal husbandry practices, and accelerated globalization, all of which have been major and inter-related drivers of the re-emergence of epidemic infectious diseases (Gubler, 1998a).

The first evidence of the re-emergence of infectious disease occurred in the 1970s, but the process greatly accelerated in the latter two decades of the twentieth century. Old diseases that were once effectively controlled began to reappear in epidemic form – for example, dengue, Japanese encephalitis, West Nile Virus, epidemic polyarthritis, yellow fever, measles, plague, cholera, tuberculosis, leishmaniasis, malaria, etc. In addition, numerous newly recognized diseases began to cause epidemics, such as HIV/AIDS, the hemorrhagic fevers (Marburg, Ebola, Lassa, hantavirus, Crimean-Congo, arenaviruses, dengue and yellow fever), avian influenza, Hendra and Nipah encephalitis, severe acute respiratory syndrome (SARS), Lyme disease, ehrlichiosis, and others. In addition to those factors mentioned above, resistance of bacterial pathogens to antibiotics, drug resistance in malaria parasites, insecticide resistance in mosquitoes, new medical technology (e.g. organ transplantation) and immunosuppression by drugs and disease (AIDS), and ecologic encroachment by humans and animals have all played a role in the emergence/re-emergence of infectious diseases as a global public health problem (Gubler, 1998a, 2002; Smolinski *et al.*, 2003). In 2002, an estimated 26 percent of deaths worldwide were attributable to infectious and parasitic diseases (Fauci *et al.*, 2005); 24 percent of the global burden of disease, as measured by disability adjusted life years (DALYs), was caused by infectious diseases (World Health Organization, 2004).

A unique feature of the twentieth century re-emergence of infectious diseases has been the rapid global spread of some infectious agents, such as SARS, avian influenza, West Nile Virus, and dengue. This global spread is tied directly to modern transportation and globalization, both of which are directly dependent on the major urban centers of the world. In the past 50 years the global human population has exploded, and nearly all of that growth has occurred in the cities of the developing world. Here, the majority of the urban population typically lives in substandard housing with no electricity, water, waste management, or sewage systems. This creates ideal conditions for increased mosquito-, rodent-, water- and food-borne infectious diseases, as well as for sexually transmitted and communicable diseases. The global airline network connects these cities, providing the ideal mechanism for transporting exotic pathogens to new geographic locations. This chapter reviews the contribution of urbanization, both directly and indirectly, to the twentieth century re-emergence of infectious diseases, focusing on dengue/dengue hemorrhagic fever as a case study.

## The role of urbanization in infectious diseases

In his seminal book *Plagues and Peoples*, McNeil (1976) described how the development of major urban centers, and the successive stages of regional, and eventually global, trade linkages via new trade routes such as the Silk Road connecting the Middle East with Asia, explain historical patterns of the emergence of plagues and, indeed, the outcome of a number of pivotal events in history. Thus the relationship between urbanization and infectious diseases is an ancient one, of which the current phase is in some ways (but not all) a continuation of this story. The story is largely one of a dynamic in which the human populations of cities grow large and dense enough to insure a constant "crop" of susceptible individuals not exposed in previous outbreaks, fueling an epidemic cycle as well as providing the social-ecologic conditions for endemic persistence of a succession of infectious diseases. A significant fraction of the "susceptibles" is contributed by the flow of rural migrants, as well as urban natives born subsequent to previous outbreaks – a demographic pattern that today involves much higher numbers of migrants. This dynamic is elaborated upon by the expansion of networks of cities linked by land and sea through the movement of people, their associated animals, and commodities, providing new opportunities for pathogen dispersal. Cities, at least where adequate resources and political stability exist, have responded with the development of social institutions (hygienic laws, customs, and behaviors) and physical infrastructure (health-care, sanitation and waste management systems, etc.). Thus, the historic as well as the present-day dynamics of human infectious diseases largely reflect the ecological and evolutionary interplay of microparasites, the proportion of susceptible individuals in a population, and cultural adaptation. An underlying complication is the continuous evolution (and co-evolution) of humans, along with their domestic animals, and pathogens – with the occasional addition of new pathogens that have jumped the species' barrier (e.g. between apes and humans as in the case of HIV) and become successfully established in the human population.

Adding to the complexity today, and resulting in a new era of infectious disease emergence qualitatively different from that of the past, is the fact that the ecological theater in which this age-old co-evolutionary play is performed has been changing at an unprecedented pace. As a result, not only is there a need to restore investment in infectious disease prevention; new perspectives and models are also needed to understand and predict epidemic disease emergence, and to develop preventive measures that take into account the new social ecology of modern urbanization. A "social ecological systems" perspective, in which cities and their surroundings are seen as so-called "coupled human-natural systems," is one new way of thinking that has been helpful to explain patterns of disease emergence in relation to urbanization (Wilcox and Colwell, 2005; Wilcox and Gubler, 2005). A graphical representation of this perspective is shown in Figure 4.1, which illustrates the linkage of human systems on a regional scale with natural systems, such as ecological communities and ultimately host–parasite complexes, on successively smaller ecological scales. Changes at the level of the regional environment, such as population growth, cascade down through these successively smaller scales to facilitate the emergence or re-emergence of

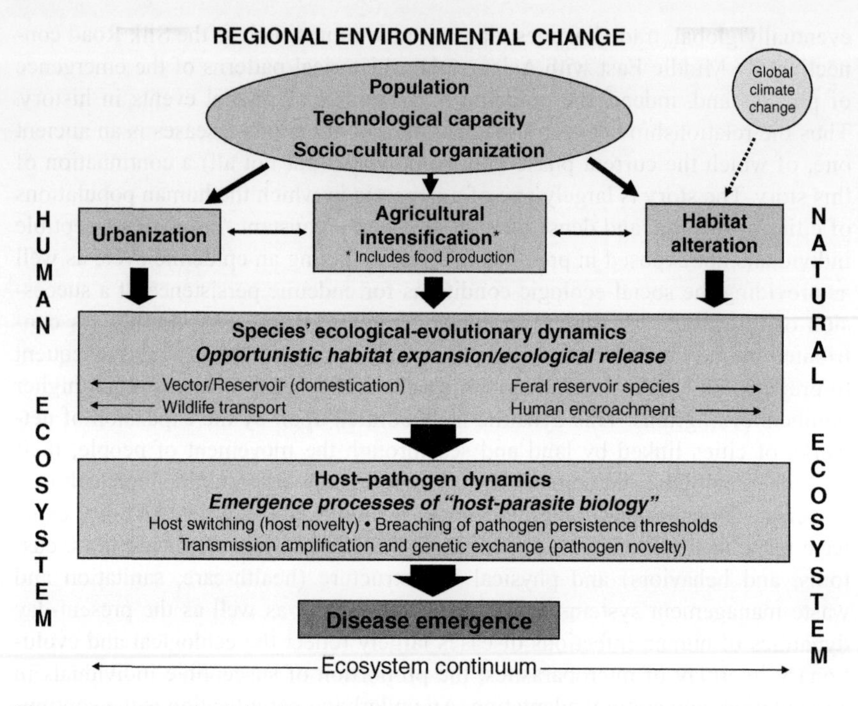


**Figure 4.1** Causal relationship of regional scale environmental drivers on disease emergence represented as a "coupled human–natural system." Modified from Wilcox and Gubler (2005).

infectious diseases. This coupled system thus "sends" the diseases in the opposite direction, up through the system to potentially impact public health on a regional or even global scale.

## Current urban demographic trends

While the process of urbanization has been going on for all of recorded human history, its pace has dramatically increased in the past hundred years. From 1900 to 1950 the world's urban population increased from approximately 220 million to 732 million, and then from 1950 to 2005 to about 3.2 *billion*. By 2030, demographers project the number will be about 4.9 billion. Sometime in 2005, for the first time in history, there were more people living in cities than in the countryside. By 2030, about 60 percent of the world's population will be living in cities (United Nations, 2006). In 1800, Beijing was the only city with a population of over a million people. From 1800 to 1990, the average size of

**Table 4.1 World megacities* 1975, 2000, and 2015 (projected): population in millions**

| 1975 | 2000 | 2015 |
|---|---|---|
| Tokyo (19.8) | Tokyo (26.4) | Tokyo (26.4) |
| New York (15.9) | Mexico City (18.1) | Mumbai (26.1) |
| Shanghai (11.4) | Mumbai (18.1) | Lagos (23.2) |
| Mexico City (11.2) | São Paolo (17.8) | Dhaka (21.1) |
| São Paolo (10) | Shanghai (17) | São Paolo (20.4) |
| | New York (16.6) | Karachi (19.2) |
| | Lagos (13.4) | Mexico City (19.2) |
| | Los Angeles (13.1) | New York (17.4) |
| | Kolkata (12.9) | Jakarta (17.3) |
| | Buenos Aires (12.6) | Kolkata (17.3) |
| | Dhaka (12.3) | Delhi (16.8) |
| | Karachi (11.8) | Metro Manila (14.8) |
| | Delhi (11.7) | Shanghai (14.6) |
| | Jakarta (11) | Los Angeles (14.1) |
| | Osaka (11) | Buenos Aires (14.1) |
| | Metro Manila (10.9) | Cairo (13.8) |
| | Beijing (10.8) | Istanbul (12.5) |
| | Rio de Janeiro (10.6) | Beijing (12.3) |
| | Cairo (10.6) | Rio de Janeiro (11.9) |
| | | Osaka (11.0) |
| | | Tianjin (10.7) |
| | | Hyderabad (10.5) |
| | | Bangkok (10.1) |

*Cities with populations ⩾10 million
Source: United Nations Population Fund (2001).

the world's hundred largest cities grew from around 200,000 to over 5 million (Hardoy *et al.*, 2001). There now are more than 40 cities with populations of at least 5 million, and 19 with more than 10 million. The latter are referred to as *megacities*, the list of which has grown and will continue to do so dramatically in the coming decades (see Table 4.1).

These are actually no longer discrete metropolitan areas surrounded by well-defined rural areas, but *urban agglomerations* that typically include the original city, now represented as a central urban zone, surrounded by a mix of suburbs, semi-urban, and semi-rural areas, all of which are interlinked. These are in turn connected (via the central city) to a global transportation network facilitating the rapid flow of people, vectors, and pathogens globally. Some urban agglomerations, like Tokyo, Mexico City, New York, and the Ruhr, are composed of more than one central city and one municipal government, but in many ways function like a single social ecological entity. The population of Tokyo, Yokohama, Kawasaki, and Saitama is more than 34 million. The population of Mexico City, Nezahualcoyotl, Ecatepec, and Naucalpan is more than 22 million. Seoul can be thought of as including Bucheon, Goyan, Incheon, Seongnam, and Suweon, with a total population of about 22 million. Not only has uncontrolled urbanization produced "cities" with population sizes unimaginable two generations ago, it has also created a new geography in which large, small, and medium human settlements across large regions have coalesced to create regional landscapes qualitatively different from those of the past. A similar growth pattern is unfolding in hundreds of smaller urban areas throughout the developing and developed world.

Thus, cities like Bangkok are now either referred to as the original municipality, currently with around 6 million people, or as "greater" Bangkok, which encompasses the surrounding districts with which the urban "habitat" of the original administrative unit is now contiguous. Such contiguous municipalities effectively constitute a single pool of humans – and a potential disease reservoir – which, in the case of Bangkok, now exceeds 10 million people. Even if the physical infrastructure and thus urban human density are not contiguous, it often becomes effectively so from a human pathogen's standpoint. The connectivity and mixing of people made possible by the modern transportation infrastructure and commuter lifestyle makes this so.

Nearly 500 cities now have population sizes approaching or exceeding 500,000. This is just above what mathematical epidemiology has found to be the critical population size for disease persistence and recurrent epidemics – a key transition point beyond which the trend of increasing infectious disease re-emergence or emergence is much more difficult to reverse, as is described further below.

Most of the fastest growing of these cities are in the developing tropical zones, where climate, environmental, and social ecological conditions are favorable for the transmission of pathogens responsible for the vast majority of old and new infectious diseases. While these conditions and the disease patterns are an integral part of the history of the development of human settlements and civilization, that the scale and magnitude of the present era of human-induced environmental change is unprecedented is starkly illustrated by the figures on urban population growth in developing countries during the past half-century. From 1950 to 2005, the urban population of the developing nations increased from just over 308 million to about 2.25 *billion* (United Nations, 2006). Most of that growth has resulted from immigration from rural areas, with many people bringing their rural lifestyle to the city. Figure 4.2 illustrates the growth of urban populations from 1950 to 2030, and shows that the largest increase has been in the low- and moderate-income countries of the developing world.

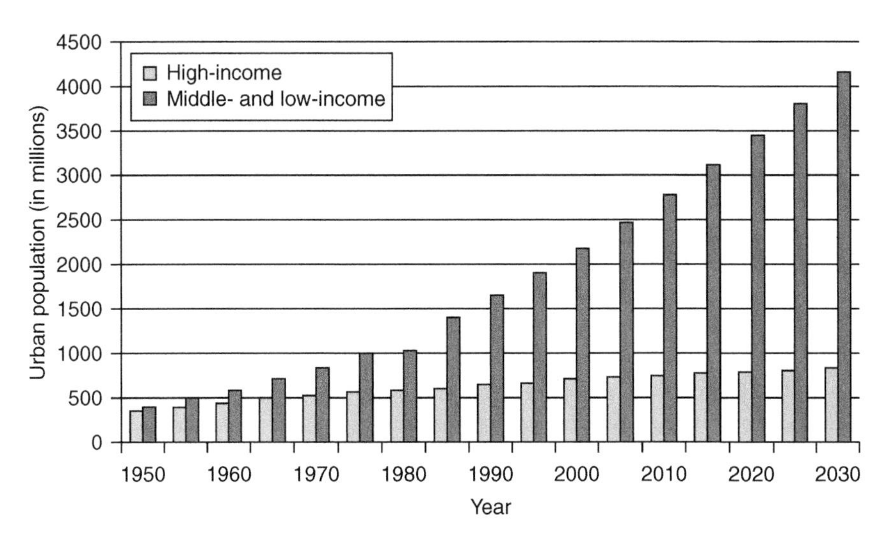


**Figure 4.2** Growth of urban populations, 1950–2030. Source: United Nations (2002); World Bank (2002).

## Risk parameters associated with urbanization

How urban social ecology affects emerging disease risk, especially the underlying mechanisms and dynamics involving environmental, human behavioral, and other factors, is poorly understood. However, a number of the parameters affecting risk are known. The risk for urban infectious disease outbreaks is greatest not only where the population density is highest, but also where people, public infrastructure, and public services are poor, and where access to medical care and basic public health programs does not keep pace with population growth. This includes cities in many of the same countries in the low- to moderate-income category just described, which typically are severely overcrowded. Poor neighborhoods lack safe and adequate housing, as well as reliable clean water, sewage disposal, and waste management. Public health is usually underfunded, and surveillance is non-existent or primitive. Generally, the supply of trained medical and public health professionals is insufficient to meet basic public needs. And because all the districts and neighborhoods in a large urban area are linked by modern transportation systems, pathogens can circulate with ease. Ironically, the construction of modern transportation systems intended to support modernization and economic development ensures the mixing of infected and susceptible people at a historically unprecedented rate. Many large urban complexes in developing Asia, for example, where highway and mass transit systems were lacking a decade ago, either now have them or soon will. Moreover, these are linked to transportation networks connecting even the most distantly separated urban centers. So the possibility now exists that an infectious disease outbreak within a neighborhood will spread readily not only throughout the city and to the surrounding areas, but also across the country and beyond national borders. This is of course exactly the scenario describing the near pandemic of SARS in 2003. Once this novel pathogen, whose source may have been fruit bats in Southern China, emerged near the rapidly growing Guangzhou in Guangdong Province (China), it easily spread to Hong Kong by surface transportation, and to Singapore, Vietnam, and even Toronto by infected air travelers, in barely a few days. In 2004, approximately a billion people traveled by air, with an ever-increasing number of them going to and from fast-growing cities in the tropical developing world. From 1983 to 1994, the number of air passengers leaving the United States doubled from about 20 million to 40 million, and more than half went to tropical countries. With modern laboratory analyses, it has been possible to track these epidemics across the urban hubs of common world travel and trade routes.

Although the global spread of epidemic pathogens tends to occur episodically, serious disease-causing pathogens now regularly erupt in cities and spread outward to the surrounding less densely populated areas of the country. In today's booming urban economic centers of Asia, like Guangzhou (approximately 9.5 million in the metropolitan area), Ho Chi Minh City (about 5.4 million), and Bangkok, much of the labor force is almost constantly on the move to and from cities and the country. The magnitude of human movement in and out of cities in developing countries cannot be overstated. A significant proportion (often the majority) of the urban population is of non-native origin and migrants. These people immigrating to the urban centers seeking economic opportunity still call "home" the small cities, towns, and villages of their origin, often because it is where their wives, children, parents, and grandparents live. On holidays like the New Year they travel from the city to their homes *en masse*. Each individual who travels home is capable of spreading a disease contracted in the city, and *vice versa*. Not long after the SARS outbreak in Southern China, and when the pathogen was apparently not yet contained, tens of millions of people took a holiday from work to travel home for several weeks during the Chinese New Year. Even with its highly centralized authority, the Chinese Government was largely powerless to control this annual mass migration. Fortunately, in this case, the disease was nonetheless contained or self-limiting.

As already mentioned, a critical population size is required to sustain an epidemic and for diseases to become endemic, the size depending on characteristics intrinsic to a pathogen. For example, this figure is about 250,000 for measles (Black, 1966; Anderson *et al.*, 1992). For this and a number of other common diseases, relatively isolated cities of less than a few hundred thousand or fewer people cannot generate a sufficiently large and constant flow of susceptible (immunologically naïve) people to fuel epidemics. Those that do arise as the result of an imported pathogen quite simply "burn out" quickly. For example, a number of Pacific Islands with small populations and limited tourism have had only limited outbreaks of dengue even though they have the *Aedes* mosquito vectors. Of course today, given the exponential growth in human mobility provided by modern transportation, even geographically isolated cities or human populations are becoming less and less biologically isolated. As this level of connectivity lowers the barriers to pathogen dispersal, the population size thresholds that once limited their continual transmission effectively cease to exist.

Even with effective family planning programs in place, demographic momentum ensures growth in most developing countries will continue until the middle of this century and, as illustrated by the Chinese New Year holiday example above, the capacity to restrict human movement is quite limited in most countries even after an epidemic has started, let alone before. Human population growth and mobility will remain a potent factor underlying disease emergence and re-emergence, as it has for the past three decades. This is clearly suggested for dengue in most Asian countries, where the mosquito vectors have been generally widespread and abundant.

Improved vector control could alter this situation in the future. However, eliminating the mosquitoes like *A. aegypti*, the primary vector for the pathogens responsible for several important emerging diseases, across large geographic areas or even single cities, likely is no longer possible. Today's urban conglomerations consist of tens of thousands and in some cases even millions of households, virtually every one potentially harboring this highly domestic mosquito. Just as with human mobility, the movement of commodities – between villages, provinces, cities, and countries – consisting of materials capable of harboring eggs, larvae, or adult mosquitoes has grown exponentially. Even if a mosquito population can be extinguished within a semi-urban village on a city's outskirts, or eliminated from an urban district, its absence will be only temporary. The constant influx of mosquito propagules insures the "empty patch" of domestic habitat will not stay unoccupied and unexploited for long. For example, a recent study of *A. aegypti* movement in Thailand and Puerto Rico (Harrington *et al.*, 2005) showed that individual mosquitoes commonly dispersed actively only as far as adjacent households. Inter-village dispersal (that is, active flight of up to half a kilometer) was found to be rare. Yet passive long-distance dispersal via hitch-hiking on human transport was deduced to be common, based on genetic evidence demonstrating the ecological connectivity between populations.

In light of these and other recent research findings, it's clear today's sprawling and globally interconnected urban landscapes, and the limited effectiveness and ecological risks associated with insecticides, requires new approaches to mosquito vector control. Fortunately, disease control, and even prevention, can be accomplished without completely eliminating the vector, so long as its abundance and biting success is reduced. Keeping mosquito abundance below certain levels by managing the environment and influencing the frequency with which humans are bitten through various efforts can decrease epidemic frequency and severity, as well as preventing pathogens like dengue from expanding its endemic cycle geographically. Disease ecology theory suggests the rate of spread of an infection, including whether this rate is high enough to initially spark an epidemic, is sensitive to both mosquito density and herd immunity in the human population. In disease endemic regions, pathogens like dengue viruses circulate via "silent transmission" in the human population, erupting episodically due to the interaction of a variety of dynamic factors, including the proportion of susceptible people, seasonal changes in mosquito abundance and survival, and viral evolution.

## Dengue and other emerging arbovirus diseases

About 177 pathogens are recognized as re-emerging or emerging, of which 73 percent are estimated to be zoonotic (Woolhouse and Gowtage-Sequeria, 2005) – that is, they are maintained in transmission cycles that involve domestic and/or wild animals, but can infect humans. Many, like dengue, are arboviral diseases, a term that describes a virus that requires a blood-sucking (hematophagous) arthropod, like a mosquito or tick, to complete its lifecycle. Except for dengue, which has fully adapted to an *A. aegypti*–human–*A. aegypti* cycle in tropical urban settings, all arboviruses have a non-human reservoir host, like a bird, rodent, or monkey. Of the more than 534 registered zoonotic and arboviruses, about 130 have been documented to cause illness in humans. Those of public health significance belong to three families: *Flaviviridae* (e.g. dengue fever, yellow fever, West Nile Virus), *Togaviridae* (e.g. Chikungunya fever, Ross River virus), and *Bunyaviradae* (Rift Valley fever, California encephalitis) (Gubler, 2002). The past few decades have seen a significant increase in the frequency, geographic spread, and virulence of a number of arboviral diseases. Table 4.2 provides a selective list of urban arboviral diseases that have public health importance. As vector-borne diseases require warm climates and moisture to thrive, they emerge more readily in tropical climates. However, many arthropod vectors are active during the summer but can "over-winter" in temperate climates.

### Dengue fever as a classic case study of the impact of urbanization

Dengue fever is an old disease, and is the classic case study of the recent re-emergence of a globally significant disease that originated as a zoonosis. Its historical pattern of emergence provides many lessons for containing the global spread of other, more recently recognized, arboviral diseases with the potential for becoming major urban public health threats.

The first reports of illness clinically compatible with dengue date back to a Chinese medical encyclopedia first published in AD 265 and last edited during

**Table 4.2 Urban emerging infectious diseases of public health importance**

| Family/virus | Vector | Vertebrate host | Ecology | Disease in humans | Geographic distribution |
|---|---|---|---|---|---|
| **Togaviridae** | | | | | |
| *Chikungunya* | Mosquitoes | Human, primates | U, S, R | SFI | Africa, Asia |
| *Ross River* | Mosquitoes | Human, primates | R, S, U | SFI | Australia, South Pacific |
| *Mayaro* | Mosquitoes | Birds | R, S, U | SFI | South America |
| **Flaviviridae** | | | | | |
| *Dengue 1–4* | Mosquitoes | Human, primates | U, S, R | SFI, HF | Worldwide in tropics |
| *Yellow fever* | Mosquitoes | Human, primates | R, S, U | SFI, HF | Africa, South America |
| *Japanese encephalitis* | Mosquitoes | Birds, pigs | R, S, U | SFI, ME | Asia, Pacific |
| *St Louis encephalitis* | Mosquitoes | Birds | R, S, U | SFI, ME | Americas |
| West Nile Virus | Mosquitoes | Birds | R, S, U | SFI, ME | Africa, Asia, Europe, US |
| **Bunyaviridae** | | | | | |
| *Oropouche* | Midges | ? | R, S, U | SFI | Central and South America |

U, urban; S, suburban; R, rural; SFI, systemic febrile illness; ME, meningoencephalitis; HF, hemorrhagic fever
Source: Gubler (2002).

the Northern Sung Dynasty in AD 992. Epidemics of dengue-like illness were reported in 1635 in the Caribbean, and in 1699 in Panama. By the end of the eighteenth century the viruses and their vectors apparently had a worldwide distribution, with epidemics of dengue-like illness being reported in Batavia (Jakarta) in Indonesia (1779), Cairo in Egypt (1779), and Philadelphia in Pennsylvania, USA (1780) (Gubler, 1997). It should be noted that these were not confirmed dengue epidemics, because the viruses were not isolated until 1943–44. However, clinically and epidemiologically, the disease was compatible with dengue.

The historical evolution of dengue to become the most important arboviral disease of humans as we enter the twenty-first century is closely tied to the evolution of urbanization and commerce (globalized trade). The primitive cycle of dengue viruses involved canopy-dwelling mosquitoes and non-human primates in the rainforests of Asia, and possibly Africa. Humans who entered the forests to hunt, cut wood, or for other activities were exposed to the viruses through the bite of infected mosquitoes. With an incubation period of up to 14 days, people became ill (and infectious) after they returned to their village outside the forest, thus exposing peridomestic mosquitoes in the villages to the virus. These latter mosquitoes, such as *A. albopictus*, transmitted epidemics, but because of the small human populations in the villages the infections soon died out and transmission ceased until another virus was introduced. These epidemics were thus very infrequent and sporadic.

In the seventeenth, eighteenth, and nineteenth centuries, as global trade and the shipping industry developed, port cities sprouted on all continents, followed by the building of inland cities and larger port cities. The water barrels carried on sailing vessels were frequently infested with mosquitoes, and it was not uncommon for the ships to maintain active transmission of diseases like dengue and yellow fever among the mosquitoes and crew members (Gubler, 1997). When the ships docked at a port city, both the mosquitoes and the viruses went ashore with the crew. Thus were mosquitoes and viruses imported to and established in port cities around the world.

The history of the spread of *A. aegypti* provides a classic example. Although a feral mosquito in Africa, it was introduced to the villages and cities of West Africa, where it adapted to breeding in stored water containers. From there it was taken, along with yellow fever, to the Americas during the slave trade in the sixteenth and seventeenth centuries, infesting port and inland cities. Even the temperate United States was infested, with the mosquito being maintained in the port cities of the Gulf Coast during the winter and expanding up the rivers and waterways to inland cities during the summer months (Gubler, 1997). The epidemics of dengue and yellow fever in cities like Philadelphia were the direct result of this kind of commerce. From the Americas, *A. aegypti* spread to the Pacific and Asia by the same means. This mosquito ultimately became highly adapted to humans and the urban environment, infesting most tropical cities of the world, and became the most efficient epidemic vector of urban dengue and yellow fever (Gubler, 1989, 1998b).

As noted above, by the late eighteenth century dengue viruses had a worldwide distribution in the tropics. Because the viruses were dependent on sailing vessels for geographic spread, however, epidemics were infrequent, often with periods of 10–40 years with no epidemic activity. Once a virus was introduced to a new region, however, it would move from country to country within that region at a much faster pace. This was the status of the disease at the beginning of World War II.

The war in the Pacific and Asian Theaters initiated the twentieth-century pandemic of dengue (Gubler, 1998b). Both the Allied and Japanese armies put hundreds of thousands of susceptible troops into the area. The movement of those troops, along with war materials, was responsible for all four dengue virus serotypes and *A. aegypti* mosquitoes being spread throughout the region. By the end of the war, dengue was hyperendemic (the co-circulation of multiple virus serotypes) in most countries of Asia.

In the years following World War II, an economic boom began in Asia that is continuing today. It was this dramatic economic development, combined with unprecedented population growth, that was the primary driving force of uncontrolled urbanization that has occurred in most Asian cities in the past 50 years. The influx of people, primarily from rural areas, led to rapid and uncontrolled urban growth. Forced to live in inadequate housing in areas where there was no water, sewage, electricity, or waste management, people had to store water in containers, which made ideal larval habitats for *A. aegypti* mosquitoes. The large mosquito populations living in intimate association with crowded human populations similarly provided ideal conditions for epidemic transmission of the dengue virus. It was in this setting in the 1950s and 1960s that the much more serious and sometimes fatal form of dengue, dengue hemorrhagic fever (DHF), emerged in epidemic form. By 1970, DHF was a leading cause of hospitalization and death among children in Southeast Asia. In the latter two decades of the twentieth century, epidemic DHF spread throughout Asia, east to China and Taiwan, and west to the Indian subcontinent.

Urbanization was occurring in other parts of the world as well, especially in the Americas. Fortunately, however, dengue and yellow fever had been effectively controlled in the 1950s and 1960s in the Americas by the *A. aegypti* eradication program initiated in 1946, which focused on larval mosquito control using a combination of environmental management and DDT. Because there were no epidemics of dengue and yellow fever, however, this program was disbanded in the early 1970s (Gubler, 1989; Gubler and Trent, 1994). Thus began the reinvasion of tropical American countries by *A. Aegypti* – but this time there were much larger cities to host them. By the beginning of the twenty-first century, Mexico and most of the Caribbean, South and Central American countries had been re-colonized by this mosquito.

The era of jet travel and modern transportation began in the 1960s, but accelerated in the 1970s and 1980s. This provided the ideal mechanism for the hyperendemic dengue melting pot of Southeast Asia to seed the rest of the world with dengue viruses. The viruses first moved into the Pacific Islands in the early 1970s, and into the Americas in the late 1970s. The 1980s and 1990s saw the whole of the tropical world become hyperendemic, resulting in greatly increased frequency of epidemic dengue fever and the emergence of DHF in the Pacific and Americas (Gubler, 1997). As shown by the maps in Figure 4.3, in 1970 dengue was either hypoendemic with only one virus serotype circulating, or non-endemic in most countries of South and Central America, the Caribbean and West Africa; only Southeast Asia was hyperendemic with all four serotypes co-circulating. Today, the whole of the tropical world is hyperendemic, with all four virus serotypes co-circulating throughout the Americas, across tropical Africa, South Asia, Southeast Asia, Australasia, and Oceania. As a result, the epidemics have became more frequent, and larger, on a global level. In 2006, approximately 2.5–3 billion people live in areas at risk for dengue, which infects an estimated 50–100 million

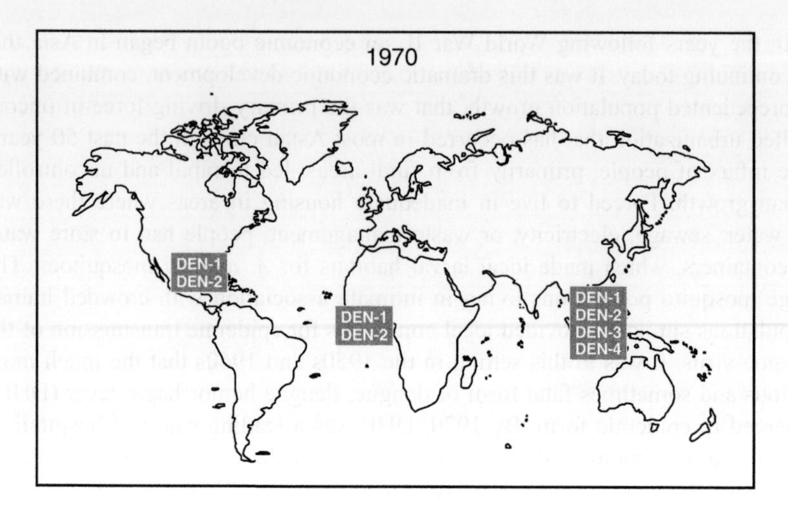


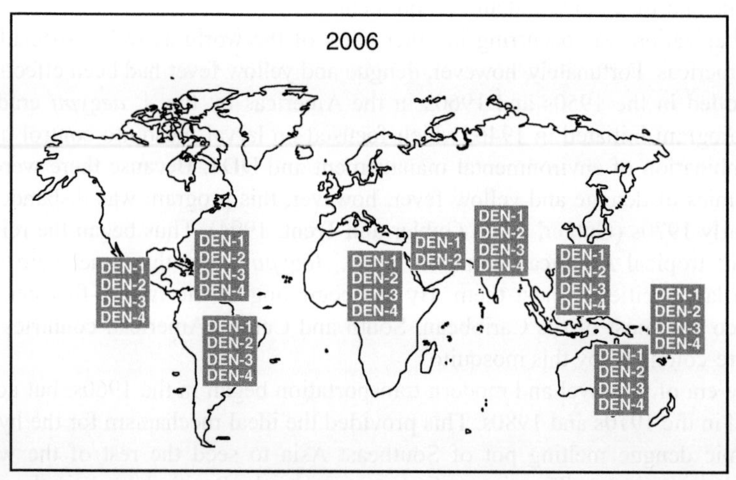


**Figure 4.3** Global distribution of dengue virus serotypes in 1970 and 2006. Source: Gubler (1998b).

persons per year, with 500,000 cases of DHF and 20,000–25,000 deaths (Gubler, 1998b; World Health Organization, 1999).

As can be seen in Figure 4.4, the increased incidence of DF/DHF in the past 50 years closely tracks global population growth, most of which is urban population growth. In Thailand, the annual number and frequency of dengue cases closely tracks the historic population increase across the country and in Bangkok.

Figure 4.5 shows that the increase in dengue cases in Bangkok closely tracks population growth, and can therefore be projected to increase for at least several decades under current conditions. Moreover, dengue frequency in terms of

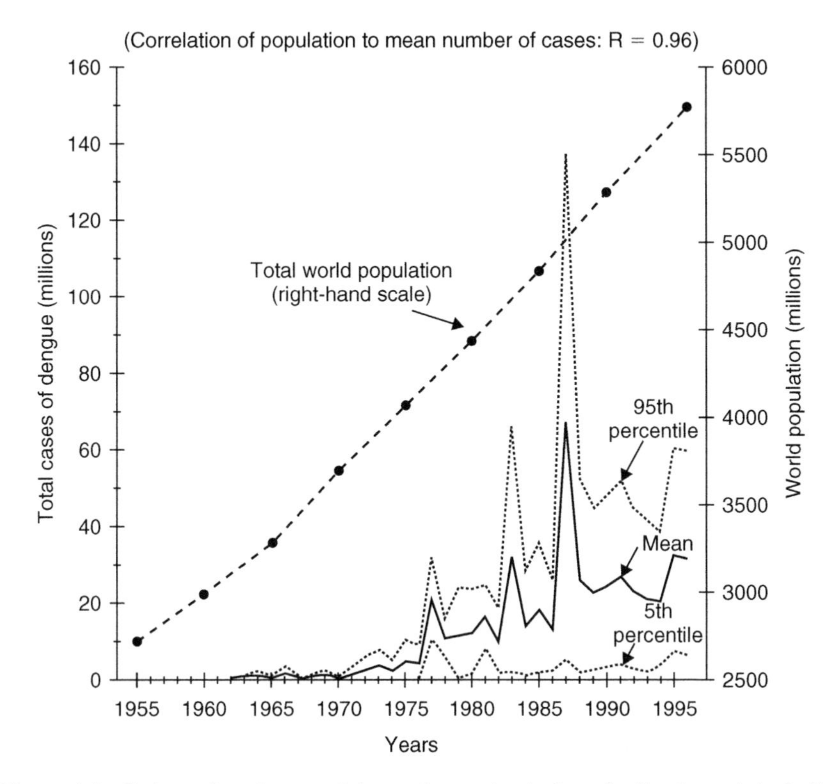


**Figure 4.4** Estimated total cases of dengue hemorrhagic fever for Southeast Asia, India, China, Latin America, and the Caribbean. Source: Gubler and Meltzer (1999).

the proportion of months with reported cases tends to increase sharply for provinces exceeding about 500,000 in population size (Wearing and Rohani, 2006). Population growth is a surrogate measure of urbanization and all its attendant social-ecological factors that facilitate disease emergence.

One recent study documented how dengue epidemics travel in a wave out from Bangkok at an average rate of 148 km per month (Cummings *et al.*, 2004). Bangkok serves as a regional "epicenter" for major epidemics in Thailand on a 3- to 5-year cycle (Nisalak *et al.*, 2003). These patterns, being uncovered through the accumulation of increasingly detailed data and more sophisticated molecular and statistical research tools, are probably representative of what is occurring in all the cities of tropical developing countries. These tropical urban centers are the spawning grounds for epidemic dengue (Gubler, 2004a).

The key to understanding the recent major resurgence of dengue, along with most of the other 177 or more emerging infectious diseases, requires an appreciation of dynamics of a "coupled human–natural system" mentioned earlier, seen as

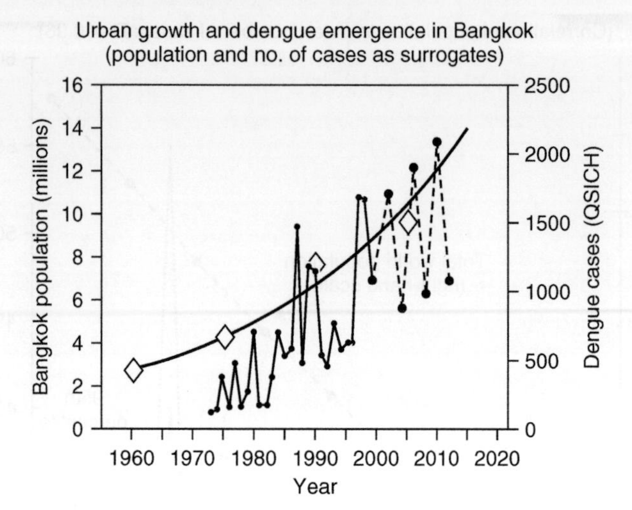


**Figure 4.5** Historical and projected growth in dengue cases and urban population in Bangkok. Population growth serves as a surrogate or indicator of a wide range of social-ecological factors accompanying urbanization. Dashed line represents projected dengue cases assuming current circumstances, such as *per capita* levels of vector-control efforts, remain constant. Source: Wilcox (unpublished); based on historic and projected population size of greater Bangkok and dengue case data for the Queen Sirikit National Institute of Child Health in Bangkok, published in Nisalak *et al.* (2003). Projected future cases, year 2000 on, were estimated by linear extrapolation from least-squares fitted regression for peak years and trough years as a function of population size.

an inherent characteristic of urban ecosystems when viewed from a social ecological perspective (Wilcox and Colwell, 2005). Thus, this complex situation can be simplified somewhat by considering it from the standpoint of how human society – in this case in the form of poorly or ill-guided public policy with regard to public health, urbanization, and globalization – and nature interact. "Nature" here refers to the ecological and associated evolutionary processes represented by viral (and vector) dispersal, genetic change, inter-serotype and serotype–host interaction occurring across spatial scales involving virus–mosquito–human interactions in a single village or urban neighborhood to the regional and global level, with urban expansion and global transport as the dominant influences. Based on this perspective, the re-emergence of dengue and similar diseases can be described as follows.

In the first half of the twentieth century public health measures focused on vector control, and were very effective in controlling dengue and other *A. aegypti*-borne diseases such as yellow fever. Beginning in the second half of the twentieth century, and especially during the latter 30 years, rapid, uncontrolled urbanization in tropical regions of the developing world combined with exponentially increasing global transport of people, animals, and commodities developed

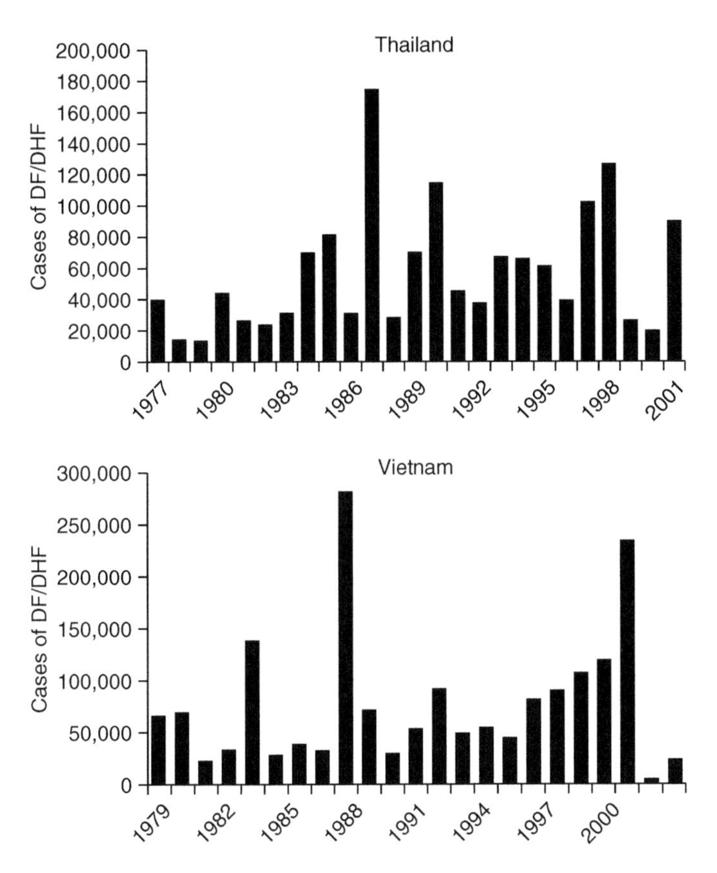


**Figure 4.6** The dynamics of dengue. Dengue outbreaks in Thailand and Vietnam now occur on a 3- to 5-year cycle instead of the 10- to 40-year prior historical pattern. Source: Gubler (2004a).

into dominant social-ecological forces, as already noted above. A growing lack of effective mosquito control in crowded urban centers and the increasing movement of viruses via modern transportation facilitated increasing hyperendemicity in large urban centers in the tropics. The result has been epidemic cycles shortened from a 10- to 40-year to a 3- to 5-year cycle, as the case data for Vietnam and Thailand show in Figure 4.6.

Finally, dengue provides the classic model of how the geographic spread of an infectious disease, a principal characteristic by which many viral diseases are classified as emerging, can be revealed by the tracking of a specific viral genetic strain on a map. Before 1989 DHF was common in Southeast Asia, but rare on the Indian subcontinent, despite the circulation of all four serotypes. After 1989, regular epidemics of DHF were reported on the Indian subcontinent and Sri Lanka.

The change did not appear to be due to a general increase in viral transmission, but to a change in virus subtype (Lanciotti *et al.*, 1994; Messer *et al.*, 2003). The majority of people with severe disease were infected with a new subtype of DENV-3, which was clearly derived from the pre-DHF epidemic DENV-3 strain, most likely via genetic drift and selection (Bennett *et al.*, 2003). While the exact processes by which epidemic DHF arose in Sri Lanka are not fully understood, it is clear that the DENV-3 strain associated with DHF in Sri Lanka was derived from the strain previously circulating in Sri Lanka, and was not the same as the DENV-3 circulating in the Southeast Asia region. It appears this Indian subcontinent subtype then spread from South Asia to East Africa (Gubler *et al.*, 1986; Messer *et al.*, 2003). Genetic studies show the DENV-3 subtype III viruses currently found in Latin America are also closely related to isolates found in East Africa and South Asia. Figure 4.7 illustrates the most likely route of the global spread of DENV-3, subtype III from South Asia to East Africa and Central and South America, and, although based on global population movement data, it is likely that the American DENV-3 was introduced from Asia.

Two other arboviral diseases, yellow fever and Chikungunya fever, whose emergence appears to be following a pattern disturbingly similar to the early re-emergence of dengue, also have a transmission cycle in urban areas similar to dengue, being transmitted by *A. aegypti* and *A. albopictus*. Changes in the transmission dynamics of both are also associated with social ecological changes accompanying urbanization, along with regional environmental change and globalization.

Chikungunya is Swahili for "that which bends up," and the name comes from the muscle and joint symptoms of the diseases, which can be debilitating and last for weeks or months. People with Chikungunya experience a range of other symptoms, such as fever, headache, fatigue, nausea, vomiting, and skin rash. Like dengue and yellow fever, the Chikungunya virus exists in a natural cycle involving mosquitoes and monkeys in the rain forests of Africa. It was first isolated in Tanzania in 1953, and has since been identified in epidemics in western, central and southern Africa, and in a number of Asian countries, such as Indonesia, the Philippines, Thailand, Myanmar, and India. In 2005 and 2006 there were numerous epidemics in India and the islands off the east coast of Africa. A large epidemic in the heavily populated Indian state of Andhra Pradesh spread to neighboring states with fatalities in a number of cities, including Udaipur, Chittorgarh, and Bhilwara. By the time this chapter was going to press 1.39 million cases had been reported across 13 Indian states (NVBDCP, 2006). Currently, Chikungunya fever is most likely spread by infected travelers, but it has been endemic in Asia for decades, and has the potential to become an urban disease globally.

Yellow fever was the most important urban infectious disease in the Americas until the twentieth century. The *A. aegypti* eradication program noted above eliminated the mosquito and the disease throughout most of the Americas (Gubler, 2004b). Endemic in Africa and South America, the first recorded outbreak of

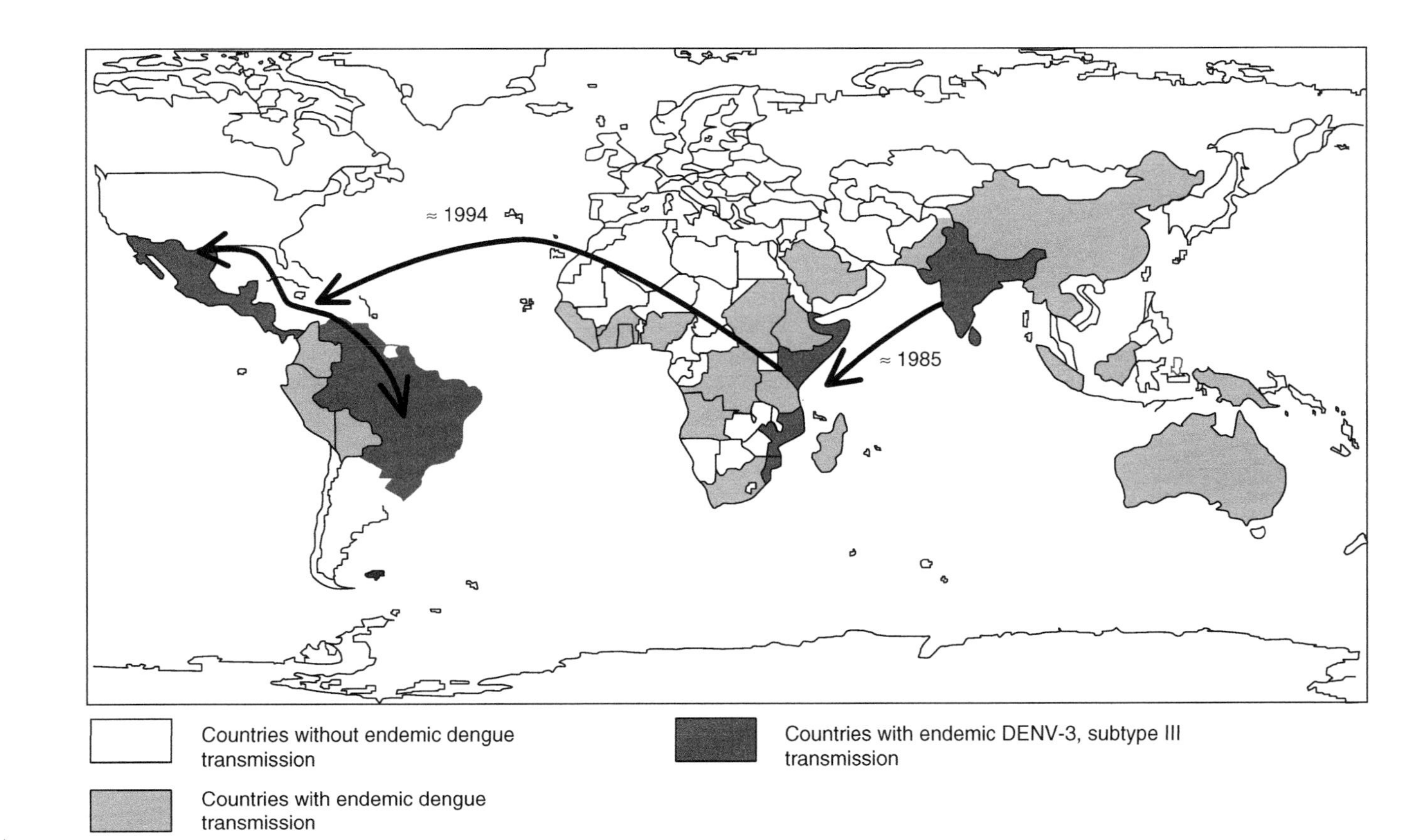


**Figure 4.7** Global spread of dengue virus 3 (DENV-3). Subtyping studies show the likely spread of a single subtype of dengue virus 3 from its origin on the Indian subcontinent to East Africa and Latin America. Source: Messer *et al.* (2003).

yellow fever in the Western Hemisphere occurred in 1648, and over the next 400 years epidemics were recorded across much of South and Central America, and as far north as New York City (Carter and Frost, 1931). Once Walter Reed and his colleagues had determined that *A. aegypti* spread yellow fever outbreaks, control focused on destroying the mosquito in the larval stage in domestic water-storage containers, and killing adult mosquitoes with insecticides – usually DDT. In 1901, William Gorgas developed the first effective control program in Havana, Cuba, which in 1904 was replicated in Panama. Over the next few years, programs were initiated in Rio de Janeiro in Brazil, Vera Cruz in Mexico, and Guayaquil in Ecuador. In 1937 a live-attenuated vaccine was developed, which was used in West Africa but not in the Americas. Nevertheless, successful mosquito control in the Americas worked to eliminate urban epidemics of yellow fever (Gubler, 2004b). With the re-expansion of *A. aegypti* across its former geographic range in Latin America, however, it has been dengue that has re-emerged most dramatically.

Today, yellow fever persists in three kinds of transmission cycles (Table 4.3). These cycles illustrate the process by which social-ecological factors, such as settlement patterns (including the urban expansion into rural zones and agriculture communities into forests), produce a landscape continuum from natural habitat to urban habitat. Arboviruses, because of their capability for relatively

**Table 4.3 Transmission cycles for yellow fever**

| Type | Transmission cycle |
|---|---|
| Sylvatic or jungle yellow fever | This is a disease of the rainforest in which the virus is transmitted between monkeys and wild mosquitoes; it is seen only rarely in people, in those working in logging or other activities in the rainforest |
| Intermediate yellow fever | This occurs in "zones of emergence," like savannah areas of Africa during the rainy season, where there is increased contact between humans and semi-domestic mosquitoes; even if a number of villages are involved simultaneously, outbreaks affect only relatively small populations |
| Urban yellow fever | This involves domestic *A. aegypti* mosquitoes and produces the largest and most dangerous epidemics in cities of tropical Africa (between 15° north and 10° south of the equator); it is less common than dengue, possibly because both viruses compete for the same vector and hosts |

rapid evolution (aided by a parallel domestication process exhibited by mosquitoes like *A. aegypti*), have the potential over time to "move" across this landscape to become major public health threats for urban areas.

At present, 33 countries in Africa with a total population of 468 million are at risk for yellow fever. Since 2000, 18 countries in Africa have reported yellow fever outbreaks, 13 of them in West Africa (World Health Organization, 2006). Given the inadequacies of the health system in these countries, it is likely that the number of reported cases is well below the actual number. Most people infected with yellow fever have no symptoms or only mild symptoms, and are not likely to see a physician who would then report their case to public health authorities. In endemic areas there also is a shortage of laboratories capable of performing virologic analyses. The major threat of epidemic yellow fever, however, is in the Americas, where over 300 million people live in urban areas infected with *A. aegypti* (Gubler, 2004b). So far yellow fever has not taken hold in Asia, but if urban yellow fever epidemics begin to occur, as in the Americas, a major global public health emergency will occur, because all of the Asia-Pacific region is at high risk (Gubler, 2004b).

## What the future holds

What can we learn from historical and recent experience, and what does a social-ecological perspective reveal that can help with the challenges ahead in controlling diseases in an increasingly urbanized world? First, one-dimensional vector control measures and technological quick fixes will not work. The *Aedes* mosquito cannot be completely eliminated from any of today's cities in the tropics. Nor can effective vaccines be developed for every infectious disease that is a potential public health problem. Integrated multi-level disease prevention and control programs will be necessary (Gubler, 1989). However, if complacency prevails after these approaches achieve success, as it did when control programs against *A. aegypti* were disbanded and merged with other public health programs, prevention programs will again fail (Gubler, 1989, 1998a, 2005).

Moreover, a top-down approach and methods, based on a limited or inadequate understanding of mosquito ecology, evolution, and urban social ecology, will fail. This lesson can be learned from the flawed strategy of using ultra-low volume (ULV) insecticides to kill adult mosquitoes that replaced previous successful programs that targeted mosquito larvae in water containers (Gubler, 1989). ULV is a top-down reactive strategy, which relies heavily on prompt physician reporting of cases before widespread spraying is initiated. Space-spraying requires direct contact of pesticide with the adult mosquito. It does not penetrate inside houses or kill mosquito larvae in water containers, where the mosquitoes thrive. Within a few hours of spraying, *A. aegypti* is again feeding on and infecting humans.

Surveillance and public education should be constant and based on a bottom-up community participation and community ownership approach. Between outbreaks, it should be expected that the index of suspicion for primary care medical providers and public awareness will fall (Gubler, 1989). Public health professionals should stress the professional obligation to be aware of and report cases. Public education health messages should continue when there are no outbreaks, otherwise epidemics will proceed without public health attention until they are at near peak level. In the case of a mosquito-borne disease like dengue, at that point spraying is not likely to be effective. The consequence of this public health laxness in the Americas was that *A. aegypti* and dengue both reappeared and spread in the 1970s and 1980s in more frequent and longer epidemics (Gubler, 1989, 2005).

Prevention and control begin with recognizing the potential for a particular arthropod-borne disease in a given environment and understanding the specific conditions that promote its transmission and spread. For about a century it has been known that *A. aegypti* thrives in water containers, and public health prevention effectively targeted open-water reservoirs in population centers. That is still the case today, but the difference is that human population centers are an order of magnitude larger, resulting in more frequent and larger epidemics. Controlling breeding sites today requires the help of the community and the people who live in the homes where transmission occurs (Gubler, 1989). These interventions thus require public planning and education, as well as investment. Even financially-strapped local governments can do a better job through public education and outreach, while they work on and invest in more long-term expensive projects to improve water and sewage infrastructure.

Another lesson is to continue effective immunization programs, even when case incidence is low. African countries did not maintain their mass vaccination programs for yellow fever. There are many reasons for this, including financial constraints and civil instability, but keeping up and modernizing those programs might have prevented the resurgence of yellow fever.

Continuous, active public health surveillance with the assistance of modern laboratory virology must be the mainstay for effective tracking of dengue and other emerging arboviral diseases (Gubler, 1998). Active surveillance usually depends on mandatory reporting of suspected cases by diligent primary care health providers to public health agencies. Active surveillance also requires modern laboratory diagnostic methods, which may not be readily available in many developing countries. In these instances, a productive strategy is to develop ongoing partnerships with scientists and laboratories in developed nations. Fortunately, some of these features are beginning to be integrated within region-wide programs employing interdisciplinary approaches based on a largely social-ecological perspective, also called an "ecosystem approach" or "eco-bio-social approach," as in the case of a recent initiative aimed at controlling dengue in tropical developing countries (IDRC, 2006).

There is an immediate need for frontline medical professionals to be better educated in emerging diseases, and for modern laboratory-based epidemiological surveillance in endemic areas. This includes training primary care medical professionals to accurately diagnose patients that present with the symptoms of infectious diseases and promptly notify cases to public health departments. Tropical areas need up-to-date public health laboratories capable of accurately diagnosing disease. This includes the capability of doing genetic subtyping, which requires sophisticated training of laboratory personnel and is relatively costly. There also needs to be better sharing of surveillance information; this is a foundation stone for coordinating the prevention of epidemics across regions at-risk, and it helps link local government agencies to international NGOs. After being free of the disease since 1981, even Cuba, with its strong, centralized government health system, has experienced a re-emergence of dengue in the past 10 years. So while national attention to public health is essential, top-down planning and funding has limitations. Community partnerships with government agencies are not only a key part of the solution to overcoming short resources, they are also essential to sustaining disease control activities over the long term.

## References


Anderson, R.M., May, R.M. and Anderson, B. (1992). *Infectious Diseases of Humans: Dynamics and Control*. New York: Oxford University Press.

Bennett, S.N., Holmes, E.C., Chirivella, M. *et al.* (2003). Selection-driven evolution of emergent dengue virus. *Molecular Biology and Evolution* **20**(10), 1650–1658.

Black, F.L. (1966). Measles endemicity in insular populations: critical community size and its evolutionary implication. *Journal of Theoretical Biology* **11**, 207–211.

Carter, H.R. and Frost, W.H. (1931). Yellow fever. In: L.A. Carter and W.H. Frost (eds), *An Epidemiological and Historical Study of its Place of Origin*. Baltimore: Waverly Press.

Cummings, D.A.T., Irizarr, R.A., Huang, N.E. *et al.* (2004). Travelling waves in the occurrence of dengue hemorrhagic fever in Thailand. *Nature* **427**, 344–347.

Fauci, A.S., Touchette, N.A. and Folkers, G.K. (2005). Emerging infectious diseases: a 10-year perspective from the National Institute of Allergy and Infectious Diseases. *Emerging Infectious Diseases* **11**(4), 519–525.

Gubler, D.J. (1989). *Aedes aegypti* and *Aedes aegypti*-borne disease control in the 1990s: top down or bottom up? *American Journal of Tropical Medicine and Hygiene* **40**, 571–578.

Gubler, D.J. (1997). Dengue and dengue hemorrhagic fever: its history and resurgence as a global public health problem. In: D.J. Gubler and G. Kuno (eds), *Dengue and Dengue Hemorrhagic Fever*. London: CAB International, pp. 1–22.

Gubler, D.J. (1998a). Dengue and dengue hemorrhagic fever. *Clinical Microbiology Reviews* **11**(3), 480–496.

Gubler, D.J. (1998b). Resurgent vector-borne diseases as a global health problem. *Emerging Infectious Diseases* **4**, 442–450.

Gubler, D.J. (2002). The global emergence/resurgence of arboviral diseases as public health problems. *Archives of Medical Research* **33**, 330–342.

Gubler, D.J. (2004a). Cities spawn epidemic dengue viruses. *Nature Medicine* **10**(2), 129–130.
Gubler, D.J. (2004b). The changing epidemiology of yellow fever and dengue, 1900 to 2003: full circle? *Comparative Immunology Microbiology and Infectious Diseases* **27**(5), 319–330.
Gubler, D.J. 2005. The emergence of epidemic dengue fever and dengue hemorrhagic fever in the Americas: a case of failed public health policy. *Pan-American Journal of Public Health* **17**(4), 221–224.
Gubler, D.J. and Meltzer, M. (1999). The impact of dengue/dengue hemorrhagic fever on the developing world. *Advances in Virus Research* **53**, 35–70.
Gubler, D.J. and Trent, D.W. (1994). Emergence of epidemic dengue/dengue hemorrhagic fever as a public health problem in the Americas. *Infectious Agents and Disease* **2**, 383–393.
Gubler, D.J., Sather, G.E., Kuno, G. and Cabral, A.J.R. (1986). Dengue 3 transmission in Africa. *American Journal of Tropical Medicine and Hygiene* **35**, 1280–1284.
Hardoy, J.E., Mitlin, D. and Satterthwaite, D. (2001). *Environmental Problems in an Urbanizing World: Finding Solutions for Cities in Africa, Asia, and Latin America*. London: Earthscan Publications Ltd.
Harrington, L.C., Scott, T.W., Lerdthusness, K. *et al.* (2005). Dispersal of the dengue vector *Aedes aegypti* within and between rural communities. *American Journal of Tropical Medicine and Hygiene* **72**(2), 209–220.
IDRC (2006). *Eco-biosocial Research on Dengue in Asia: Understanding Ecosystem Dynamics for Better-Informed Dengue Prevention*. Available at: the International Development Research Center website, http://www.idrc.ca/en/ev-88137-201-1-DO_TOPIC.html (accessed 21 December 2006).
Lanciotti, R.S., Lewis, J.G., Gubler, D.J. and Trent, D.W. (1994). Molecular evolution and epidemiology of dengue-3 viruses. *Journal of General Virology* **75**, 65–75.
McNeil, W.H. (1976). *Plagues and Peoples*. New York: Doubleday.
Messer, W.B., Gubler, D.J., Harris, E. *et al.* (2003). Emergence and global spread of a dengue serotype 3, subtype III virus. *Emerging Infectious Diseases* **9**(7), 800–809.
NVBDCP (2006) *Chikungunya Fever Situation in the Country during 2006 (as on 03.01.2007)*. Available at: the National Vector-Borne Disease Control Program, Government of India website, http://www.namp.gov.in/Chikun-cases.html (accessed 5 January 2007).
Nisalak, A., Endy, T.P., Nimmanitya, S. *et al.* (2003). Serotype-specific dengue virus circulation and dengue disease in Bangkok, Thailand, from 1973 to 1999. *American Journal of Tropical Medicine and Hygiene* **68**, 191–202.
Patlak, M. (1996). Book reopened on infectious diseases {electronic version}. *FDA Consumer Magazine* **30**(3).
Smolinski, M.S., Hamburg, M.A. and Lederberg, J. (eds). (2003). *Microbial Threats to Health: Emergence, Detection, and Response*. Washington, DC: Institute of Medicine, National Academies Press.
United Nations (2002). *World Urbanization Prospects: The 2001 Revision*. New York: United Nations.
United Nations, Department of Economic and Social Affairs (2006). *World Urbanization Prospects: The 2005 Revision Population Database*. Available at: the United Nations, Department of Economic and Social Affairs website, http://esa.un.org/unup/ (accessed 21 December 2005).
United Nations Population Fund (2001). *The State of World Population 2001*. Available at: the United Nations Population Fund website, http://www.unfpa.org/swp/2001/english/tables.html (accessed 21 December 2006).

Wearing, H.J. and Rohani, P. (2006). Ecological and immunological determinants of dengue epidemics. *Proceedings of the National Academy of Sciences* **103**(31), 11,802–11,807.
Wilcox, B.A. and Colwell, R.R. (2005). Emerging and re-emerging infectious diseases: biocomplexity as an interdisciplinary paradigm. *EcoHealth* **2**(4), 244–257.
Wilcox, B.A. and Gubler, D.J. (2005). Disease ecology and the global emergence of zoonotic pathogens. *Environmental Health & Preventive Medicine* **10**(5), 263–272.
Woolhouse, M.E.J. and Gowtage-Sequeria, S. (2005). Host range and emerging and reemerging pathogens. *Emerging Infectious Diseases* **11**(12), 1842–1847.
World Bank (2002). *World Development Report 2002: Building Institutions for Markets.* New York: Oxford University Press for the World Bank.
World Health Organization (2004). *The World Health Report 2004 – Changing History.* Geneva: World Health Organization.
World Health Organization (2006). *Epidemiological Trends and Current Situation of Yellow Fever.* Available at: the World Health Organization website, http://www.who.int/csr/disease/yellowfev/surveillance/en/index.html (accessed 29 October 2006).
World Health Organization, Regional Office for Southeast Asia, New Delhi (1999). *Prevention and Control of Dengue and Dengue Hemorrhagic Fever: A Comprehensive Guideline.* Geneva: World Health Organization, pp. 1–137.

# Suburbanization in developed nations

# 5

**Richard C. Falco, Gary P. Wormser and Thomas J. Daniels**

The suburbs. Mention the word, and many people think of nice houses along tree-lined streets, nestled into pristine wooded areas that serve as the perfect backdrop to an almost idyllic existence. The suburbs are often seen as a retreat, far enough away from the hustle and bustle of the city, yet usually a short commute to urban job centers.

Since World War II, the massive expansion of suburbs has been the way that metropolitan areas grow. As suburbs were created, lawns and parks replaced forests and farms, and entire neighborhoods were built into woodlots. However, the exodus from the nation's older cities to previously unspoiled lands has come at a cost. The suburbs, with their typically wooded properties and increasing human population, have become an important environment for interaction between people and arthropod vectors, particularly ticks. In the United States, the emergence of Lyme disease provides the archetypal example of just such a process.

Lyme disease was unheard of in the United States before 1977, and likely would have gone unrecognized for many more years if not for the efforts of two mothers living in suburban towns in Connecticut. They independently reported what appeared to be outbreaks of juvenile rheumatoid arthritis, sometimes associated with a rash, to the Connecticut State Health Department in 1975, which eventually led to the "discovery" of Lyme disease as a distinct clinical entity (Steere *et al.*, 1977a, 1977b; Aronowitz, 1989). While it is probable that the agent of Lyme disease was present in the northeastern United States hundreds of years ago, it likely went unnoticed due to the often more serious, and sometimes fatal, effects of other diseases afflicting North Americans throughout the nineteenth century (Barbour and Fish, 1993). Presently, Lyme disease is the most common tick-borne disease in the United States and Europe (O'Connell *et al.*, 1998; Orloski *et al.*, 2000; CDC, 2004, 2005). In the United States, over the decade 1994 to 2004, Lyme disease case numbers increased, with an average of almost 17,000 cases reported annually (CDC, 2006a; see also Figure 5.1).

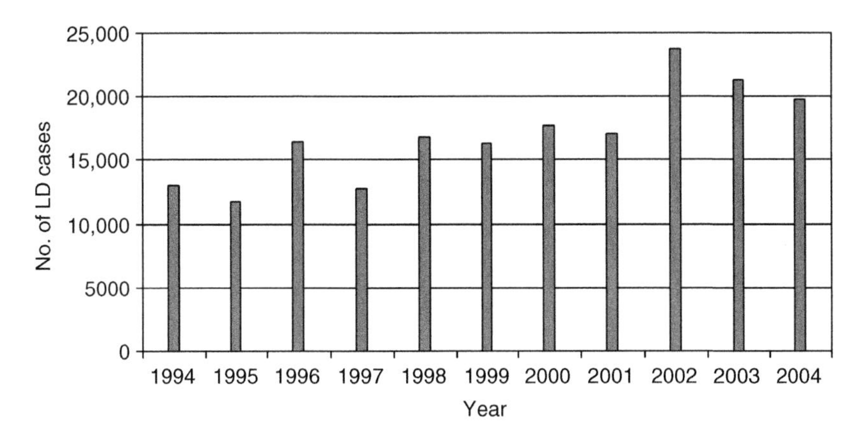


**Figure 5.1** Bar chart of US Lyme disease cases. Data from CDC (2006a).

The histories of Lyme disease and suburbanization are intertwined. The spread of suburbia is a relatively new phenomenon in human history, and it has had a tremendous impact on how we interact with nature. It is hoped that the lessons learned from Lyme disease will be applicable to other infectious diseases where the local landscape is a critical factor in the transmission of the pathogen.

## Lyme disease – overview

Lyme disease is the most prevalent tick-borne disease in the world (CDC, 2006b). In the United States, the *Ixodes scapularis* tick is the most important vector for Lyme disease. Lyme disease is caused exclusively by the spirochete *Borrelia burgdorferi* in North America, whereas in Europe this infection is caused by *B. burgdorferi*, *B. afzelii*, or *B. garinii*, and, rarely, other borrelial species (Nadelman and Wormser, 1998; Wang *et al.*, 1999; Steere, 2001; Stanek and Strle, 2003).

### Clinical picture

The clinical picture of Lyme disease can be quite complex, consisting of the early skin manifestation known as erythema migrans (EM), as well as potentially more serious sequelae that result from hematogenous dissemination of the spirochete (Wormser, 2006). The more serious clinical sequelae involve the joints, nervous system, and/or heart (Nadelman and Wormser, 1998; Steere, 2001; Stanek and Strle, 2003). Knowledge of the broad clinical spectrum of Lyme disease is an essential component for proper diagnosis and, consequently, effective treatment.

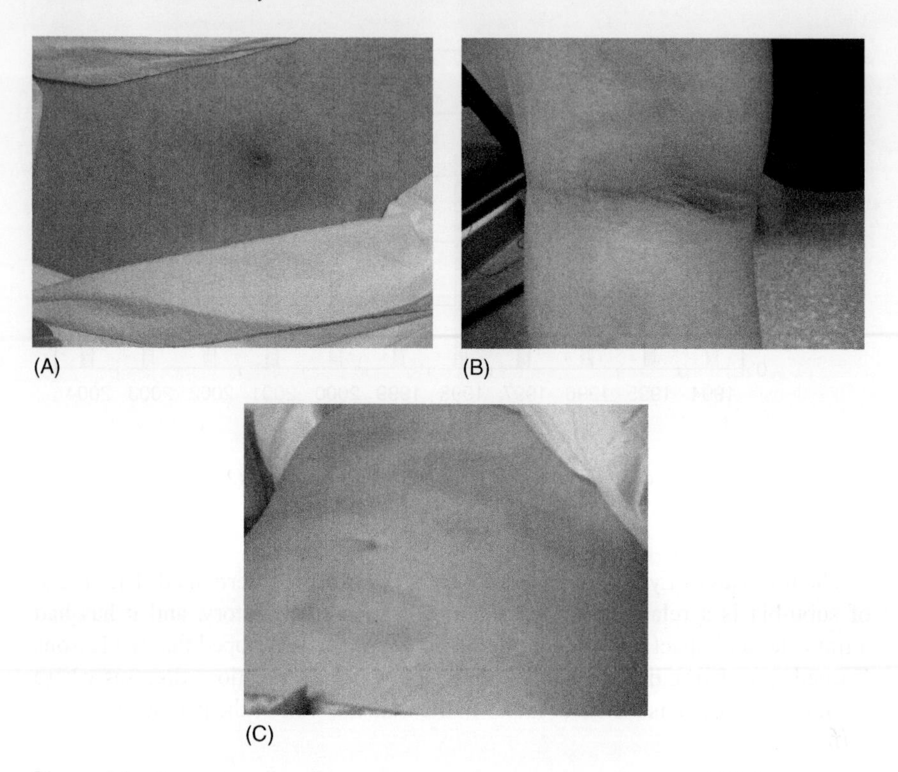


**Figure 5.2** Examples of erythema migrans. Reprinted from Wormser (2006), with permission from the Massachusetts Medical Society. (A) Patient from New York State with culture-confirmed Lyme disease. The patient had a single erythema migrans lesion of $8.5 \times 5.0$ cm on the abdomen. The lesion is homogeneous in color except for a prominent central punctum (presumed site of preceding tick bite). (B) Patient from New York State with culture-confirmed Lyme disease. The patient had a single erythema migrans lesion of $11.5 \times 7.5$ cm in the popliteal fossa of the left leg. More intense erythema is found to the right of the center of the lesion. (C) A patient from New York State with culture-confirmed Lyme disease. The patient had over 40 erythema migrans lesions. Note the prominent central clearing of the lesions present on the abdomen.

EM typically develops 7–14 days (range 3–30 days) after tick detachment, and is characterized by a rapidly expanding, erythematous skin lesion (Steere *et al.*, 1983; Berger, 1989; Nadelman and Wormser, 1995; Nadelman *et al.*, 1996, Wormser, 2006; see also Figure 5.2). Over one-half of United States patients with erythema migrans have concomitant systemic symptoms such as fatigue, arthralgias, headache, or neck pain (Steere *et al.*, 1983; Nadelman and Wormser, 1995, 1998; Nadelman *et al.*, 1996; Wormser, 2006).

About 20–25 percent of United States patients with erythema migrans in recent studies have had more than a single skin lesion (Nadelman *et al.*, 1996;

Wormser *et al.*, 2005; Wormser, 2006). The secondary lesions are believed to arise by hematogenous dissemination from the site of primary infection at the tick bite site (Wormser *et al.*, 2005).

## Tick ecology and Lyme disease risk

Lyme disease is closely associated with local ecology (Fish, 1993; Gern and Falco, 2000). Therefore, understanding the dynamic nature of Lyme disease risk requires an understanding of the ecology of vector ticks, including their seasonal and annual activity patterns, as well as the ecology of *B. burgdorferi* with respect to ticks and reservoir hosts.

There are four tick species that are primary vectors for Lyme disease: *I. scapularis* (= *I. dammini*) (Oliver *et al.*, 1993) in the eastern and midwestern US and Canada, *I. pacificus* in the western US, *I. ricinus* in Europe, and *I. persulcatus* in Eurasia (Gern and Falco, 2000). Due to the high prevalence of Lyme disease in the eastern US and the fact that suburbanization in the eastern US as it relates to Lyme disease risk is the primary focus of this chapter, we will use the life history of *I. scapularis*, commonly called the "black-legged tick" or "deer tick," as a model for our discussion of tick ecology.

### *Lifecycle and seasonal activity patterns of* I. scapularis

Ticks are obligate parasites. All active stages of *I. scapularis* must take a blood meal in order to survive, with each stage feeding on a host animal. These ticks begin their life as eggs, which hatch into sexually immature, six-legged larvae that are about the size of a grain of sand, approximately 0.5 mm in length (Fish, 1993; see also Figure 5.3). Larval *I. scapularis* ticks prefer to feed on small animals like mice or birds. They feed for 3–4 days, taking in blood until they become fully engorged, then drop off the host, usually into the leaf litter. In the northeastern US, larval host-seeking activity peaks during the months of August and September (Fish, 1993; see also Figure 5.4).

Engorged larvae molt into sexually immature, eight-legged nymphs that are about the size of a poppy seed, approximately 1 mm in length (Fish, 1993). The nymphs take a blood meal, usually preferring to feed on small or medium-sized mammals such as mice, chipmunks, and raccoons. Nymphs usually remain attached to their host for 4–5 days while feeding to repletion. Nymphs are most active during the late spring and early summer in the northeastern US, although some nymphs have been collected in the field as late as October (Fish, 1993).

The engorged nymph will drop off the host and eventually molt into a sexually mature, eight-legged adult in the late summer or early fall. The adult female ticks, approximately 2.5 mm in length (Fish, 1993), feed on large mammals, usually white-tailed deer (*Odocoileus virginianus*) (see pp. 148 for further details on the

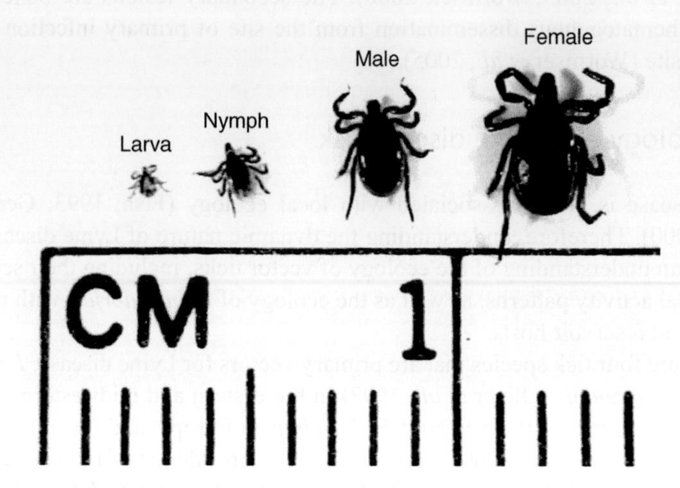


**Figure 5.3** Active stages of *I. scapularis*. Photograph courtesy of J. Vellozzi.

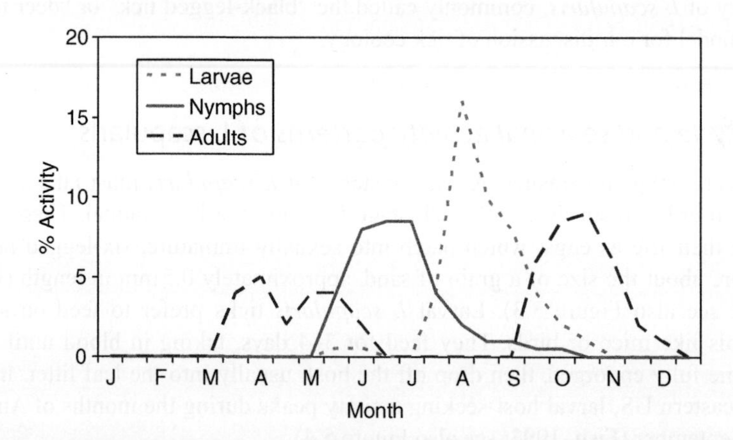


**Figure 5.4** Seasonal activity *of I. scapularis* in Westchester County, NY, based on sampling conducted from 1990 to 2005.

role of deer in Lyme disease ecology). The smaller adult males are also found on deer, but they feed only intermittently, while females remain attached and will feed for about a week. Mating typically takes place on the host animal, where tick densities are higher than in the environment. Adult ticks in the northeastern and midwestern US have two activity peaks during the year, with a primary peak in October and November and a smaller peak in early spring (Fish, 1993; Figure 5.4). Mark–release–recapture studies demonstrate that these two activity peaks represent the same cohort of adult ticks (Daniels *et al.*, 1989). Those adults that do not find a host in the fall will become active again in the spring, although host-seeking during the winter months can occur on warmer days if a threshold temperature is reached; this has been reported to vary between −0.6ºC (Schulze *et al.*, 2001) and 4.0ºC (Duffy and Campbell, 1994).

Engorged adult female ticks will drop off the host and lay approximately 2500 eggs per tick in the spring; these will hatch into larvae which are most active in late summer, and the cycle will begin again (Daniels *et al.*, 1996). The whole process takes two years in the northeastern US, although the duration of the life-cycle may vary in other parts of the geographic range for *I. scapularis* (Clark *et al.*, 1998; Lindsay *et al.*, 1998).

## B. burgdorferi *in host animals and ticks*

*I. scapularis* has a wide host range, with immature stages feeding on over 30 species of mammals, 49 species of birds, and several species of lizards (Magnarelli *et al*, 1986; Anderson, 1988). A few relatively common species, such as the white-footed mouse (*Peromyscus leucopus*), may feed the majority of individual ticks (Wilson, 1998). This generalist granivore (Nupp and Swihart, 1996) is common in woodlands of the eastern US. Its significance in the transmission cycle of tick-borne pathogens lies in its role as a reservoir host that is capable of becoming infected with *B. burgdorferi*, *Anaplasma phagocytophilum*, and *Babesia microti* from ticks that feed on it, maintaining the infections for at least several weeks, and transmitting the pathogens to new ticks it picks up in the environment (Piesman and Spielman, 1979). Although white-footed mice are considered the principal reservoir for *B. burgdorferi* in nature (Levine *et al.*, 1985; Donahue *et al.*, 1987), other animals can serve to infect ticks with this pathogen and may be important reservoirs in certain regions. These include medium-sized mammals such as raccoons (*Procyon lotor*) and skunks (*Mephitis mephitis*) (Fish and Daniels, 1990), as well as smaller mammals such as chipmunks (Mather *et al.*, 1989), voles, and gray squirrels (Hanincova *et al.*, 2006), which play an important role in Lyme disease ecology in residential areas. Birds are important as carriers of ticks and some, such as American robins (*Turdus migratorius*), serve as reservoirs of *B. burgdorferi* (Anderson *et al.*, 1986; Battaly *et al.*, 1987; Richter *et al.*, 2000). Birds are implicated in the emergence of Lyme disease in previously unaffected areas (Richter *et al.*, 2000).

Interestingly, although white-tailed deer can serve as hosts for both immature and adult *I. scapularis*, they do not appear to be competent reservoirs for *B. burgdorferi*. In one study, only 1 percent of nymphal ticks derived from larvae collected from deer were infected – significantly less than the 23 percent infection rate of field-collected nymphal ticks in that geographic area (Telford *et al.*, 1988).

The *B. burgdorferi* infection rate of ticks varies considerably among stages. Transovarial transmission of *B. burgdorferi*, defined as infection of the maternal oocytes by pathogens, with passage to the progeny (Harwood and James, 1979), is not very efficient in either *I. scapularis* or the closely-related *I. pacificus* tick species (Magnarelli *et al.*, 1987; Schoeler and Lane, 1993). Therefore, ticks in the larval stage typically have very low infection rates (<1 percent) and are not considered epidemiologically important vectors for *B. burgdorferi*.

Nymphal *I. scapularis*, on the other hand, have infection rates that usually average 20 to 25 percent in endemic areas, although there can be tremendous site-to-site variation in these rates (Maupin *et al.*, 1991; Schwartz *et al.*, 1997; Daniels *et al.*, 1998). Adult *I. scapularis* have had an additional blood meal and are therefore afforded an increased opportunity to become infected with *B. burgdorferi*. Infection rates in adult ticks typically average 45 to 50 percent in endemic areas (Maupin *et al.*, 1991; Schwartz *et al.*, 1997).

The two-year lifecycle of *I. scapularis* in the northeastern US is an important factor in maintaining *B. burgdorferi* in nature. Host-seeking larvae are uninfected with *B. burgdorferi*, therefore they must feed on infected reservoir host animals to become infected and pass the spirochete transstadially to the nymphal stage. This is possible because the larvae of one population cohort feed after the nymphs of the previous cohort in any one year (Figure 5.4). Since approximately 25 percent of the nymphs in that previous cohort (with peak activity in June and July) are infected, they serve to infect hosts, such as white-footed mice, which then serve to infect the larvae that subsequently feed on them in August and September, and the spirochete is maintained through the enzootic cycle. This two-year lifecycle, and the resulting seasonal activity patterns of immature *I. scapularis*, are the keys to both the efficient transmission of *B. burgdorferi* to the next generation of ticks (Spielman *et al.*, 1985; Wilson and Spielman, 1985) and the high risk of human exposure to spirochete-infected nymphs (Fish, 1993).

It should be noted, however, that since the two-year lifecycle may not occur in all parts of the range of *I. scapularis*, zoonotic transmission of *B. burgdorferi* may not always occur in this way. For example, in the southern US, larval activity may precede and/or occur along with that of nymphs, and thus amplification of the spirochete in *I. scapularis* is attenuated (Clark *et al.*, 1998).

## Lyme disease risk and tick activity

Human risk for Lyme disease is a function of two important environmental factors: seasonal weather and light conditions, which dictate the time of year each stage in the lifecycle is most active; and the abundance of ticks in a given location. Although there are three active stages in the deer tick lifecycle, it is the nymphal tick that has the greatest impact on public health in terms of risk of Lyme disease. This is demonstrated by comparing the seasonal activity of nymphal and adult *I. scapularis* with the temporal distribution of Lyme disease cases. For example, in a study in Westchester County, New York, an endemic area for Lyme disease, there was a significant correlation between seasonal activity of nymphal *I. scapularis* and the onset of EM, the hallmark of early Lyme disease (Falco *et al.*, 1999; see also Figure 5.5). In this study, 74.2 percent of all EM's

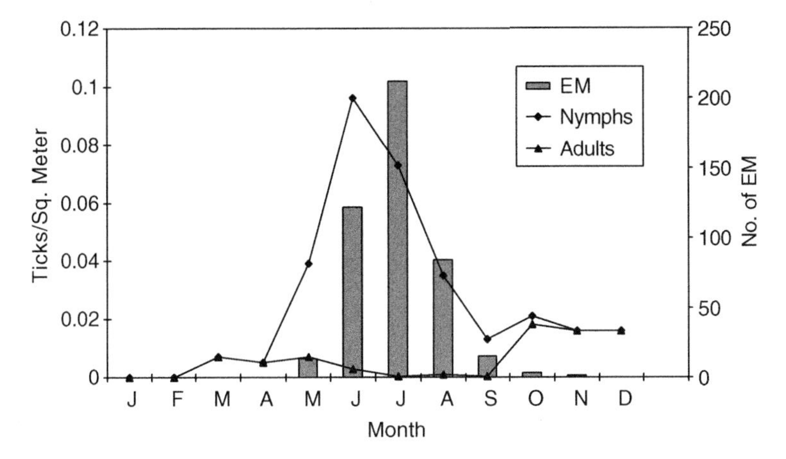


**Figure 5.5** Monthly abundance of nymphal and adult *I. scapularis* measured in central Westchester County, NY, and the monthly distribution of EM diagnosed at WCMC, 1991–1996. Taken from Falco *et al.* (1999).

over a six-year period were diagnosed during the months of June and July, when nymphal ticks are most active. In contrast, EMs were rarely diagnosed during times of the year when just adult ticks are active, and there was no significant correlation between the incidence of EM and adult tick abundance.

The importance of nymphal *I. scapularis* in the epidemiology of Lyme disease is further supported by studying temporal changes in tick numbers and Lyme disease case reports. In Westchester County, New York, annual fluctuations in nymphal tick abundance, as measured by mark–release–recapture studies of ticks, were shown to be directly correlated with the numbers of patients with EM diagnosed at a local Lyme disease clinic (Falco *et al.*, 1999; Figure 5.6). Similar results were obtained in Connecticut comparing field-collected tick numbers and Lyme disease cases reported to the state health department (Stafford *et al.*, 1998). These studies demonstrate that nymphal *I. scapularis* serve to drive the seasonal and annual patterns of Lyme disease incidence.

Why are nymphal *I. scapularis* more important vectors of *B. burgdorferi* than adult ticks, when the nymphal infection rate is roughly one-half of the adult tick infection rate? This is likely due to three factors:

1. The seasonal activity pattern of the nymphal stage, which peaks during the spring and summer months (Figure 5.4), coincides with the time of year in which outdoor human activity is typically high in the northeastern and midwestern US. This affords greater opportunity for contact between people and nymphal ticks. Additionally, *I. scapularis* nymphs are active during the warmer months, when people often dress more casually with clothing that is

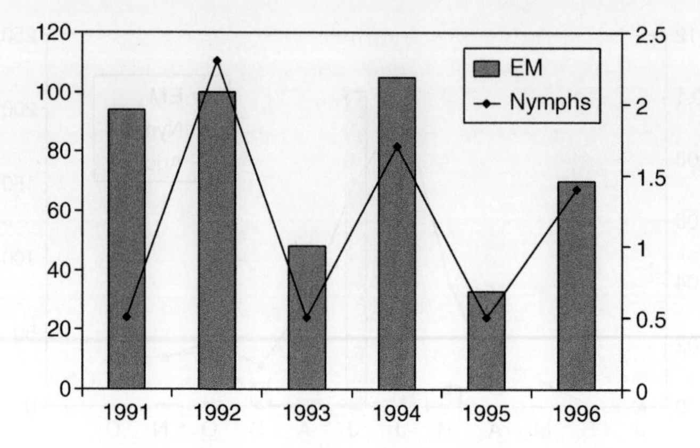


**Figure 5.6** Annual nymphal *I. scapularis* density measured in central Westchester County, NY, and annual numbers of EM cases diagnosed at WCMC, 1991–1996. Taken from Falco *et al.* (1999).

less likely to act as a barrier to host-seeking ticks than clothing worn during colder seasons.

2. Due to interstadial (stage-to-stage) mortality, nymphal abundance is much higher than that of adults, with one study in Westchester County, New York, showing the number of host-seeking nymphs to be almost 300 percent higher than the number of adult ticks (Daniels *et al.*, 2000). Thus, there are far more nymphs available to parasitize humans and transmit *B. burgdorferi* than adults.
3. The tiny size of the nymphal tick (Figure 5.3) makes it more difficult to detect and remove them from the body than the larger adult tick (Yeh *et al.*, 1995; Falco *et al.*, 1996). This size disparity becomes even more important when considering the fact that transmission of *B. burgdorferi* usually occurs a minimum of 48 hours after the tick attaches to the human host (Piesman *et al.*, 1987, 1991; Piesman, 1989). The larger adult tick, therefore, is easier to find and remove before transmission occurs, while the smaller nymph has a greater chance of staying undetected and remaining attached for more than the critical 48-hour period (Yeh *et al.*, 1995; Falco *et al.*, 1996).

## Social determinants of Lyme disease risk

### Historical distribution of Lyme disease vectors in the US

Since the late 1980s, three major foci of Lyme disease in the US have been identified and account for the vast majority of cases reported each year (Barbour and Fish, 1993). The most significant in terms of case numbers is in the northeastern and mid-Atlantic regions, followed by the upper midwestern region of

Wisconsin and Minnesota. Disease transmission in both areas is due to the presence of *I. scapularis*. A third focus, encompassing northern California, Oregon, and Washington, is characterized by a transmission cycle involving a different vector, *I. pacificus*, the western black-legged tick, which has a relatively low level of infection with *B. burgdorferi*. *I. pacificus* feeds frequently on the western fence lizard (*Sceloperus occidentalis*), which is an incompetent reservoir for the spirochete (Lane *et al.*, 1991), so transmission of the agent to ticks is infrequent. Furthermore, there is little overlap of cohorts (Padgett and Lane, 2001) during the tick's lifecycle – a trait that may impede spirochete transmission from one generation to the next. Since the number of reported cases of Lyme disease in this focus tends to be only about 1 percent of the total cases nationally (CDC, 2004), the following discussion will not address *I. pacificus*.

In reviewing data on northern populations of *I. scapularis*, Spielman *et al.* (1993) noted that relict populations of this tick originally could be found along the terminal moraines that formed as glaciers pushed their way south during the last ice age, then receded about 15,000 years ago (Davis, 1983). The remaining tick populations were isolated and limited to coastal portions of the northeastern US and to the northern Great Plains. In Europe and Asia, terminal moraines also marked areas primarily affected by Lyme disease on those continents (Spielman *et al.*, 1993; Steere, 1994), though the vector tick species are different in these locations.

It is unclear whether *I. scapularis* populations might once have been more widely distributed in the United States than at present, possibly expanding and declining with white-tailed deer numbers (Spielman *et al.*, 1985), but as recently as the early twentieth century populations of this tick were small. Studies at the time revealed the presence of *I. scapularis* and *I. muris* in just a few sites in the northeastern US (Bishopp and Smith, 1937; Cobb, 1942; Smith and Cole, 1943). Later work by Hyland and Mathewson (1961) and Good (1972, 1973) indicated that tick numbers were relatively low and probably had remained localized. For instance, Larrouse *et al.* (1928) found *I. scapularis* on Naushon Island in 1926 when they attempted to control American dog ticks (*Dermacentor variabilis*) there using a parasitic wasp, *Ixodiphagus* (*Hunterellus*) *hookeri*, though *I. scapularis* was apparently absent, or at least very rare, on the neighboring islands of Martha's Vineyard and Nantucket (Hertig and Smiley, 1937; Spielman *et al.*, 1979). Several museum specimens of *I. scapularis* collected from New Haven, Connecticut, and from various sites on Long Island, New York, in the 1930s and 1940s also suggest that the species may have been fairly widespread geographically, but with small, isolated populations (Anastos, 1947; Collins *et al.*, 1949; Ginsberg, 1993).

Although cases of Lyme disease likely occurred before the outbreak that began in the 1970s, they were relatively few in number and would have represented localized phenomena, given the highly focal distribution of *I. scapularis*. Persing *et al.* (1990) documented the presence of *B. burgdorferi* DNA in museum specimens of *I. scapularis* collected on Montauk Point, New York, in the 1940s, and cases of "Montauk knee" and "Montauk spider bite" were probably due to tick bites (Ginsberg, 1993). Marshall *et al.* (1994) found evidence of *B. burgdorferi* DNA in mouse specimens collected in Massachusetts during the 1890s, suggesting the presence of an enzootic cycle that would have resulted in infected *I. scapularis* ticks. Ecological conditions, however, were not right for a widespread increase in Lyme disease cases until the 1970s, though the stage was being set for just such an event centuries earlier as humans opened up the wilderness that marked the landscape of early America. Because the emergence of Lyme disease in the northeastern and midwestern US is so closely associated with the presence of white-tailed deer (Wilson and Childs, 1997; Piesman, 2002), the role of deer in Lyme disease ecology should be reviewed.

## Agriculture, land-use patterns, and deer

The remarkable landscape changes that occurred in the northeastern US over the last four centuries involving deforestation, intensive agriculture, farm abandonment, reforestation, and human population increase have had tremendous impacts on wildlife assemblages (Foster *et al.*, 2002). With the decline in agriculture that typified the eastern US in the mid-1800s, for instance, the forest expanded, providing habitat for numerous wildlife species that had previously been restricted. Most obvious was the change in population-size and distribution of white-tailed deer in the eastern US.

White-tailed deer in the eastern half of the US underwent four distinct population phases from 1500 to 1900 (McCabe and McCabe, 1997) that coincided with the landscape changes made by a growing and mobile human population in the US. These phases included a massive harvest of deer from 1500 to 1800, a regrowth of the white-tailed deer herds as harvest limits were imposed from 1800 to 1865, the "exploitation era" from 1850 to 1900 when deer were under extreme hunting pressure, and a period of regrowth of the deer population that began in 1900 and continues to this day (McCabe and McCabe, 1997). The emergence of Lyme disease is associated with this last phase.

However, Ginsberg (1993) cautioned against assuming that the relationship between deer and tick abundance is entirely straightforward. The spread of white-tailed deer has occurred concurrently with changes in a whole range of environmental conditions, some of which may have been responsible for increases in both tick and deer populations in recent decades (as opposed to deer population increases directly causing tick population increases). Furthermore, on a broad scale the distribution of *I. scapularis* is poorly correlated with that of deer. Deer are common in the Adirondack Mountains (Severinghaus and Brown, 1956) and in northwestern New Jersey (Schulze *et al.*, 1984), for instance, where *I. scapularis* is uncommon. While ticks may eventually be abundant in those areas, other environmental variables may be important in dictating suitable habitat (Telford, 2002). Therefore, it is important to distinguish the role that deer have as vehicles for the geographic spread of ticks from their role in regulating populations of ticks in endemic areas (Ginsberg, 1993).

The key to understanding how Lyme disease became the most important vector-borne illness in North America lies in the realization that human-mediated changes in land-use patterns impact how wildlife reservoirs and disease vectors interact with humans sharing that land. Lyme disease occurs wherever the vector is abundant.

## Human influence on deer and tick populations

### *Suburbanization and landscape changes*

The growth of *I. scapularis* tick populations and the deer herd on which they depend was a consequence of landscape changes that began in the nineteenth century (Cronon, 1983) and continue to this day. As abandoned farmland became available for forest regeneration in the 1800s, it also became available for homesteads, built increasingly distant from the urban centers that marked much of the nation's early settlement. Suburbanization may not be a uniquely American trend, but it is one that has been wholeheartedly embraced in the US. Indeed, it could be argued that standards for the impact of suburbanization on society and the land, both good and bad, have been set in the US. For instance, nowhere in Europe do urban areas sprawl as much as in the United States. Less than a quarter of the US population lived in suburbia in 1950, but today, well over one-half does (Nivola, 1999).

Although suburbanization is largely regarded as a post-World War II phenomenon, it actually began much earlier. On the North American continent, Boston, Philadelphia, and New York City established suburbs well before the Revolutionary War, and suburbanization accelerated in the mid-1800s; in Europe, the pace of suburbanization in London was nearly as rapid (Jackson, 1985). The popularity of suburbs in the United States grew in the twentieth century, beginning in the 1920s, and was most rapid from then through the 1950s (Staley, 1999). By the 1950s, however, the suburbs were growing at a median level of 30 percent faster than the cities they surrounded. The underlying cause was the introduction and popularity of the automobile, facilitated by huge federal outlays for highway construction. Before then, people had to live within walking distance of work or near a rail or streetcar line. They needed to live close to local schools, merchants, doctors, family, etc., and densely settled cities solved these problems (Rappaport, 2005). Other important factors contributing to suburbanization included rapid population growth, large increases in wealth as the US economy did not falter after World War II (unlike the situation in Europe and Asia), high urban crime rates, and policies of the federal government that subsidized home ownership (Nivola, 1999; Rappaport, 2005), all of which made emigration from the cities not just feasible but also attractive.

The 1970 census showed that more people moved to the suburbs than to central cities or rural districts that year – a trend that continued as the suburbs grew and the populations of cities declined (Baumgartner, 1988). By 1990, over 60 percent of the population of 320 metropolitan areas lived in the suburbs (Rusk, 1995). Although the pace of suburbanization since the 1990s slowed to just above 10 percentage points per decade (Rappaport, 2005), the net effect was that households increasingly chose to live in low density, vehicle-dependent suburbs (Kahn, 2000). In 1950, the share of metropolitan area residents who lived in central cities was 57 percent; by 1990, that had fallen to 37 percent (Mieszkowski and Mills, 1993; Langdon 1994).

In addition to the sheer number of people moving out of the cities, the pattern of development in which single homes are placed on quarter-acre lots carved out of the forest leads inevitably to habitat fragmentation. However, our tendency to build out of proportion to our population growth leads to sprawl, so that much more land is put under development than is needed to address issues of human density. Chicago has grown just 4 percent in population over the past 20 years, yet there has been a 46 percent increase in land consumption. The city of Los Angeles is now larger than two states combined: Delaware and Rhode Island. The increase in sprawl is reflected in the number of vehicle miles traveled (VMT) each year. In the US, VMT has increased 65 percent over the past 20 years while population growth has grown just 21 percent (Environmental & Energy Study Institute, 2000). The current trend toward hyperfragmentation (why limit ourselves to disrupting just one acre when five are available?) is conducive to the spread and maintenance of this tick-borne disease cycle.

## *Lyme disease and the suburbs*

Suburbanization has greatly altered the epidemiological landscape. The steady rise in human Lyme disease incidence in the northeast over the last three decades is related to landscape modification through suburbanization (Maupin *et al.*, 1991; Barbour and Fish, 1993; Frank *et al.*, 1998). Eastern deciduous forest communities were fragmented by suburban development, resulting in a habitat matrix that is ideal for deer and some small rodents, especially the white-footed mouse (Daily and Ehrlich, 1995). Although the net amount of forest has not changed significantly in the past two decades (Brownstein *et al.*, 2003), continued building has created a higher number of forest patches interspersed with residential development. The proximity of humans, wildlife, and ticks in a habitat that is suitable for all three virtually ensures exposure of people to tick bites and the agents of Lyme disease, human granulocytic anaplasmosis, and babesiosis (Falco and Fish, 1988a; Ehrlich and Ehrlich, 2002). Furthermore, deer are often protected from hunting – a situation that not only permits the maintenance of a large tick population, but has also led to increased tension among residents who do and do not advocate controlling the number of deer in areas where Lyme disease is endemic (see below). While suburban growth has fostered *I. scapularis* population expansion, the spread is not uniform and the resultant tick distribution is patchy, with varying densities in adjacent areas. In general, all landscapes can be thought of as mosaics composed of discrete bounded patches that are differentiated by a number of biotic (e.g. vegetation features, host abundance) and abiotic (e.g. weather) factors (Pickett and Cadenasso, 1995). Patches can arise because of natural or human-caused disturbance, fragmentation of the land, regeneration of a habitat type, or persistent differences in environmental resources (Forman and Godron, 1986). In the case of Lyme disease, the deciduous forest matrix contains small habitat patches, some of which may be suitable for *I. scapularis* and some not. The relationship among habitat patches across a range of spatial scales provides important information on factors influencing the distribution of *I. scapularis*.

Past studies have used a landscape approach to identify habitat features associated with the presence of *I. scapularis*. One, conducted in a residential area of Westchester County, New York, focused on a single neighborhood and found considerable variation among sampled properties. Although Maupin *et al.* (1991) noted a general pattern in which progressively fewer ticks were found on moving from the woodland to the wooded edge and onto residential lawns, it appears that the edge or ecotone habitat probably was underestimated. Edge effects may extend some distance into adjacent woodlots (Harris, 1984), though Maupin *et al.* (1991) limited consideration of edge to "non-ornamental, unmaintained edge which abutted woodlots…" (p. 1106), and this may have resulted in too few ticks being counted.

Dister *et al.* (1993) conducted a regional analysis of habitat features throughout Westchester County, New York, using LANDSAT Thematic Mapper imagery and a GIS (geographic information system) relating land-cover composition to canine seroprevalence for antibodies to *B. burgdorferi*. They noted that the percentage of land cover represented by deciduous forest was the variable most highly correlated with canine seroprevalence, a measure of Lyme disease risk (Daniels *et al.*, 1993; Falco *et al.*, 1993). However, Dister *et al.* (1993) concluded that further analysis of spatial context with finer-scaled databases was needed to understand the influence that surrounding land cover classes may have on Lyme disease risk.

This approach offers a systematic way of discerning which ecological features of the landscape influence tick abundance, and how temporal changes in those features play a role at the population level. Because the suburban landscape is a disjunct pattern of lawns, homes, and woodland defined by human activity, anthropogenic effects on habitat structure are important to consider. Among the anthropogenic effects that may have a strong influence on tick distribution is the creation of edge habitat or ecotone, which is characteristic of the fragmentation that accompanies the building of homes in previously uninhabited woodland (Temple and Wilcox, 1985; Verner, 1985). Such areas appear to be particularly important for adult *I. scapularis*. For example, Stafford and Magnarelli (1993) noted that adult ticks were recovered predominantly from lawn and ecotone areas, and Schmidtmann *et al.* (1994) found adults most commonly at the woodland-pasture interface (i.e., edge) on their study sites. Kramer *et al.* (1993) observed that drag-sampling along trails in parkland (essentially narrow corridors bordered by edge) yielded relatively high numbers of adults. Using county case reports, Glass *et al.* (1995) found that in Baltimore County, Maryland, the risk of human Lyme disease decreased with increasing distance from forest edge (Dennis *et al.*, 1998).

As noted earlier, the significance of edge habitats lies in their attractiveness to white-tailed deer (Harlow, 1984) and a variety of rodent species (see, for example, Diffendorfer *et al.*, 1995), all of which are fed upon routinely by *I. scapularis* (Fish and Dowler, 1989). When wildlife hosts utilize these habitats, engorged ticks may be dropped that will later molt to host-seeking ticks with the potential to transmit pathogens to their next host – possibly a human. This is especially relevant in the case of *P. leucopus*, which commonly inhabits ecotones. Differences in the abundance of white-footed mice from one site to another may be related to the distribution of forest fragments as a result of land development (Krohne and Hoch, 1999).

Although fragmentation of the landscape has been shown to increase both tick density and infection prevalence (the percentage of ticks infected with a specific pathogen, *B. burgdorferi* in this case) (Ostfeld and Keesing, 2000; Schmidt and Ostfeld, 2001), two components of the entomologic risk of Lyme disease, work by Brownstein *et al.* (2005) indicates that the incidence of human disease is actually lower in fragmented contexts. These investigators speculate that human contact with infected ticks may be lower with increased fragmentation because of a reduction in forest patch size and an increase in isolation. In that case, fewer properties will contain or adjoin woodlots that support tick populations, resulting in an overall reduction in peridomestic exposure (Brownstein *et al.*, 2005). It might be concluded that although development will increase the risk of Lyme disease to a point, further development resulting in increased isolation of a residential property (farther distance from a woodland patch) will then reduce risk. Since it is unlikely that most people residing in Lyme disease-endemic areas would support increased development as a means of evading infection, the way that landscape structure affects risk should be considered in future residential planning.

## Environmental and behavioral factors impacting risk

On the local level, risk for Lyme disease is not homogeneous. Even in areas considered endemic, nymphal tick abundance can vary significantly. For example, in New York State, which has reported more Lyme disease cases over the

12-year period from 1993 to 2004 than any other state in the US (CDC, 2006a), tick abundance can vary significantly not only within the state (Daniels *et al.*, 1998) but also within a county (Falco and Fish, 1992; Falco *et al.*, 1995), and even between neighborhoods (Maupin *et al.*, 1991). Such disparity in the distribution of ticks is likely due to several factors, including host activity patterns and changes in microhabitat that impact tick survival, both of which are strongly influenced by the local landscape features. In the northeastern US, risk for Lyme disease based on landscape features can be largely divided into two categories: residential and recreational.

The peridomestic nature of Lyme disease was first described in 1988, when studies in southern New York State of both Lyme disease cases and *I. scapularis* ticks removed from humans suggested that tick bites and *B. burgdorferi* infections often occurred around the home as a result of activities on the lawn (Falco and Fish, 1988a, 1988b). Subsequent investigations have provided additional evidence that residential exposure to tick bites is a major factor contributing to the development of Lyme disease. For example, in a study of lawns and adjacent woodlots in residential areas of southeastern Connecticut, 26.5 percent of *I. scapularis* nymphs and 36.4 percent of *I. scapularis* adults were collected directly from lawns, although the risk of exposure to infected nymphs varied spatially with the type of landscape and with each individual residence (Stafford and Magnarelli, 1993). In a residential subdivision in Westchester County, an attempt was made to determine the spatial distribution of *I. scapularis* in all habitats within the residential environment (Maupin *et al.*, 1991). Most nymphs were collected from the woods (67.3 percent) and ecotone (unmaintained edge) (21.6 percent), with smaller numbers found on ornamental plants (9.1 percent) and lawns (2.0 percent) (Figure 5.7). However, the presence of ticks on lawns and ornamental plants was of concern because most

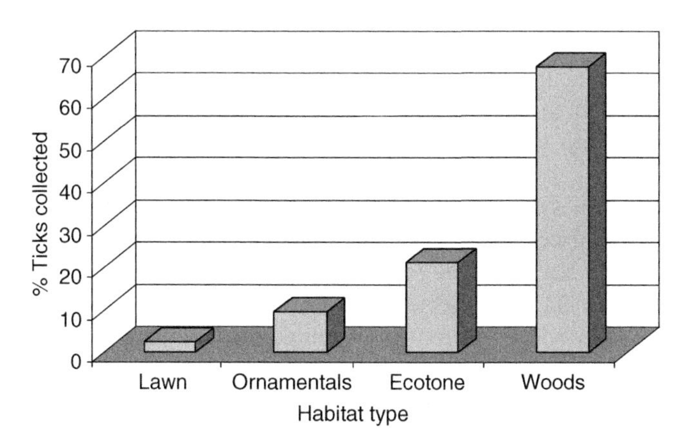


**Figure 5.7** Abundance of *Ixodes scapularis* ticks (all stages) collected by drag sampling in four habitat types on 67 properties, Armonk, NY. Data from Maupin *et al.* (1991).

residents spent the majority of their outdoor time involved in recreational and lawn-maintenance activities in these two habitats, with lawns that are adjacent to woodlots of special concern (Maupin *et al.*, 1991; Duffy *et al.*, 1994). Perhaps the most important reason that exposure to infected ticks in Westchester County, New York, is largely peridomestic is the amount of edge that typifies many properties. Mowed backyard lawns frequently abut woodland along a strip of unmaintained edge habitat harboring ticks. The extremely high level of residential risk for Lyme disease may also explain the propensity for young children to acquire tick bites, since children under 10 years of age are most likely to spend much of their outdoor recreational time in their own back yard (Falco and Fish, 1988b).

While epidemiologically less important than residential risk, recreational activities also provide an important opportunity for tick exposure in the northeastern US. In particular, recreational parks may present a high risk for human exposure to tick bites, as tick abundance can be quite high (Falco and Fish, 1989). Additionally, serologic studies have shown that outdoor workers in parks and other areas of largely undisturbed woodland have an elevated Lyme disease risk when compared to control groups without such exposure (Bowen *et al.*, 1984; Schwartz and Goldstein, 1990). It should be noted that outside of the northeastern US, where there tends to be less development and a more rural landscape, recreational areas may play an even more important epidemiologic role in the risk of exposure to *I. scapularis* ticks by humans.

Parks are also likely to play a role in human exposure to Lyme disease in urban areas. Although the suburban and rural landscapes are far more important contributors to the overall burden of Lyme disease cases in the US, evidence suggests there is at least some risk in more developed and populated urban areas. For example, in Westchester County, New York, studies utilizing a serosurvey of canine exposure to *B. burgdorferi* and remote sensing technology to analyze land-cover composition demonstrated that there was a higher than expected risk of exposure in the southern, urban region (Dister *et al.*, 1993; Falco *et al.*, 1993). While exposure was lower in the urbanized area compared to the suburban/rural northern region (67.3 percent vs 17.3 percent seroprevalence in dogs, respectively), there was at least some local exposure to ticks in the urban landscape (Falco *et al.*, 1993).

Additional studies have demonstrated that *I. scapularis* ticks infected with *B. burgdorferi* can be found in and near urban settings. For example, adult *I scapularis* infected with *B. burgdorferi* were collected within 3 km of Philadelphia, Pennsylvania (Anderson *et al.*, 1990), and in Bridgeport, Connecticut, collections of ticks and blood from white-tailed deer revealed both *I. scapularis* infected with *B. burgdorferi*, as well as seropositive deer. In the latter study, the authors concluded that foci for Lyme disease can occur in forested urban settings as well as in rural areas if there are ticks, rodents, birds, and large mammals present (Magnarelli *et al.*, 1995). In northern New York City (Bronx borough), mouse trapping and drag sampling conducted in Van Cortlandt Park resulted in the collection of *I. scapularis* infected with *B. burgdorferi* or with *A. phagocytophilum* (Daniels *et al.*, 1997). Although tick populations were low, the presence of infected ticks suggested that *I. scapularis* can exist at low populations in urban areas when deer are present even on an intermittent, seasonal basis, as apparently occurs in this park. Unlike the peridomestic exposure to ticks infected with *B. burgdorferi* that is typical of suburban and rural areas, tick populations in urban areas likely result in more focal exposure, restricted to woodland habitat "islands" which exist primarily as parkland (Daniels *et al.*, 1997).

## Social and political barriers to Lyme disease prevention

### The limitations of personal protection and public education

Personal protection measures have long been recommended for the prevention of Lyme disease and other tick-borne diseases. Such measures are typically divided into three categories: behavior modification, repellent use, and prompt removal of attached ticks (White, 1993; see also Table 5.1). Lyme disease prevention and educational programs in endemic areas have stressed the use of such personal protective measures (Dennis, 1995; CDC, 2000), and advice regarding personal protection against tick bites has sometimes been the major focus of Lyme disease prevention efforts on the part of public health officials (Williams *et al.*, 1986; Sigal and Curran, 1991). This advice serves to shift responsibility for

**Table 5.1 Personal protection (from White, 1993)**

1. Behavior modification
    - Wear light-colored clothing
    - Tuck long pants into socks
    - Conduct frequent clothing checks when outside
    - Conduct full-body exam before going to sleep
2. Use repellents
    - DEET-based products can be used on skin and clothing to repel ticks
    - Permethrin-based products can be used as clothing treatments to kill and repel ticks
3. Promptly remove all feeding ticks with forceps (tweezers) or tick-removal tools
    - Apply antiseptic
    - Bring tick to health-care provider within 72 hours of removal to discuss whether single-dose doxycycline is appropriate for prophylaxis (Nadelman *et al.*, 2001; Wormser, 2006)
    - Contact physician if rash, fever, 'flu-like illness, or unexplained joint pain occurs.

Lyme disease prevention from the governmental health agencies to the individual, in effect minimizing the government's role in actively reducing Lyme disease incidence.

However, reliance on personal protection as the cornerstone of prevention efforts is problematic. While it is likely that strict adherence to personal protection recommendations will reduce an individual's risk of acquiring tick bites, and therefore Lyme disease, there is also convincing evidence on a population level that this strategy does not significantly reduce the overall burden of Lyme disease. The continued rising incidence of Lyme disease in the United States, despite the importance placed on personal protection measures, suggests that either this approach is not very effective or, more likely, that too few individuals are engaging in these practices consistently enough for it to be very effective (Hayes *et al.*, 1999). For example, studies in Westchester County, New York, where Lyme disease has been endemic since 1982 and active public education has been implemented (Williams *et al.*, 1986), show that for the six-year period from 1991 to 1996 there was a significant correlation between Lyme disease cases (as defined by the presence of EM) and the abundance of nymphal ticks (Falco *et al.*, 1999; see Figure 5.6). Similar results were obtained in Connecticut (Stafford *et al.*, 1998), where Lyme disease public education efforts also have been implemented (Herrington *et al.*, 1997). These data suggest that it is the abundance of host-seeking nymphal *I. scapularis* that determines annual fluctuations in Lyme disease cases, regardless of the public education effort put forth.

While the diligent use of repellents likely reduces the risk of tick bites (Schreck *et al.*, 1986; Stafford, 1989), this measure is not being used extensively by the public (Herrington *et al.*, 1997; Shadick *et al.*, 1997). That conclusion seems to be particularly true of residents in suburban areas where ticks are encountered virtually daily during the spring and summer months, compared to those whose risk of exposure to ticks is occasional, such as through recreational activities. A study in Pennsylvania found that twice as many people took protective measures against tick bites before outdoor employment compared with those who ventured into a yard or other property associated with their home (Smith *et al.*, 2001). Cartter *et al.* (1989) found that while 90 percent of Connecticut high-school students surveyed believed tick avoidance behavior could prevent Lyme disease, fewer than 50 percent reported practicing any preventive behaviors. Even thorough training in personal protection measures is not always sufficient to prevent Lyme disease if exposure to ticks is on a daily basis. This may be because of the failure of the prevention method, but is more likely due to reduced diligence on the part of the individual (Falco and Daniels, 1993).

Thus, altering public behavior to prevent tick bites has been largely unsuccessful and public education efforts should not be the centerpiece of future Lyme disease prevention plans. Rather, an integrated approach focusing on the source of risk, namely vector ticks, would likely have the greatest chance for success.

## The controversy over chemical use and deer control

The most effective way for a suburban homeowner to reduce risk for tick bites significantly has been, and still is, the environmental application of insecticides (Barbour and Fish, 1993; Mount, 1993; Gern and Falco, 2000). However, there has been reluctance on the part of residents and some public health officials to advocate and implement chemical control due to environmental concerns (Stafford, 1991; Sonenshine, 1993; Golaine, 1992). The aversion to the use of chemicals, even those that have been approved by the federal Environmental Protection Agency for use in residential areas, has had a negative impact on efforts to prevent Lyme disease in endemic areas. However, due to the high prevalence of Lyme disease in suburban areas of the northeast, the benefit of reduced tick abundance through annual insecticide applications to lawns may outweigh potential environmental concerns (Barbour and Fish, 1993).

Although chemical control continues to be the most effective way of killing *I. scapularis* ticks and reducing Lyme disease risk in the residential landscape, it will do little to curb risk at a population level if a large segment of the at-risk population does not use it. Changing people's attitudes towards the use of pesticides, even relatively safe and effective ones used to mitigate a credible public health risk, is a challenge that goes beyond the scope of vector control. Environmental concerns regarding chemical use can be expected to continue and even accelerate, with intense pressure on government, the private sector, and the scientific community to develop environmentally sound, practical alternatives (Sonenshine, 1993).

Another controversial approach to preventing Lyme disease involves the reduction of tick populations through deer management. Controlling *I. scapularis* by reducing or eliminating deer beyond traditional deer management practices, such as routine hunting, has been studied and may be effective in some circumstances (Wilson *et al.*, 1984, 1988; Deblinger *et al.*, 1993, Telford, 1993). However, it would be difficult to maintain deer densities at a low enough level to have a significant impact on tick populations (Stafford *et al.*, 2003). Furthermore, such practices have become increasingly socially distasteful, and their large-scale implementation is unlikely, particularly in suburban residential settings (Stafford, 1993; Wilson and Deblinger 1993). As long as the "hunt" versus "no hunt" debate centers on the fundamental values of each side rather than on biology, as it so often does (McShea *et al.*, 1997), Lyme disease prevention through any type of host control is unlikely.

## Future strategies for Lyme disease prevention

Although the prospect for a new human Lyme disease vaccine remains, and research continues in that direction (Thomas and Fikrig, 2002), future efforts to prevent Lyme disease will likely move toward more efficient tick-control strategies, through either enhancing present strategies or developing new ones. These approaches will be based on the understanding that most emerging diseases are zoonotic in origin, with an ecology that should be investigated and considered before informed control practices can be implemented.

Not surprisingly, several approaches targeting white-tailed deer are under investigation. Application of insecticide directly to deer has shown promise in reducing the abundance of *I. scapularis* in the suburban environment. One such method uses a "four-poster device" – essentially a large central bin holding corn that spills out into two feeding troughs, one on each side of the corn bin. On the two corners of each trough are vertical posts that have been treated with the insecticide. As deer maneuver to feed from the trough, they rub their head and neck against the vertical posts and are passively treated with insecticide, typically a permethrin compound (Pound *et al.*, 2000a, 2000b). This method has successfully reduced the abundance of host-seeking ticks by 69–100 percent over a three-year period on study sites that were between 2.55 and 10.1 $km^2$ in area (Carroll *et al.*, 2002; Solberg *et al.*, 2003). This approach has great potential for control of *I. scapularis* tick populations, and should be considered by communities that want to reduce Lyme disease risk over a large area. More widespread use of the four-poster device may be achieved in coming years, particularly with the use of newer chemical acaricides that provide a longer-lasting residual treatment on deer. Consequently, less labor will be involved for those charged with maintaining the devices.

Much recent research and discussion (see, for example, DeNicola *et al.*, 1996, 1997; Muller *et al.*, 1997; Rudolph *et al.*, 2000) has focused on the possibility of controlling urban/suburban deer populations with immunocontraceptives – essentially vaccines that produce an antibody response to proteins that are involved in reproduction. This approach is appealing because it theoretically offers the potential to control deer in a humane fashion, without the distasteful outcome of "terminal management". It would also have an advantage over methods like surgical sterilization (which requires direct capture and handling of the deer), or the administration of synthetic steroid hormones, where there are some restrictions regarding the status of the doe at the time (pregnant or not) and which are not currently licensed for use with deer (Warren, 2000).

While fertility control may seem to be a logical alternative for reducing deer numbers, and ultimately reducing the risk of exposure from ticks that depend on deer, practical and logistical difficulties of administering the agents have prevented this method from being used routinely (Warren, 2000). Among the problems that researchers have yet to overcome with immunocontraceptives are the following:

1. The lack of a commercial source for the vaccine antigen (PZP: porcine zona pellucida)
2. Difficulty in obtaining FDA approval for a vaccine that uses Freund's Complete Adjuvant (FCA) in its formulation, since this interferes with tuberculosis testing of animals

3. The need to treat deer in spring and summer, before the breeding season but at a time when attracting deer to bait stations so they can be darted is difficult because of abundant natural foods
4. The tendency for treated females to display recurrent estrous cycles, effectively prolonging the breeding season and disrupting normal behavior
5. The fact that incomplete vaccinations can result in a drop in vaccine antibody titers, resulting in late-born fawns that have difficulty surviving
6. The process is time-consuming and costly – Rudolph *et al.* (2000) estimated the expense at $800–$1100 per treated doe
7. There may not be a population decline unless every female of reproductive age is vaccinated (Warren, 2000).

Work on this front is ongoing, but Telford (2002) cautions that even if an effective immunocontraceptive is developed and it is practical to deliver it to large numbers of animals, herds will almost certainly have to be reduced first, probably to fewer than six to eight deer per square mile (Telford, 1993), by hunting before this measure might be expected to work.

The possibility of a vaccine targeting *B. burgdorferi* in wildlife, particularly white-footed mice, is intriguing. Tsao *et al.* (2001) administered a recombinant outer surface protein A (OspA) vaccine to mice that were experimentally infected with *B. burgdorferi*, and found reduced transmission of the agent to uninfected ticks that were allowed to feed on them. After three vaccinations, infection prevalence in ticks was reduced by 99 percent. Field trials of the vaccine showed a significant reduction in *B. burgdorferi*-infected host-seeking nymphs the year after vaccination (Tsao *et al.*, 2004). However, the vaccination coverage (percentage of mice vaccinated) was estimated to be only about 55 percent – a difficulty that would be expected in any program requiring the administration of the vaccine by syringe. Results also highlighted the role that other vertebrates may play a part as reservoirs of infection. Tsao *et al.* (2004) concluded that alternative hosts probably have a greater role in infecting larvae than previously believed. Future work using an oral vaccine (Scheckelhoff *et al.*, 2006) delivered to a variety of host species in the field could resolve these problems.

The use of tick assembly and arrestment pheromones to enhance the efficacy of pesticides (Sonenshine *et al.*, 2003) and the use of biological control agents such as entomopathogenic fungi (see, for example, Benjamin *et al.*, 2002) also show promise in controlling *I. scapularis* in the field, and these avenues should be investigated further. Both approaches can be used as part of an effective integrated pest management (IPM) program that reduces the risk not only of Lyme disease, but of other tick-borne diseases as well.

On a larger and long-term scale, changes in land-use and development patterns that minimize the creation of ecotones by preserving forest stands intact rather than fragmenting the habitat will go far toward alleviating the future burden of tick-borne diseases. However, our past experience with Lyme disease prevention and control has taught us that the level of active participation and acceptance on the part of community members and government officials (Hayes *et al.*, 1999) must also be considered before a truly effective strategy is developed. In this case, the importance of informing the public about the consequences of current practices and cultivating the will to act in a manner that will improve public health in the future cannot be overstated.

## Conclusions

Not very long ago, a list of health challenges for the twenty-first century included sanitation and hygiene, vaccination, antibiotics and other antimicrobials, technological advances in detecting and monitoring diseases, serologic testing, viral isolation and tissue culture, and molecular techniques that can help diagnose and track the transmission of new threats (CDC, 1999). The paper went on to discuss the value of molecular genetics and how the US public health system must prepare to address diverse challenges including emergence of new infectious diseases, the re-emergence of old diseases, large food-borne outbreaks, and acts of bioterrorism. Not until the last sentence was the need for research into environmental factors that facilitate disease emergence mentioned. A mere seven years later, with a growing understanding of the threats that climate change and habitat destruction pose to ecosystems and public health, it is unlikely that such an oversight would be made. The fact is that we are facing the emergence and re-emergence of diseases caused by environmental changes of anthropogenic origin.

As we have seen, the proliferation of suburban development and fragmented landscapes in the northeastern and midwestern US has generated an ideal habitat for people emigrating from the cities, for commensal animal species such as deer that proliferate in the presence of humans, and for the parasites those animals host. This mixture of large numbers of vectors and people brings with it an increased potential for diseases such as Lyme disease, babesiosis, and human granulocytic anaplasmosis. Cases of Lyme disease are on the rise throughout the nation, notably from regions that have experienced similar alterations of landscape, have rapidly growing deer herds, and are becoming increasingly suburbanized (Spielman *et al.*, 1993; Wormser *et al.*, 2007).

The number of tools available to analyze ecological parameters that affect vector distribution has grown considerably in the last two decades – for example, satellite imagery and weather records from around the world support robust analyses of population data that could not be considered in the past. Much of this progress comes as a result of the interest in global warming, a critically important phenomenon that cannot be ignored any longer. Climate constrains the range of many infectious diseases, and weather affects the timing and intensity of outbreaks (Epstein, 1999). A strong argument can be made that diseases transmitted by mosquitoes – malaria, dengue fever, yellow fever, West Nile Virus infection, and several types of encephalitis, for instance – will be the ones that gain most from global warming (Epstein, 2001). This is mainly a function of vector mosquitoes benefiting from larger geographic ranges and faster development times. The future is less clear for ticks in a warmer world. Tick-borne diseases are also sensitive to climatic conditions, but favor cooler temperatures. In Africa, Rogers and Randolph (2000) found mean monthly maximum temperature to be the strongest predictor of tick occurrence at the margins of endemic zones. Only 2°C determined the difference between areas where ticks were present or absent in southeastern Africa. In the southern US, Rocky Mountain Spotted Fever may decline due to the vector tick's (*Dermacentor variabilis*) intolerance of high temperatures and diminished humidity (Haile, 1989; Patz *et al.*, 1996). Brownstein *et al.* (2003) concluded that *I. scapularis*' distribution is limited to those areas with a mean minimum temperature of −7°C in winter opening the possibility that global warming can increase the range of this tick in the future. Coupled with the ongoing creation of suitable habitat for ticks and their hosts resulting from suburbanization, the likelihood is high that tick-borne diseases will remain a significant threat in developed countries.

The means of combating ticks and reducing risk are limited in number. Host-targeted approaches are laborious, but environmentally more friendly than the classical use of chemical insecticides. To use them to their full advantage, however, requires knowledge of the ecology of a disease system that is frequently difficult to obtain and which takes time to understand; vector–host–pathogen lifecycles are complex and not easily unraveled. Furthermore, the public at risk needs to understand the value of such efforts. This is a formidable task for a suburbanized human population that is generally poorly informed about nature and wildlife dynamics (Foster *et al.*, 2002) that drive vector-borne diseases. Most important will be acknowledging the role that humans have in dictating which wildlife species have a place in suburbia, and the types of habitats we permit them to share with us. We need to recognize the direct and indirect cultural control we have over the modern landscape (Foster *et al.*, 2002).

## References

Anastos, G. (1947). Hosts of certain New York ticks. *Psyche* **54**, 178–180.

Anderson, J.F. (1988). Mammalian and avian reservoirs for *Borrelia burgdorferi*. *Annals of the New York Academy of Science* **539**, 180–191.

Anderson, J.F., Duray, P.H. and Magnarelli, L.A. (1990). *Borrelia burgdorferi* and *Ixodes dammini* in the greater Philadelphia area. *Journal of Infectious Diseases* **161**, 811–812.

Anderson, J.F., Johnson, R.C., Magnarelli, L.A. and Hyde, F.W. (1986). Involvement of birds in the epidemiology of the Lyme disease agent *Borrelia burgdorferi*. *Infection and Immunity* **51**, 394–396.

Aronowitz, P.B. (1989). A Connecticut housewife in Francis Bacon's court: Polly Murray and Lyme disease. *Pharos Alpha Omega Alpha Honor Medical Society* **52**(4), 9–12.

Barbour, A.G. and Fish, D. (1993). The biological and social phenomenon of Lyme disease. *Science* **260**, 1610–1616.

Battaly, G.R., Fish, D. and Dowler, R.C. (1987). The seasonal occurrence of *Ixodes dammini* and *Ixodes dentatus* (Acari: Ixodidae) on birds in a Lyme disease endemic area of New York State. *Journal of the New York Entomological Society* **95**, 461–468.

Baumgartner, M.R. (1988). *The Moral Order of a Suburb*. New York: Oxford University Press.

Benjamin, M.A., Zhioua, E. and Ostfeld, R.S. (2002). Laboratory and field evaluation of the entomopathogenic fungus *Metarhizium anisopliae* (Deuteromycetes) for controlling questing adult *Ixodes scapularis* (Acari: Ixodidae). *Journal of Medical Entomology* **39**, 723–728.

Berger, B.W. (1989). Dermatologic manifestations of Lyme disease. *Reviews of Infectious Diseases* **11**(Suppl. 6), S1475–1481.

Bishopp, F.C. and Smith, C.N. (1937). A new species of *Ixodes* from Massachusetts. *Proceedings of the Entomological Society of Washington* **39**, 133–138.

Bowen, G.S., Schulze, T.L., Hayne, C. and Parkin, W.E. (1984). A focus of Lyme disease in Monmouth County, New Jersey. *American Journal of Epidemiology* **120**, 387–394.

Brownstein, J.S., Holford, T.R. and Fish, D. (2003). A climate-based model predicts the spatial distribution of the Lyme disease vector *Ixodes scapularis* in the United States. *Environmental Health Perspectives* **111**, 1152–1157.

Brownstein, J.S., Skelly, D.K., Holford, T.R. and Fish, D. (2005). Forest fragmentation predicts local scale heterogeneity of Lyme disease risk. *Oecologia* **146**, 469–475.

Carroll, J.F., Allen P.C., Hill, D.E. *et al.* (2002). Control of *Ixodes scapularis* and *Amblyomma americanum* through use of the "4-poster" treatment device on deer in Maryland. *Experimental and Applied Acarology* **28**, 289–296.

Cartter, M.L., Farley, T.A., Ardito, H.A. and Hadler, J.L. (1989). Lyme disease prevention – knowledge, beliefs, and behaviors among high school students in an endemic area. *Connecticut Medicine* **53**, 354–356.

Centers for Disease Control and Prevention (1999). Achievements in public health, 1900–1999: control of infectious diseases. *Morbidity and Mortality Weekly Report* **48**(29), 621–629.

Centers for Disease Control and Prevention (2000). Surveillance for Lyme disease – United States, 1992–1998. *Morbidity and Mortality Weekly Report* **49**, 1–11.

Centers for Disease Control and Prevention (2004). Lyme disease – United States, 2001–2002. *Morbidity and Mortality Weekly Report* **53**, 365–369.

Centers for Disease Control and Prevention (2005). Summary of notifiable diseases – United States, 2003. *Morbidity and Mortality Weekly Report* **52**(Suppl.), 1–85.

Centers for Disease Control and Prevention (2006a). *Number of Reported Cases of Lyme Disease by State, 1994–2003*. Available at: http://www.cdc.gov/ncidod/dvbid/lyme/ld_rptdLymeCasesbyState94_03.htm (accessed 10 April 2006).

Centers for Disease Control and Prevention (2006b). *Lyme Disease*. Available at: http://www.cdc.gov/ncidod/dvbid/lyme/who_cc/index.htm (accessed 20 July 2006).

Clark, K.L., Oliver, J.H., McKechnie, D.B. and Williams, D.C. (1998). Distribution, abundance, and seasonal activities of ticks collected from rodents and vegetation in South Carolina. *Journal of Vector Ecology* **23**, 89–105.

Cobb, S. (1942). Tick parasites on Cape Cod. *Science* **95**, 503.

Collins, D.L, Nardy, R.V. and Glasgow, R.D. (1949). Some host relationships of Long Island ticks. *Journal of Economic Entomology* **42**, 110–112.

Cronon, W. (1983). *Changes in the Land: Indians, Colonists, and the Ecology of New England*. New York: Hill and Wang.

Daily, G.C. and Ehrlich, P.R. (1995). *Development, Global Change, and the Epidemiological Environment* (available at: http://dicoff.org).

Daniels, T.J., Fish, D. and Falco, R.C. (1989). Seasonal activity and survival of adult *Ixodes dammini* (Acari: Ixodidae) in southern New York State. *Journal of Medical Entomology* **26**, 610–614.
Daniels, T.J., Fish, D. and Schwartz, I. (1993). Reduced abundance of *Ixodes scapularis* (Acari: Ixodidae) and Lyme disease risk by deer exclusion. *Journal of Medical Entomology* **30**, 1043–1049.
Daniels, T.J., Falco, R., Curran, K.L. and Fish, D. (1996). Timing of *Ixodes scapularis* (Acari: Ixodidae) oviposition and larval activity in southern New York. *Journal of Medical Entomology* **33**, 140–147.
Daniels, T.J., Falco, R.C., Schwartz, I. *et al.* (1997). Deer ticks (*Ixodes scapularis*) and the agents of Lyme disease and human granulocytic ehrlichiosis in a New York City park. *Emerging Infectious Diseases* **3**, 353–355.
Daniels, T.J., Boccia, T.M., Varde, S. *et al.* (1998). Geographic risk for Lyme disease and Human Granulocytic Ehrlichiosis in southern New York state. *Applied Environmental Microbiology* **64**, 4663–4669.
Daniels, T.J., Falco, R.C. and Fish, D. (2000). Estimating population size and drag sampling efficiency for the black-legged tick, *Ixodes scapularis* (Acari: Ixodidae). *Journal of Medical Entomology* **37**, 357–363.
Davis, M.B. (1983). Holocence vegitational history of the Eastern United States. In: H.E. Wright Jr. (ed.), *Late-Quaternary Environments of the United States*, Vol. 2: *The Holocene*. Minneapolis: University of Minnesota Press, pp. 166–179.
Deblinger, R.D., Wilson, M.L., Rimmer, D.W. and Spielman, A. (1993). Reduced abundance of immature *Ixodes dammini* (Acari: Ixodidae) following incremental removal of deer. *Journal of Medical Entomology* **30**, 144–150.
DeNicola, A.J., Swihart, R.K. and Kesler, D.J. (1996). The effect of remotely delivered gonadotropin formulations on reproductive function of white-tailed deer. *Drug Development and Industrial Pharmacy* **22**, 847–850.
DeNicola, A.J., Kesler, D.J. and Swihart, R.K. (1997). Remotely delivered prostaglandin F-2alpha implants terminate pregnancy in white-tailed deer. *Wildlife Society Bulletin* **25**, 527–531.
Dennis, D.T. (1995). Lyme disease. *Dermatology Clinics* **13**, 537–551.
Dennis, D.T., Nekomoto,T.S., Victor, J.C. *et al.* (1998). Reported distribution of *Ixodes scapularis* and *Ixodes pacificus* (Acari: Ixodidae) in the United States. *Journal of Medical Entomology* **35**, 629–638.
Diffendorfer, J.E., Gaines, M.S. and Holt, R.D. (1995). Habitat fragmentation and movements of three small mammals (*Sigmodon*, *Microtus*, and *Peromyscus*). *Ecology* **76**, 827–839.
Dister, S., Beck, L., Wood, B. *et al.* (1993). The use of GIS and remote sensing technologies in a landscape approach to the study of Lyme disease transmission risk. In: *Proceedings of the Seventh Annual Symposium in Geographic Information Systems in Forestry, Environmental, and Natural Resource Management*. Vancouver, BC, Canada.
Donahue, J.G., Piesman, J. and Spielman, S. (1987). Reservoir competence of white-footed mice for Lyme disease spirochetes. *American Journal of Tropical Medicine and Hygiene* **36**, 92–96.
Duffy, D.C. and Campbell, S.R. (1994). Ambient air temperature as a predictor of activity of adult *Ixodes scapularis* (Acari: Ixodidae). *Journal of Medical Entomology* **31**, 178–180.
Duffy, D.C., Clark, D.D., Campbell, S.R. *et al.* (1994). Landscape patterns of abundance of *Ixodes scapularis* (Acari: Ixodidae) on Shelter Island, New York. *Journal of Medical Entomology* **31**, 875–879.
Ehrlich, P.R. and Ehrlich, A.H. (2002). Population, development, and human natures. *Environment and Development Economics* **7**, 158–170.

Environmental & Energy Study Institute (2000). *Sprawl in Europe and the United States: Contrasting Patterns of Urban Development and their Policy Implications* (available at: www.eesi.org/briefings/publications/06.21.99sprawl.pdf).
Epstein, P.R. (1999). Climate and health. *Science* **285**, 347–348.
Epstein, P.R. (2000). Is global warming harmful to health? *Scientific American* **2000**, 36–43.
Falco, R.C. and Daniels, T.J. (1993). Lyme disease control and tick management. In: H. Ginsberg (ed.), *Ecology and Environmental Management of Lyme Disease*. New Brunswick: Rutgers University Press, pp. 167–171.
Falco R.C. and Fish D. (1988a). Prevalence *of Ixodes dammini* near the homes of Lyme disease patients in Westchester County, New York. *American Journal of Epidemiology* **127**, 826–830.
Falco, R.C. and Fish, D. (1988b). Ticks parasitizing humans in a Lyme disease endemic area of southern New York State. *American Journal of Epidemiology* **128**, 1146–1152.
Falco, R.C. and Fish, D. (1989). Potential for exposure to tick bites in recreational parks in a Lyme disease endemic area. *American Journal of Public Health* **79**, 12–15.
Falco, R.C. and Fish, D. (1992). A comparison of methods for sampling the deer tick, *Ixodes dammini*, in a Lyme disease endemic area. *Experimental and Applied Acarology* **14**, 165–173.
Falco, R.C., Smith, H., Fish D. *et al.* (1993). Distribution and prevalence of canines seropositive for *Borrelia burgdorferi* antibodies in a Lyme disease endemic area. *American Journal of Public Health* **83**, 1305–1310.
Falco, R.C., Daniels, T.J. and Fish, D. (1995). Increase in abundance of immature *Ixodes scapularis* (Acari: Ixodidae) in an emergent Lyme disease endemic area. *Journal of Medical Entomology* **32**, 522–526.
Falco, R.C., Fish, D. and Piesman, J. (1996). Duration of tick attachment in a Lyme disease endemic area. *American Journal of Epidemiology* **143**, 187–192.
Falco R.C., McKenna D.F., Daniels T.J. *et al.* (1999). Temporal relation between *Ixodes scapularis* abundance and risk for Lyme disease associated with erythema migrans. *American Journal of Epidemiology* **149**, 771–776.
Fish, D. (1993). Population ecology of *Ixodes dammini*. In: H.S. Ginsberg (ed.), *Ecology and Environmental Management of Lyme Disease*. New Brunswick: Rutgers University Press, pp. 99–117.
Fish, D. and Daniels, T.J. (1990). The role of medium-sized mammals as reservoirs of *Borrelia burgdorferi* in southern New York. *Journal of Wildlife Diseases* **26**, 339–345.
Fish, D. and Dowler, R.C. (1989). Host associations of ticks (Acari: Ixodidae) parasitizing medium-sized mammals in a Lyme disease endemic area of southern New York. *Journal of Medical Entomology* **26**, 200–209.
Forman, R.T.T. and Godron, M. (1986). *Landscape Ecology*. New York: Wiley.
Foster, D.R., Motzkin, G., Bernardos, D. and Cardoza, J. (2002). Wildlife dynamics in the changing New England landscape. *Journal of Biogeography* **29**, 1337–1357.
Frank, D.H., Fish, D. and Moy, F.H. (1998). Landscape features associated with Lyme disease risk in a suburban residential environment. *Landscape Ecology* **13**, 27–36.
Gern, L. and Falco, R.C. (2000). Lyme disease. *Revue scientifique et technique de l'Office international des epizooties* **19**, 121–135.
Ginsberg, H.S. (1993). Geographical spread of *Ixodes dammini* and *Borrelia burgdorferi*. In: H.S. Ginsberg (ed.), *Ecology and Environmental Management of Lyme Disease*. New Brunswick: Rutgers University Press, pp. 63–82.
Glass, G.E., Schwartz, B.S., Morgan, J.M. *et al.* (1995). Environmental risk factors for Lyme disease identified with geographic information systems. *American Journal of Public Health* **85**, 944–948.

Golaine, A. (1992). Fear of pesticides thwarts Lyme disease prevention effort. *Priorities* (National Council for Science and Health), Spring Issue, 39–41.
Good, N.E. (1972). Tick locality and host records from Long Island and southeastern New York State. *Entomological News* **83**, 165–168.
Good, N.E. (1973). Ticks of eastern Long Island: notes on host relations and seasonal distribution. *Annals of the Entomological Society of America* **66**, 240–243.
Haile, D.G. (1989). Computer simulation of the effects of change in weather patterns on vector-borne disease transmission. In: J.R. Smith and D.A. Tirpak (eds), *The Potential Effects of Global Climate Change in the United States*. Washington, DC: USEPA.
Hanincova, K., Kurtenbach, K., Diuk-Wasser, M. *et al.* (2006). Epidemic spread of Lyme borreliosis, northeastern United States. *Emerging Infectious Diseases* **12**, 604–611.
Harlow, R.F. (1984). Habitat evaluation. In: L.K. Halls (ed.), *White-tailed Deer: Ecology and Management*. Harrisburg: Stackpole Books, pp. 601–628.
Harris, L.D. (1984). *The Fragmented Forest*. Chicago: University of Chicago Press.
Harwood, R.F. and James, M.T. (1979). *Entomology and Human Health*. New York: Macmillan.
Hayes, E.B., Maupin, G.O., Mount, G.A. and Piesman, P. (1999). Assessing the prevention effectiveness of local Lyme disease control. *Journal of Public Health Management Practice* **5**, 84–92.
Herrington, J.E., Campbell, G.L., Bailey, R.E. *et al.* (1997). Predisposing factors for individuals' Lyme disease prevention practices: Connecticut, Maine, and Montana. *American Journal of Public Health* **87**, 2035–2038.
Hertig, M. and Smiley, D. (1937). The problem of controlling woodticks on Martha's Vineyard. *Vineyard Gazette*, Edgartown, MA, 15 January.
Hyland, K.E. and Mathewson, J.A. (1961). The ectoparasites of Rhode Island mammals. I. The ixodid tick fauna. *Wildlife Diseases* **11**, 1–14.
Jackson, K.T. (1985). *Crabgrass Frontier: The Suburbanization of the United States*. New York: Oxford University Press.
Kahn, M.E. (2000). The environmental impact of suburbanization. *Journal of Policy Analysis & Management* **19**, 569–586.
Kramer, V.L., Carper, E.R. and Beesley, C. (1993). Mark and recapture of adult *Ixodes pacificus* (Acari: Ixodidae) to determine the effect of repeated removal sampling on tick abundance. *Journal of Medical Entomology* **30**, 1071–1073.
Krohne, D.T. and Hoch, G.A. (1999). Demography of *Peromyscus leucopus* populations on habitat patches: the role of dispersal. *Canadian Journal of Zoology* **77**, 1247–1253.
Lane, R.S., Piesman, J. and Burgdorfer, W. (1991). Lyme borreliosis: relation of its causative agent to its vectors and hosts in North America and Europe. *Annual Review of Entomology* **36**, 587–609.
Langdon, P. (1994). *A Better Place to Live: Reshaping the American Suburb*. Amherst: University of Massachusetts Press.
Larrouse, F., King, A.G. and Wolbach, S.B. (1928). The overwintering in Massachusetts of *Ixodiphagus caucertei*. *Science* **67**, 351–353.
Levine, J.F., Wilson, M.L. and Spielman, A. (1985). Mice as reservoirs of the Lyme disease spirochete. *American Journal of Tropical Medicine and Hygiene* **34**, 355–360.
Lindsay, R.L., Marker I.K., Surdeiner, G.A. *et al.* (1998). Survival and development of the different life stages of *Ixodes scapularis* (Acari: Ixodidae) held within four habitats on Long Point, Ontario, Canada. *Journal of Medical Entomology* **35**, 189–199.
Magnarelli, L.A., Anderson, J.F., Apperson, C.S. *et al.* (1986). Spirochetes in ticks and antibodies to *Borrelia burgdorferi* in white-tailed deer from Connecticut, New York State, and North Carolina. *Journal of Wildlife Diseases* **22**, 178–188.

Magnarelli, L.A., Anderson, J.F. and Fish, D. (1987). Transovarial transmission of *Borrelia burgdorferi* in *Ixodes dammini* (Acari: Ixodidae). *Journal of Infectious Diseases* **156**, 234–236.

Magnarelli, L.A., Denicola, A., Stafford, K.C. and Anderson, J.F. (1995). *Borrelia burgdorferi* in an urban environment: white-tailed deer with infected ticks and antibodies. *Journal of Clinical Microbiology* **33**, 541–544.

Marshall, W.F. III, Telford, S.R. III, Rys, P.N. *et al.* (1994). Detection of *Borrelia burgdorferi* DNA in Museum Specimens of Peromyscus leucopus. *Journal of Infectious Diseases* **170**, 1027–1032.

Mather, T.N., Wilson, M.L., Moore, S.I. *et al.* (1989). Comparing the relative potential of rodents as reservoirs of the Lyme disease spirochete (*Borrelia burgdorferi*). *American Journal of Epidemiology* **130**, 143–150.

Maupin, G.O., Fish, D., Zultowsky, J. *et al.* (1991). Landscape ecology of Lyme disease in a residential area of Westchester County, New York. *American Journal of Epidemiology* **133**, 1105–1113.

McCabe, T.R. and McCabe, R.E. (1997). Recounting whitetails' past. In: W.J. McShea, H.B. Underwood and J.H. Rappole (eds), *The Science of Overabundance*. Washington, DC: Smithsonian Institution Press, pp. 11–26.

McShea, W.J., Underwood, H.B. and Rappole, J.H. (eds) (1997). *The Science of Overabundance*. Washington, DC: Smithsonian Institution Press.

Mieszkowski, P. and Mills, E. (1993). The causes of metropolitan suburbanization. *Journal of Economic Perspectives* **7**, 135–147.

Mount, G.A. (1993). Lyme disease control and tick management. In: H.S. Ginsberg (ed.), *Ecology and Environmental Management of Lyme Disease*. New Brunswick: Rutgers University Press, pp. 171–175.

Muller, L.I., Warren, R.J. and Evans, D.L. 1997. Theory and practice of immunocontraception in wild mammals. *Wildlife Society Bulletin* **25**, 504–514.

Nadelman, R.B. and Wormser, G.P. (1995). Erythema migrans and early Lyme disease. *American Journal of Medicine* **98** (Suppl. 4A), S15–24.

Nadelman, R.B. and Wormser, G.P. (1998). Lyme borreliosis. *Lancet* **352**, 557–565.

Nadelman, R.B., Nowakowski, J., Forseter, G. *et al.* (1996). The clinical spectrum of early Lyme borreliosis in patients with culture positive erythema migrans. *American Journal of Medicine* **100**, 502–508.

Nadelman, R.B., Nowakowski, J., Fish, D. *et al.* (2001). Prophylaxis with single-dose doxycycline for the prevention of Lyme disease after an *Ixodes scapularis* tick bite. *New England Journal of Medicine* **345**, 79–84.

Nivola, P. (1999). Are Europe's cities better? *The Public Interest*, **137** (reprinted at: www.uli.org).

Nupp, T.E. and Swihart, R.K. (1996). Effect of forest patch area on population attributes of white-footed mice (*Peromyscus leucopus*) in fragmented landscapes. *Canadian Journal of Zoology* **74**, 467–472.

O'Connell, S., Granstrom, M., Gray, J.S. and Stanek, G. (1998). Epidemiology of European Lyme borreliosis. *Zentralblatt fur Bakteriologie* **287**, 229–240.

Oliver, J.H., Owsley, M.R., Hutcheson, H.J. *et al.* (1993). Conspecifity of the ticks *Ixodes scapularis* and *I. dammini* (Acari: Ixodidae). *Journal of Medical Entomology* **30**, 54–63.

Orloski, K.A., Hayes, E.B., Campbell, G.L. and Dennis, D.T. (2000). Surveillance for Lyme disease – United States, 1992–1998. *Morbidity and Mortality Weekly Report, CDC Surveillance Summary* **49**, SS3, 1–11.

Ostfeld, R.S. and Keesing, F. (2000). Biodiversity and disease risk: the case of Lyme disease. *Conservation Biology* **14**, 722–728.

Padgett, K.A. and Lane, R.S. (2001). Lifecycle of *Ixodes pacificus* (Acari: Ixodidae): timing of developmental processes under field and laboratory conditions. *Journal of Medical Entomology* **38**, 684–693.
Patz, J.A., Epstein, P.R., Burke, T.A. and Balbus, J.M. (1996). Global climate change and emerging infectious diseases. *Journal of the American Medical Association* **275**, 217–223.
Persing, D.H., Telford, S.R. III, Rys, P.N. *et al.* (1990). Detection of *Borrelia burgdorferi* DNA in museum specimens of *Ixodes dammini* ticks. *Science* **249**, 1420–1423.
Pickett, S.T.A. and Cadenasso, M.L. (1995). Landscape ecology: spatial heterogeneity in ecological systems. *Science* **269**, 331–334.
Piesman, J. (1989). Transmission of Lyme disease spirochetes (*Borrelia burgdorferi*). *Experimental and Applied Acarology* **7**, 71–80.
Piesman, J. (2002). Ecology of *Borrelia bugdorferi* sensu lato in North America. In: J.S. Gray, O. Kahl, R.S. Lane and G. Stanek (eds), *Lyme borreliosis: Biology, Epidemiology, and Control*. New York: CABI Publishing.
Piesman, J. and Spielman, A. (1979). Host associations and seasonal abundance of immature *Ixodes dammini* in southeastern Massachusetts. *Annals of the Entomological Society of America* **72**, 829–832.
Piesman, J., Mather, T.N., Sinsky, R.J. and Spielman, A. (1987). Duration of tick attachment and *Borrelia burgdorferi* transmission. *Journal of Clinical Microbiology* **25**, 557–558.
Piesman, J., Maupin, G.O., Campos, E.G. and Happ, C.M. (1991). Duration of adult female *Ixodes dammini* attachment and transmission of *Borrelia burgdorferi*, with description of a needle aspiration isolation method. *Journal of Infectious Diseases* **163**, 895–897.
Pound, J.M., Miller, J.A., George, J.E. and Lemeilleur, C.A. (2000a). The "4-poster" passive treatment device to apply acaricide for controlling ticks (Acari: Ixodidae) feeding on white-tailed deer. *Journal of Medical Entomology* **37**, 588–594.
Pound, J.M., Miller, J.A. and George, J.E. (2000b). Efficacy of amitraz applied to white-tailed deer by the "4-poster" topical treatment device in controlling free-living lone star ticks (Acari: Ixodidae). *Journal of Medical Entomology* **37**, 878–884.
Rappaport, J. (2005). The shared fortunes of cities and suburbs. *Federal Reserve Bank of Kansas City Economic Review* (available at: www.kc.frb.org/publicat/econrev/pdf/3qtrapp.pdf).
Richter, D., Spielman, A., Komar, N. and Matuschka, F. (2000). Competence of American robins as reservoir hosts for Lyme disease spirochetes. *Emerging Infectious Diseases* **6**, 133–138.
Rogers, D.J. and Randolph, S.E. (2000). The global spread of malaria in a future, warmer world. *Science* **289**, 1763–1766.
Rudolph, B.A., Porter, W.F. and Underwood, H.B. (2000). Evaluating immunocontraception for managing suburban white-tailed deer in Irondequoit, New York. *Journal of Wildlife Management* **64**, 463–473.
Rusk, D. (1995). *Cities without Suburbs*, 2nd edn. Washington, DC: Woodrow Wilson Center Press.
Scheckelhoff, M.R., Telford, S.R. and Hu, L.T. (2006). Protective efficacy of an oral vaccine to reduce carriage of *Borrelia burgdorferi* (strain N40) in mouse and tick reservoirs. *Vaccine* **24**, 1949–1957.
Schmidt, K.A. and Ostfeld, R.S. (2001). Biodiversity and the dilution effect in disease ecology. *Ecology* **82**, 609–619.
Schmidtmann, E.T., Carroll, J.F. and Potts, W.J.E. (1994). Host-seeking of black-legged tick (Acari: Ixodidae) nymphs and adults at the woods-pasture interface. *Journal of Medical Entomology* **31**, 291–296.

Schoeler, G.B. and Lane, R.S. (1993). Efficiency of transovarial transmission of the Lyme disease spirochete, *Borrelia burgdorferi*, in the western black-legged tick, *Ixodes pacificus* (Acari: Ixodidae). *Journal of Medical Entomology* **30**, 80–86.

Schreck, C.E., Snoddy, E.L. and Spielman, A. (1986). Pressurized sprays of permethrin or DEET on military clothing for personal protection against *Ixodes dammini* (Acari: Ixodidae). *Journal of Medical Entomology* **23**, 396–399.

Schulze, T.L., Bowen, G.S., Bosler, E.M. *et al.* (1984). *Amblyomma americanum*: a potential vector of Lyme disease in New Jersey. *Science* **224**, 601–603.

Schulze, T.L., Jordan, R.A. and Hung, R.W. (2001). Effects of selected meterological factors on diurnal questing of *Ixodes scapularis* and *Amblyomma americanum* (Acari: Ixodidae). *Journal of Medical Entomology* **38**, 318–324.

Schwartz, B.S. and Goldstein, M.D. (1990). Lyme disease in outdoor workers: risk factors, preventive measures, and tick removal methods. *American Journal of Epidemiology* **131**, 877–885.

Schwartz, I., Fish, D. and Daniels, T.J. (1997). Prevalence of the Rickettsial agent of human granulocytic ehrlichiosis in ticks from a hyperendemic focus of Lyme disease. *New England Journal of Medicine* **337**, 49–50.

Severinghaus, C.W. and Brown, C.P. (1956). History of white-tailed deer in New York. *New York Fish and Game Journal* **3,** 129–167.

Shadick, N.A., Daltroy, L.H., Phillips, C.B. *et al.* (1997). Determinants of tick-avoidance behaviors in an endemic area for Lyme disease. *American Journal of Preventive Medicine* **13**, 265–270.

Sigal, L.H. and Curran, A.S. (1991). Lyme disease: a multifocal worldwide epidemic. *Annual Review of Public Health* **12**, 85–109.

Smith, C.N. and Cole, M.M. (1943). Studies of parasites of the American dog tick. *Journal of Medical Entomology* **36**, 569–572.

Smith, G., Wiley, E.P., Hopkins, R.B. *et al.* (2001). Risk factors for Lyme disease in Chester County, Pennsylvania. *Public Health Reports*, **116**(Suppl. 1), 146–156.

Solberg, V.B., Miller, J.A., Hadfield, T. *et al.* (2003). Control *of Ixodes scapularis* (Acari: Ixodidae) with topical self-application of permethrin by white-tailed deer inhabiting NASA, Meltsville, Maryland. *Journal of Vector Ecology* **28**, 117–134.

Sonenshine, D.E. (1993). Lyme disease control and tick management. In: H.S. Ginsberg, (ed.), *Ecology and Environmental Management of Lyme Disease*. New Brunswick: Rutgers University Press, pp. 175–181.

Sonenshine, D.E., Adams, T., Allan, S.A., McLaughlin, J., Webster, F.X. (2003). Chemical composition of some components of the arrestment pheromone of the black-legged tick, *Ixodes scapularis* (Acari: Ixodidae) and their use in tick control. *Journal of Medical Entomology* **40**, 849–859.

Spielman, A., Clifford, C.M, Piesman, J. and Corwin, M.D. (1979). Human babesiosis on Nantucket Island, USA: Description of the vector, *Ixodes* (*Ixodes) dammini* n. sp. (Acarina: Ixodidae). *Journal of Medical Entomology* **15**, 218–234.

Spielman, A., Wilson, M., Levine, J. and Piesman, J. (1985). Ecology of *Ixodes dammini*-borne human babesiosis and Lyme disease. *Annual Review of Entomology* **30**, 439–460.

Spielman, A., Telford, S.R. III and Pollack, R.J. (1993). The origins and course of the present outbreak of Lyme disease. In: H.S. Ginsberg (ed.), *Ecology and Environmental Management of Lyme Disease*. New Brunswick: Rutgers University Press, pp. 83–96.

Stafford, K.C. (1989). Lyme disease prevention: personal protection and prospects for tick control. *Connecticut Medicine* **52**, 347–351.

Stafford, K.C. (1991). Effectiveness of carbaryl applications for the control of *Ixodes dammini* (Acari: Ixodidae) nymphs in an endemic residential area. *Journal of Medical Entomology* **28**, 32–36.

Stafford, K.C. (1993). Lyme disease control and tick management. In: H.S. Ginsberg (ed.), *Ecology and Environmental Management of Lyme Disease*. New Brunswick: Rutgers University Press, pp. 78–81.
Stafford, KC. and Magnarelli, L.A. (1993). Spatial and temporal patterns of *Ixodes scapularis* (Acari: Ixodidae) in southeastern Connecticut. *Journal of Medical Entomology* **30**, 762–771.
Stafford, K.C., Cartter, M.L., Magnarelli, L.A. *et al.* (1998). Temporal correlations between tick abundance and prevalence of ticks infected with *Borrelia burgdorferi* and increasing incidence of Lyme disease. *Journal of Clinical Microbiology* **36**, 1240–1244.
Stafford, K.C., DeNicola, A.J. and Kilpatrick, H.J. (2003). Reduced abundance of *Ixodes scapularis* (Acari: Ixodidae) and the tick parasitoid *Ixodiphagus hookeri* (Hymenoptera: Encyrtidae) with reduction of white-tailed deer. *Journal of Medical Entomology* **40**, 642–652.
Staley, S.R. (1999). The sprawling of America: in defense of the dynamic city. *Policy Study* **251**, Reason Public Policy Institute (available at: www.npca.org/ba/ba287.html).
Stanek, G. and Strle, F. (2003). Lyme borreliosis. *Lancet* **362**, 1639–1647.
Steere, A.C. 1994. Lyme disease: a growing threat to urban populations. *Proceedings of the National Academy of Science* **91**, 2378–2383.
Steere, A.C. (2001). Lyme disease. *New England Journal of Medicine* **345**, 115–125.
Steere, A.C., Malawista, S.E., Hardin, J.A. *et al.* (1977a). Erythema chronicum migrans and Lyme arthritis: the enlarging clinical spectrum. *Annals of Internal Medicine* **86**, 685–698.
Steere, A.C., Malawista, E.E., Snydman, D.R. *et al.* (1977b). Lyme arthritis: an epidemic of oligoarticular arthritis in children and adults in three Connecticut communities. *Arthritis & Rheumatism* **20**, 7–17.
Steere, A.C., Bartenhagen, N.H. and Craft, J.E. (1983). The early clinical manifestations of Lyme disease. *Annals of Internal Medicine* **99**, 76–82.
Telford, S.R. (1993). Lyme disease control and tick management. In: H.S. Ginsberg (ed.), *Ecology and Environmental Management of Lyme Disease*. New Brunswick: Rutgers University Press, pp. 164–167.
Telford, S.R. III. (2002). Deer tick-transmitted zoonoses in the eastern United States. In: A.A. Aguirre, R.S. Ostfeld, G.M. Tabor *et al.* (eds), *Conservation Medicine: Ecological Health in Practice*. New York: Oxford University Press, pp. 310–324.
Telford, S.R., Mather, T.N., Moore, S.I. *et al.* (1988). Incompetence of deer as reservoirs of the Lyme disease spirochete. *American Journal of Tropical Medicine and Hygiene* **39**, 105–109.
Temple, S.A. and Wilcox, B.A. (1985). Introduction: predicting effects of habitat patchiness and fragmentation. In: J. Verner, M.L. Morrison and C.J. Ralph (eds), *Wildlife 2000: Modeling Habitat Relationships of Terrestrial Vertebrates*. Madison: University of Wisconsin Press, pp. 261–262.
Thomas, V. and Fikrig, E. (2002). The Lyme disease vaccine takes its toll. *Vector-borne Zoonotic Diseases* **2**, 217–222.
Tsao, J., Barbour, A.G., Luke, C.J. *et al.* (2001). OspA immunization decreases transmission of *Borrelia burgdorferi* spirochetes from infected *Peromyscus leucopus* mice to larval *Ixodes scapularis* ticks. *Vector-borne Zoonotic Diseases* **1**, 65–74.
Tsao, J.I., Wootton, J.T., Bunikis, J. *et al.* (2004). An ecological approach to preventing human infection: vaccinating wild mouse reservoirs intervenes in the Lyme disease cycle. *Proceedings of the National Academy of Science* **101**, 18,159–18,164.
Verner, J. (1985). Summary: predicting effects of habitat patchiness and fragmentation – the researcher's viewpoint. In: J. Verner, M.L. Morrison and C.J. Ralph (eds), *Wildlife*

*2000: Modeling Habitat Relationships of Terrestrial Vertebrates*. Madison: University of Wisconsin Press, pp. 327–329.

Wang, G., van Dam, A.P., Schwartz, I. and Dankert, J. (1999). Molecular typing of *Borrelia burgdorferi* sensu lato: taxonomic, epidemiological, and clinical implications. *Clinical Microbiology Reviews* **12**, 633–653.

Warren, R.J. (2000). Overview of fertility control in urban deer management. In: *Proceedings of the 2000 Annual Conference of the Society of Theriogenology, December 2000, San Antonio, TX* (available at: http://coryi.org/florida_panther/deerfertilitycontrol.pdf).

White, D.J. (1993). Lyme disease surveillance and personal protection against ticks. In: H.S. Ginsberg (ed.), *Ecology and Environmental Management of Lyme Disease*. New Brunswick: Rutgers University Press, pp. 99–117.

Williams, C.L., Curran, A.S., Lee, A.C. and Sousa, V.O. (1986). Lyme Disease: Epidemiologic Characteristics of an Outbreak in Westchester County, NY. *American Journal Public Health* **76**, 62–65.

Wilson, M.L. (1998). Distribution and abundance of *Ixodes scapularis* (Acari: Ixodidae) in North America: ecological processes and spatial analysis. *Journal of Medical Entomology* **35**, 446–457.

Wilson, M.L. and Childs, J.E. (1997). Vertebrate abundance and the epidemiology of zoonotic diseases. In: W.J. McShea, H.B. Underwood and J.H. Rappole (eds), *The Science of Overabundance: Deer Ecology and Population Management*. Washington, DC: Smithsonian Institution, pp. 224–248.

Wilson, M.L. and Deblinger, R. (1993). Vector management to reduce the risk of Lyme disease. In: H.S. Ginsberg (ed.), *Ecology and Environmental Management of Lyme Disease*. New Brunswick: Rutgers University Press, pp. 126–156.

Wilson, M.L. and Spielman, A. (1985). Seasonal activity of immature *Ixodes dammini*. *Journal of Medical Entomology* **22**, 408–414.

Wilson, M.L., Levine, J.F. and Spielman, A. (1984). Effect of deer reduction on the abunbdance of the deer tick (*Ixodes dammini*). *Yale Journal of Biology and Medicine* **57**, 697–705.

Wilson, M.L., Telford, S.R., Piesman, J. and Spielman, A. (1988). Reduced abundance of immature *Ixodes dammini* (Acari: Ixodidae) following elimination of deer. *Journal of Medical Entomology* **25**, 224–228.

Wormser, G.P. (2006) Early Lyme disease. *New England Journal of Medicine* **354**, 2794–2801.

Wormser, G.P., McKenna, D., Carlin, J. *et al.* (2005). Brief communication: hematogenous dissemination in early Lyme disease. *Annals of Internal Medicine* **142**, 751–755.

Wormser, G.P., Dattwyler, R.J., Shapiro, E.D. *et al.* (2007). Infectious Diseases Society of America practice guidelines for clinical assessment, treatment and prevention of Lyme disease, human granulocytic anaplasmosis and babesiosis. *Clinical Infectious Diseases* **42**, 1089–1134.

Yeh, M.-T., Bak, J.M., Hu, R. *et al.* (1995). Determining the duration of *Ixodes scapularis* (Acari: Ixodidae) attachment to tick-bite victims. *Journal of Medical Entomology* **32**, 853–858.

# *The social ecology of infectious disease transmission in day-care centers*

# 6

**Robert F. Pass**

There are many examples of outbreaks of infectious diseases that are linked to child day care. This is not surprising. Provision of day care to preschool age children creates a convergence of several factors that promote transmission of infection. The fact that group care of young children increases the likelihood that they will acquire infections due to common viruses and bacteria that cause acute illnesses, and bring these infections home to their family members is both intuitive and well documented (Pickering and Cordell, 2003).

Cytomegalovirus (CMV) is different from the infectious agents that are frequent causes of acute illnesses. It is common to think of viral infections as self-limited events during which virus is shed for days or perhaps a few weeks, corresponding to the time when symptoms are present and when virus can be transmitted to others. The effector mechanisms of the host innate or adaptive immune systems result in clearance of the virus and resolution of illness. Infection results in immunologic memory that will provide some degree of protection against future encounters with the same agent. Cytomegalovirus infection rarely causes any signs of illness in healthy adults or children. However, CMV is shed in multiple body fluids for months to years in spite of the host immune responses. Cytomegalovirus is not cleared even when viral shedding is no longer detectable; like other members of the herpesvirus family, CMV becomes latent and can reactivate with intermittent productive infection with virus in blood cells or secretions. Further distinguishing CMV from many other viruses is the fact that vertical transmission (transmission from mother to offspring) plays a very important role in both the epidemiology and medical significance of CMV.

## CMV as a model

From the point of view of a virologist, day-care centers provided a natural "experiment" through which the key role of young children in spread of CMV was clearly demonstrated and the impact of social customs that affect child-rearing practices on the epidemiology of CMV infection was revealed.

### Discovery of cytomegalovirus

The history of the discovery of CMV is closely linked to the study of fetuses and newborns who died and had pathologic findings of multiorgan involvement suggestive of congenital infection (see review in Riley, 1997). Features consistent with CMV cytopathology were described more than 50 years prior to the isolation of CMV, including intranuclear inclusions, though the viral etiology of these changes was not at first recognized. Cytopathologic changes similar to those seen in autopsies of fetuses and newborns were subsequently described in salivary glands and other organs of infants and children. The term "cytomegalia" was first used in 1921, and the presence of this cytopathology in stillbirths was considered evidence of a prenatal insult. The possibility that the responsible agent might be more common than the congenital disease was suggested by a report of cells with intranuclear and intracytoplasmic inclusions in 14 percent of salivary glands from 183 infant postmortem examinations (Farber and Wolbach, 1932). In the early 1950s, the term "generalized cytomegalic inclusion disease" was used to describe the infants with multiorgan involvement (Wyatt *et al.*, 1950). Detection of inclusion-containing cells in urine was described as an early diagnostic technique, and developmental delay and cerebral palsy were reported in a surviving infant prior to the isolation of CMV (Fetterman, 1952; Margileth, 1955).

Weller, one of the discoverers of human CMV, reviewed the circumstances and chronology of the initial tissue culture isolation of CMV which occurred in three laboratories at roughly the same time in the mid-1950s (Weller, 1970). The first isolation of CMV from tissues of a living human was made serendipitously in 1955, when Weller's laboratory was attempting to isolate *Toxoplasma* from a liver biopsy taken from an infant with hepatosplenomegaly, cerebral calcification, and chorioretinitis; the focal cytopathic effect that developed in tissue culture was typical of cytomegaloviruses, and Weller and colleagues concluded that they had recovered the agent of cytomegalic inclusion disease. Shortly afterwards, it was possible to prepare antigens for serologic testing. Study of sera from adults showed that the majority were antibody-positive, showing that CMV infections are common (Rowe *et al.*, 1956).

### The virus

Cytomegalovirus is a herpesvirus and shares physical features with other members of this family, including a linear, double-stranded DNA genome packaged in an icosahedral nucleopsid which is surrounded by tegument enveloped in amorphous lipid bilayer derived from the host cell endoplasmic reticulum–Golgi complex. CMV is large, around 200 nm in diameter, with roughly 230 kbp of DNA that encode upwards of 150 gene products. In human fibroblast tissue culture CMV grows slowly, producing foci of enlarged, rounded, refractile cells. Although there are many cytomegaloviruses, they are species-specific; human CMV infects only humans, and grows exclusively in human cells in the laboratory. In its human host, CMV infects a variety of cell types (hematopoietic, endothelial, epithelial, stromal) in multiple organs, including salivary gland, gastrointestinal tract, kidney, liver, spleen, lung, adrenal gland, reproductive organs, vasculature, brain, inner ear, and eye. An important biologic feature of CMV is its ability to establish latency and persist indefinitely in the host. Hematopoietic progenitor cells, monocyte-derived macrophages, and dendritic cells appear to play a key role in harboring CMV genome in the latent state. Reactivation of CMV from latency is stimulated by activation and differentiation of these cells. Another key feature of human CMV is the large number of viral genes that code for proteins, which have the potential to interact with host inflammatory responses or interfere with host innate and adaptive immune responses. These include gene products which interfere with host cell apoptosis, mimic host cytokines or their receptors, interfere with innate immune mechanisms, and evade host T-cell responses by interfering with antigen processing in the context of the major histocompatibility complex. The complex biology of CMV has been recently reviewed (Mocarski *et al.*, 2007). The large number of human CMV genes aimed at host cell functions is indicative of a virus that is remarkably well adapted to its host.

## CMV infection in the normal host

Approximately 95 percent of initial CMV infections in healthy pregnant women are clinically silent (Peckham *et al.*, 1983; Griffiths and Baboonian, 1984; Stagno *et al.*, 1986; see also Table 6.1). The proportion that is symptomatic may be

**Table 6.1 Primary CMV infection in pregnant women is usually clinically silent**

| Study* | Sample size | No. primary CMV infections | No. with symptoms (%) |
|---|---|---|---|
| Stagno, 1986 | 5,199 | 63 | 3 (4.8%) |
| Griffiths, 1984 | 10,847 | 58 | 2 (3.4%) |
| Peckham, 1983 | 14,789 | 28 | 0 |

*The combined enrollment of the listed studies included nearly 30,000 women (including seronegative and seropositive participants). There were 149 primary infections identified during pregnancy, but symptoms occurred in only five women with primary infection.

lower in the general population than in pregnant women, considering the fact that almost everyone acquires CMV, yet symptomatic primary infection is uncommon. When initial CMV infection is symptomatic, it causes a mononucleosis-like syndrome similar to that associated with Epstein-Barr virus. Symptomatic CMV infection in the normal host is characterized by high fever that persists for more than a week, pharyngitis and lethargy; anorexia, adenopathy, and enlargement of the liver and spleen are also sometimes present. Laboratory abnormalities include anemia, thrombocytopenia, and elevated hepatic transaminases.

## Medical importance of CMV infection

Cytomegalovirus disease occurs mainly as a result of infections in fetuses and immunocompromised hosts; in each of these patient groups, CMV is a leading (possibly *the* leading) infectious cause of morbidity. Among immunocompromised hosts, the probability of experiencing CMV disease is related to the degree of impairment of T-lymphocyte function. Prior to the use of highly active antiretroviral agents, CMV was a frequent opportunistic infection in AIDS patients. Cytomegalovirus continues to be a major problem for transplant patients; the possibility of CMV infection in such patients has made screening of organ donors, recipients, and blood products used for them standard care. Use of preventive strategies involving antiviral agents for CMV in the first few months after transplant is commonly practiced, and has dramatically reduced the frequency of CMV disease in organ transplant patients. However, aside from the chance transmission of CMV from a young child to a susceptible transplant patient (an event that has rarely been recognized), the problems posed by CMV for transplant medicine are not germane to the current discussion. In contrast, child care may be the source of virus for many congenital CMV infections.

It has been estimated that congenital CMV infection occurs in 0.5–1.5 percent of live births. The majority of the approximately 40,000 infants born in the US each year with congenital CMV infection have no clinically evident abnormalities at birth (i.e. are asymptomatic). Around 5–10 percent of newborns with congenital CMV infection have physical signs typical of congenital infection, such as petechiae, jaundice, enlarged spleen and liver, small size for gestational age, and chorioretinitis; some have clear evidence of involvement of the central nervous system, such as microcephaly, hearing loss, lethargy, hypotonia, or seizures (Istas *et al.*, 1995). The public health impact of congenital CMV infection is the result of central nervous system sequelae. Among infants who are symptomatic at birth, 50–90 percent will have cognitive impairment, motor abnormalities (cerebral palsy), sensorineural hearing loss, or impaired vision. Among the larger group that are asymptomatic at birth, hearing loss is the most frequently encountered adverse outcome, and it occurs in 7–15 percent of them. Congenital CMV infection is a leading cause of sensorineural hearing loss and mental retardation in young children. An expert panel from the Institute of Medicine of the National Academy of Sciences estimated that from 5000 to 8000 infants born in the US each year will have cognitive, motor, or sensory deficits due to congenital CMV infection (Stratton *et al.*, 2001). The same panel also concluded that, based on health-care costs and quality adjusted life years lost, a vaccine aimed at prevention of congenital CMV should be a top priority in the US.

## The social epidemiology of CMV infection

The most significant congenital CMV infections are those that result from initial maternal infection during pregnancy, and day-care centers have the potential to increase the probability that young women who are, or intend to become, pregnant will have close contact with a child who is shedding CMV.

### Reservoir and routes of transmission

Humans are the only source of human CMV infection. Transmission of CMV from person to person appears to require direct contact with body fluids. Cytomegalovirus is present in saliva, urine, semen, cervico-vaginal secretions, tears, milk, and blood, and CMV is, predictably, transmitted in settings in which susceptible persons have contact with body fluids from persons with CMV infection. Infectious aerosols and airborne virus particles do not appear to be important in transmission of CMV. Compared with respiratory viruses or varicella, agents that are often transmitted by these routes, CMV is not highly contagious. Cytomegalovirus can also be transmitted by blood products and organs transplanted from CMV seropositive donors to CMV seronegative recipients.

### Prevalence

Results of a recent study that used sera and data from a sample representative of the US population found that 58 percent of the US population have had CMV infection (are CMV-IgG antibody-positive; Staras *et al.*, 2006). Prevalence increased with age, ranging from 36 percent in 6- to 11-year-olds to 91 percent in subjects older than 80 years. Prevalence was higher among African Americans and Hispanics than among white non-Hispanics; it was also higher among females (63.5 percent) than males (54.1 percent). Low income and residence in the southern US were also associated with higher CMV prevalence. Numerous studies have found similar differences by racial/ethnic group and similar trends in age-related prevalence in convenience samples of blood donors, patient groups, hospital workers, and day-care workers. Comparison of results of seroepidemiologic studies from diverse geographic areas shows that in general, CMV infection is acquired more frequently and earlier in life in developing countries than in the developed nations of Europe and North America. CMV infection has been found in every human population that has been studied; among some indigenous peoples and populations from developing countries, infection levels reach 100 percent in childhood. In contrast, in the US and Europe initial CMV infection commonly occurs after childhood, and it is common for women to reach reproductive age prior to acquisition of CMV.

## Incidence of CMV infection

Population-based incidence data for CMV infection are not available. Cohort studies in pregnant women, in blood donors, and in hospital employees suggest that the incidence of CMV infection in young adults is around 2 percent per year, with higher rates in low-income young women (Balfour and Balfour, 1986; Stagno *et al.*, 1986; Balcarek *et al.*, 1990). Cohort studies performed in groups thought to be at higher risk for CMV infection show markedly higher rates (see Table 6.2). Patients in sexually transmitted disease clinics have very high rates of CMV infection. The incidence of CMV infection is also increased in day-care workers and parents of young children.

**Table 6.2 Incidence of CMV infection in various groups**

| Group | Rate* (%/year) | Reference |
|---|---|---|
| Blood donors | 1.57 | Balfour and Balfour, 1986 |
| Hospital employees: | | |
| • Richmond, VA | 2 | Adler, 1989 |
| • Birmingham, AL | 2.2 | Balcarek *et al.*, 1990 |
| Pregnant women: | | |
| • Middle income | 2.5 | Stagno *et al.*, 1986 |
| • Low income | 6.8 | |
| Women in STD clinic | 37 | Chandler *et al.*, 1985 |
| Parents of CMV-shedding child: | | |
| • 0–12 months | 47 | Yeager, 1983 |
| • ≤18 months | 32 | Pass and Hutto, 1986 |
| • 19 months–6 years | 13 | Pass and Hutto, 1986 |
| Day-care workers: | | |
| • Iowa | 7.9 | Murph *et al.*, 1991 |
| • Richmond | 11 | Adler, 1989 |
| • Toronto | 12.5 | Ford-Jones *et al.*, 1996 |
| • Birmingham | 20 | Pass *et al.*, 1990 |

*Percent per year who seroconverted from antibody negative to antibody positive. Reproduced, with permission, from *Fields Virology*, 4th edn, p. 2683 (Pass, 2001).

## Vertical transmission

Cytomegalovirus is transmitted from mother to baby by three routes – transplacental, intrapartum contact with maternal secretions, and through breast milk (see Table 6.3). Although transplacental transmission is of great medical significance because of the morbidity associated with congenital CMV infection, it is infrequent (around 0.5–1.5 percent) compared with intrapartum and breastmilk transmission, which occur in a sizeable proportion of babies born to seropositive mothers. If CMV is in the cervico-vaginal canal at the time of delivery, the transmission rate to the newborn is around 50 percent (Reynolds *et al.*, 1973). Overall, approximately 10 percent of women have CMV in the cervix or vagina at the time of delivery, but rates of shedding as high as 28 percent have been observed (Stagno *et al.*, 1982a). Epidemiologically, breast milk is the most important route for transmission of CMV from mother to baby. Numerous studies of human milk have found CMV (or viral DNA) in milk from seropositive mothers. Older studies using virus culture recovered CMV from the milk of from 30 to 70 percent of seropositive mothers with repeated sampling (Dworsky *et al.*, 1983; Ahlfors and Ivarsson, 1985). More recent studies using polymerase chain reaction (PCR) to detect CMV DNA report finding virus in around 95 percent of milk samples (Hotsubo *et al.*, 1994; Asanuma *et al.*, 1996; Vochem *et al.*, 1998; Jim *et al.*, 2004). Virolactia is uncommon in colostrum or after more than a few months post-partum; a Swedish study reported no detection (by virus culture) of CMV in mothers' milk before nine days or later than three months post-partum (Ahlfors and Ivarsson, 1985). Transmission of CMV to nursing infants is related to duration of nursing, and the presence or quantity of

**Table 6.3 Transmission of CMV from mother to infant: three routes of vertical transmission**

| Setting | Proposed mechanism | Maternal infection | Estimated rate of infection in infant |
|---|---|---|---|
| Pregnancy | Viremia/ placental infection | Unknown (population sample) | 1% |
| Primary infection during pregnancy | Viremia/placental infection | Seroconversion or CMV-IgM positive | 20%–40% |
| Birth, intrapartum | Virus in genital tract | Seropositive | 5% |
| Birth, intrapartum | Virus in genital tract | Positive cervical or vaginal culture | 50% |
| Breastfeeding | Virolactia | Seropositive mother | 25% |
| Breastfeeding | Virolactia | Milk positive for virus by culture or PCR | 65%–70% |

virus in milk. A cohort study of infants nursed by CMV seropositive mothers found that those nursed for less than one month were not infected, compared with a CMV infection rate of 39 percent in infants nursed for more than one month; 25 percent of infants nursed by seropositive mothers acquired CMV by one year of age (Dworsky *et al.*, 1983). When CMV was isolated from milk, 69 percent of infants were infected, compared with 10 percent of infants nursed by seropositive mothers whose milk was negative by virus culture. Review of rates of CMV infection during the first year of life from diverse geographic areas showed a consistent correlation between high prevalence of CMV at one year of age and high rates of both maternal seropositivity and breast feeding (Stagno *et al.*, 1980).

Infants who acquire CMV vertically shed the virus in their body fluids for years (Pass and Hutto, 1986). They will amplify the epidemiologic effect of vertical transmission because they can transmit CMV to susceptible age peers with whom they have contact. In addition, they will spread CMV to their susceptible caregivers, including their parents and child-care workers.

### Nosocomial CMV transmission

Spread of CMV to patients in hospitals through contact with staff or from other patients appears to be a rare event. Transmission of CMV from patients to workers also appears to be uncommon. A review of studies of CMV infection among hospital workers and a meta-analysis of studies each found little evidence for increased risk of CMV infection for health-care workers (Flowers *et al.*, 1988; Pass and Stagno, 1988). Handwashing and use of procedures to prevent contact with body fluids are routines in the health-care setting that are probably effective in preventing transmission of CMV. Transmission of CMV through medical procedures that transfer cells from donor to patient, such as blood transfusion, organ transplantation, and assisted reproductive technology, is a concern that has been addressed with prevention strategies in each of these areas. Although of medical importance, these CMV infections have little impact on the prevalence or transmission of CMV infection in the general population.

## Day care as microbiological experiment

Care of children in institutions in which sizeable numbers are grouped by age and held in relatively close contact for 8–12 hours per day could be viewed as an experiment designed to allow testing of hypotheses regarding transmission of infectious agents. Not surprisingly, increased rates of infection and disease due to a number of agents have been associated with day care; these include invasive infections due to *Hemophilus influenzae* b (pre-vaccine era), *Streptococcus pneumoniae* and *Neisseria meningitidis*, rotavirus, Salmonella, Shigella, *Giardia*

**Table 6.4 Features of day-care center based care of preschool children that promote spread of infections**

| Child/developmental features | Environmental/social factors |
|---|---|
| Naïve to many infections – no immunity from past infection | Crowding |
| Age/developmental susceptibility to certain infections | Often inadequate facilities for sanitation |
| Lack of control of body fluids/excreta | Grouping children by age |
| Unable to manage hygiene independently – handwashing, oral and nasal secretions, toileting, feeding | Cohort effect – new susceptibles enter center |
| Oral behavior – everything goes in the mouth | Family linkage |

*lamblia*, *E. coli*, respiratory viruses, varicella, hepatitis A, and parvovirus B 19. In addition to infections that can be attributed to a specific agent, there is evidence that children in day-care settings experience more frequent diarrhea, otitis media, and respiratory tract infections than children kept at home (Churchill and Pickering, 1997).

There are a number of features of day care that favor occurrence and transmission of infections (Table 6.4); these features are attributable to the developmental characteristics of young children and to the peculiar social features of these institutions. Pre-school age children have not yet encountered many common causes of infection, and they lack the adaptive immune responses that result from prior infection. When infected with an agent for the first time, they are more likely to have an illness and more likely to infect others in contact with them than an immune host would be. Hands are a key step in the transmission of many infectious agents, and the importance of hand hygiene and interrupting hand contact with body fluids in preventing spread of infections is well known. Pre-school age children do not have control of body fluids; they often have runny noses or drool, and the majority of those less than three years of age are incontinent of feces and urine. Developmentally, young pre-school aged children are not able to learn and reliably perform the routine procedures of hygiene that reduce contact with body fluids (their own as well as others), or to independently keep their hands clean. They use their mouths to explore their environment, and thus contaminate their surroundings with their oral secretions. Observation of toddlers in day care has shown that they place an object or hand in their mouths as often as 60 times per hour (Hutto *et al.*, 1986).

Features of day care that are peculiar to these institutions include crowding, the cohort effect, and family linkage. Very commonly in day-care centers children are grouped by age, and those of similar age are kept in a single room for most of the day. Even if the room is very spacious (and often it is not), the frequency with which a toddler will have contact with another toddler is dramatically increased compared with a child not in a day-care center. In addition, grouping children by age can result in many (even 20 or more) children of approximately the same age having daily contact with each other – a truly remarkable social change compared with family care. The cohort effect is the result of the fact that each year children reach school age and age out of day care, and a new group of infants or toddlers enters. This new group is likely to be naïve with regard to any infections that are endemic in that particular center, as well as to the common community-acquired respiratory and enteric infections that are spread in day-care centers. Thus this age-cohort effect means that there will be a regular supply of new susceptible children in day-care centers. Each child in day care lives with a family, and thus establishes a two-way microbiological relationship between the day-care center and the family. Children can introduce an infectious agent, such as CMV, which they acquired in their family to a day-care center. Children can bring infections home from day care, infecting family members and facilitating further spread of infection in the community; hepatitis A virus infection is the classic example (Hadler and McFarland, 1986).

Ironically, the occurrence of infectious diseases in childcare has created significant public health problems, but investigation of day-care associated infections has improved knowledge of the epidemiology of many infections. Studies of infections in day care have helped establish the following:

- the importance of young children as a source of hepatitis A virus infection
- the ability of intervention with human immune globulin to interrupt a hepatitis A outbreak
- the importance of carriage of pathogenic strains in the occurrence of invasive disease due to *Streptococcus pneumoniae* and *Hemophilus influenzae* b
- the impact of varicella vaccine on the occurrence of chickenpox in young children
- the relationship between antibiotic use and carriage of resistant bacteria
- the natural history of primary herpes simplex virus gingivostomatitis and spread of herpes among young children
- the importance of outbreaks of parvovirus B 19 infection in young children as a source of maternal infections
- the importance of young children as a source of maternal CMV infections.

## Child care and the epidemiology of CMV infection

Prior to studies of CMV infection in day-care centers, two features of CMV epidemiology were well accepted: prevalence of infection increases directly with increasing age, and the main source of CMV infections in young infants and toddlers is their mother. Accordingly, it might be expected that in any population, prevalence of CMV infection would be higher in adults than in children, and it would be very uncommon to find CMV infection in young children whose mothers had never had CMV. However, with rigorous study it became clear that the primary social environment for breeding infection was child day care-centers rather than the home.

## Surprise: CMV infection is more prevalent in children than in parents

A 1982 study of CMV infections in a day-care center that served a population comprised mostly of middle-income families with college-educated parents found that the prevalence of CMV infection in the pre-schoolers attending the center was higher than in their parents, and there were numerous examples of toddlers who were shedding CMV and whose mothers had never had CMV infection (Pass *et al.*, 1982). This paradox in related prevalence of CMV infection suggested that horizontal transmission between children was occurring, and also that widespread use of day-care centers could change the epidemiology of CMV infection in the US.

Subsequent studies in day-care centers confirmed horizontal transmission of CMV from young children, and raised concerns about transmission of CMV from children to their mothers or day-care workers. Further evidence of transmission of CMV from child to child came from a cohort study of children in day care which showed that only around 10 percent of children less than one year of age had CMV, and that the majority of children acquired CMV during their second year of life (Pass *et al.*, 1984). In one day-care center, 100 percent of children in a class group of children aged 1–2 years were shedding CMV. This pattern of infection was incompatible with a maternal source but was compatible with horizontal transmission during the second year of life, when ambulatory toddlers have frequent opportunities for contact with each other and for sharing of toys – which are often mouthed by children of this age. Studies that tracked transmission of CMV by comparing restriction fragment length polymorphism (RFLP) of DNA ("DNA fingerprinting") from CMV strains isolated from children in day-care centers confirmed horizontal transmission of virus between children (Adler, 1985).

## Public health concern from CMV in day-care centers

Cytomegalovirus infection did not appear to cause illness in the children; investigators who studied large numbers of young children who acquired CMV in day-care centers did not identify symptomatic infections (Adler, 1988a; Pass,

1990). However, the fact that CMV infection and shedding of virus were so common in children in day-care centers raised concern about the possibility that they would transmit virus to their caregivers (mothers and day-care workers), who were mostly women of childbearing age, and that increased CMV infection rates in these young women would result in increased occurrence of congenital CMV infection. A markedly increased rate of CMV infection was found in parents of children in day care compared with parents of children in home care, and infection in the parents was clearly related to whether or not their toddler was shedding CMV (Pass *et al.*, 1986). As with child-to-child transmission, RFLP analysis confirmed transmission of CMV from children who acquired virus in day care to parents (Adler, 1986). Subsequent studies showed that day-care workers had increased rates of CMV infection ranging from around 8 percent to 20 percent per year, compared with a rate of around 2 percent in other adults (Adler, 1989; Pass *et al.*, 1990; Murph *et al.*, 1991). In addition, study of CMV infection in day-care workers and RFLP analysis of viral DNA from CMV isolated from children and workers showed that the children were the source of virus for the day-care workers (Adler, 1988b). The connection between day-care center CMV infections and congenital infection was confirmed in family studies that used RFLP analysis of CMV DNA to demonstrate that a child who acquired CMV in day care transmitted virus to its mother, and that this virus was transmitted transplacentally to a fetus (Pass *et al.*, 1987).

## Social organization and ecology of human CMV infection

In societies in which all or almost all mothers are seropositive to CMV and breastfeeding is widely practiced, it might be expected that half or more of children would acquire CMV during infancy and these children would infect their age peers during childhood. As a result, the proportion of women of childbearing age who had never had CMV infection would be very low, and disease due to congenital CMV infection would be uncommon. Observations in countries in which almost all young adults are seropositive and infants are routinely breastfed support this conclusion (Stagno *et al.*, 1982b; Sohn *et al.*, 1992). It is likely that this was also the situation in the US and Western Europe prior to the twentieth century. During the twentieth century rates of breastfeeding in the US decreased, and it is likely that this led to lower rates of CMV infection in infants and less opportunity for spread of CMV among young children (Wright and Schanler, 2001). In addition, most young children were cared for within the family, where they had relatively limited opportunity (compared with center care) for contact with other children. The use of group day-care centers increased dramatically in the US in the latter part of the twentieth century; the proportion of children in day-care centers increased from 6 percent in 1965 to 30 percent in 1993, and it has continued to rise (Hofferth, 1996). In the latter part of the twentieth century, many young mothers in the US had been raised under circumstances that allowed them to escape CMV infection during childhood; they had not been breastfed and they had not been in a group-care situation as young children. These CMV-susceptible young mothers were placing their own children in day-care centers, where they would very commonly acquire CMV, shed virus for years and possibly infect their mothers – who might well be pregnant. A recent case–control study of maternal risk found that contact with young children significantly increased the risk of having a newborn with congenital CMV infection (Fowler and Pass, 2006).

Although population-based studies aimed at determining whether use of day care has increased the prevalence of CMV infection in young children, or the rate of congenital CMV infection, have not been conducted, it is certainly plausible that widespread use of day-care centers could change the epidemiology of CMV infection in the US. Breast-milk transmission is the means by which CMV is transmitted from one generation to the next. In the US, day-care centers could be viewed as an amplifier of breast milk-acquired CMV infection. Perhaps 10 percent of children acquire CMV through breast milk, and these children will shed virus in their body fluids for years. If 10 percent of children who enter day care were shedding CMV, it might be expected that in most day-care centers there would be at least one infected child in each entering age cohort. Daily contact between toddler-age children would almost certainly increase the proportion of children infected several-fold, and children infected in the day-care center will also shed virus for years, continuing to infect their family members and others who have close contact with them.

## Conclusions

Cytomegalovirus is extremely well adapted to persist in, and be spread among, its human hosts. The proportion of the population that acquires CMV in childhood is very dependent on child-rearing practices, particularly breastfeeding and group care of children. Since disease due to congenital CMV infection results mainly from infections in susceptible women during pregnancy, it could be speculated that this public health problem is largely the result of changes in social behavior (bottle-feeding and child-rearing practices) that occurred in the twentieth century in the US and other developed countries. Although current widespread use of day-care centers probably increases the occurrence of congenital CMV infection, it might be expected that over time the increasing infection rate among pre-school children may eventually lead to a higher proportion of the population acquiring CMV prior to reproductive age, and therefore fewer problems with sequelae of congenital CMV infection. Although vaccines for prevention of congenital CMV infection are in development, for the time being the only measures available for prevention of maternal and congenital CMV infection are traditional public health approaches. The Centers for Disease Control and Prevention recommend that day-care workers and pregnant women be informed about the risk of acquiring CMV infection from a young child; procedures to prevent contact with body fluids are recommended for prevention (information available at: www.cdc.gov/CMV/index.htm, accessed June 5, 2006).

## References

Adler, S.P. (1985). The molecular epidemiology of cytomegalovirus transmission among children attending a day care center. *Journal of Infectious Diseases* **152**, 760–768.

Adler, S.P. (1986). Molecular epidemiology of cytomegalovirus: evidence for viral transmission to parents from children infected at a day care center. *Pediatric Infectious Diseases* **5**, 315–318.

Adler, S.P. (1988a). Cytomegalovirus transmission among children in day care, their mothers and caretakers. *Pediatric Infectious Diseases Journal* **7**, 279–285.

Adler, S.P. (1988b). Molecular epidemiology of cytomegalovirus: viral transmission among children attending a day care center, their parents, and caretakers. *Journal of Pediatrics* **112**, 366–372.

Adler, S.P. (1989). Cytomegalovirus and child day care. Evidence for an increased infection rate among day-care workers. *The New England Journal of Medicine* **321**, 1290–1296.

Ahlfors, K. and Ivarsson, S.A. (1985). Cytomegalovirus in breast milk of Swedish milk donors. *Scandinavian Journal of Infectious Diseases* **17**, 11–13.

Asanuma, H., Numazaki, K., Nagata, N. *et al.* (1996). Role of milk whey in the transmission of human cytomegalovirus infection by breast milk. *Microbiology and Immunology* **40**, 201–204.

Balcarek, K.B., Bagley, R., Cloud, G.A. and Pass, R.F. (1990). Cytomegalovirus infection among employees of a children's hospital: no evidence for increased risk associated with patient care. *Journal of the American Medical Association* **263**, 840–844.

Balfour, C.L. and Balfour, H.H. (1986). Cytomegalovirus is not an occupational risk for nurses in renal transplant and neonatal units. *Journal of the American Medical Association* **256**, 1909–1914.

Chandler, S., Holmes, K.K., Wentworth, B.B. *et al.* (1985). The epidemiology of cytomegaloviral infection in women attending a sexually transmitted disease clinic. *Journal of Infectious Diseases* **152**, 597–605.

Churchill, R.B. and Pickering, L.K. (1997). Infection control challenges in child-care centers. *Infectious Disease Clinics of North America* **11**, 347–365.

Dworsky, M., Yow, M., Stagno, S. *et al.* (1983). Cytomegalovirus infection of breast milk and transmission in infancy. *Pediatrics* **72**, 295–299.

Farber, S. and Wolbach, S.B. (1932). Intranuclear and cytoplasmic inclusions ("protozoan-like bodies") in the salivary glands and other organs of infants. *American Journal of Pathology in Children* **8**, 123–126.

Fetterman, G.H. (1952). New laboratory aid in clinical diagnosis of inclusion disease of infancy. *American Journal of Clinical Pathology* **22**, 424–425.

Flowers, R.H. III, Torner, J.C. and Farr, B.M. (1988). Primary cytomegalovirus infection in pediatric nurses: a meta-analysis. *Infection Control and Hospital Epidemiology* **9**, 491–496.

Ford-Jones, E.L., Kitai, I., Davis, L. *et al.* (1996). Cytomegalovirus infections in Toronto child-care centers: a prospective study of viral excretion in children and seroconversion among day-care providers. *Pediatric Infectious Diseases Journal* **15**, 507–514.

Fowler, K.B. and Pass, R.F. (2006). Risk factors for congenital cytomegalovirus infection in the offspring of young women: exposure to young children and recent onset of sexual activity. *Pediatrics* **118**, 286–292.
Griffiths, P.D. and Baboonian, C. (1984). A prospective study of primary cytomegalovirus infection during pregnancy: final report. *British Journal of Obstetrics and Gynaecology* **91**, 307–315.
Hadler, S.C. and McFarland, L. (1986). Hepatitis in day care centers: epidemiology and prevention. *Reviews of Infectious Diseases* **8**, 548–557.
Hofferth, S.L. (1996). Child care in the United States today. *The Future of Children* **6**, 41–61.
Hotsubo, T., Nagata, N., Shimada, M. *et al.* (1994). Detection of human cytomegalovirus DNA in breast milk by means of polymerase chain reaction. *Microbiology and Immunology* **38**, 809–811.
Hutto, C., Little, E.A., Ricks, R. *et al.* (1986). Isolation of cytomegalovirus from toys and hands in a day care center. *Journal of Infectious Diseases* **154**, 527–530.
Istas, A.S., Demmler, G.J., Dobbins, J.G. and Stewart, J.A. (1995). Surveillance for congenital cytomegalovirus disease: a report from the National Congenital Cytomegalovirus Disease Registry. *Clinical Infectious Diseases* **20**, 665–670.
Jim, W.T., Shu, C.H., Chiu, N.C. *et al.* (2004). Transmission of cytomegalovirus from mothers to preterm infants by breast milk. *Pediatric Infectious Diseases Journal* **23**, 848–851.
Margileth, A.M. (1955). The diagnosis and treatment of generalized cytomegalic inclusion disease of the newborn. *Pediatrics* **15**, 270–283.
Mocarski, E.S., Shenk, T. and Pass, R.F. (2007). Cytomegaloviruses. In: D. Knipe and P. Howley (eds), *Fields Virology*, 5th edn. Philadelphia: Lippincott, Williams & Wilkins.
Murph, J.R., Baron, J.C., Brown, K. *et al.* (1991). The occupational risk of cytomegalovirus infection among day care providers. *Journal of the American Medical Association* **265**, 603–608.
Pass, R.F. (1990). Day care centers and transmission of cytomegalovirus: new insight into an old problem. *Seminars in Pediatric Infectious Diseases* **1**, 245–251.
Pass, R.F. (2001). Cytomegaloviruses. In: D. Knipe and P. Howley (eds), *Fields Virology*, 4th edn. Philadelphia: Lippincott, Williams & Wilkins.
Pass, R.F. and Hutto, C. (1986). Group day care and cytomegaloviral infections of mothers and children. *Reviews of Infectious Diseases* **8**, 599–605.
Pass, R.F. and Stagno, S. (1988). Cytomegalovirus. In: L.G. Donowitz (ed.), *Hospital-Acquired Infection in the Pediatric Patient*. Baltimore: Williams & Wilkins, pp. 174–187.
Pass, R.F., August, A.M., Dworsky, M.E. and Reynolds, D.W. (1982). Cytomegalovirus infection in a day care center. *New England Journal of Medicine* **307**, 477–479.
Pass, R.F., Hutto, C., Reynolds, D.W. and Polhill, R. B. (1984). Increased frequency of cytomegalovirus in children in group day care. *Pediatrics* **74**, 121–126.
Pass, R.F., Hutto, S.C., Ricks, R. and Cloud, G. A. (1986). Increased rate of cytomegalovirus infection among parents of children attending day care centers. *New England Journal of Medicine* **314**, 1414–1418.
Pass, R.F., Little, E.A., Stagno, S. *et al.* (1987). Young children as a probable source of maternal and congenital cytomegalovirus infection. *New England Journal of Medicine* **316**, 1366–1370.
Pass, R.F., Hutto, C., Lyon, M.D. and Cloud, G. (1990). Increased rate of cytomegalovirus infection among day care center workers. *Pediatric Infectious Diseases Journal* **9**, 465–470.
Peckham, C.S., Chin, K.S., Coleman, J.C. *et al.* (1983). Cytomegalovirus infection in pregnancy: preliminary findings from a prospective study. *Lancet* **1**, 1352–1355.

Pickering, L.K. and Cordell, R.L. (2003). Infectious diseases associated with out-of-home childcare. In: S.S. Long, L.K. Pickering and C.G. Prober (eds), *Principles and Practice of Pediatric Infectious Diseases*, 2nd edn. Philadelphia: Churchill Livingstone, pp. 27–33.

Reynolds, D.W., Stagno, S., Hosty, T.S. *et al.* (1973). Maternal cytomegalovirus excretion and perinatal infection. *New England Journal of Medicine* **289**, 1–5.

Riley, H.D. (1997). History of the cytomegalovirus. *Southern Medical Journal* **90**, 184–190.

Rowe, W.P., Hartley, J.W., Waterman, S. *et al.* (1956). Cytopathogenic agent resembling human salivary gland virus recovered from tissue cultures of human adenoids. *Proceedings of the Society for Experimental Biology and Medicine* **92**, 418–424.

Sohn, Y.M., Park, K.I., Lee, C. *et al.* (1992). Congenital cytomegalovirus infection in Korean population with very high prevalence of maternal immunity. *Journal of Korean Medical Science* **7**, 47–51.

Stagno, S., Reynolds, D.W., Pass, R.F. and Alford, C.A. (1980). Breast milk and the risk of cytomegalovirus infection. *New England Journal of Medicine* **302**, 1073–1076.

Stagno, S., Dworsky, M.E., Torres, J. *et al.* (1982a). Prevalence and importance of congenital cytomegalovirus infection in three different populations. *Journal of Pediatrics* **101**, 897–900.

Stagno, S., Pass, R.F., Dworsky, M.E. and Alford, C.A. (1982b). Maternal cytomegalovirus infection and perinatal transmission. *Clinical Obstetrics and Gynecology* **25**, 563–576.

Stagno, S., Pass, R.F., Cloud, G. *et al.* (1986). Primary cytomegalovirus infection in pregnancy. Incidence, transmission to fetus, and clinical outcome. *Journal of the American Medical Association* **256**, 1904–1908.

Staras, S.A.S., Dollard, S.C., Radford, K. *et al.* (2006). Seroprevalence of cytomegalovirus infection in the United States, 1988–1994. *Clinical Infectious Diseases* **43**, 1143–1151.

Stratton, K., Durch, J. and Lawrence, R. (2001). *Vaccines for the 21st Century: A Tool for Decisionmaking*. Washington, DC: National Academy Press.

Vochem, M., Hamprecht, K., Jahn, G. and Speer, C. P. (1998). Transmission of cytomegalovirus to preterm infants through breast milk. *Pediatric Infectious Diseases Journal* **17**, 53–58.

Weller, T.H. (1970). Cytomegaloviruses: the difficult years. *Journal of Infectious Diseases* **122**, 532–539.

Wright, A.L. and Schanler, R.J. (2001). The resurgence of breastfeeding at the end of the second millennium. *Journal of Nutrition* **131**, S421–425.

Wyatt, J.P., Lee, R.S. and Pinkerton, H. (1950). Generalized cytomegalic inclusion disease. *Journal of Pediatrics* **36**, 271–294.

Yeager, A.S. (1983). Transmission of cytomegalovirus to mothers by infected infants: another reason to prevent transfusion-acquired infections. *Pediatric Infectious Diseases* **2**, 295–297.

# Protecting blood safety 7

**Peter L. Page and H.F. Pizer**

It is estimated that each year more than 81 million units of blood are collected worldwide, but the availability of this blood for medical use is not evenly distributed across the globe. The average number of blood donations per 1000 population is 12 times greater in high-income countries and 3 times greater in medium-income countries than in low-income countries (WHO, 2005a). Only about 27 million of the 81 million units (approximately 33 percent) are collected in low- and medium-income countries, where 82 percent of the world's population resides. In developed nations about one patient in ten admitted to the hospital receives blood or some type of blood product, while in developing countries the number is much smaller because of financial constraints, limited access to advanced medical procedures, and a lack of a modern, effective blood-banking system. It is estimated that 70 percent of blood transfusions in Africa go to children with malaria and to treat women with post-partum hemorrhage, and that perhaps 100,000 people die annually because of unsafe transfusion practices (Heyns *et al.*, 2005). Still, the demand grows for safe blood products in developing nations.

About 8 million people in the United States voluntarily donate about 12 million units of blood each year. These units are then processed into about 20 million units of blood components, such as red blood cells, platelet, and plasma units (The National Blood Data Resource Center of the United States, 2005; US General Services, Administration, 2006). Modern medical care would not be possible without a safe and dependable blood supply to treat leukemia and certain cancers; to perform complex surgical procedures like heart operations, liver transplants and joint replacement; and to treat blood diseases like sickle cell anemia and thalassemia. Blood replacement is essential in acute trauma care, including accidents, burns, and battlefield injuries. The demand for blood is also growing in response to the world's aging demographics – for example, in the United States people over the age of 69 make up about 10 percent of the population, but receive about 50 percent of components.

This chapter is primarily about components of volunteer blood donations that are each transfused as a single unit, usually "as is." A single donation of a pint of whole blood is virtually always separated by centrifugation into a unit of red blood cells and a unit of plasma; a unit of platelets or a unit of cryoprecipitated plasma can also be prepared, but this is done for less than half of whole blood donations, since patients' clinical needs for platelets and cryoprecipitated plasma are much less than patients' needs for red blood cell transfusion. Each of these units is labeled, and regulated as a drug, following FDA regulations. While today virtually all of these blood components from whole blood donation are from volunteer donors, the FDA would permit paying these donors, but requires that the component's label be marked conspicuously with the words "PAID DONOR" as opposed to "VOLUNTEER DONOR." Whole blood donation is permitted no more often than once every eight weeks (to prevent development of iron deficiency in the donor).

Most of the plasma separated from red blood cell units is not needed for direct transfusion to patients, and so it is provided as "Recovered Plasma" to large multi-national for-profit companies that pool and fractionate it into plasma derivatives. The major source of plasma for plasma derivative manufacture, however, is "Source Plasma" collected by plasmapheresis (not whole blood collection). In the US, as well as many other countries, this Source Plasma is collected from donors who are paid (i.e. not volunteer), and the units of Source Plasma are labeled "Paid Donor." Plasmapheresis today is an automated procedure in which blood is removed from a donor, and the red cells are returned but the plasma is retained outside the body. Donors may undergo this procedure as often as twice a week. Source Plasma from tens of thousands of donations is pooled by the plasma fractionators who then, utilizing a series of complex processes, chemically and physically separate the plasma into lots of several plasma derivatives, including albumin, gamma globulin, and sometimes Factor VIII and Factor IX concentrates (used in developing parts of the world to treat patients with hemophilia A and B respectively), as well as alpha 1 anti-trypsin. Currently, the process of plasma fractionation includes several purifying and sterilizing steps which render these derivatives essentially safe, even though the source material is from pools of plasma from thousands of paid donors (each donation is, though, still tested for hepatitis B and C, as well as HIV). Prior to the availability of HIV and hepatitis C testing, and prior to the addition of sterilizing steps in plasma fractionation in the 1980s, large numbers of patients with hemophilia were regularly exposed to hepatitis C and HIV, which resulted in a number of national scandals and law suits (see p. 208; Negligence, human error, and failed oversight). Today in the US most patients with hemophilia A and B are treated with synthetic preparations of Factor VIII and Factor IX (respectively), which are viewed as essentially virally safe but also much more expensive. The sterilizing steps used in plasma fractionation are not suited to the single blood components used for direct transfusion, since these processes would damage red blood cells, platelets, and whole plasma, such that these components would not be efficacious.

Testing of blood donations for infectivity is viewed as part of the manufacturing process for the pharmaceutical drug (red blood cells, platelets, or plasma), and the test is to "determine the suitability for human use" of the blood component. This is the "purpose" for this testing, rather than for making a clinical diagnosis in the person donating the blood (who is presumed to be healthy and not a "patient") or screening the blood donor population for syphilis or HIV. Accordingly, it is very important that the false-negative rate in testing be as low as it can be (to prevent disease transmission by transfusion) – this would be good "sensitivity." However in doing this, the false-positive rate increases – this would be bad "specificity." Since there is a false-positive rate, and since blood centers view it as their obligation to inform donors of unsuitable test results, even false-positive test results make donors ineligible for subsequent blood donation. More important, though, is for donors to know if they are truly positive, in which case counseling is required to prevent future transmission to others (e.g. to sex partners), and referral to a health-care facility to initiate treatment, if appropriate, is required. It is therefore important that there be a "confirmatory" test to complement each reactive screening test result. To use anti-HIV testing as an example, the screening assay is an automated enzyme-linked immunoassay (EIA); if the EIA is repeatedly reactive then the donor is deferred in any event, but a Western blot (the confirmatory assay) is performed; if the WB is positive, the donor is considered infected and needs to be counseled to avoid exposing others, and to seek treatment for HIV. The more difficult counseling, however, is of the donor whose screening assay is repeat-reactive and whose confirmatory assay is negative – a false positive. This person is healthy and not infected, but is not permitted to donate blood any more (by federal regulation; there are complicated "re-entry" algorithms available which some blood centers use).

As we shall discuss, the social ecology for maintaining a safe blood supply is complex, but one point cannot be overemphasized: voluntary donation is its cornerstone. Where blood is in short supply and nations are poor, it is common to pay donors. Time and again this practice has been shown to increase the risk of introducing potentially lethal infectious agents. To date, the Government of Malawi is one of only a few in developing nations to establish a national voluntary blood donation system. It took two years to set up the system and its benefits were rapidly seen: the death rate from malarial anemia dropped by 60 percent and pregnancy-related mortality fell by more than 50 percent (Heyns *et al.*, 2005). Even with a purely voluntary donor system, blood must be screened for transmissible infectious agents. The strategies for optimal screening are based on the human social environment, which include factors related to demography, behavior, and geography. For example, in South Africa and Zimbabwe it is estimated that the overall rate of HIV infection is greater than 20 percent of the population. Based on the risk factors known to be associated with HIV transmission, both countries instituted appropriate donor-screening procedures and reduced the rate of HIV positivity in blood donations to below 0.5 percent (and these units are discarded, not used). This has not been accomplished throughout sub Saharan Africa. The World Health Organization (WHO) estimates that HIV-contaminated blood still accounts for 5 percent of HIV infections throughout the region (Heyns *et al.*, 2005).

## A brief history of improving blood safety

In 1937, Dr Bernard Fantus established the first blood bank in the United States at Cook County Hospital in Chicago. He is also credited with coining the term "blood bank" to describe a laboratory capable of preserving and storing blood. In just a few years, blood-banking spread across the United States and Europe. In 1940 the process of fractionation for breaking down plasma into albumin and gamma globulin was discovered, and these products soon were available for clinical use. World War II created an immediate demand for blood, and around this time the American Red Cross started using the vacuum bottle to collect and store donated blood. Whole blood donations were tested for ABO and Rh type, as well as syphilis. Meanwhile, the United States Government set up a nationwide program for collecting blood, and the "Plasma for Britain" program to aid the British war effort. During the war, albumin was used to treat shock, the Coombs test was discovered, and acid citrate dextrose (ACD) solution came into use to prolong shelf-life – thereby making more blood available for transfusion.

During World War II and the succeeding years, blood centers were established across the United States and other developed nations. The American Association of Blood Banks (AABB) was established in 1947, with a mission to promote safe blood-banking and improve public and professional education. Fueled by advances in medical care, like open-heart surgery and trauma care, there was steady increase in the demand for blood during the 1950s. In 1970, component therapy came into medical practice – first with red blood cells and plasma, and later with platelets and cryoprecipitated antihemophilic factor (AHF) for transfusion. Today, blood is transfused to a patient essentially only as one of the components of whole blood – i.e. red cells, platelets, or plasma – while whole blood is rarely transfused as such. This approach is beneficial in two respects: it is clinically better for the individual patient to receive only the component needed (and not be exposed to potential adverse effects of the other, unrelated component), and it also permits several patients to benefit from each unit of donated whole blood. In addition, the optimal storage conditions are different for the different blood components – red cells are refrigerated, plasma is frozen, and platelets are agitated at room temperature. Up to four different components may be derived from one unit of blood, but nowadays rarely more than three are prepared.

Giving blood is relatively painless and, in all but very rare circumstances, free of serious adverse consequences. The fluid lost is usually replaced naturally within 24 hours, but it can take up to two months to replace the lost red blood

**Table 7.1 Risk of transfusion-related hepatitis B infection**

| Year | Risk/unit | Comment |
|---|---|---|
| 1970 | 1:855 | Based on CEP rate |
| 1979 | 1:562 | TTV study |
| 1979 | 1:2809 | TTV (today's tests) |
| 1995 | 1:250,000 | Test sensitivity |
| 1996 | 1:63,000 | Window |
| 1999 | 1:138,700 | Window, 1989 incidence |

Source: Adapted from Dodd (2001), with permission.

**Table 7.2 Current transfusion risks***

| Agent | Point estimate | 95% Confidence level |
|---|---|---|
| HIV | 1:493,000 | 1:202,000–1:2,778,000 |
| Hepatitis B | 1:63,000 | 1:31,000–1:147,000 |
| Hepatitis C | 1:103,000 | 1:28,000–1:288,000 |
| HTLV 1 and II | 1:641,000 | 1:256,000–1:2,000,000 |

*Per unit of blood that is negative in laboratory testing.
Adapted from Schreiber *et al.* (1996).

cells. Whole blood can be donated once every eight weeks (56 days). Multiple units of platelets can be collected as frequently as every two weeks by a procedure called platelet pheresis. Two or three units of plasma can be collected at a time by plasmapheresis, even more frequently. However, two units of red blood cells can safely be donated only once every 16 weeks, also by plasmaapheresis.

Over the past four decades there has been enormous improvement in the safety of the blood supply. From 1970 to 2000, the overall risk of acquiring a transfusion-transmitted infection in the United States dropped from one in 70,000 to one in 11.15 million units transfused. This is mirrored in the risk of acquiring specific infectious agents, like hepatitis B and C – for example, from 1970 to 1999 the risk of acquiring hepatitis B virus dropped from 1:855 to 1:138,700 units transfused (Dodd, 2001). Table 7.1 illustrates the risk of acquiring HBV between 1970 and 1999, and Table 7.2 illustrates current estimated risk of acquiring HIV, hepatitis B and C, and HTLV I and II via transfusion. Table 7.3 estimates the risk of acquiring these agents, comparing it with other life risks such as sports, accidents, and other illnesses.

**Table 7.3 Risk assessment: infection from blood transfusion versus other life risks**

| Risk of new infection with transfusion-transmitted disease | Per unit transfused |
|---|---|
| Hepatitis B | 63,000 |
| Hepatitis C | 103,000 |
| HIV-1 | 676,000 |
| HIV-2 | <10,000,000,000 |
| HIV-1, Subtype O | <167,000,000 |
| **Activity** | **Odds of death per person per year** |
| Motorcycling | 50 |
| Rock climbing | 7150 |
| Playing football/soccer | 5000 |
| Struck by auto | 20,000 |
| Influenza | 5000 |
| Leukemia | 12,500 |
| Earthquakes (in California) | 588,000 |

Source: Reproduced from McCullough (1993), with permission from the *Journal of the American Medical Association*, © 1993; all rights reserved.

A person is far more likely to die in a motorcycle accident or from rock climbing than from receiving a blood transfusion. Until 1999 the tests for infectivity mainly involved measuring antibody to the virus. An important recent addition to blood safety came in 1999 with Nucleic Acid Amplification Testing (NAT), which employs a new technology that specifically can detect very small amounts of the genetic materials of viruses like HCV, HIV, and West Nile Virus (WNV).

## The social ecology of blood safety

Most infectious agents transmissible by transfusion have been around for many years, so in general it is prior donor activity that determines whether a unit of blood is infected or not. Occasionally a new microbe, like the human immunodeficiency virus (HIV) or *Babesia microti*, is introduced. And, while West Nile Virus may have been in Africa for some time, it was first identified only recently in the United States. Regardless whether it is an ancient or new agent, similar human activities are likely to underpin their entry into and transmission along the medical blood supply. They include intravenous drug use with needle-sharing, high-risk or multiple partner sexual activity, exposure to insect vectors, travel, population migration, and medical negligence, and each may be influenced by economic, political, and social conditions. These subjects are discussed elsewhere in this volume, so this chapter will focus on their specific impact on blood safety.

By far the greatest challenge to blood safety today is in the developing world, primarily in sub Saharan Africa and parts of Asia, as a consequence of inadequate resources and dated technology in the health sector. The result has been epidemics of transfusion-related HIV, hepatitis, and other infectious diseases. Failing to test all donated blood for infectious agents is thought to have caused up to 16 million HBV infections, 5 million HCV infections, and 160,000 cases of HIV worldwide (Heyns *et al.*, 2005). In 1975, a World Health Assembly (WHA) resolution called for countries to adopt nationally coordinated, voluntary blood transfusion services. So far, fewer than 30 percent of the signatory states have done so. In contrast, developed nations have made a deliberate policy to ignore cost in order to emphasize safety. The result has been steady increases in transfusion safety and cost. In the United States, the Food and Drug Administration (USFDA) has been regulating blood-banking for several decades. However, around 1990 the FDA also started applying the regulations for pharmaceutical manufacture to blood-banking. Also, in its mission to protect the public health, it is not permitted to consider cost when promulgating rules and conducting oversight – so, even though proven not to be cost-effective, the FDA approved, on an interim basis, p-24 antigen testing for HIV. The FDA at the time was looking ahead to when a more effective NAT test would be available. Still, one study estimated it costs about $2 million to prevent one case of transfusion-related HIV (Heyns *et al.*, 2005). Currently, NAT is routinely performed to detect HIV-1, hepatitis C Virus (HCV), and West Nile virus. Figure 7.1 illustrates the steady rise in the price of blood-banking services, due in part to increased testing. It should also be noted that in earlier years blood centers traditionally had underpriced the cost of providing red cells for transfusion.

The challenge is to make widespread testing commonplace in countries where cost is a limiting factor. Currently, advanced testing for HIV and HCV, and bacterial culture tests on platelets, are not available in most developing countries.

## Effective public health strategies: incentives, education, screening and procedures

Relatively low-cost, low-technology public health strategies continue to be the first-line and most important defenses to protect blood safety. They include

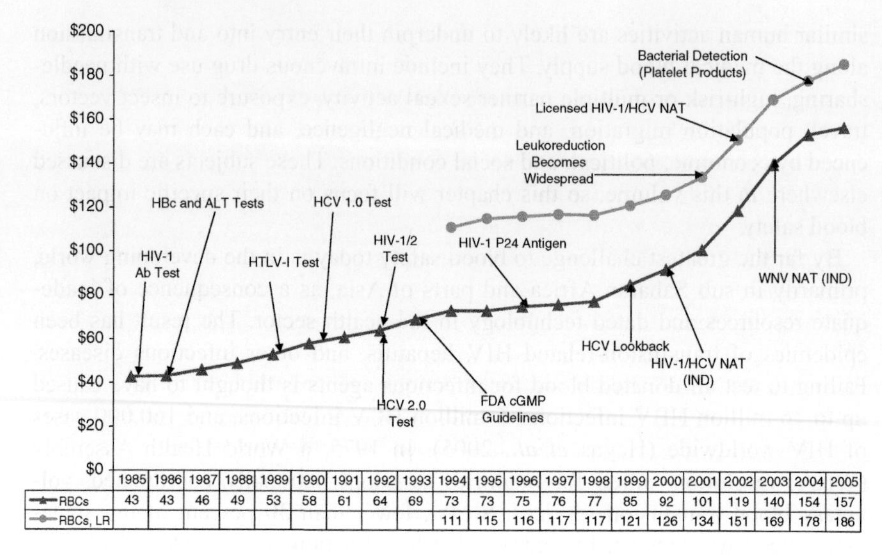


| | 1985 | 1986 | 1987 | 1988 | 1989 | 1990 | 1991 | 1992 | 1993 | 1994 | 1995 | 1996 | 1997 | 1998 | 1999 | 2000 | 2001 | 2002 | 2003 | 2004 | 2005 |
|---|---|---|---|---|---|---|---|---|---|---|---|---|---|---|---|---|---|---|---|---|---|
| RBCs | 43 | 43 | 46 | 49 | 53 | 58 | 61 | 64 | 69 | 73 | 73 | 75 | 76 | 77 | 85 | 92 | 101 | 119 | 140 | 154 | 157 |
| RBCs, LR | | | | | | | | | | 111 | 115 | 116 | 117 | 117 | 121 | 126 | 134 | 151 | 169 | 178 | 186 |

**Figure 7.1** Graph of safety measures and mean red cell fees, 1985–2005. Reproduced with permission of America's Blood Centers.

establishing an altruistic system for blood donation, effective public education, rigorous professional training, and continuous epidemiological surveillance and consistent front-line screening (not testing) of prospective donors *before they give blood*. While the array of accurate laboratory screening tests continues to grow, it is probably fair to predict that there never will be a fully effective laboratory test for every infectious agent that can be transmitted via blood transfusion. The environment of infectious microbes is always changing, as is the ecology of human activities that can spread them. It is therefore realistic to expect that the next infectious agent will be present in asymptomatic blood donors before there is a test for it. Other sources of risk include the donor that is infectious but not yet positive by available testing; an infection that is immuno-silent; lab error; and the eruption of a new infectious agent or new variant of a known agent. With this reality in mind, continuous epidemiological surveillance to uncover changing disease patterns and a willingness to be creative, diligent, and rapid in adapting to new challenges are the keys to maintaining public health. In addition, as we learned from the outset of the AIDS epidemic, public health measures will have to adapt to ethical and legal parameters of pluralistic societies. The challenges to maintaining sufficient blood supplies for life-saving medical needs are constantly changing while demand is always increasing. Not long ago, there were concerns related to discriminating against vulnerable and stigmatized groups. Today, a concern is rejecting people that have traveled to or lived in Western Europe.

## The voluntary donor system

Voluntary donation continues to be the foundation for maintaining blood safety. To date, only about 40 countries in the world have established systems for fully voluntary blood donation. This includes only about half of the 52 WHO European Member States, with 17 more in the process of developing one (WHO, 2005b). The problem is primarily in Eastern Europe. Africa is less advanced, with only about 15 percent of the 46 Member States in the African Region having a voluntary system, and only 30 percent have plans on the table to institute such a policy. It is estimated that about 60 percent of transfusions come from family donors, who are known to be more likely than voluntary donors to have a transmissible infectious agent (WHO, 2002).

The United States has a fully voluntary system. Most collection centers in the United States stopped paying for blood in the late 1960s, and with that came a clear reduction in transmission of hepatitis by transfusion. In the early 1970s in the United States there was a risk of approximately 6–8 percent of contracting some form of hepatitis from a transfusion. The risk now of becoming infected with hepatitis B is about 1 in 250,000, and the risk of acquiring hepatitis C is less than 1 in 3300 units transfused. This remarkable improvement in reducing post-transfusion hepatitis may be considered a marker for overall blood safety that occurred from the mid-1960s onward with the US national shift to voluntary blood donation. To appreciate the role of voluntary donation in this improvement, it is helpful to note that at about this time Australia Antigen testing (HBsAg Assay) for hepatitis B was introduced. However, the Australian Antigen test does not detect hepatitis C (previously called Non A/Non B), and HCV transmission declined over these years. It seems reasonable to conclude that voluntary donation works. After 1970, only a very few institutions in the United States continued to pay certain blood donors, and the last few stopped the practice in the past 15 years. Since 2004, we believe that all blood components transfused in the US come from volunteer sources.

In summary, the most critical area for change is in the developing world, where well-established voluntary donor systems still do not exist. The consequence is a much higher rate of transfusion-related infection in these countries, due to agents like HIV, hepatitis B and C, and Chagas' disease, where it is found only in the Americas (WHO, 2004). In response to this public health problem, the World Health Organization strongly advocates that "Member States promote national blood transfusion services, based on voluntary non-remunerated donations, and promulgate laws to govern their operation" (see also the May 1975 World Health Assembly resolution WHA 28.72). And the problem is not isolated to Africa. WHO Europe (2006) notes:

> The spread of the HIV/AIDS epidemic makes this issue {voluntary donation} of primary concern for many countries of the WHO European Region. The latest data show that in Western Europe, where blood donations are mainly non-remunerated,

HIV prevalence has declined regularly over time to 1.3 per 100,000 donations (2002). In Eastern Europe, where the national blood supplies are mainly based on paid or family replacement donors, HIV prevalence has increased alarmingly during the last years, up to more than 40 times when compared to some Western European countries.

## Public education and health screening

Another element in the first-line defense system is effective public education and health screening. In the United States, it is estimated that each year less than 5 percent of individuals eligible to donate blood actually do so. A donor must be in good health, weigh at least 110 pounds, and be at least 17 years of age (some states permit donors under the age of 17 with parental approval). Most blood banks have no upper age limit, and an increasing number of seniors give blood. Getting more voluntary donors requires appealing to both individual self-interest and community spirit. The American Red Cross public education message points out that most of us will "face a time of great vulnerability in which we will need blood. And that time is all too often unexpected" (American Red Cross, 2006a).

An effective public media campaign also involves educating individuals about the need for voluntary exclusion, so that people in a high-risk group or in poor health do not try to donate blood. At present in the United States, guidelines for ineligibility include (Food and Drug Administration, 2002):

- a prior history of illegal intravenous drug use
- a man who has had sexual contact with another man (MSM) since 1977; however, this is currently being reconsidered, the proposal being to accept MSM only if there has been no MSM activity in the last year
- any history of receiving clotting-factor concentrates
- a positive HIV test
- a history of engaging in sex for money or drugs since 1977
- a history of hepatitis since age 11
- a history of babesiosis or Chagas' disease
- ever having taken Tegison for psoriasis
- risk factors for Crueutzfeldt-Jakob disease (CJD), which include a family member with CJD, receipt of a dura mater transplant or administration of human pituitary derived growth hormone, and/or transmission via reusable instruments employed during brain surgery
- risk factors for variant Creutzfeldt-Jacob Disease or vCJD, which include three months or more spent in the United Kingdom from 1980 through 1996, or five years spent in Europe from 1980 to the present.

Public health prevention also includes identifying and refraining from collecting blood at high-risk locations, such as prisons and mental health facilities, where there is a high-risk for hepatitis; and, even though there is an effective test for HIV contamination, avoiding community settings with a relatively high concentration of men who have sex with men, because of the risk of HIV infection. To some, this policy is controversial.

Educational materials provided in advance contain clear and specific instructions that list the exclusions for giving blood, such as recent travel to a geographic location where malaria is endemic, or residence in Europe where variant Creutzfeldt-Jacob Disease (vCJD) is present (see below). Upon entering the blood collection center, prospective donors are given written materials with information on the risks and symptoms associated with infectious diseases transmitted by blood transfusion. They are asked questions and then given a form to sign indicating they have read and understood the material, and have provided accurate personal information. Questions also are asked about donor safety, such as a medical history of heart disease, current fever, or other sign of infectious disease. At that point, they can elect to leave without giving blood.

Donors then are asked clear and specific questions about behaviors that increase the risk of carrying an infectious agent that can be transmitted by transfusion, such as injecting drugs, sexual activity, recent tattooing, and travel. These questions are updated as needed to account for changing infectious disease epidemiology. For example, in response to the newly emerging AIDS epidemic of the early 1980s, blood collection centers began asking men about sexual contact with other men. Even though a virus had not yet been isolated and determined to cause AIDS, population-based data were showing the new disease to be behaving like hepatitis B, which already was well known to be caused by a virus and spread by sexual activity, by sharing intravenous needles during recreational drug use, and by blood transfusion. In general, the public health approach is to continue an effective screening measure until population-based data points conclusively in another direction. It is rare to discontinue the strict application of a screening protocol until there is convincing data over a sustained period of time. Thus, it was several years before Haitians in the United States stopped being considered a separate risk group for HIV.

In the United States, the policy generally is to screen for individual behavior, not group ethnicity. The first instructions for identifying AIDS were promulgated in January 1983. In March 1985, a laboratory test became available to screen for HIV in donated blood. The test was developed within a year, and was made available for donor screening prior to large-scale experience with the screening test and its confirmatory test, the Western Blot. Accordingly, as experience developed, with millions of blood donations being tested, information regarding the false-positive rate became available and the test was modified over the next few years to improve its specificity. Still, the screening questions were not changed until solid epidemiological information became available. To do this, donors with positive test results were interviewed and specific criteria for Western Blot positivity began to be developed by looking at individuals who had been considered Western Blot-positive but had no risk factors for disease. Then the exclusion policy was revised to ask donors specific questions about behavior. Guidelines recently were revised for tattooing, now done on a state-by-state basis based on each state's own regulation of tattooing facilities.

## Epidemiological surveillance

Each month, the American Red Cross monitors the true positive rate of each of its infectious disease markers. Since the beginning of HIV testing, there has been a clear downward trend in units that confirm positive for HIV (Dodd, 1994). This reflects improvement in donor screening. There has also been new and more active coordination between agencies and organizations through weekly surveillance of West Nile Virus positivity in blood donors. This positivity actually precedes the CDC's identification of human cases of West Nile Virus disease, so in this situation the coordination of blood-banking and general epidemiology has positive synergistic impacts for public health. Since all positive tests on blood donors are followed up individually by the local blood region, a local increase in confirmed positives suggesting a local epidemic can then be further followed up by local public health entities. And local public health departments notify the local Red Cross about changes in disease epidemiology, so blood collection centers can refrain from blood collection drives in potential epidemic situations.

## Laboratory testing

In the past 15 years there has been a proliferation of new and better laboratory tests for infectious diseases that can be transmitted by transfusion. Before 1985, blood products were tested only for antibodies to *Treponema pallidum* (the bacterium that causes syphilis), and the Australian Antigen for hepatitis B surface antigen (HBsAg). Additional tests were developed for hepatitis B and C, HIV 1 and 2, and HTLV-I. Testing for parasites has been discussed for many years, but the FDA has yet to license a test to screen for malaria or Chagas' disease. With the risk of contamination from viral agents diminished, there is now a greater emphasis on developing assays for parasites.

In the United States, the policy is to screen each unit of blood as soon as a new assay is licensed; this currently includes the following laboratory tests (see Table 7.4):

- syphilis
- hepatitis B (HBV)
- hepatitis C (HCV)
- human immunodeficiency virus (HIV-1 and HIV-2)
- human T-lymphotropic virus (HTLV-I and -II)

**Table 7.4 How blood is tested**

| Disease | Test | Implemented | Discontinued |
|---|---|---|---|
| HIV/AIDS | HIV-I Antibody test | 1985 | |
| | HIV-1/2 Antibody test | 1992 | |
| | HIV-I p24 Antigen test | 1996 | 2003 |
| Hepatitis B | Hepatitis B surface antigen | 1971 | |
| | Hepatitis B core antibody | 1987 | |
| Hepatitis C | Anti-HCV | 1990 | |
| Hepatitis | ALT | 1986 | 2003 |
| Syphilis | Serologic test for syphilis – TP or RPR | 1948 | |
| Human T-cell Lymphotropic Virus (HTLV) | HTLV-I Antibody test | 1989 | |
| | HTLV-I/II Antibody test | 1998 | |
| Hepatitis C and HIV/AIDS | Nucleic acid testing (NAT) | 1999 | |
| West Nile Virus | Nucleic acid testing (WNV-NAT) | 2003 | |

Note: CMV testing is performed on some units of blood for patients who require CMV-negative blood – for example, neonates weighing less than 1500 g, and immuno-compromised or immune-suppressed patients.
Source: American Red Cross, 2006b (http://www.pleasegiveblood.org/education/blood_tested.php)

- West Nile Virus (WNV)
- Testing of platelet donations for bacterial contamination.

Once approved, a laboratory test will be conducted on all donations until either a new and better test is licensed, or population-based data indicate that it is no longer productive. For example, ALT testing for hepatitis continued until 2003, and was discontinued only when a more specific and sensitive test became available to detect HCV. Similarly, HIV1 p24 antigen testing was discontinued in 2003, but only when a specific and sensitive nucleic acid amplification test (NAT) was licensed for HIV1 and added to routine donor screening. From 15 March 1996 to 27 March 1999, HIV-1 p24 antigen testing was performed on approximately 45,000,000 units. Even with additional neutralization testing, there were many false-positive test results. Still, testing continued, and during this period only five donors were found to be HIV-antigen positive and HIV-antibody negative – which was about 1 in 9 million units tested. Currently, the United States is the only country to test for HIV-RNA (the United Kingdom and others test for HCV-RNA, but not for HIV-RNA). It is estimated that RNA testing reduces the risk of HIV and HCV transmission to about 1 in 2 million blood units.

It should be noted that there can be a difference between a laboratory test available for blood testing, and that licensed for the clinical assessment of patients. For example, there are diagnostic clinical tests for Chagas' disease and malaria, but still none licensed for screening blood. In contrast, the laboratory test for anti-HIV1 was licensed for blood-donor screening before it was approved for patient use.

## High-risk human behavior

On 13 May 1981, John Paul II was shot and critically wounded in an assassination attempt. During the five-hour surgical procedure at a Rome hospital to repair his wounds, the Pope was given six pints of blood. On 20 June 1981, he was hospitalized with a high fever and inflammation of the right lung. The Pope was tested and found to be infected with cytomegalovirus (CMV), a herpes-type virus that can be transmitted by sexual contact and blood transfusion. He had acquired CMV from the blood administered during surgery. At the time, neither Italian law nor standard medical practice required the transfused units to be tested for CMV (Catholic News Service, 2005). A number of years later, filtering the white blood cells' blood components came into common use, which reduces the risk of CMV from transfusion. Pope John Paul's experience illustrates the link between human behavior and transfusion risk. There are a number of infectious agents associated with human high-risk sexual behavior and injecting drug use that are of special concern for safety of the blood supply. The first to be identified was syphilis, for which testing began just after World War II, at a time when the rate of syphilis infection was much higher in the general population. Then and now, however, the risk of acquiring syphilis via blood transfusion, if present at all, is exceedingly small – in fact, in the United States there have been no recorded cases transmitted by transfusion in many years, and only one report in the literature (in the 1970s) of a poorly documented case of syphilis transmitted by transfusion. Moreover, the spirochete that causes syphilis is fragile and becomes inactive during the first few days of refrigerated blood storage. The same cannot be said for hepatitis B, HIV, HTLV, and CMV, however, as these infections are spread sexually and can also be transmitted by transfusion.

### Impact of injection drug use and sexual activity

Since it was recognized in late 1982 that blood transfusions could transmit HIV, there has been increased international attention on the link between sexual activity, injecting illicit drugs, and blood safety. Epidemics related to high-risk sexual activity, such as sex work and trading sex for drugs, are discussed in Chapter 2, and epidemics related to intravenous drug use are discussed in Chapter 3. Both types of behavior are known to impact the safety of the blood supply, especially when donors are paid. Blood is an ideal medium for harboring, growing, and transmitting the agents that are spread by these behaviors, such as HIV, the hepatitis B and C viruses, HTLV, CMV, and others. To combat this problem, the Federation of Red Cross and Red Crescent Societies has called for expanded drug abuse prevention, harm reduction, and public education. They believe that "National Societies, with their unique combination of volunteer support and community membership alongside their status as auxiliaries to the public authorities in the humanitarian field, have a special contribution to make to the best approach to the challenges at the national and community level" (Kopketzky, 2005). For more than two decades international attention has been on the AIDS epidemic, where HIV, like the hepatitis B virus, is known to spread by sexual contact and IVDU. Of note is that five countries in the former Soviet Union and Asia that have combined populations of almost 2 billion are seeing HIV epidemics of more than 50,000 registered cases per country (Wolfe and Malinowska-Sempruch, 2007). These epidemics are associated with high-risk human sexual activity and injection drug use, and these nations also have limited safety controls on blood banking. Even with improved laboratory screening, the front-line defense in these states, as in developed nations, is effective public health screening. The US guidelines that relate most directly to uncovering high-risk sexual activity and IVDU also exclude any individual with AIDS or a positive HIV test. In addition, they exclude individuals that have:

- a history of injecting illicit drugs, including steroids and other medications not prescribed by a physician
- any man who since 1977 has had sex with another man, even once
- anyone who since 1977 who has ever taken money, drugs, or other payment for sex
- anyone who since 1977 was born in, lived in, or received a transfusion or medical treatment in Cameroon, Central African Republic, Chad, Congo, Equatorial Guinea, Gabon, Niger, or Nigeria, or *had sexual contact* in the past 12 months with anyone described above.

Type O is a rare form of HIV found primarily in West Africa. As of this writing, there is no laboratory test approved in the United States to detect Type O in donated blood. In 1996, the United States Food and Drug Administration promulgated rules that prohibit blood donation by individuals born in these West African States after 1977, and persons that have had sexual contact with someone born in one of these nations since 1977.

## Impact of migration, travel, and geography

While travel, migration, and trade are expanding rapidly over the globe, the movement between countries of blood components for transfusion is still relatively uncommon, and to date the blood supply of developed nations has remained remarkably immune from infection by travel and migration. For example, it is estimated that 300–500 million people have malaria, but perhaps only a handful of cases each year in the United States are due to infection via the blood supply. Similarly, while Chagas' disease is endemic in Mexico, Central and South America, and millions of people from these countries travel annually to America, the reported incidence of Chagas' disease due to blood transfusion in the United States is very, very low (Chamberland *et al.*, 1998). Nevertheless, concern is real because blood is an extraordinarily effective vehicle for transmitting infectious agents (Chamberland *et al.*, 1998). Since a fully effective laboratory test for every blood-transmissible infectious disease in the world is not on the horizon, the current first line of defense will have to remain effective health screening and public education *before* individuals give blood. It also is hopeful to note that new laboratory tests to screen human blood are being developed. As noted already, in the wake of improved screening tests for infections transmissible by transfusion, we are seeing a greater emphasis on developing screens for parasites (e.g. Chagas' disease and malaria) that can infect the blood supply. Even with these population shifts, the blood supply in developed nations remains remarkably safe, as illustrated in Table 7.5, which shows the low rate of transfusion-related infection due to malaria and to *T. cruzi* infections not endemic in the United States, as well as babesiosis, which is endemic in the US.

**Table 7.5 Estimates of the risk of infectious complications of blood transfusion: Babesia, malaria, *T. cruzi***

| Agent | Frequency per million units (per actual unit) | Clinical disease | | |
|---|---|---|---|---|
| | | Acute per million units | Chronic per million units | Deaths per million units |
| Babesia | >1 (>1/100,000) | 1.0 | 0 | 0.25 |
| Malaria | >4 (>1/250,000) | 4.0 | 0 | 0.00 |
| *T. cruzi* | 24–80 (1/12,500–1/50,000) | 0.0–1.0 | 0 | 0.00 |

Source: Adpated from Dodd (1994).

## Malaria

Between 1958 and 1998, the Centers for Disease Control recorded 103 cases of transfusion-transmitted malaria in the United States. Malaria is not endemic in North America, so these cases most likely were the result of blood donated by people who were asymptomatic carriers. In the United States, potential donors are screened by questionnaire and asked to refrain from giving blood until one year after visiting a malarial area, three years after completing treatment for malaria, or three years after living in an area where it is endemic. While very rare in the US, malaria is epidemic in tropical areas and causes serious health consequences, including death. There is no practical test available to screen donors. Table 7.5 illustrates that the US estimate of malaria contamination to the blood supply currently is small, at about just over 1/250,000 units with no fatalities. However, the problem is wholly different in tropical countries, particularly in sub Saharan Africa. Worldwide malaria is the most significant cause of death due to parasitic infection. The World Health Organization estimates there are at least 300 million acute cases of malaria each year globally, which lead to more than a million deaths. Probably 90 percent of the deaths occur in Africa, mostly in young children. Blood transfusion is used to treat life-threatening malaria in young children, but also poses the risk of acquiring HIV infection (Obonyo *et al.*, 1998; WHO, 2006a).

## Creutzfeldt-Jacob Disease (CJD) and variant Creutzfeldt-Jacob Disease (vCJD)

CJD is a rare degenerative, fatal disorder of the central nervous system. Worldwide, its incidence is about one person per million per year. CJD can affect humans in three ways: sporadic CJD, which has no known risk factors and accounts for 85 percent of CJD cases; hereditary CJD, which occurs in individuals with a family history of the disease and tests positive for specific genetic mutations; and acquired CJD, which is transmitted by exposure to brain or nervous system tissue. Acquired CJD accounts for less than 1 percent of CJD cases, and has occurred most in individuals that received repeated injections of human pituitary gland growth hormone. This preparation was prepared from pools of pituitary glands from a number of cadavers. Through the 1950s it was used to treat congenital dwarfism, although this practice was subsequently supplanted by a synthetic growth hormone preparation. CJD also has been transmitted to patients who have undergone brain surgery, including transplant of the dura mater – the covering of the brain and spinal cord. This material was harvested from human cadavers and then used in some brain operations. The dura mater had been stored in a vat along with dura mater from other cadavers that had contaminated material. Affected individuals may take decades to develop symptoms, and then progress rapidly to dementia, severe loss of coordination, and death. While the cause of CJD remains uncertain, the suspicion now is that it occurs in response to abnormal changes in the shape of brain prions. Currently, there is no screening test for the disease, and while blood transfusions have never been shown to transmit CJD, as a precaution the Food and Drug Administration (FDA) prohibits blood donation by individuals who may be at risk. These include potential donors who have received injections of human-derived pituitary hormone, those with a family history of CJD, or those who have had surgeries that involved transplanted dura mater.

Similar to CJD, and an issue with regard to travel, is variant Creutzfeldt-Jacob Disease (vCJD). It also is a rare, degenerative and very similar fatal disorder of the central nervous system, thought to occur after humans have eaten beef contaminated with bovine spongiform encephalopathy (BSE, or "mad cow" disease). In 1996 the first cases of vCJD were reported in the United Kingdom, and there soon was concern on the part of blood banks and public health officials that it could contaminate the blood supply. To date the problem has remained quite rare, with only cases in the United Kingdom (UK) and a few in France and Italy. There has been one case in the United States, but it was in a person who had lived in the UK during the period of greatest risk. So far there have recently been two cases, both in the UK and presumed to have been transmitted via blood transfusion. The USFDA policy seeks to strike a reasonable balance between guarding against the risk of spreading vCJD through blood-banking, and preserving the supply of blood products for medical use. Currently, the policy relates to travel and residence in Europe, and recommends the following individuals be deferred indefinitely:

- those that spent a total of three months or more in the United Kingdom (UK) from the beginning of 1980 through the end of 1996; or who have spent a total of five years or more in Europe from 1980 to the present
- current or former US military personnel, civilian military employees, and their dependents that resided at US military bases in Northern Europe (Germany, UK, Belgium, and the Netherlands) for a total of six months or more from 1980 through 1990; or that resided elsewhere in Europe, such as Greece, Turkey, Spain, Portugal, or Italy, from 1980 through 1996
- those that received any blood or blood component transfusions in the UK between 1980 and the present
- persons that injected bovine insulin since 1980 from cattle raised in the United Kingdom, unless it is possible to confirm the product was not manufactured after 1980 from cattle in the UK.

Guidelines also were issued in 2001 by the US Department of Defense (DoD). They also relate to travel and residence in Europe, and recommend active-duty military personnel, civil service employees, and their family members defer indefinitely from giving blood if:

- they traveled or resided in the UK for a cumulative total of three months or more at any time from 1980 through the end of 1996
- they received a blood transfusion in the UK at any time from 1980 to the present
- they traveled to or resided anywhere in Europe for a cumulative total of six months or more at any time from 1980 through the end of 1996; or traveled to or resided anywhere in Europe for a cumulative total of five years or more at any time from 1 January 1997 to the present.

These policies are under constant review by the FDA in light of new information about vCJD and BSE.

There has been much discussion about the development of a test for vCJD that could be used to screen blood donors, but there has been little real progress. In order to be sensitive enough to identify virtually all potentially infectious donations, almost all screening tests will have a false-positive rate. To deal with this it is necessary to have a suitable confirmatory test prior to beginning donor screening. The rationale is that, with the risk of transmission by transfusion being so small, it would be unacceptable to be in a situation of notifying a fair number of volunteer blood donors that they have a test result suggesting they may later develop a fatal disease for which there is no treatment, and when there is uncertainty about the meaning of the test result.

## Chagas' disease

Almost 100 years ago, the Brazilian physician Carlos Chagas discovered the parasite *Trypanosoma cruzi*, which causes Chagas' disease. Today, it infects as many as 18 million people worldwide. Once established in the body the infection is lifelong, and several thousand South and Central Americans die annually of heart and digestive problems caused by the parasite. Up to 20 percent of infected people never exhibit symptoms. Over the years, Chagas' disease has often been transmitted by transfusion in South America. Prior to the availability of a test, and since, there has been no treatment for Chagas' disease. Earlier, gentian violet was added directly to units of blood prior to transfusing them. More recently, testing for Chagas' disease has become routine in much of South America. While the infection remains rare outside of South and Central America, it has come to be considered a threat to the blood supply because of global travel and migration, especially from South America to North America. To date, there have been only seven cases of transfusion-transmitted Chagas' disease reported in North America (two of them in Canada), but it is estimated that several million people from countries where Chagas' disease is present now reside in the United States, and 100,000 or more may be infected with *T. cruzi* (Kirchoff *et al.*, 1987). In several investigational studies of US blood donors with confirmed sensitive tests, so far there has been only a handful of cases of blood donors with confirmed positive tests, but not all have been tracked to donors from countries where Chagas' disease is endemic. As a precaution, the American Association of Blood Banks' guidelines prohibit blood donation from anyone who has had Chagas' disease. Laboratory and screening tests are now under development. It is expected that, when approved, such testing will be implemented routinely on all donations.

## Babesiosis

Babesiosis is a parasitic infection of red blood cells caused by the protozoa *Babesia microti*, similar to malaria, and carried by the white-footed mouse and transmitted by deer tick bites. It appears primarily in the northeastern United States, in coastal areas that are home to the white-footed mouse. Cases also have been identified in the Upper Midwest and Pacific Northwest. Approximately 60 transfusion-associated cases have been reported in the United States. While babesiosis often is quite mild, some patients (including those without a spleen, the elderly, or the immunocompromised) may be at risk of serious illness. Occasionally babesiosis is misdiagnosed as malaria, but when correctly diagnosed appropriate antibiotic therapy is effective. There are no useful tests available for screening blood donors, although testing strategies are being discussed. The American Association of Blood Banks (AABB) requires that all donors be asked if they have a history of babesiosis, and individuals with a history of the disease are deferred from donating blood.

## Lyme disease

The social ecology of Lyme disease is related to the migration of the population in developed nations to suburban areas (see Chapter 5). It is associated with the bite of the same species of deer tick as is the vector for babesiosis, and can cause an illness that affects many systems within the body. Despite a number of researchers looking for examples of Lyme disease transmitted by transfusion, none has been found; and this is probably because it is present in blood and potentially capable of being transmitted by transfusion for only a very short period of time. There is one interesting case of an individual thought to have been positive for both Lyme disease and babesiosis whose blood donation transmitted only babesiosis but not Lyme disease (Cable *et al.*, 1993; McQuiston *et al.*, 2000). While transfusion-related cases have not been reported, public health agencies and the AABB are monitoring this disease because of the remote chance that it could affect transfusion safety. People with a history of Lyme disease can give blood provided they have completed a full course of antibiotic treatment and no longer have any symptoms.

## Ehrlichiosis

First discovered in 1994, human ehrlichiosis (HE) is a bacterial infection caused by several types of rickettsiae that spread by tick vectors from dogs and other animals to humans. Like Lyme disease, ehrlichiosis is occurring in a context of widening suburban sprawl in North America, Europe, and Japan. Once in the bloodstream, the bacteria invade and kill white blood cells. There are two types of ehrlichiosis: human granulocytic (HGE) and human monocytic ehrlichiosis (HME). The threat to the blood supply appears to be primarily from HGE. From 1986 to 1997, there were only 449 cases of HGE reported in the United States. Symptoms include relatively mild and self-limited fever, headache, and malaise. However, 10–20 percent of people with HGE go on to suffer encephalitis, acute respiratory distress syndrome, and opportunistic infections, and up to 5 percent of infections may be fatal. Treatment with antibiotics such as tetracylines is effective. The incidence of HGE is not well understood among blood donors, with perhaps only one presumptive case in a blood recipient. Preliminary studies indicate that HGE can survive in refrigerated blood for up to 18 days (Walker and Dumler, 1997; Leiby *et al.*, 2002).

## Severe acute respiratory syndrome (SARS)

SARS is a respiratory infection that can produce serious complications. Most cases identified have been in Asia, but there have been outbreaks elsewhere, including the United States and Canada. To date, there has been no evidence of SARS transmission through blood transfusion. However, the potential exists, because the virus associated with SARS is present in the blood of people who are sick. If it were possible for the virus to be present in blood before an individual becomes ill and for that person to donate blood, then that individual potentially could introduce the virus into the blood supply. To prevent this from occurring, blood collection facilities ask potential donors orally or in writing about recent travel to a SARS-affected country, a history of SARS infection, or possible exposure to SARS. Because the risk of contracting SARS through a blood transfusion theoretically exists during periods of SARS epidemics anywhere in the world, anyone who has traveled to a SARS-infected area or who has had close contact with a person with SARS or someone suspected of having SARS is asked to refrain from donating blood for at least 14 days after arriving in the United States. "Close contact" is defined as having cared for, lived with, or had direct contact with the respiratory secretions and/or body fluids of a person known to have, or to have had, SARS. Anyone who has been ill with SARS or suspected SARS should refrain from donating blood for 28 days from the last date of treatment and symptoms. In addition, all donors are asked to call the blood center after donating if they then become ill. If this call occurs prior to the blood being transfused, the unit(s) can be intercepted and destroyed. As long as a donor is and remains well, no other measures are necessary.

## Negligence, human error and failed oversight

In the United States and other developed nations, errors in blood component preparation and testing are nowadays very rare – primarily due to bar-coding of all donor registration forms, blood donations, components, and sample tubes, as well as to large-scale automation of viral testing and testing for blood type, with computerized assimilation of all these test results to each pertinent blood donation. One type of error that can still occur is at the blood collection site, in the labeling of the form, the donation and/or the tubes of collected blood. Below are two examples of how human errors such as this might occur.

1. While each blood donor is handled separately from the others, a staff person evaluating the donor's history can turn the donor and the form over to another staff person, and then to the venipuncture technician to collect the blood and the tubes for testing. As a check, a second staff person is tasked to ask the donor to provide his name again so it can be checked with the paperwork before collecting the unit. If this check is not performed correctly, one donor's identifying information can become associated with another donor's unit of blood. While the donated unit will have all of the test results associated with it, if the donor subsequently calls back (this is an infrequent event) saying that he has developed a fever or has just learned of recent prior exposure to hepatitis, the wrong unit of blood will be recalled, and the one that should be recalled could be transfused.
2. A more ominous error would be a mix-up between the "whole blood number," with its associated bar code of one unit of blood, with the numbering on the tubes for testing from another donation. The bar code and number are manually placed on each of three or four tubes for testing, and on the unit of blood itself. If a staff person is interrupted during the placing of these bar codes on tubes and bags, or if the wrong tubes are picked up for filling just before or after collection of the unit of blood, then the test results will be associated with the wrong unit for transfusion. If one donor's test results are positive, the unit could still be labeled as acceptable due to clear test results from the other donor, and transfused. Since 80 percent of blood donations are from repeat donors and the blood type on prior donations is checked before labeling the current donation, the blood center most probably will know an error has occurred, will investigate, and will prevent the labeling and release of all

potentially affected units if the blood types on the two donors are different (e.g. one is A positive and another is O negative). Finding such a discrepancy is a very rare event, due to the emphasis on training collection staff about the importance of donor identity and proper handling of blood samples and unit labeling. In addition, hospitals are required to re-check the ABO and Rh type of each unit of labeled blood they receive from a blood center before issuing it for transfusion to a patient. The consequence of adhering to strict procedures is that finding such a discrepancy at the hospital is a very, very rare event. Nevertheless, the practice of continuing this last check at the hospitals continues. If a test tube or unit does not have a proper number with bar code on it, it is not tested and not used, and the donation is discarded.

A number of high-profile scandals have occurred in blood-banking in developed nations, most notably during the first years of the AIDS epidemic, but also related to hepatitis C. One occurred in Canada, where in 1984 the Canadian Red Cross Society was importing plasma donations from the United States. At the time it was widely known that the AIDS epidemic in the United States had been associated with blood transfusion, especially among people with hemophilia. Meanwhile, Canada was not fully self-sufficient in its supply of blood products and, in particular, did not have adequate supplies of Factor VIII concentrate (the plasma derivative) for people with hemophilia. The Canadian Red Cross was faced with the difficult problem of weighing the risk of importing potentially tainted blood products against that of putting hemophiliacs at risk because of insufficient supplies of clotting agent. The Canadian Red Cross chose to continue importing US supplies. However, while the US Centers of Disease Control had in 1984 approved a test for HIV in blood, the Canadian Red Cross did not implement testing until late 1985.

A national scandal erupted when critics accused the government and the Canadian Red Cross of denial and negligence. In 1988, the Canadian Hemophilia Society requested compensation for victims. They were ignored. The victims persisted. In 1993, the Federal Government appointed a Commission of Inquiry, headed by Justice Horace Krever, to investigate the spread of HIV and hepatitis C through the blood-banking system. They concluded that Red Cross and Government oversight had led to failure to act in a timely manner to prevent the spread of HIV and hepatitis through the blood supply, that the system had been under-funded, and that Canada lacked a strong national blood policy with clear lines of responsibility across organizations and agencies involved in the entire process of blood-banking. Victims and families were awarded compensation. The Commission also concluded that donated blood is a national public resource, and safety is paramount. A national independent board was established, national standards were promulgated, and in 1998 the Red Cross was replaced by the new Canadian Blood Services to manage the blood-banking system. In 2004, federal, provincial, and territorial governments announced the country's first national standards to govern the handling of blood and blood products from "vein-to-vein." In May 2005, the Canadian Red Cross apologized and pleaded guilty to distributing tainted blood products. Dr Roger Perrault, the National Medical Director of the Canadian Red Cross at the time of the scandal, three other physicians, and Armour Pharmaceuticals (located in the United States) were put on trial for criminal negligence. That proceeding continues. It is estimated that perhaps 20,000 people were infected with hepatitis C, and about 1000 with HIV.

As in Canada, epidemics of contaminated Factor VIII concentrate products for hemophiliacs also occurred in France and Japan during the 1980s. Scandals also erupted when it was recognized that high-level government officials either knew or should have known that the products were unsafe, and were culpable for permitting them to remain in use. In France, perhaps 4000 individuals were infected. Three French officials were charged: the former Premier, Laurent Fabius; his superior at the time, the Social Affairs Minister Georgina Dufoix; and the Health Minister responsible for oversight, Edmond Hervé. At the core of the charge of negligence was that the three had delayed introducing the US blood-screening test into France until a rival French product was ready to go on the market. Hervé was convicted of negligence for HIV acquired by two recipients of the products, but the court failed to hand down punishment. Still, at least there was public recognition of official responsibility (BBC News, 1999). In Japan, more than 1400 hemophiliacs were exposed to HIV and at least 500 died as a result. The scandal raised a national outcry about whether the Japanese Government's health bureaucracy was too tied to the pharmaceutical industry, and thereby put profits over people. Akihito Matsumara, head of the Government Ministry responsible for handling blood and blood products, was charged with negligence and given a suspended one-year sentence for the death of a patient who in 1986 contracted AIDS from a contaminated transfusion. By that time, heat treatment to sterilize blood replacement products from HIV had been in place in the United States for two years (BBC News, 2001).

A much larger epidemic of HIV, hepatitis, and possibly other infectious diseases may have occurred in China as a result of unsafe practices by unscrupulous private blood collection companies too closely linked with government officials. These enterprises operated in Henan, Anhui Shaanxi, and Hebei provinces, where mostly poor rural Chinese were paid to donate blood. They re-used non-sterile blood collection equipment between donors and re-infused repeat donors with potentially tainted blood in order to reduce their symptoms of anemia and get them to donate more often. So far there has never been a full surveillance program to document the extent of the epidemic, but it could be very large, because the combined population of these provinces is about equal to that of Western Europe. Not only are the donors at risk, but also their sexual partners. Activists who protested that there was a Government cover-up were beaten and jailed. In February 2002, proof of official Government participation in the blood donor scheme appeared in the form of a videotape sent to the United Nations, the PRC Ministry of Health, and the news media. It showed 20 villagers' blood donor cards that had been issued by the Henan Ministry of Health, and that the individuals had been allowed to be repeat donors. It was not until March 2004, approximately seven years after the beginning of the epidemic, that the national Ministry of Health acknowledged that transmission of HIV had occurred in the 1990s due to blood collection. To date, no individual Chinese official has been held responsible (Agency France Press, 2004; Wan and Li, 2007).

## Responding to the challenges ahead

Maintaining a safe blood supply will require constant vigilance on the part of blood-banking and public health officials. No one could have predicted the enormous challenges created by the AIDS pandemic, and no doubt new challenges will arise in the years ahead. One such relatively new concern is terrorism (see Chapter 12). Officials in the United States are concerned about the possibility that terrorists could obtain access to the smallpox virus and use it against the public. Before 1972, vaccinia, the live virus used in smallpox vaccinations, was routinely administered to Americans. Smallpox can be highly infectious – for example, a scab at the inoculation site can contain infectious virus, so it is possible for a scab that spontaneously separates from the skin to inadvertently infect close contacts that touch the vaccination site or dressing. Another area of concern is blood-banking. In response to this potential threat, the United States Department of Health and Human Services (HHS) has been working with state and local governments to strengthen preparedness. Part of the strategy is to expand the national stockpile of smallpox vaccine, which is highly effective when administered shortly after exposure. To make certain the virus is not transmitted through a blood transfusion, potential donors will be asked at blood collection facilities about a history of vaccination or close contact with anyone who has been vaccinated. A vaccine recipient who has had no complications may donate after the vaccination scab has spontaneously separated, or 21 days after vaccination, whichever is the later. Individuals who receive a smallpox vaccination may be asked to refrain from donating for an interval of two months if a scab was pulled off or knocked off (i.e. did not spontaneously separate), and to refrain for 14 symptom-free days if there has been a complication or contact with a vaccine recipient that developed skin lesions or other complications. If smallpox vaccinations were to be administered rapidly to a large section of the population, this would have a substantial negative effect on the blood supply, particularly platelets. The current FDA recommendation is for individuals to refrain from donating blood for one month after smallpox vaccination.

Another future consideration for blood safety is "pathogen inactivation." Several large corporations have spent hundreds of millions of dollars exploring this. There are different approaches to pathogen inactivation of red cells, platelets, and plasma; since the functionality of red cells, platelets, and plasma proteins are different in their fragility, the function of each must be preserved to be effective. These approaches involve exposing each single blood component to a chemical, sometimes along with UV light, such that viruses, bacteria, and parasites would be inactivated and no longer infectious. These processes require addition of inactivating chemicals to the blood, then washing out these chemicals so the patient is not exposed to the chemical. To perform this inactivation on each component would be expensive and require very large laboratories, and would generate huge volumes of hazardous waste (the wash solution). Due to the increasing safety of blood for transfusion that has occurred in the last 20 years, it is questioned whether this (now) small incremental (theoretic) improvement in safety is worth the huge expense (and risk to staff in handling all these chemicals). Clinical trials of these processes have been terminated (red cells), though some continue (platelets).

Traditionally a conservative group, blood-bankers, sensitive to patient safety, public perception, and regulatory agencies (e.g. FDA), have been slow to change longstanding processes developed earlier to improve safety. Following this tradition, more and more tests will be added to each donation, more and more questions will be asked of all prospective donors, and more and more donors will be deferred – such that the blood supply will dwindle. To offset this, major initiatives to encourage donations from new donors, and more frequent donations from existing suitable donors, will continue to be needed.

For some time it has been a popular lay belief that "artificial blood" or "synthetic blood" will solve this problem. Truly synthetic blood is represented by perfluorocarbons that can carry and release oxygen. Several trials have been discontinued due to safety concerns. Other problems include that the beneficial effect is brief, administration of 100 percent oxygen is required during its use, and the recipient patient's blood is then milky, interfering with the interpretation of many clinical laboratory test results. These perfluorocarbons have no role other than oxygen delivery, and cannot supplant the need for platelet or plasma transfusion.

Some refer to stroma-free hemoglobin solutions as artificial blood, but the source is still blood. Advantages include extended shelf-life and more lax storage conditions, no need to match blood type, and presumed lack of infectious risk. Some clinical trials with this have been halted due to safety concerns as well, and the beneficial effect in the recipient is short-lived – again, it is just a bridge until transfusion of traditional red cells becomes available. A bovine-sourced hemoglobin continues to face multiple challenges in getting through regulatory and other hurdles.

There is no artificial substitute for platelets on the horizon. Several isolated plasma proteins have been synthesized (Factor VIII and Factor IX for example), and others can be (albumin) but it wouldn't be cost-effective. No substitute for gamma globulin is expected, since this requires the panoply of antibodies from >1000 donors to all the things these donors have been exposed to. This changes as the population is exposed to new infectious and other agents to which they make antibodies.

Multiple attempts to grow blood cells in tissue culture and propagate them have not yet been fruitful.

While pathogen inactivation may help to maintain safety for additional infectious threats, it would be expensive. It might eventually, however, permit more individuals to be eligible as donors, and various approaches to pathogen inactivation appear applicable to red cells, platelets, and plasma.

The bottom line is that for quite some time we will continue to be dependent on altruistic voluntary donations from more and more people, and more and more often. If eligible, please give blood. In the United States you can make an appointment at a site convenient to you by calling 1-800-GIVE LIFE. If not eligible, encourage friends, family and acquaintances that may be eligible to donate.

## References


Agency France Press (2004). Officials say most Chinese provinces may have AIDS from selling blood. Agency France Presse, 3 March.

American Red Cross (2006a). *Help Now Give Blood.* Available at: http://www.redcross.org/donate/give/ (accessed 9 May 2006).

American Red Cross (2006b). *Blood Donation Eligibility Guidelines.* Available at: http://www.redcross.org/services/biomed/0,1082,0_557_,00.html (accessed 9 May 2006).

BBC News (1999). World: Europe Blood scandal ministers walk free. BBC News, 9 March.

BBC News (2001). Japan blood scandal official convicted. BBC News, 28 September.

Cable, R.G., Krause, P., Badon, S. *et al.* (1993). Acute blood donor co-infection with *Babesia microti* (Bm) and *Borrelia burgdorferi* (Bb). *Transfusion* **33**(Suppl.), S50.

Catholic News Service, 2005. Available at: http://www.catholicnews.com/jpii/stories/story02.htm (accessed 6 January 2006).

Chamberland, M.E., Epstein, J., Dodd, R.Y. *et al.* (1998). Blood safety. *Emerging Infectious Diseases* **4**(3). Available at: http://www.cdc.gov/ncidod/eid/vol4no3/chambrln.htm (accessed 15 May 2006).

Dodd, R.Y. (1994). Adverse consequences of blood transfusion: quantitative risk. In: S.T. Nance (ed.), *Blood Supply: Risks, Perceptions and Prospects for the Future.* Bethesda: American Association of Blood Banks, pp. 1–24.

Dodd, R.Y. (2001). Germs, gels and genomes: a personal recollection of 30 years in blood safety testing. In: S.L. Stramer (ed.), *Blood Safety in the New Millenium.* Bethesda: American Association of Blood Banks, pp. 97–122.

Heyns, A.D.P., Benjamin, R.T., Swanevelder, J.P.R. *et al.* (2005). Blood supply and demand. *Lancet* **365**(9478), 2151.

Kirchoff, L.V., Gam, A.A. and Gilliam, F.C. (1987). American trypanosomiasis (Chagas' disease) in Central American immigrants. *American Journal of Medicine* **82**, 915–920.

Kopetzky, W. (2005). *Statement by Wolfgang Kopetzky, Special Representative of the International Federation and Secretary General of the Austrian Red Cross, at the UN Commission on Narcotic Drugs, in Vienna, 8 March.* Available at: http://www.ifrc.org/docs/news/speech05/wk080305.asp (accessed 1 January 2006).

Leiby, D.A., Chung, A.P.S., Cable, R.G. *et al.* (2002). Relationship between tick bites and the seroprevalence of *Babesia microti* and *Anaplasma phagocytophila* (previously *Ehrlichia* sp.) in blood donors. *Transfusion* **42**(12), 1585.

McCullough, J. (1993). The nation's changing blood supply system. *Journal of the American Medical Association* **269**, 2239–2245.

McQuiston, J.H., Childs, J.E., Chamberland, M.E. and Tabor, E. (2000). Transmission of tick-borne agents of disease by blood transfusion: a review of known and potential risks in the United States. *Transfusion* **40**, 274–284.

Obonyo, C.O., Steyerberg, E.W., Oloo, A.J. and Habbema, J.D. (1998). Current and emerging infectious risks of blood transfusions. *American Journal of Tropical Medicine and Hygiene* **59**(5), 808–812.

The National Blood Data Resource Center of the United States (2005). Available at: http://www.aabb.org/All_About_Blood/FAQs/aabb_faqs.htm (accessed 27 December 2005).

Schreiber, G.B., Busch, M.P., Kleinman, M.D. and Korelitz, J.J. (1996). The risk of transfusion-transmitted viral infections. *New England Journal of Medicine* **334**(26), 1685–1690.

United States Food and Drug Administration (2002). *Guidance for Industry: Revised Preventive Measures to Reduce the Possible Risk of Transmission of Creutzfeldt-Jakob Disease (CJD) and Variant Creutzfeldt-Jakob Disease (vCJD) by Blood and Blood Products*. January 2002. Available at: http://www.fda.gov/cber/gdlns/cjdvcjd.htm (accessed 15 May 2006).

US General Services, Administration (2006). *Progress in Blood Supply Safety* (by Monica Revelle). Available at: http://www.pueblo.gsa.gov/cic_text/health/blood-ss/blood2.htm.

Walker, D.H. and Dumler, J.S. (1997). Human monocytic and granulocytic ehrlichioses: discovery and diagnosis of emerging tick-borne infections and the critical role of the pathologist. *Archives of Pathology and Laboratory Medicine* **121**, 785–791.

Wan, Y.H. and Li, X. (2007). Consequences of stalled response. In: C. Beyrer and H.F. Pizer (eds), *Public Health and Human Rights: Evidenced-based Approaches*. Baltimore: Johns Hopkins University Press.

WHO Regional Office for Europe (2005). Press release EURO/11/05, Copenhagen, 13 June. Available at: http://www.euro.who.int/mediacentre/PR/2005/20050613_1 (accessed 6 April 2006).

Wolfe, D. and Malinowska-Sempruch, K. (2007). Seeing double: mapping contradictions in HIV prevention and illicit drug policy. In: C. Beyrer and H.F. Pizer (eds), *Public Health and Human Rights: Evidenced-Based Approaches*. Baltimore: Johns Hopkins University Press.

World Health Organization (2002). *Blood Safety: A Strategy for the African Region*. Brazzaville: World Health Organization Regional Office for Africa.

World Health Organization (2004) *Global Database on Blood Safety Report 2001–2002*. Available at: http://www.who.int/bloodsafety/GDBS_Report_2001-2002.pdf (accessed 8 August 2006).

World Health Organization (2005a). *Fact Sheet*. Available at: URO/05/05.http://www.euro.who.int/Document/Mediacentre/fs0505e.pdf (accessed 10 June 2005).

World Health Organization (2005b). *Supply of Safe Blood can Help Combat HIV/AIDS Epidemic: voluntary, unpaid blood donations cut the risk of infection through transfusion*. Geneva: WHO.

World Health Organization (2006a). *Roll Back Malaria*. Available at: http://www.rbm.who.int/cmc_upload/0/000/015/370/RBMInfosheet_3.htm (accessed 6 January 2006).

World Health Organization (2006b). *Voluntary Blood Donation*. Available at: http://www.who.int/bloodsafety/voluntary_donation/en/ (accessed 6 January 2006).

World Health Organization Europe (2006). *First Ever World Blood Donor Day Campaigns for Boosting Voluntary, Non-remunerated Blood Donations*. Available at: http://www.euro.who.int/aids/prevention/20040614_3 (accessed 9 May 2006).

# Food safety in the industrialized world 8

Marguerite A. Neill

Implied in the concept of food "safety" is that there is a risk of illness associated with its ingestion. Transmission of disease by food has been recognized since Biblical times, and the then current concept of food safety was straightforward and simple – namely, exposure avoidance. It has been a relatively modern construct that food could be made safer and healthier – safer from specific risks to thus prevent transmission of infection, but also healthier to prevent disease and promote health. Improved nutritional content through food fortification has eliminated the major nutritional deficiencies of goiter (iodine), rickets (vitamin D), and pellagra (niacin) in the developed world, and decreases in microbial contamination have decreased specific infectious food-borne diseases as well. The availability of safer and healthier foods is considered one of the ten great public health achievements of the twentieth century in the United States (CDC, 1999a, 1999b).

Both acute and chronic illness may occur following consumption of tainted food (Tauxe and Neill, 2006). While diarrheal illness has traditionally been considered the main manifestation of food-borne disease, several other clinical manifestations are now recognized, ranging from hepatitis, sepsis, meningitis, and paralysis to chronic neurologic disease (see Table 8.1). The microbial agents that cause these illnesses are diverse, and include bacteria, mycobacteria, viruses, parasites, and probably prions. Since 1990, eleven emerging pathogens to humans have been recognized (Nipah, Hendra, Hanta, West Nile, avian influenza and SARS viruses, *Escherichia coli* O157:H7, *Vibrio cholerae* O139, Cyclospora, cryptosporidium, and variant Creutzfeld-Jacob disease, or vCJD). All but two (*V. cholerae* O139, Cyclospora) came from zoonotic sources, and of the nine with zoonotic origin three have been food-borne (*E. coli* O157:H7, Cyclospora and, probably, vCJD). While many factors converge to facilitate the emergence of a pathogen (IOM, 2003), of note are those most relevant to new food-borne pathogens (see Table 8.2). Since these factors are neither static nor self-limited, it should be anticipated that additional food-borne pathogens will emerge even

**Table 8.1 Clinical spectrum of food-borne illness and examples of common causative agents**

| Type of illness | Examples of causative agents |
|---|---|
| Acute enteric illness | |
| • Nausea and vomiting within 6 hours | *Staphyloccoccus aureus, Bacillus cereus* |
| • Vomiting and diarrhea | Norovirus, rotavirus |
| • Diarrhea and abdominal cramping | *ETEC, EPEC, Clostridium perfringens* |
| • Diarrhea and fever | Non-typhoidal Salmonellae<br>Vibrio |
| • Bloody diarrhea | *E. coli* O157:H7<br>*Campylobacter jejuni*<br>*Shigellae*<br>*Vibrio parahemolyticus* |
| Enteric fever | *Salmonella typhi*<br>Brucella |
| Acute sepsis | *V. vulnificus* |
| Acute hepatitis | Hepatitis A virus |
| Acute pseudoappendicitis | *Yersinia enterocolitica,*<br>*Y. pseudotuberculosis* |
| Acute neurological illness | |
| • Paralysis | Botulism<br>Paralytic shellfish poisoning<br>Guillain-Barré syndrome |
| • Paresthesias | Scombroid, ciguatera |
| • Meningitis | *Listeria monocytogenes* |
| Chronic enteric illness | |
| • Diarrhea >3 weeks | Giardia<br>*Cryptosporidium parvum*<br>*Cyclospora cayetensis*<br>Brainerd diarrheal syndrome |
| Chronic neurologic illness | |
| • Seizures (neurocysticercosis) | *Taenia solium* |
| • Congenital abnormalities | *Toxoplasma gondii* |
| • Encephalitis (AIDS patients) | *Toxoplasma gondii* |
| Chronic anemia | Hookworm |
| Vitamin B-12 deficiency | *Diphyllobothrium latum* |

if more traditional ones (e.g. *Salmonella typhi, Mycobacterium tuberculosis*) are controlled or eliminated. The question is not "if" new pathogens will be recognized, but when, to which microbial sector will they belong, who will be most affected, and what will be the burden of disease?

**Table 8.2 Factors in the emergence of food-borne pathogens**

- Microbial adaptation
- Human susceptibility to infection
- Human demographics and behavior
- Economic development and land use
- International travel and commerce
- Technology and industry

Reprinted with permission: Institute of Medicine (2003a).

## Global magnitude and trends

Because food-borne disease is commonly manifest as diarrhea, and death from diarrheal disease is uncommon in the industrialized countries, there is the misperception that food-borne disease is not a significant public health problem. However, the data suggest otherwise. Within the past decade, estimates of the incidence of acute gastroenteritis have been published from several industrialized countries, including the United States, Canada, England, Northern Ireland and the Republic of Ireland, the Netherlands, and Australia (Wheeler *et al.*, 1999; de Wit *et al.*, 2001; Herikstad *et al.*, 2002; Imhoff *et al.*, 2004; Majowicz *et al.*, 2004; Scallan *et al.*, 2004; Hall *et al.*, 2005). Differences in study design do not allow direct comparison of the estimates, but an overall approximation is that in these industrialized nations approximately 20 percent of the population suffers from at least one episode of acute gastroenteritis each year, and that about one-third of the cases are due to food-borne transmission.

In the United States, food-borne disease is estimated to cause 76 million illnesses yearly (approximately one case per four persons per year), leading to 325,000 hospitalizations and 5000 deaths annually (Mead *et al.*, 1999). Economic costs have been estimated at $6.9 billion for the four most common bacterial causes (Buzby *et al.*, 1996).

The purpose of all these studies is to derive reliable data on the incidence of food-borne disease (Flint *et al.*, 2005), a goal endorsed by the World Health Organization. The determination of both pathogen-specific and food-specific risks serves several useful purposes by providing data on which to base clinical recommendations, by sharpening the focus of control strategies, and by documenting the effectiveness (or lack thereof) for specific prevention and control efforts (Batz *et al.*, 2005). The primary aim of these public health studies is not the removal of trade barriers, but to direct efforts to decrease the disease burden of food-borne illness.

Globally, diarrheal disease is the second leading cause of death from infectious diseases worldwide, and is approximately 1000-fold more common in the developing, compared with the developed, nations. Most of the deaths in developing nations are in children aged less than five years, whereas in the industrialized nations most diarrhea-related deaths are in the elderly. There are multiple contributors to diarrheal disease in the developing world, including unclean water, lack of sanitation, person-to-person transmission, crowding, close exposure to animals, and contaminated environments as well as contaminated food.

## Changing profile of causative agents

One hundred years ago, the roster of food-borne pathogens was very different from that of today. *M. tuberculosis* was transmitted by milk from infected cows, typhoid fever was spread by foods contaminated by cooks and food handlers infected with *Salmonella typhi*, and botulism was a steady risk from home preserved foods (Sobel, 2005). Brucellosis was transmitted through consumption of milk, meat, and organs from infected animals (Pappas *et al.*, 2005). Trichinosis, acquired most often from consumption of inadequately cooked, infected pork products, affected one-sixth of the population (Schantz, 1983; Moorhead *et al.*, 1999). Food-borne transmission of these pathogens has been virtually eliminated in the United States and, with the exception of botulism and trichinosis, clinical cases of these infections today are most often acquired overseas.

Even as these traditional pathogens faded in importance for their food-borne transmission, other additions were being made to the list of food-borne pathogens – particularly in the past three decades. The non-typhoidal salmonellae have emerged as common and serious causes of human illness throughout the industrialized world. While still associated with foods of animal origin (beef, poultry, and dairy products), the non-typhoidal salmonellae have increasingly been associated with eggs (Braden, 2006) and fresh produce. Highly drug-resistant strains have emerged which are associated with greater clinical severity in these infections (Helms *et al.*, 2002). In a recent multi-state outbreak of *S. typhimurium* Definitive Type 104 (DT104) associated with ground beef, 41 percent of the cases in the case–control study had been hospitalized (Dechet *et al.*, 2006).

*Listeria monocytogenes* was previously known as a cause of neonatal sepsis and meningitis in immunocompromised persons, and its route of transmission was obscure. It is now well recognized as a food-borne pathogen capable of causing febrile gastroenteritis in healthy persons (Ooi and Lorber, 2005; Schlech *et al.*, 2005), in addition to severe disease in neonates and the immunocompromised host. The high-profile pathogen *Escherichia coli* O157:H7 is noteworthy for the severity of the acute colitis it causes and the post-diarrheal complication of the hemolytic uremic syndrome, the latter being the most common cause of acute renal failure in childhood (Tarr *et al.*, 2005). Raw molluscan shellfish, such as oysters, clams, and mussels, have emerged as the distinctive food risk for illness with two Vibrio species, *V. vulnificus* and *V. parahemolyticus*. An unusual cause of sepsis and meningitis in premature and low birth-weight infants, *Enterobacter sakasakii*, has emerged as a risk associated with powdered infant formula (Drudy *et al.*, 2006).

Viruses and parasites have also been newly identified as food-borne pathogens. Noroviruses, first discovered in 1972, are now recognized as probably the most common causative agent of gastroenteritis and of food-borne infections in the United States (Fankhauser *et al.*, 2002; Glass *et al.*, 2000). *Cyclospora cayetanensis* and *Cryptosporidium parvum*, both intestinal protozoa of non-human mammals, have emerged from zoonotic sources as clinically and economically important causes of food-borne disease.

As noted earlier, the list of food-borne pathogens will continue to expand. The reasons for this include improved surveillance for both sporadic and outbreak-associated illness, and more sophisticated investigation using molecular biology techniques. Pressures for cost containment in the United States have de-emphasized coprodiagnostics, such as stool cultures, fecal microscopy, and antigen detection, since many enteric illnesses are self-limited. Since any apparently sporadic case may be part of a larger outbreak of food-borne illness, greater awareness of the public health benefit of coprodiagnostic studies is needed (Guerrant *et al.*, 2001), and societal value should be accorded to finding and removing a contaminated food vehicle still in commerce.

## Large-scale food production and distribution

In 1900, 39 percent of the US population lived and worked on family farms of less than a few hundred acres. Food animals were raised in small herds or groups, and slaughtered locally. Due to the lack of refrigeration, meat, milk, and eggs were transported only a few miles from their origin, and then consumed locally and relatively quickly. Fresh produce was limited to what was in season. Overall, food production was done manually and was labor intensive. Mechanization increased productivity and promoted competition. Farms became larger in size, requiring more capital investment, and today less than 2 percent of the population lives and works on US farms (see Table 8.3). Food animals now are raised commercially in very large herds or flocks under conditions to maximize weight gain in short grow-out periods. Meat, milk, and eggs are shipped considerable distances and, due to refrigeration, are consumed later after harvest. Importation of produce and other crops has brought an end to seasonality, with many fresh items now available year round. Farming in the US is now agribusiness, characterized by more intensive land use rather than development of new farmland, commercial rather than family ownership, and a vertically integrated organizational structure. These trends have had considerable ramifications on the safety of the food supply.

**Table 8.3 Selected trends in US agriculture**

| | 1900 | 1997 |
|---|---|---|
| US population living on farms | 39% | 1.8% |
| No. of farms (millions) | 5.7 | 1.9 |
| Average farm size (acres) | 146 | 487 |
| Mechanization (wheel tractors) | 4% | 89% |
| Government payments | 0 | $5 billion |
| Farms growing or raising: | | |
| • Cattle | 85% | 55% |
| • Milk cows | 79% | 6% |
| • Chickens | 97% | 5% |
| • Corn | 82% | 23% |
| • Vegetables | 61% | 3% |
| • Soybeans | 0 | 19% |

Data from Economic Research Service, United States Department of Agriculture.

Mass production and distribution of foods permit economies of scale in production, processing, and retailing, but provide an Achilles heel should a contamination event occur. Outbreaks of food-borne illness in recent years have occurred on a size and scale not seen 50 years ago. The classic example of the old outbreak was that of staphylococcal food-poisoning at the church picnic, with a short incubation period, rapid recognition, and affecting less than 50 persons in a circumscribed group. Examples of very large outbreaks today include the thousands of cases of salmonella infections from contaminated ice cream (Hennessy *et al.*, 1996) and milk (Ryan *et al.*, 1987); *E. coli* O157:H7 infections from hamburger (Bell *et al.*, 1994), as well as radish sprouts in Japan (Michino *et al.*, 1999); and hepatitis A from green onions (Wheeler *et al.*, 2005). Each of these outbreaks occurred across multiple states or prefectures and population groups. The scale of these food-borne outbreaks was due to centralized production and distribution, and is akin to that of urban water-borne outbreaks before municipal chlorination.

Food distribution networks are now quite complex, as well as covering widespread geographic regions (Hedberg *et al.*, 1994). Both finished products and ingredients may be shipped substantial distances, passing through brokers, distributors, and other middle suppliers to food processors and retailers. Products and ingredients may be commingled with that from other sources, so that even product from a small producer can contaminate large lots – for example, contaminated mozzarella cheese from one small producer was supplied to four larger processors who shredded it into larger batches, resulting in a multi-state outbreak of salmonellosis (Hedberg *et al.*, 1992). One hamburger patty can represent meat from 600 individual cattle. These aspects impact outbreak recognition, particularly when sporadic or low-level contamination of product occurs, since the attack rate is usually low and cases are geographically dispersed, and outbreak control efforts may be thwarted by the inability to trace product back ultimately to the source.

Aquaculture as a food production sector did not exist a century ago but is growing rapidly worldwide, partly as a response to produce foods of high protein value for an expanding global population. By 2030, aquaculture farms, both marine and freshwater, will produce half of fish consumption globally (Tidwell and Alan, 2001). Although food-borne trematodiasis is an emerging disease mainly seen in the developing world (Keiser and Utzinger, 2005), international markets and trade are simply the conduit for these pathogens to non-travelers in industrialized countries. Paragonimus infections were recently noted in several Californians who had eaten raw imported crab (ProMed, 2006).

Antimicrobial drug resistance has increased among bacterial food-borne pathogens, with increased severity and increased mortality seen in human infections with drug-resistant compared with drug-susceptible strains (Molbak, 2005). Agricultural use of antibiotics for both therapeutic and non-therapeutic purposes in food animals is widely regarded as the main selection pressure for drug resistant strains. Transfer of drug-resistant pathogens to humans via the food chain has been documented (Spika *et al.*, 1987), and several lines of evidence support the linkage of agricultural use of antibiotics and human illness with drug-resistant bacterial pathogens (reviewed in Swartz, 2002). Agricultural use of antibiotics, particularly for non-therapeutic purposes (growth promotion), dwarfs human use. The topic of agricultural use of antibiotics is discussed in Chapter 9.

## Importation

Perhaps no feature so dramatically illustrates the changes in food-borne illness etiology as importation. All commodities from meat, poultry, dairy, fish, grains, produce, and processed foods have been affected to varying degrees. Greater ethnic diversity in the industrialized nations results from immigration from developing countries, with each immigrant group having its unique cultural food preferences. Nearly 50 million persons now travel yearly from the industrialized world to Asia, Africa, and Latin America, where they experience varied cuisines and novel foods. Both these trends have increased demand in the industrialized nations for exotic food products obtainable either more cheaply or only through importation.

Fresh fruits and produce are now regularly imported into the industrialized nations, following the harvest season through the year around the globe. An entire generation now expects otherwise seasonal produce items to be routinely available year round. Fruit and vegetable crops in the developing world may be grown and harvested under conditions quite different from those for produce grown in the United States. Consumption of imported contaminated foods has caused illness with common pathogens such as salmonella, or exotic pathogens such as Cyclospora (Herwaldt *et al.*, 1997) or *Vibrio cholerae* (Taylor *et al.*, 1993). Even insubstantial exposures can result in illness when the infectious dose of a pathogen is low and there is little pre-existing immunity in the population. In the outbreaks of cyclosporiasis associated with imported raspberries, the berries often were only used as a garnish yet attack rates were often greater than 80 percent (Herwaldt *et al.*, 1997).

Currently there are no regulatory standards for microbiological criteria for fresh produce in the United States. Many of these, such as salad items and fruits, are consumed raw, offering immediate opportunity for ingestion of pathogenic organisms distinctly uncommon in the developed world. It is no longer necessary to travel abroad to acquire enterotoxigenic *E. coli*, the most common cause of traveler's diarrhea (Naimi *et al.*, 2003; Beatty *et al.*, 2006; Daniels, 2006). "Don't eat the salad" is a key piece of pre-travel advice for travelers to the developing world, yet salad consumed domestically represents some risk as well. In the US, the proportion of outbreaks associated with fresh produce increased ten-fold between 1977 and 1997 (see Figure 8.1), and the implicated vehicles (lettuce, juice, melon, sprouts, and berries) all would be considered "healthy" foods (Sivapalasingam, 2004).

Preparation of fresh produce to be served raw requires extensive handling, such as peeling, slicing, chopping, or shredding. These manual operations in

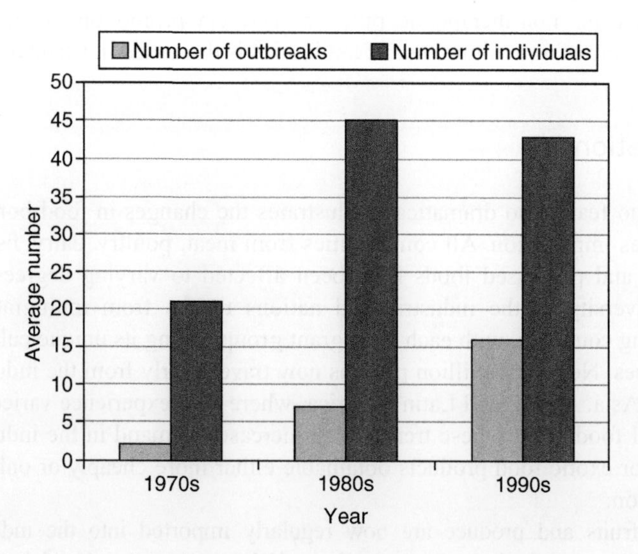


**Figure 8.1** Outbreaks of food-borne illness associated with fresh produce in the United States, 1973–1997. Reproduced with permission of the Institute of Medicine (2003a).

commercial processing create opportunities for contamination by infected food handlers in processing facilities. These workers are often poorly educated, have low wages and no benefits such as sick time or health insurance, and worker turnover may approach 50 percent per year. This confluence of factors makes it difficult to provide and reinforce training in sanitation and hygienic food preparation practices, and workers have only economic disincentives to stay out of work when ill. Mass production, chilled storage, and widespread distribution have repeatedly amplified food-handler associated outbreaks involving fresh produce.

## Changes in food processing

As cardiovascular disease, hypertension, and cancer were found to be linked to diets high in saturated fat and sodium, health promotion efforts shifted dietary consumption patterns away from the traditional meat and potatoes to less red meat, more chicken and fish, and more fruits and raw vegetables. There is greater emphasis on minimally processed foods, along with lower fat content and less sodium – characteristics attributed to a "heart healthy" diet. While these alterations to food content may be helpful in reducing cardiovascular risk, they can modify the food matrix to be less inhibitory to pathogenic as well as spoilage organisms.

Refrigeration is generally viewed as improving the storage conditions for foods that would otherwise spoil quickly at ambient temperature. However, one food-borne pathogen stands out for its capacity to grow at refrigeration temperatures: *L. monocytogenes*. In some processed foods, such as deli meats and cheeses, there can be outgrowth of very low numbers of *L. monocytogenes*, if present, during refrigerated storage. Product formulations now need to address shelf-life if the formulation permits growth of this pathogen.

Increasing emphasis on convenience and pre-cooked foods is fostered by time demands both in the home setting and in food service, schools, and restaurants. Pre-cooked ground beef has twice caused severe outbreaks of hemorrhagic colitis due to *E. coli* O157:H7 infection (Belongia *et al.*, 1991; Jay *et al.*, 2004). Packaged ready-to-eat (RTE) products now account for a substantial proportion of total food expenditures in industrialized countries. Microwave ovens are used at home by 85 percent of US households, yet many consumers do not know that uneven heating of microwaved meat and dairy products is a food safety risk rather than an organoleptic (how the taste, color, odor, and feel of a food affects the senses) preference.

Limitations to parental time and role strain from competing work and family demands influence children's nutritional choices, and may also affect obesity outcomes (McIntosh *et al.*, 2006). There are no studies focused on food-borne illness risk in relation to parental time to prepare safe meals. The Behavioral Risk Factor Surveillance System (BRFSS) included questions on food safety behaviors of the adults to whom it was administered, but did not address the preparation of food for children (CDC, 1998a). School-based health education on food-related matters is not uniform, and is mostly focused on dietary behaviors and nutrition (CDC, 1998b). It is not clear whether educational efforts directed at adults would be more effective than those directed at children to decrease risky food-handling practices, but in general child-focused education has been more successful in increasing health and safety behaviors (e.g. seat belt use, smoking cessation) by adults.

## Social and demographic influences on food preferences

The association of saturated fat consumption with cardiovascular disease and some cancers has led to changes in dietary recommendations of both the types and amounts of specific foods consumed. That recommendations have translated to actual changes in consumption in the United States is clearly evident in *per capita* consumption of selected food items over a nearly 30-year period (Table 8.4). From 1970 to 1997, the most prominent changes have included a 91 percent increase in poultry consumption and a 155 percent increase in cheese consumption. Fresh fruit and vegetable consumption increased by 32 percent and 22 percent respectively, and seafood by 25 percent. Target commodities for saturated fat have had declines in their annual *per capita* consumption, with decreases of 67 percent for whole milk, 16 percent for red meat, and 23 percent for eggs.

**Table 8.4 Changes in food consumption in the United States, 1970 and 1997**

| Food commodity | Per capita consumption (pounds) | | |
|---|---|---|---|
| | 1970 | 1997 | Change (%) |
| Red meat | 132 | 111 | −16 |
| Poultry | 34 | 65 | +91 |
| Fish | 12 | 15 | +25 |
| Dairy products | | | |
| • Milk – whole | 219 | 73 | −67 |
| • Milk – low fat or skim | 50 | 134 | +168 |
| • Cheese | 11 | 28 | +155 |
| • Yogurt | 0.8 | 5.1 | +190 |
| • Eggs (no.) | 309 | 239 | −23 |
| • Fresh fruit | 101 | 133 | +32 |
| • Fresh vegetables | 153 | 186 | +22 |
| • Flour, cereal products | 136 | 200 | +47 |
| • Caloric sweeteners | 122 | 154 | +26 |

Data from Economic Research Service, United States Department of Agriculture.

These shifts in consumption patterns have driven changes in food retailing and agriculture. The average grocery store today is often a supermarket stocked with 25,000–50,000 items from all over the world. Stores are often designed so that the produce department is the first area encountered upon entering the store, and, depending on the season of the year in the United States, as much as 80 percent of a produce item may be imported (for example, cucumbers and green onions in January–March). Seafood counters contain fresh fish flown in almost daily. Even without seasonality, economics plays its own role. The North American Free Trade Agreement enacted in 1994 has altered the dynamics of importation of agricultural products among the United States, Mexico, and Canada, which now represent one market. Livestock for meat production often have been raised in at least two of these countries before slaughter. Food consumption patterns in the industrialized world have shifted in their commodity type, amount, and source – more frequently including exotic or ethnic foods, more chicken, fish, cheese, fresh fruits and vegetables, and with different production sources for items that previously were domestically produced. The grocery store essentially has been transformed into a locally accessible but globally representative marketplace.

## Changes in spending, food preparation, and consumption

As disposable income has risen in the industrialized countries, the proportion of household income needed to purchase food has dropped, even as total spending on food has increased. In 1929 in the United States, families and individuals spent nearly one-quarter of their disposable income on food; by mid-century this had fallen to 20 percent, and in 2005 was 9.9 percent (Figure 8.2). Canada, the United Kingdom, and most of the countries in Western Europe have similar spending shares of 10–12 percent. By stark contrast, in developing countries food expenditures often account for 50 percent or more of family income.

In 2005, American spending on food was $1 billion, with families and individuals accounting for approximately 85 percent. The location where these expenditures occurred continues a remarkable and sustained trend with clear ramifications for food safety issues. In 1935, only 3 percent of total food expenditures were spent on food away from home; this has risen steadily to 48 percent today (Figure 8.3). Stated differently, in 1935, 3 cents of every food dollar was spent for food away from home; today, 48 cents of every dollar of food expenditures is spent at restaurants, fast food outlets, snack bars, catering establishments, cafeterias, and other commercial food-service locations.

Because commercial food service establishments are profit oriented, they are sensitive to consumer preference. The proliferation of salad bars and offerings of raw and/or cold items is in response to consumer preference, which in turn has prompted greater reliance on imports to meet year-round demand. To maintain profitability, the commercial food-service operators must hold down costs,

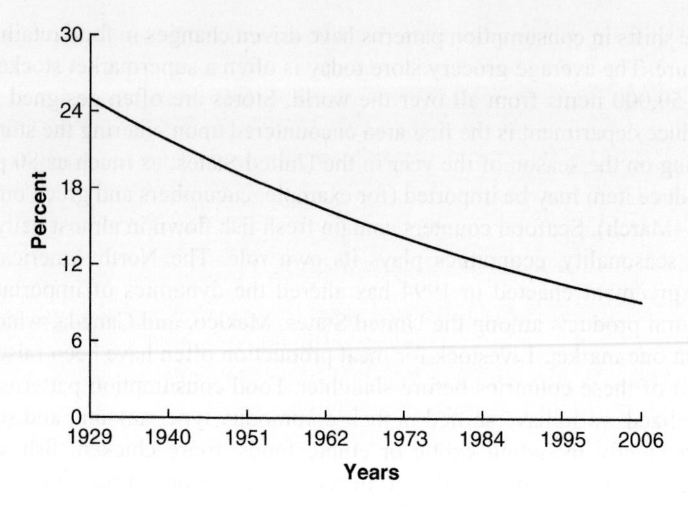


**Figure 8.2** Food expenditures as a share of disposable income for families and individuals. Source: data from the Economic Research Service, US Department of Agriculture.

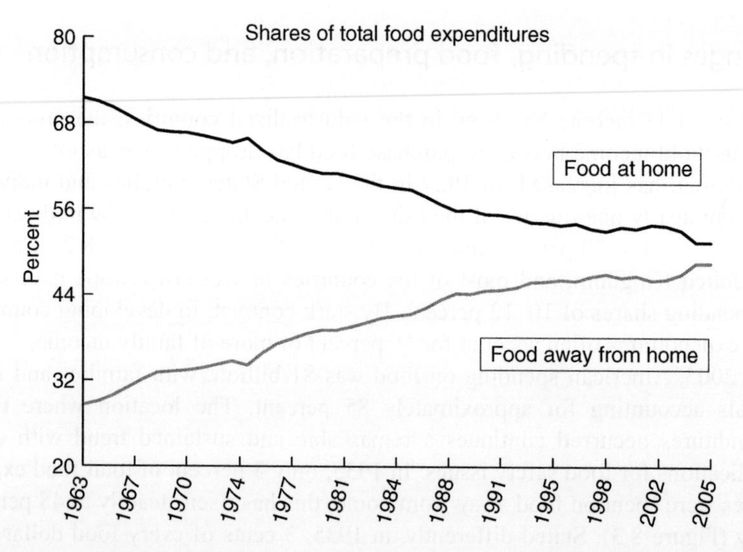


**Figure 8.3** Location of total food expenditures. Source: Economic Research Service, US Department of Agriculture.

which currently account for 86 cents of every dollar in sales, compared with an average of 76 cents for all US industry. This keeps the pressure on wages to remain low, with the aforementioned consequence of high employee turnover rates and difficulty maintaining a high degree of food safety education among food-service workers.

These trends converge to indicate that for consumers, much of their protection from food-borne illness for a large segment of their meals is less under their control and has been *de facto* delegated to other individuals and systems. An individual consumer's protection now is vested in regulatory oversight of, and adherence to, good manufacturing practices and sanitation among a complex network of food producers, processors, distributors, and commercial food-service workers.

## Demographic and cultural influences

The aging of the population in the United States and most other industrialized countries has already altered food-borne disease epidemiology, and is projected to impact food preferences and consumption patterns. By 2020 in the United States, one in five persons will be aged ⩾65 years. These 71 million individuals will represent 12.4 percent of the population. In 2000, there were 35 million persons ⩾65 years. Depending on the pathogen, older persons may be more susceptible to clinical illness if infected (e.g. listeriosis); develop more severe disease (e.g. non-typhoidal salmonellosis); be more likely to develop complications (e.g. HUS with *E. coli* O157:H7 colitis); and be more likely to die (e.g. Vibrio infections). Although hospitalization rates vary by food-borne pathogen, they are usually higher for persons over age 60 years.

In addition to older persons, other populations have distinctly increased susceptibility for food-borne illness (see Table 8.5). Immunocompetent women are at increased risk of listeriosis when pregnant. Infants less than one year of age have the highest age-specific rate of salmonellosis; pre-term infants appear most susceptible to *E. sakazakii* infection. Persons living with HIV infection and AIDS have more severe salmonella infections and more protracted diarrhea due to cryptosporidium. In the United States, there are an estimated 950,000 persons

**Table 8.5 Clinical conditions with increased risk of food-borne infection and illness**

- Extremes of age (<5 years and >65 years)
- Pregnancy
- Liver disease, cirrhosis
- Alcoholism
- HIV infection
- Transplantation (solid organ, stem cell)
- Cancer chemotherapy
- Immunosuppressive drugs – corticosteroids, monoclonal antibodies, immune modulators
- Gastric acid suppressing drugs

living with HIV infection, and 40,000 new infections occur each year, representing a sizeable vulnerable population.

Perhaps the most significant trend is that of an increasing proportion of the population that is immunocompromized. Transplant recipients are an illustrative example. In 2005, there were nearly 100,000 persons on active waiting lists for solid organ transplant, and 28,108 transplants were performed; 62 percent of these were kidney transplants (OPTN, 2006). Bone marrow transplant recipients are another growing category, with these being performed for an increasing number of hematologic and oncologic conditions. There are an estimated 10 million persons in the United States with renal disease, and the annual number of new cases has tripled over a decade, most due to hypertension and diabetes. In 1997, 360,000 persons were on dialysis or had received renal transplants, and the transplant waiting list had been expanding at 11 percent per year for the past three years (Health and Human Services Department (US), 2000). Advances in earlier detection and treatment have changed the epidemiology of cancer, making it curable for some and converting it to a chronic disease for others. In 1971, there were 3 million cancer survivors in the United States; this had more than tripled to 9.8 million in 2001 (3.5 percent of the population) (CDC, 2004). Chronic diseases such as connective tissue diseases and inflammatory bowel disease are treated with a variety of immunomodulating drugs (e.g. corticosteroids, immunosuppressive agents, anti-cytokine monoclonal antibodies) which increase susceptibility to infection.

Gastric acid is the first line of defense against food-borne pathogens, and it has been known for many decades that achlorhydria and gastrectomy increase the risk of Vibrio and salmonella infections, as well as tuberculosis. Yet treatment of common conditions such as peptic ulcer and gastro-esophageal reflux (GERD) with acid-blocking drugs neutralizes this primary defense. In 2003, there were 33 million prescriptions for H2 blockers and proton pump inhibitors – which, if administered to unique individuals, would have treated 7 percent of the population. The actual proportion of the population taking acid-blocking medications is likely higher, as these drugs are also available over the counter.

These population groups are increasing at varying rates. Some are new and exist because of medical progress, such as transplant recipients and preterm infants, while others, like the elderly, have always been present but magnify a disease trend because of their increase. Although there is a dearth of data comparing the risk of food-borne illness among these groups, it is likely they are not at equal risk. Some probably bear a disproportionate share of the burden of food-borne illness from individual pathogens. The cost of that disease burden is also measured with newer outcomes – examples being a need for renal transplantation following HUS from *E. coli* O157:H7, or the rejection of a solid organ transplant precipitated by salmonella infection.

Several cultural trends in the industrialized countries have influenced food-borne illness patterns therein. Congregate settings such as nursing homes and day-care centers have been the site of outbreaks of initially food-borne disease. An amplifier effect is seen with secondary transmission to staff and patients' family members with outbreaks due to *E. coli* O157:H7 and shigella, both low-inoculum infections capable of being transmitted from person to person in settings where hygiene is inadequate. Petting zoos have emerged as a new location for *E. coli* O157:H7 outbreaks, with mostly young children acquiring infection from direct contact with ruminant animals and their barnyard environment (CDC, 2005).

The need for behavioral change is evident in the lack of knowledge and consistent practice of food safety habits by the public, as indicated by responses in the periodic surveys of adults in the Behavioral Risk Factor Surveillance System (BRFSS). After handling raw meat, 18 percent of respondents indicated they did not wash their hands (CDC, 1998a). Videotaped observational study showed that consumers tended to underestimate the risk of specific practices, and repeatedly made food handling and sanitation errors (Anderson *et al.*, 2004). Cumulatively, these errors would increase individual risk of food-borne disease over time. One wonders whether these behavioral risks for cross-contamination during food preparation are related to increased consumption of food away from home, and a subsequent lack of familiarity with cooking and safe food preparation practices.

## Reacting, coping, and preventing

During the first several decades of the twentieth century, food safety in the industrialized world could be largely categorized as a cycle of reacting and coping – reacting to outbreaks and coping with sporadic illnesses. However, by the latter two decades a much greater emphasis was placed on prevention of all disease, both outbreak-related and sporadic. From the food safety toolbox (Table 8.6), a variety of tools have been used to decrease the morbidity, mortality, and costs of food-borne disease. Current analyses are focusing on the relative effectiveness and impact of each tool.

Sporadic infections account for the vast majority of food-borne illnesses. Of the total burden of food-borne disease in the United States, there is no identified agent in 81 percent of cases (Mead *et al.*, 1999), although many of these may be due to viral causes. While outbreaks may not be a large proportion of the total disease burden of food-borne illness, they are extremely important for the focus they bring to food safety efforts.

Food-borne outbreaks have been important catalysts of regulatory change, such as required time and temperature conditions for precooked roast beef following *Salmonella* outbreaks (IOM, 2003b), pathogen reduction steps such as pasteurization for apple cider and fresh juices because of associated *E. coli* O157:H7 (cider) and *Salmonella* (juice) infections (IOM, 2003b), and new processing requirements for ready-to-eat poultry and meat products after a large

**Table 8.6 The food safety toolbox and future needs**

| Tool | Twenty-first century need |
|---|---|
| Surveillance | More real-time data analysis |
| • Syndromal | Enhanced virtual networks for public health communication |
| • Pathogen-specific | |
| • Computer analyzed | Greater international participation |
| "Farm to fork" continuum | Tighter vertical integration |
| • Conceptual approach | |
| Regulatory framework | Eliminate fragmented statutory foundation |
| • Inspection | |
| • Microbial sampling | Creation of science-based standards |
| • Microbial standards | Regulatory flexibility |
| • Performance standards | |
| HACCP | More extensive use |
| Education and behavioral change | Improve effectiveness |
| • Consumers | Broader reach |
| • Food producers, processors | |
| Traceability | Broader application<br>Greater depth for traceback |
| Irradiation | Broader commodity approval<br>Greater public acceptance |
| Quantitative microbial risk assessment | Inform food safety objectives |
| Lawsuits and legal liability | |

multi-state outbreak of listeriosis due to turkey deli meat (Gottlieb *et al.*, 2006). The large-scale *E. coli* O157:H7 outbreak in the western United States (Bell *et al.*, 1994) spurred the Pathogen Reduction/Hazard Analysis Critical Control Program regulation by the United States Food Safety Inspection Service (FSIS). HACCP is an engineering control, aiming to prevent contamination rather than inspecting to remove defective product. Originally implemented and validated in processed foods, HACCP was implemented in 1996 for meat and poultry slaughter facilities, and modified HACCP approaches are being used in seafood and food service.

Surveillance has been, and will remain, the cornerstone of the foundation of food safety in particular, and public health in general. At one time, surveillance was pathogen-specific and almost synonymous with postcards, bulging filing cabinets, and dreary reports lacking timeliness and filled with minutiae. Today, surveillance is increasingly electronic, and not just for a specific pathogen but also for a syndrome or illness. Particularly in food-borne disease, surveillance is being used not just to find outbreaks and remove contaminated vehicles from commerce, but also to drive applied research for prevention measures. Molecular subtyping of *E. coli* O157:H7 over a two-year period in Minnesota showed multiple subtypes, identifying only four outbreaks during the period of the study (Bender *et al.*, 1997). The data were suggestive that there may be multiple "mini-outbreaks," with different sources, moving transiently and quickly through the food supply. The attendant implication is that control measures need to be multifactorial.

"FoodNet" has been a cooperative effort in the United States to conduct population-based studies of sporadic food-borne disease (Allos *et al.*, 2004). It has provided an improved, more precise estimate of actual disease occurrence – for example, that 39 symptomatic cases of *Salmonella* infection occur for every one case that is cultured and reported (Voetsch *et al.*, 2004). Now in its tenth year, FoodNet has allowed comparison of disease trends over time (CDC, 2006a), and has been used to measure the impact of regulatory changes aimed at prevention of food-borne infection. While *Campylobacter*, *E. coli* O157:H7, Listeria, and *Salmonella* infections have all decreased since 1996, seemingly correlated with the Pathogen Reduction and Hazard Analysis Critical Control Program (HACCP), most of that decrease was in the first four years after its implementation. For *E. coli* O157:H7 the incidence at FoodNet sites has been the same for the past three years, and may reflect non-food-borne transmission.

For egg-associated *Salmonella enteriditis* there has been only a partial success in the convergence of measures to control the problem, including on-farm egg quality assurance programs, continuous refrigeration of shell eggs, and education for consumers, restaurants, and retail and institutional kitchens. While the incidence has decreased from a high of 3.9 per 1,000,000 in 1994 (10 in the New England states) to 2.2 per 1,000,000 in 2003, the latter is double the incidence in the 1980s (Braden, 2006).

"PulseNet" is a national network of state public health laboratories in which molecular genetic analysis ("DNA fingerprinting") of food-borne pathogens is digitized and shared via a secured electronic network. By showing which cases have the same subtype, PulseNet has repeatedly speeded recognition of geographically diffuse, low-level outbreaks, as well as separating outbreak from sporadic background cases – in both circumstances providing direction for epidemiologic investigation and source ascertainment. Similar laboratory networks for food-borne pathogens exist in Canada and Europe ("EnterNet"). The challenges in bringing these networks together and adding additional world regions include the use of standardized protocols for molecular analysis and interpretation, transparency in data sharing, and compatibility of the computer networks.

A recent outbreak of salmonellosis illustrates the convergence of several trends to find a notable weak spot in food protection in the United States. Many case-patients were Latino or Asian – a clue that led to the identification that illness was linked to the consumption of imported mangos. This exotic fruit must be disinfested to prevent the importation of the Mediterranean fruit fly. A hot-water dip technique was used, but the water was not chlorinated or filtered; the technique had been mandated by USDA and was widely used, but had not been assessed for microbiological implications (Sivapalasingam *et al*., 2003). The hot-water dip method was implemented in order to move away from a fumigation method with carcinogenic potential, but it had not been fully evaluated for microbiological effectiveness, thus fulfilling the "law of unintended consequences" (Jones and Schaffner, 2003).

Irradiation of foods (including the aforementioned mangos) could substantially reduce food-borne infections (Tauxe, 2001; Osterholm, 2004; Osterholm and Norgren, 2004). It is supported by a wealth of data on safety, as well as by federal agencies in the United States (CDC, USDA, FDA), a variety of medical and public health organizations, food processing groups, and international organizations (WHO, FAO). Although approved for use in the United States on meat, poultry, fruits, and vegetables, it is used in $<0.002$ percent of these commodities (US General Accounting Office, 2000). The barriers to its broader use are related to ignorance, lack of consumer acceptance, and subsequent small market share for irradiated product.

Legal liability and lawsuits have had some effect on food companies, although this has been difficult to quantify. Restaurants were the targets of one-third of the lawsuits from 1988 to 1997 (Buzby *et al*., 2001). Parent companies were the next largest category targeted. Of those that went to court, only 31 percent were won by the plaintiff, with a median award of $25,560 in 1998 dollars. The data are distorted by the fact that many claims are settled out of court, the details are not publicly available, and (probably) the weaker claims go to court. As food tracing systems become deeper and more effective at traceback, the ability to identify the point of contamination improves. This increases the potential for product liability to be assigned to a specific company, and would be a direct legal incentive to produce safer foods.

An outbreak of *E. coli* O157:H7 infections in the late summer of 2006 is illustrative of the problems with the current tools for food safety when viewed in the context of modern trends in food-borne disease. Nearly 200 cases occurred, spread over more than half of the 50 states and Canada. Infection was acquired by the consumption of a "healthy" food, raw bagged spinach (CDC, 2006b). The spinach was traced under multiple brand names and through distributors to production within three counties in southern California; the ultimate source of the contamination is unknown as this chapter is going to press. This latest outbreak illustrates that problems with food-borne pathogens in produce cannot be lopsidedly assigned only to imported produce. Regulations such as microbial standards would need to be applied equally to imported as well as domestic product, if they are to be effective in reducing disease, and avoid constituting a trade barrier.

A major shortcoming in many food-borne disease outbreaks is that they represent "too much, too late." The outbreak unfolds in the rearview mirror, with retrospective construction of the chain of events that led to contamination and the spread of illness. Recalls of contaminated product not infrequently occur too late to affect the epidemic curve, because the contaminated vehicle quickly cleared through the food distribution system. Chasing the horse after it is out of the barn is a costly consumption of scarce public health resources, and re-emphasizes the primacy of prevention strategies.

What are the needs for the twenty-first century in food safety, and how should the tools in the toolbox (Table 8.6) be used? The populations in the industrialized countries will continue to grow, with more elderly persons, greater ethnic diversity in younger populations, and more people living longer with immunocompromising conditions (see Figure 8.4). The emphasis will remain on dietary patterns to prevent cardiovascular disease, cancer, osteoporosis, and obesity, with less red meat and saturated fats, and more fresh fruits, vegetables, grains, poultry, and fish. A large proportion of the foods consumed will be imported, prepared outside the home, and/or consumed at restaurants or other commercial food-service locations. Food safety in the industrialized nations will exist in a new equilibrium with the health and environmental conditions in the developing world. New pathogens will be identified and others will re-emerge into new ecological niches.

The regulatory framework in the United States has been a patchwork of statutory regulations developed using science that is now considered outdated, such as the organoleptic basis for meat inspection (IOM, 2003b). A move toward science-based standards that utilize the tools of modern molecular biology coupled with greater regulatory flexibility would keep pace with the shift in population demographics. Tighter vertical integration along the farm-to-fork continuum would foster greater awareness of microbial hazards among food producers, processors, retailers, and consumers, and would be expected to drive markets. Real-time surveillance, already feasible, would need to be coupled with real-time analysis and appropriate action. This is particularly germane to preparedness for a bioterrorist attack through the food supply.

Food safety objectives (FSOs) provide an integration of risk assessment and public health objectives translated into a numerical goal. FSOs work in reverse, from the public health objective (e.g. reducing the incidence of pathogen x-related illness to a specified level or by a specified proportion at a defined time) to a target for food processors (e.g. that no more than y cfu/gram of pathogen x can be present in the serving at the time of consumption). The FSO defines the acceptable level of protection (ALOP), a numerical value left undefined in HACCP (IOM, 2003b). The future of food safety is coming into focus in risk

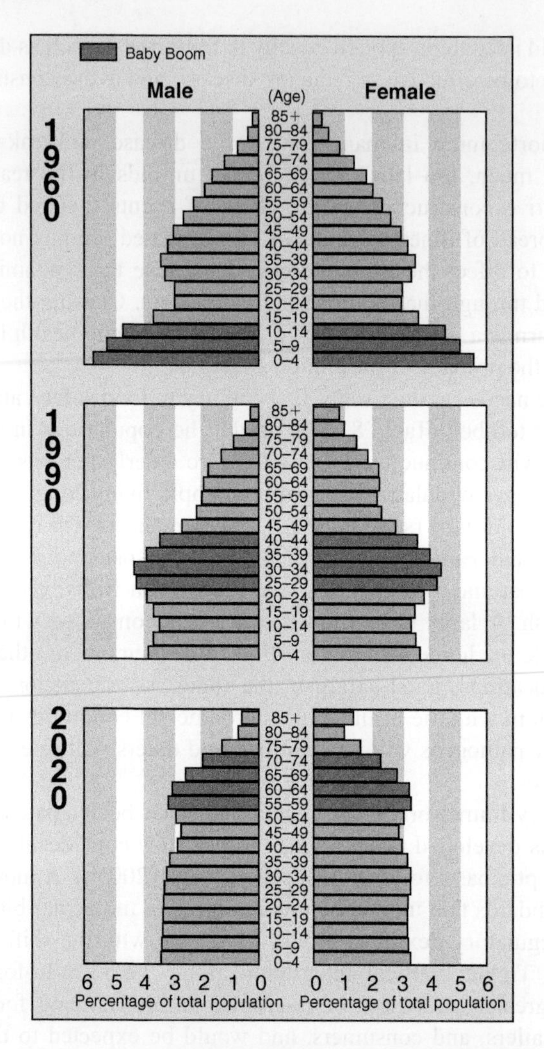


**Figure 8.4** United States population age structure 1960–2020. Source: US Bureau of the Census, www.census.gov/ipc/prod/97agewc.pdf.

management economics, an emerging field. When the cost of the intervention(s) is integrated with the benefit(s) to society as a whole, an appropriate level of protection (ALOP) can be calculated (IOM, 2003b). In this approach, as the level of food safety increases, there is greater marginal cost to society (Figure 8.5). It can also be deduced that with tighter control of the food supply for microbial pathogens, additional interventions to "enhance" food safety come with less benefit and greater cost.

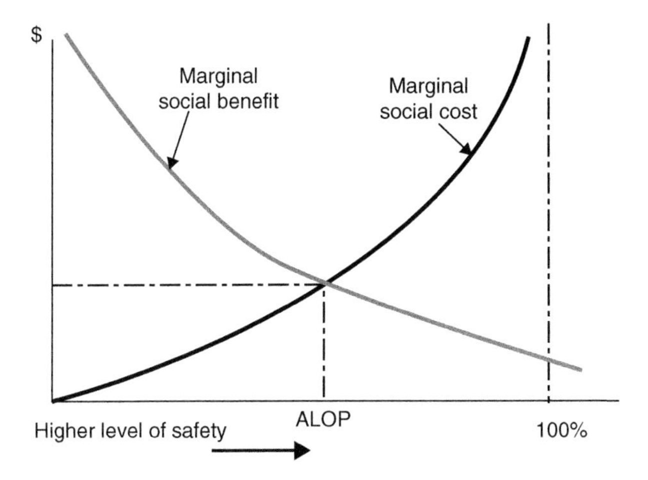


**Figure 8.5** Toward a public health goal: relating an appropriate level of protection (ALOP) to marginal social benefit and cost. Reproduced with permission of the Institute of Medicine.

## Conclusions

Food-borne disease in the industrialized world is shifting in its epidemiologic patterns of who, what, when, where, and why. Food is abundant and cheap, with less of household income being expended on it. Traditional pathogens like trichinosis, tuberculosis, and typhoid have been eliminated. Dietary changes are reflected in increases in less traditional pathogens, such as non-typhoidal salmonellae, and more produce-related illnesses. Immunocompromised persons now constitute significant proportions of the populations in industrialized countries. Nearly half of food expenditures are now on food consumed away from home, and imported foods are regularly available. New pathogens will be identified and others will re-emerge in different ecologic niches. In the global village, consumers in the industrialized countries now are directly connected to the health and environmental conditions of food producers in the developing world. Their food safety is less under their personal control, and increasingly is vested in a complex network of producers, distributors, processors, and retailers. Because the strength of such a network is measured by its weakest link, the future of food safety will be a greater emphasis on prevention and tighter control of the food supply. Regulation is the pact that society has with government to provide the safety net which markets do not provide. To keep pace with change, food safety regulations must evolve to become more science-based, have greater flexibility, and be more transparent, while society must choose the cost it is willing to pay for the benefits of improvements in public health.

## References

Allos, B.M., Moore, M.R., Griffin, P.M. *et al.* (2004). Surveillance for sporadic food-borne disease in the 21st century: the FoodNet perspective. *Clinical Infectious Diseases* **38**(Suppl. 3), S115–120.

Anderson, J.B., Shuster, T.A., Hansen, K.E. *et al.* (2004). A camera's view of consumer food-handling behaviors. *Journal of the American Dietetic Association* **104**, 186–191.

Batz, M.B., Doyle, M.P., Morris, J.G. Jr *et al.* (2005). Attributing illness to food. *Emerging Infectious Diseases* **11**, 993–999.

Beatty, M.E., Adcock, P.M., Smith, S.W. *et al.* (2006). Epidemic diarrhea due to enterotoxigenic *Escherichia coli. Clinical Infectious Diseases* **42**, 329–334.

Bell, B.P., Goldoft, M., Griffin, G.M. *et al.* (1994). A multistate outbreak of *Escherichia coli* O157:H7-associated bloody diarrhea and hemolyticuremic syndrome from hamburgers. The Washington experience. *Journal of the American Medical Association* **272**, 1349–1353.

Belongia, E.A., MacDonald, K.L., Parham, G.L. *et al.* (1991). An outbreak of *Escherichia coli* O157:H7 colitis associated with consumption of precooked meat patties. *Journal of Infectious Diseases* **164**, 338–343.

Bender, J.B., Hedberg, C.W., Besser, J.M. *et al.* (1997). Surveillance by molecular subtype for *Escherichia coli* O157:H7 infections in Minnesota by molecular subtyping. *New England Journal of Medicine* **337**, 388–394.

Braden, C. (2006). *Salmonella enterica* serotype enteritidis and eggs: a national epidemic in the United States. *Clinical Infectious Diseases* **43**, 512–517.

Buzby, J., Roberts, T., Lin, C.T. *et al.* (1996). *Bacterial Food-borne Disease: Medical Costs and Productivity Losses*. Washington, DC: Economic Research Service, US Department of Agriculture.

Buzby, J.C., Freuzen, P.D. and Rasco, B. (2001). *Product Liability and Microbial Food-borne Illness* (available at: www.ers.usda.gov/Publications/aer799).

CDC (1998a). Multistate surveillance for food-handling, preparation, and consumption behaviors associated with food-borne diseases: 1995 and 1996 BRFSS food-safety questions. *Morbidity and Mortality Weekly Report* **47**(SS-4), 33–57.

CDC (1998b). Characteristics of health education among secondary schools – school health education profiles, 1996. *Morbidity and Mortality Weekly Report* **47**(SS-4), 1–31.

CDC (1999a). Ten great public health achievements – United States, 1900–1999. *Morbidity and Mortality Weekly Report* **48**, 241–243.

CDC (1999b). Achievements in public health, 1900–1999. *Morbidity and Mortality Weekly Report* **48**, 905–943.

CDC (2004). Cancer survivorship – United States, 1974–2001. *Morbidity and Mortality Weekly Report* **52**, 526–529.

CDC (2005). Outbreaks of *Escherichia coli* O157:H7 associated with petting zoos – North Carolina, Florida, and Arizona. *Morbidity and Mortality Weekly Report* **54**, 1277–1280.

CDC (2006a). Preliminary FoodNet data on the incidence of infection with pathogens transmitted commonly through food – 10 States, United States, 2005. *Morbidity and Mortality Weekly Report* **55**, 392–395.

CDC (2006b). Ongoing multistate outbreak of *Escherichia coli* serotype O157:H7 infections associated with consumption of fresh spinach – United States. *Morbidity and Mortality Weekly Report Dispatch* 26 September. Available at: www.cdc.gov/mmwr (accessed 29 September 2006).

Daniels, N.A. (2006). Enterotoxigenic *Erscherichia coli*: traveler's diarrhea comes home. *Clinical Infectious Diseases* **42**, 335–336.

Dechet, A.M., Scallan, E., Gensheimer, K. *et al.* (2006). Outbreak of multidrug-resistant *Salmonella enterica* serotype *typhimurium* Definitive Type 104 infection linked to commercial ground beef, northeastern United States, 2003–2004. *Clinical Infectious Diseases* **42**, 747–752.

de Wit, M.A.S., Koopmans, M.P.G., Kortbeek, L.M. *et al.* (2001). Sensor, a population-based cohort study on gastroenteritis in The Netherlands: incidence and aetiology. *American Journal of Epidemiology* **154**, 666–674.

Drudy, D., Mullane, N.R., Quinn, T. *et al.* (2006). *Enterobacter sakazakii*: an emerging pathogen in powdered infant formula. *Clinical Infectious Diseases* **42**, 996–1002.

Fankhauser, R.L., Monroe, S.S., Noel, J.S. *et al.* (2002). Epidemiologic and molecular trends of "Norwalk-like viruses" associated with outbreaks of gastroenteritis in the United States. *Journal of Infectious Diseases* **186**, 1–7.

Flint, J., VanDuynhoven, Y.T., Angulo, F.J. *et al.* (2005). Estimating the burden of acute gastroenteritis, food-borne disease, and pathogens commonly transmitted by food: an international review. *Clinical Infectious Diseases* **41**, 698–704.

Glass, R.I., Noel, J., Ando, T. *et al.* (2000). The epidemiology of enteric caliciviruses from humans: a reassessment using new diagnostics. *Journal of Infectious Diseases* **181**(Suppl. 2), S254–261.

Gottlieb, S.L., Newbern, E.C. Grivvin, P.M. *et al.* (2006). Multistate outbreak of listeriosis linked to turkey deli meat and subsequent changes in US regulatory policy. *Clinical Infectious Diseases* **42**, 29–36.

Guerrant, R., VanGilder, T., Steiner, T.S. *et al.* (2001). Practice guidelines for the management of infectious diarrhea. *Clinical Infectious Diseases* **32**, 331–350.

Hall, G.V., Kirk, M.D., Becker, N. *et al.* (2005). Estimating food-borne gastroenteritis, Australia. *Emerging Infectious Diseases* **11**, 1257–1263.

Health and Human Services Department (US) (2000). Chronic kidney disease: issues and trends. In: *Healthy People 2010*. Available at: www.healthypeople.gov/Document/HTML/Volume1/04CKD.htm.

Hedberg, C.W., Korlath, J.A., D'Aoust, J-Y. *et al.* (1992). A multistate outbreak of *Salmonella javiana* and *Salmonella oranienburg* infections due to consumption of contaminated cheese. *Journal of the American Medical Association* **268**, 3303–3307.

Hedberg, C.W., MacDonald, K.L. and Osterholm, M.T. (1994). Changing epidemiology of food-borne disease: a Minnesota perspective. *Clinical Infectious Diseases* **18**, 671–680.

Helms, M., Vastrup, P., Gerner-Smidt, P. and Molbak, K. (2002). Excess mortality associated with antimicrobial drug-resistant *Salmonella typhimurium*. *Emerging Infectious Diseases* **8**, 490–495.

Hennessy, T.W., Hedberg, C.W., Slutsker, L. *et al.* (1996). A national outbreak of *Salmonella enteritidis* infections from ice cream. *New England Journal of Medicine* **334**, 1281–1286.

Herikstad, H., Yang, S., van Gilder, T. *et al.* (2002). A population-based estimate of the burden of diarrheal illness in the United States: FoodNet 1996–1997: are the rates of diarrhea increasing? *Epidemiology and Infection* **129**, 9–17.

Herwaldt, B. and Ackers, M.L., with the Cyclospora Working Group (1997). An outbreak in 1996 of cyclosporiasis associated with imported raspberries. *New England Journal of Medicine* **336**, 1548–1556.

Imhoff, B., Morse, D., Shiferaw, B. *et al.* (2004). Burden of self-reported acute diarrheal illness in FoodNet surveillance areas, 1998–1999. *Clinical Infectious Diseases* **38**(Suppl. 3), 219–226.

IOM (Institute of Medicine) (2003a). *Microbial Threats to Health: Emergence, Detection and Response* (S. Smolinski, M.A. Hamburg and J. Lederberg (eds). Washington, DC: National Academies Press.

IOM (2003b). *Scientific Criteria to Ensure Safe Food.* Committee on the Review of the Use of Scientific Criteria and Performance Standards for Safe Food, Food and Nutrition Board, Board on Agriculture and Natural Resources. Washington, DC: National Academies Press.

Jay, M.T., Garrett, V., Mohle-Boetani, J.C. *et al.* (2004). A multistate outbreak of *Escherichia coli* O157:H7 infection linked to consumption of beef tacos at a fast-food restaurant chain. *Clinical Infectious Diseases* **39**, 1–7.

Jones, T.F. and Schaffner, W. (2003). *Salmonella* in imported mangos: shoeleather and contemporary epidemiologic techniques together meet the challenge. *Clinical Infectious Diseases* **37**, 1591–1592.

Keiser, J. and Utzinger, J. (2005). Emerging food-borne trematodiasis. *Emerging Infectious Diseases* **10**, 1507–1514.

Majowicz, S.E., Dore, K., Flint, J.A. *et al.* (2004). Magnitude and distribution of acute, self reported gastrointestinal illness in a Canadian community. *Epidemiology and Infection* **132**, 607–617.

McIntosh, A., Davis, G., Nayqa, R. *et al.* (2006). Parental time, role strain, and children's fat intake and obesity-related outcomes. *USDA CCR No. 19* (available at: http://www.ers.usda.gov/Publications/CCR19/ccr19.pdf).

Mead, P.S., Slutsker, L., Dietz, V. *et al.* (1999). Food-related illness and death in the United States. *Emerging Infectious Diseases* **5**, 607–625.

Michino, H., Araki, K., Minami, S. *et al.* (1999). Massive outbreak of *Escherichia coli* O157:H7 infection in school children in Sakai City, Japan, associated with consumption of white radish sprouts. *American Journal of Epidemiology* **150**, 787–796.

Molbak, K. (2005). Human health consequences of antimicrobial drug-resistant *Salmonella* and other food-borne pathogens. *Clinical Infectious Diseases* **41**, 1613–1620.

Moorhead, A., Grunenwald, P.E., Dietz, V.J. and Schantz, P.M. (1999). Trichinellosis in the United States, 1991–1996: declining but not gone. *American Journal of Tropical Medicine and Hygiene* **60**, 70–74.

Naimi, T.S., Wicklund, J.H., Olsen, S.J. *et al.* (2003). Concurrent outbreaks of *Shigella sonnei* and enterotoxigenic *Escherichia coli* infections associated with parsley: implications for surveillance and control of food-borne illness. *Journal of Food Protection* **66**, 535–541.

National Institutes of Health. Chronic kidney disease: issues and trends. In: *Healthy People 2010* (available at: www.healthypeople.gov/Document/HTML/Volume1/04CKD.htm).

Ooi, S.T. and Lorber, B. (2005). Gastroenteritis due to *Listeria monocytogenes. Clinical Infectious Diseases* **40**, 1327–1332.

OPTN (Organ Procurement and Transplantation Network) (2006). *Transplant Year (2004–2005) by Organ Based on Data as of 22 September* (available at: www.optn.org).

Osterholm, M.T. (2004). Food-borne disease: the more things change, the more they stay the same. *Clinical Infectious Diseases* **39**, 8–10.

Osterholm, M.T. and Norgren, A.P. (2004). The role of irradiation in food safety. *New England Journal of Medicine* **350**, 1898–1901.

Pappas, G., Akritidis, N., Bosilkovski, M. *et al.* (2005). Brucellosis. *New England Journal of Medicine* **352**, 2325–2336.

ProMed (2006). *Parigonimus-US(CA). 20060820.2337.* Available at: www.promedmail.org (accessed 25 August 2006).

Ryan, C.A., Nickels, M.K., Hargrett-Bean, N.T. *et al.* (1987). Massive outbreak of antimicrobial-resistant salmonellosis traced to pasteurized milk. *Journal of the American Medical Association* **258**, 3629–3674.

Scallan, E., Fitzgerald, M., Collins, C. *et al.* (2004). Acute gastroenteritis in Northern Ireland and the Republic of Ireland: a telephone survey. *Communicable Disease and Public Health* **7**, 761–767.

Schantz, P.M. (1983). Trichinosis in the United States – 1947–1981. *Food Technology* **March**, 83–86.

Schlech, W.F. III., Schlech, W.F. IV., Haldane, H. *et al.* (2005). Does sporadic *Listeria* gastroenteritis exist? A 2-year population-based survey in Nova Scotia, Canada. *Clinical Infectious Diseases* **41**, 778–784.

Sivapalasingam, S., Barrett, E., Kimura, A. *et al.* (2003). A multistate outbreak of *Salmonella enterica* serotype newport infection linked to mango consumption: impact of water-dip disinfestation technology. *Clinical Infectious Diseases* **37**, 1585–1590.

Sivapalasingam, S. (2004). Fresh produce: a growing cause of outbreaks of food-borne illness in the United States, 1973 through 1997. *Journal of Food Protection* **67**, 2342–2353.

Sobel, J. (2005). Botulism. *Clinical Infectious Diseases* **41**, 167–173.

Spika, J., Waterman, S., Hoo, G. *et al.* (1987). Chloramphenicol-resistant *Salmonella Newport* traced through hamburger to dairy farms: a major persisting source of human salmonellosis in California. *New England Journal of Medicine* **316**, 565–570.

Swartz, M.N. (2002). Human diseases caused by foodborne pathogens of animal origin. *Clinical Infectious Diseases* **34**(Suppl. 3), S111–122.

Tarr, P.I., Gordon, C.A. and Chandler, W.L. (2005). Shiga-toxin-producing *Escherichia coli* and haemolytic uraemic syndrome. *Lancet* **365**, 1073–1086.

Tauxe, R.V. (2001). Food safety and irradiation: protecting the public from food-borne infections. *Emerging Infectious Diseases* **7**, 516–521.

Tauxe, R.V. and Neill, M.A. (2006). Food-borne infections and food safety. In: B.A. Bowman and R.M. Russell (eds), *Present Knowledge in Nutrition*, 9th edn. Washington, DC: International Life Sciences Press.

Taylor, J.L., Tuttle, J., Pramukul, T. *et al.* (1993). An outbreak of cholera in Maryland associated with imported commercial frozen fresh coconut milk. *Journal of Infectious Diseases* **167**, 1330–1335.

Tidwell, J. and Alan, G.L. (2001). Fish as food, aquaculture's contribution. Ecological and economic impacts and contributions of fish farming and capture fisheries. *EMBO Journal* **2**, 958–963.

US Census Bureau. *Aging in the United States*. Available at: www.census.gov/ipc/prod/97agewc.pdf (accessed 24 May 2007).

US General Accounting Office (2000). Food irradiation: available research indicates that benefits outweigh risks. *Publication no. GAO/RCED-00-217*. Washington, DC: US General Accounting Office.

Voetsch, A.C., van Gilder, T.J., Angulo, F.J. *et al.* (2004). FoodNet estimate of the burden of illness caused by nontyphoidal *Salmonella* infections in the United States. *Clinical Infectious Diseases* **38**(Suppl. 3), S127–134.

Wheeler, J.G., Sethi, D., Cowden, J.M. *et al.* (1999). Study of infectious intestinal disease in England: rates in the community, presenting to general practice, and reported to national surveillance. *British Medical Journal* **318**, 1046–1050.

Wheeler, C., Vogt, T.M., Armstrong, G.L. *et al.* (2005). An outbreak of hepatitis A associated with green onions. *New England Journal of Medicine* **353**, 890–897.

## Recommended reading for clinicians

Acheson, D.W.K. and Fiore, A.E. (2004). Preventing food-borne disease – what clinicians can do. *New England Journal of Medicine* **350**, 437–440.

CDC (2004). Diagnosis and management of food-borne illnesses: a primer for physicians and other healthcare personnel. *Morbidity and Mortality Weekly Report* **53** (RR-4), 1–33.

Kendall, P.A., Hillers, V.V. and Medeiros, L.C. (2006). Food safety guidance for older adults. *Clinical Infectious Diseases* **42**, 1298–1304.

# Antibiotic resistance and nosocomial infections

# 9

**Ronald Jay Lubelchek and Robert A. Weinstein**

The mid-twentieth century discovery and subsequent mass-production of antimicrobial agents ushered in a period during which previously deadly infectious diseases, such as *Streptococcus pneumoniae* meningitis and *Staphylococcus aureus* endocarditis, could at last be cured. Optimism abounded during the early days of the antibiotic era. By 1967, the US Surgeon General reportedly stated that we could "close the book" on infectious diseases (Fauci, 2001). Yet even as these apocryphal words were spoken, bacteria – driven by the selection pressure of antimicrobial exposure and enabled by an astounding degree of genetic mutability – had begun to develop antibiotic resistance, thus undermining the era's "magic bullets." Today, despite our extensive armamentarium of antimicrobial agents, infectious diseases represent the second leading cause of death and a primary cause of disability worldwide (Fauci, 2001). Between 1980 and 1992, deaths attributed to infectious diseases increased by 58 percent. Though the HIV/AIDS epidemic comprises half of the increase, other emerging infections, including those due to antibiotic-resistant bacteria, contribute significantly to the escalation in infectious disease-related mortality (Pinner *et al.*, 1996).

## Case study

A 40-year-old woman with diabetes and end-stage renal disease requiring hemodialysis presented in March 2001 with non-healing metatarsal ulcers. Amputation was required; wound cultures grew methicillin-susceptible *Staphylococcus aureus*. In May 2001, wound cultures from a recurrent foot ulcer grew both vancomycin-susceptible *Enterococcus fecalis* and methicillin-resistant *S. aureus* (MRSA). From January 2002 to March 2002 the patient received multiple courses of systemic antibiotics, including vancomycin, for recurrent lower extremity infections (Chang *et al.*, 2003).

During an April 2002 hospitalization for further amputation, the patient developed an MRSA dialysis fistula abscess and bacteremia. In May 2002, cultures from the exit site of a new dialysis catheter grew MRSA, requiring further treatment with vancomycin and replacement of the catheter. In June 2002, cultures from a suspected catheter exit site infection grew the world's first-ever reported vancomycin-resistant clinical isolate of *S. aureus* (defined as a MIC $>32\,\mu g/ml$). The catheter tip culture also grew a garden-variety vancomycin-resistant *Enterococcus (VRE) fecalis*. DNA sequencing revealed the presence of the *vanA* gene, which conferred high-level vancomycin resistance in both strains and apparently had "jumped" from the enterococcal to the staphylococcal strain (Chang *et al.*, 2003).

The emergence of vancomycin-resistant *S. aureus* (VRSA) represents a cautionary tale. Although *S. aureus* displayed exquisite susceptibility to penicillin in the 1940s, it soon acquired a penicillin-hydrolyzing penicillinase enzyme, rendering it penicillin resistant. Only a year after the introduction of methicillin, a penicillinase-resistant penicillin derivative, nosocomial methicillin-resistant isolates of *S. aureus* emerged, and in the later 1990s MRSA began its run as a major community threat (Deresinski, 2005; see also Figure 9.1). From vancomycin's introduction in 1956 until 2002 this drug remained the agent of choice, and in some cases sole bactericidal option, for the treatment of invasive MRSA infections. Even as VRSA becomes a reality – five cases have been reported – *S. aureus* has also evolved resistance to novel agents such as quinupristin/dalfopristin, linezolid, and daptomycin (Malbruny *et al.*, 2002; Hayden *et al.*, 2005; Peeters and Sarria, 2005). Other microbes, such as *Acinetobacter* and *Pseudomonas*, also have developed resistance to all mainline antimicrobials (Manikal *et al.*, 2000; Deplano *et al.*, 2005). The story of *S. aureus* illustrates microbes' ability to adapt. The seemingly relentless progression toward multi-drug antibiotic resistance, coupled with an anemic new drug pipeline, evokes the possibility that we may be entering a post-antibiotic era during which we may struggle vainly to deal with increasingly resistant pathogens (Cohen, 1992; CDC, 2004).

The genetic events that drive the evolution of drug resistance do not occur in a vacuum. Whether drug-resistant mutants inch forward by gradually accumulating uncorrected nucleotide substitutions that occur an estimated $10^{-6}$ to $10^{-9}$ times per replicative cycle per gene, or bound ahead by whole-scale interspecies transmission of fully-formed genetic elements, human behavior dramatically affects the selection pressures that drive drug resistance (Livermore, 2003). In this chapter we will examine the social forces that contribute to antibiotic resistance, including antibiotic over-use in health care and agriculture; characterize

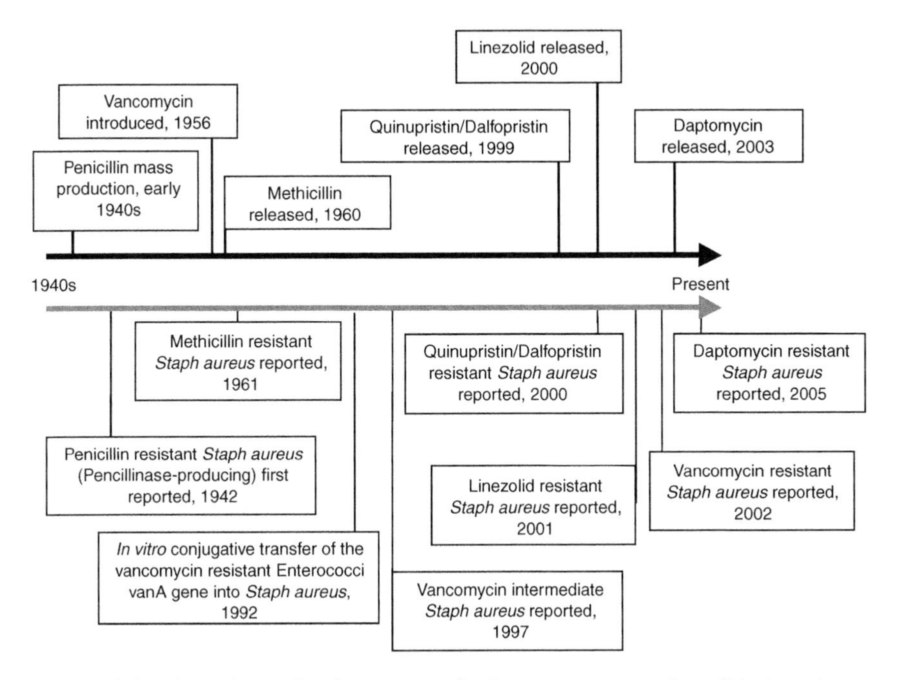


**Figure 9.1** New drug development and the emergence of antibiotic-resistant *Staphylococcus aureus*. Reproduced from Low (2005), with permission.

the epidemic of nosocomial infections (Figure 9.2); consider the impact of socio-cultural factors, such as the increased acuity of inpatient medicine, population aging, and use of invasive and prosthetic devices; and finally address potential remedies, such as antibiotic stewardship, improved application of surveillance and infection control programs, enhanced regulatory oversight, and public reporting.

## Social determinants of antibiotic resistance

### Overview: global scope and cost of antibiotic resistance

Before discussing the social determinants of antibiotic resistance, we must place the problem into a context that considers its global scale, clinical importance, and economic impact. Gram-negative bacteria containing extended spectrum beta-lactamases (ESBLs), that hydrolyze third-generation cephalosporins and most other beta-lactam antibiotics, have a global presence. In 2003, the Study for Monitoring Antimicrobial Resistance Trends (SMART) collected intra-abdominal wound culture isolates from 74 medical centers, located in 23 different countries comprising five geographic regions. ESBLs were most prevalent within nosocomial *Enterobacter* isolates from the Asia/Pacific region (Figure 9.3; Paterson

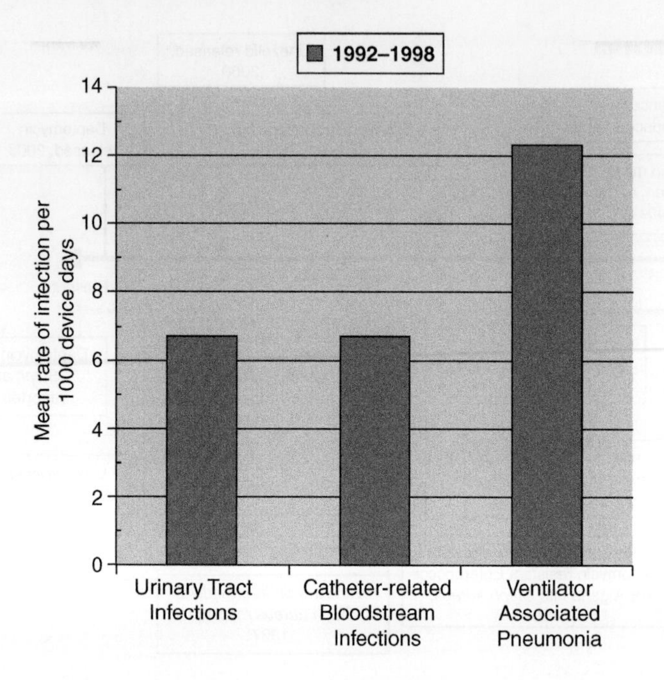


**Figure 9.2** United States intensive care unit nosocomial infection rates. Adapted from NNIS System (1998).

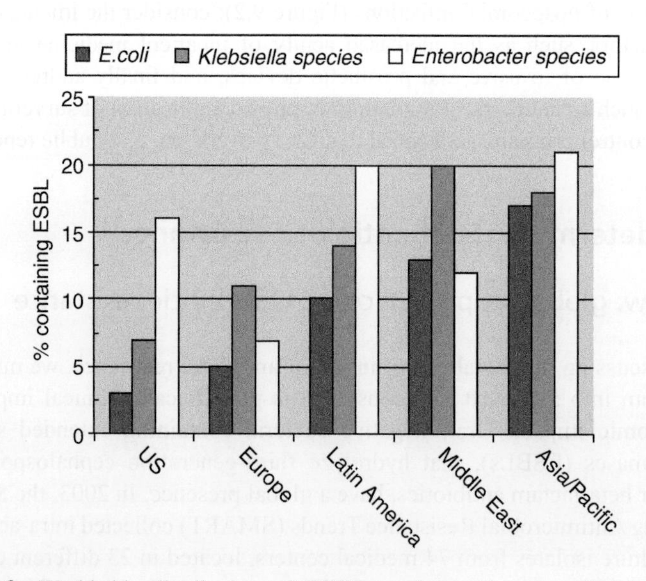


**Figure 9.3** Worldwide distribution of ESBL-containing Gram-negative bacilli from intra-abdominal sources. Adapted from Paterson *et al.* (2005).

*et al.*, 2005). In 1997, SENTRY, a similar global resistance surveillance program, examined isolates from 48 medical centers throughout Canada, the US, and Latin America; the highest rate of ESBL-containing bloodstream isolates, 33 percent, was from Latin America (Diekema *et al.*, 1999). Lautenbach and colleagues conducted a case–control study of clinical and economic impacts of infection with ESBL-containing Gram-negative bacilli. Compared to 66 matched controls, with non-ESBL containing Gram-negative infections, 33 ESBL-infected case patients had longer median lengths of stay (11 vs 7 days), longer duration until appropriate antibiotic administration (72 hours vs 11.5 hours), lower clinical response rates (76 percent vs 83 percent), and higher hospital costs ($66,590 vs $22,231) (Lautenbach *et al.*, 2001).

Significant global rates of resistance and clinical and economic impact also have been demonstrated among Gram-positive bacteria. SENTRY reported susceptibility results for over 4900 enterococcal clinical isolates collected between 1997 and 1999, documenting a 17 percent US rate of vancomycin resistance for 1999 (Low *et al.*, 2001). In 2003, the European Antimicrobial Resistance Surveillance System (EARSS), which surveys antibiotic resistance rates from 1300 hospitals in 28 European countries, found high rates of vancomycin resistance in invasive *Enterococcus faecium* infections in Portugal (50 percent), Italy (25 percent), Greece (23 percent), and Ireland (19 percent) (EARSS Management Team, 2004).

Although clinicians regularly question enterococcal virulence and impact on morbidity, an extensive meta-analysis has linked vancomycin resistance among enterococci with increases in mortality. DiazGranados and colleagues evaluated 1614 enterococcal bloodstream infections (683 vancomycin-resistant vs. 931 vancomycin-susceptible), from nine separate studies. They found that bacteremia with VRE led to a mortality rate 2.5 times that associated with vancomycin-sensitive enterococcal (VSE) bacteremia (DiazGranados *et al.*, 2005). A smaller, single-site study, comparing VRE bacteremia ($n = 21$) and VSE bacteremia ($n = 32$), showed that patients with VRE bacteremia had higher mortality (76 percent vs 41 percent), longer lengths of stay (35 vs 17 days) and, on average, $27,000 higher hospital costs per episode (Stroser *et al.*, 1998).

These examples largely reflect health-care associated infections, but the problem of antibiotic resistance pervades the community as well. *Streptococcus pneumoniae* represents the leading cause of community-acquired bacterial pneumonia, meningitis, and middle ear infections worldwide. Before the first report of reduced susceptibility in 1967, pneumococci were uniformly sensitive to penicillin (Low, 2005). Since then, the Alexander Project, set up in 1992 to monitor antibiotic resistance among respiratory tract pathogens worldwide, has revealed rising rates of pneumococcal resistance – for example, in 1998–2000, worldwide resistance was nearly 32 percent. The highest rates were in the Far East (56 percent), the Middle East (54 percent), and Africa (52 percent) (Jacobs *et al.*, 2003). Rates of penicillin resistance in pneumococci have also increased

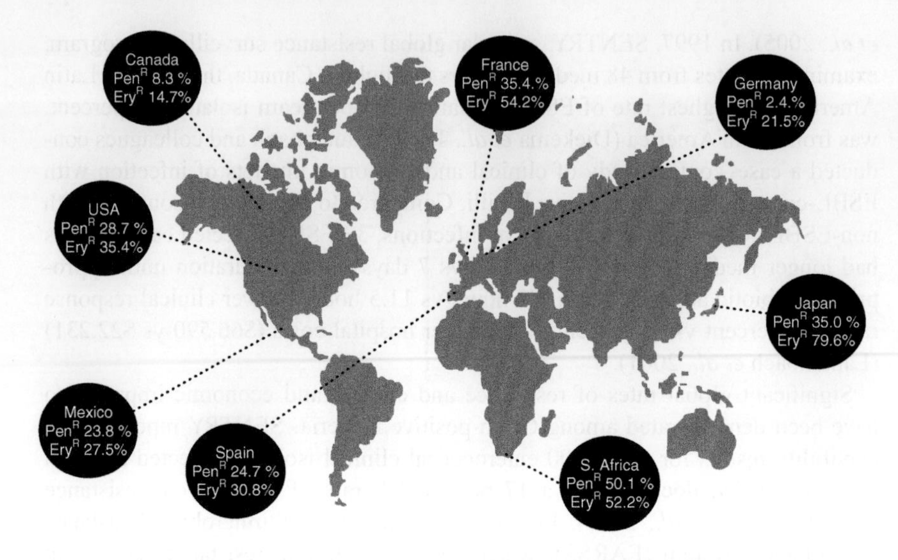


**Figure 9.4** Worldwide penicillin and erythromycin resistance, 2002–2003. Rates of macrolide and penicillin resistance in *Streptococcus pneumoniae* from Prospective Organism Tracking and Epidemiology for the Ketolide telighromycin, for 2002–2003, with penicillin resistance defined as MIC $\geqslant 2\,\mu$g/ml, and erythromycin resistance defined as MIC $\geqslant 1\,\mu$g/ml. Reproduced from Low (2005), with permission.

in the US, from 7 percent in 1992–1993 to 22 percent in 1999–2000 (Jacobs, 2003). More recent data have confirmed high rates of penicillin and macrolide resistance worldwide (Low, 2005; see Figure 9.4). Numerous reports have documented penicillin treatment failure in cases of meningitis caused by pneumococci with reduced sensitivity to penicillin (Paris *et al.*, 1995). Moreover, a study specifically looking at the effects of higher-level penicillin resistance found an association between penicillin resistance and pneumococcal pneumonia-related mortality (Feiken *et al.*, 2000).

Much of the morbidity and mortality associated with antibiotic resistance results from ineffective empiric treatment. When initial treatment for intra-abdominal sepsis failed to "cover" resistant organisms, patients experienced two times more complications (Mosdell *et al.*, 1991). Similarly, intensive care unit patients given ineffective versus effective initial antibiotic therapy had a 42 percent versus 17 percent in-hospital, infection-related mortality rate (Kollef *et al.*, 1999; Figure 9.5).

Fear of such treatment failure spurs the use of newer, often more costly, agents. For instance, if concern about pneumonia due to penicillin-resistant pneumococci led to a 10 percent market share increase for the fluoroquinolone levofloxacin, spending on empiric therapy could increase more than 100 percent based on 1999 prices for a course of amoxicillin of ($6) versus levofloxacin of ($69)

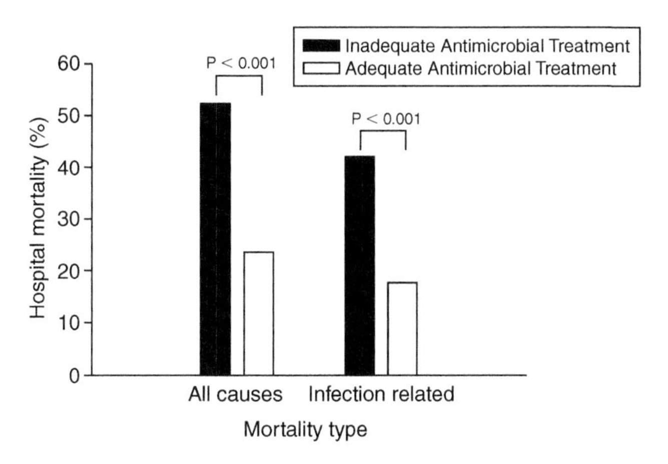


**Figure 9.5** ICU infection-related mortality based on appropriateness of initial antibiotic therapy. Reproduced from Kollef *et al.* (1999), with permission.

(Howard *et al.*, 2003). Antibiotic resistance concerns have driven an estimated $20 million increase in therapeutic choices for otitis media (Howard and Scott, 2005).

Estimating the overarching economic burden of antibiotic resistance has proven difficult. The now defunct US Congressional Office of Technology Assessment (OTA) estimated annual costs related to hospital-acquired infection caused by six antibiotic-resistant microbes at $1.5 billion (US Congress, Office of Technology Assessment, 1995). A study of "all costs" related to antibiotic resistance, including the economic impact of longer hospital stays, premature death, and the prescribing of more expensive drugs, estimated the cost at $100 million to $300 billion annually (Phelps, 1989).

## Antibiotic use and drug resistance

In this section, we review the evidence supporting a link between antibiotic use and resistance, explore inappropriate clinical uses of antibiotics, and evaluate the ecologic impact of the agricultural use of antimicrobials as "growth promoters."

### *Antibiotic use and resistance: confirming the connection*

A substantial body of data supports the connection between antibiotic use and resistance. The Meropenem Yearly Susceptibility Test Information Collection (MYSTIC) Program, a global surveillance network initiated to detect changes in carbapenem-resistance rates, reported antibiotic usage and resistance among Gram-negative isolates from 10–15 US institutions from 1999 to 2001. There was a clear population-level connection between drug use and increased resistance to ciprofloxacin for *Pseudomonas aeruginosa* and *Enterobacteriaceae*. For instance, ciprofloxacin use doubled to 6.4 defined daily doses (DDD) per 100 patient days by 2001, while Pseudomonal ciprofloxacin resistance doubled to 22.1 percent (Mutnick *et al.*, 2004; Figure 9.6).

Data linking VRE rates – nearly 29 percent in the CDC's latest National Nosocomial Infection Surveillance (NNIS) system – to antibiotic use paint a more nuanced picture than that described above for antibiotic resistance in Gram-negative bacteria (NNIS System, 2004). Results of studies linking antibiotic use with resistant pathogens, such as VRE, depend on factors such as when the study was conducted within the timeline of the pathogen's local emergence, whether the study is prospective or retrospective, patient-level or population-level, and whether it focuses on infection or colonization. Clearly, there has been a temporal relation between vancomycin use, which has increased as much as 20-fold in response to rising MRSA rates, and the emergence of VRE (Ena *et al.*, 1993). For example, as part of the Intensive Care Antimicrobial Resistance Epidemiology (ICARE) project, researchers charted rates of VRE and antibiotic consumption for 120 US ICUs from 1996 to 1999. In their multivariate regression analysis, they found a significant population-level link between vancomycin use and the rates of VRE (Fridkin *et al.*, 2001). Conversely, a meta-analysis of 20 case–control studies failed to find a clear patient-level connection between vancomycin use and VRE colonization or infection. Specifically, after controlling

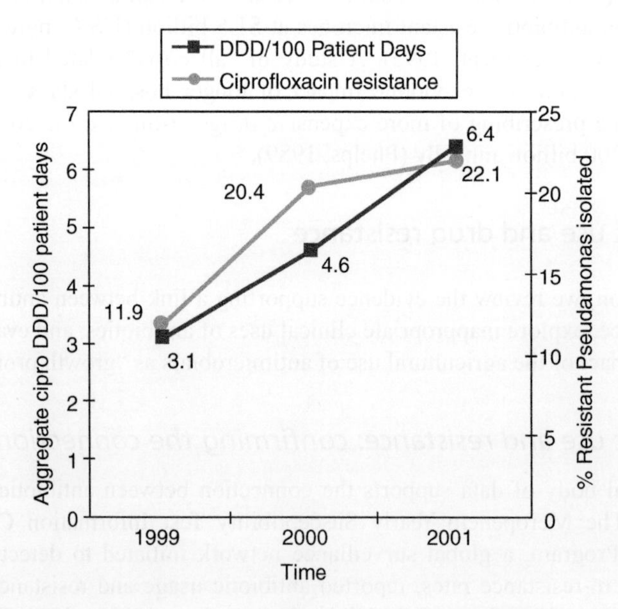


**Figure 9.6** Correlation between ciprofloxacin use and *Pseudomonas* resistance from MYSTIC Program, 1999–2001. Adapted from Mutnick *et al.* (2004).

the data for length of hospital stay, these authors noted that any significant connection between vancomycin use and VRE dissolved (Carmeli *et al.*, 1999). A large case–control study that compared 233 VRE cases with 647 well-matched controls failed to identify antecedent vancomycin use as a risk factor for VRE infection or colonization (Carmeli *et al.*, 2002).

While VRE may owe its initial emergence to vancomycin use, multi-drug resistance ensures that many antibiotic regimens can now promote VRE. The link between third-generation cephalosporin use and cases of VRE has been especially strong. Broad-spectrum third-generation cephalosporins effectively kill Gram-negative colonic micro-flora, leaving the intrinsically-resistant enterococci to multiply. A case–control study of ventilator-associated pneumonia (VAP) patients with VRE ($n = 13$) versus VAP patients without VRE ($n = 25$) found a link between VRE and exposure to third-generation cephalosporins, but not vancomycin (Bonten *et al.*, 1996). Also, a meta-analysis of 19 studies found a strong association between previous third-generation cephalosporin exposure and VRE colonization or infection, as well as significant, though weaker, connections between metronidazole and fluoroquinolone exposure and VRE (Harbarth *et al.*, 2002a).

The connection between antibiotic use and resistance also has been confirmed for community-acquired bacterial pathogens. Group A Streptococci (GAS) cause a wide range of serious infections, including pharyngitis, necrotizing soft-tissue infections, and bacteremia, along with immunologic complications such as rheumatic fever and glomerulonephritis. While GAS have remained sensitive to penicillin, macrolide-resistance has been well documented. The frequency of erythromycin-resistant isolates in Finland increased from 5 percent in 1988 to 19 percent in 1993, during a period of increasing erythromycin consumption (Seppala *et al.*, 1992, 1997; Figure 9.7).

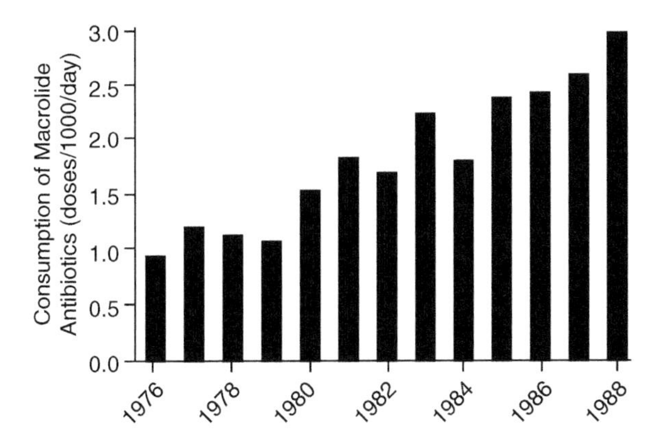


**Figure 9.7** Consumption of erythromycin by outpatients in Finland from 1976 through 1988, expressed in doses per 1000 people per day. Adapted from Seppala *et al.* (1997), with permission.

In addition to crude rates of antibiotic consumption, antibiotic use in relation to other demographic parameters may influence rates of resistance. For instance, in examining *E. coli* and enterococcal isolates from stool samples from healthy volunteers from three cities, researchers found that antibiotic resistance correlated more strongly with drug consumption as a function of population density (expressed as DDD per 1000 inhabitants per day multiplied by the number of inhabitants per $km^2$, or the DDD per $km^2$ per day) than with overall use (DDD per 1000 inhabitants per day) (Bruinsma *et al.*, 2003; Figure 9.8).

Selection pressure from residual antibiotics found in the environment, as a byproduct of either medicinal or agricultural use, also may have an impact on resistance rates. A group in Spain analyzed bacteria from water samples at various sites along the Arga River, downstream of Pamplona. They found that the urban effluent was associated with tetracycline, beta-lactam, and trimethoprim/sulfamethoxazole resistance among *Enterobacteriaceae* and *Aeromonas* isolates, and that this resistance progressively decreased at sampling sites farther from the urban center (Goni-Urriza *et al.*, 2000).

In both the hospital and the community, interactions between individuals – direct person-to-person, or indirect contact via fomites or health-care workers (HCWs) – also drive resistance rates beyond the effects of antibiotic consumption. For example, in hospitals "colonization pressure" – the number of colonized and infected patients – is a major risk factor for acquisition of VRE, MRSA, and *Pseudomonas aeruginosa* (Bonten *et al.*, 1998, 1999; Merrer *et al.*, 2000).

## Antibiotic misuse and the emergence of resistance

Despite the link between use and resistance, the appropriate use of antibiotics is one of humankind's most essential weapons against disease. So, interventions need to target inappropriate patterns of use, specifically those that have contributed most significantly to the development of resistance. In this section, we examine antibiotic misuse and explore its root causes.

Common colds – acute inflammatory changes, mostly mediated by viruses, anywhere along the continuum of the upper respiratory tract – lead to 110 million outpatient visits and cost an estimated $40 billion annually in the US. On average, common colds afflict each US adult 2.2 times per year, and child 3 times per year (Fendrick *et al.*, 2003). Despite the fact that viruses cause >90 percent of these maladies, physicians regularly prescribe antibacterial agents to treat acute upper respiratory tract infections (URIs).

Researchers used the National Ambulatory Medical Care Survey (NAMCS), an annual sampling of reasons people seek outpatient medical care, to approximate the quantity and cost of antibiotic use for URIs. NAMCS data revealed that in 1998, 84 million office visits for URIs led to 45 million antibiotic prescriptions. Using historical data to extrapolate bacterial infection rates, they concluded that 55 percent of the antibiotics were prescribed inappropriately for

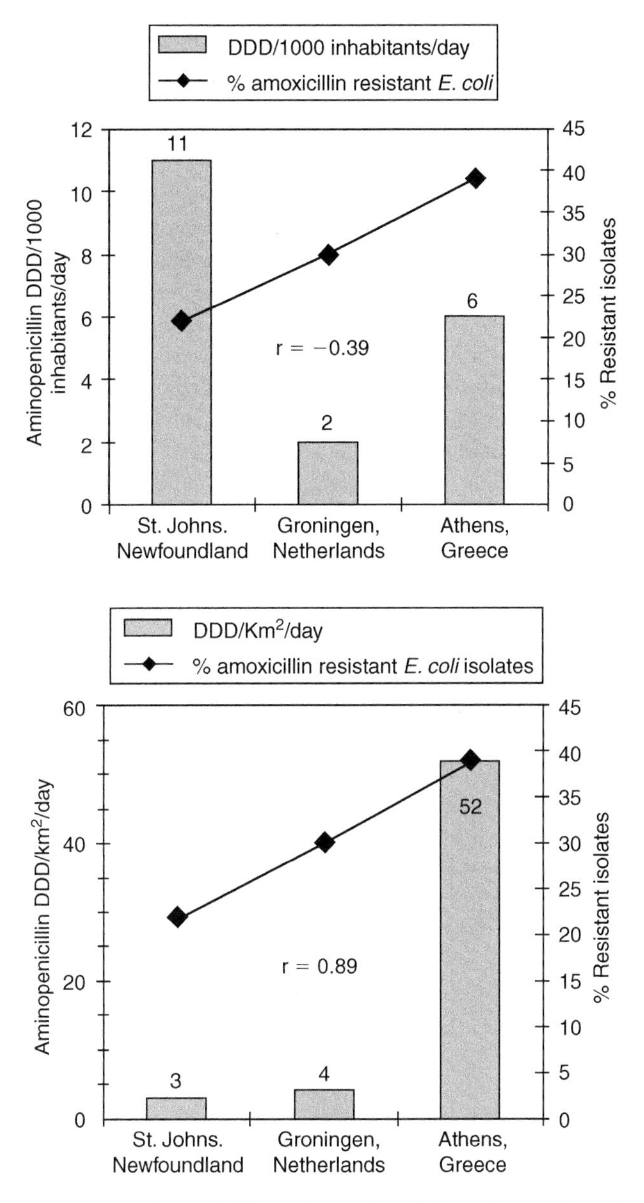


**Figure 9.8** Prevalence of amoxicillin-resistant *E. coli* from three cities, plotted against aminopenicillin consumption (DDD/1000 per day) and consumption as a function of population density (DDD/km$^2$ per day). Adapted from Bruinsma *et al.* (2003) with permission.

illnesses of viral origin. Antibiotics prescribed for URIs led to costs of $1.32 billion, of which an estimated $726 million was for unneeded antibiotics (Gonzales *et al.*, 2001; Figure 9.9). More recently, researchers estimated that, annually, $1.1 billion was being spent on 41 million unnecessary antibiotic prescriptions for viral respiratory tract infections (Fendrick *et al.*, 2003).

Similar misuse of antibacterial agents for URIs has been documented in Europe. A study of 2899 acute respiratory infections from 10 Spanish hospitals between 1994 and1995 showed that antibiotics had been prescribed for 83 percent; 41 percent of the prescriptions were considered inappropriate based on use for diseases of likely viral origin (Ochoa *et al.*, 2000). National health survey data demonstrated similar trends in France, where the proportion of acute respiratory tract infections of presumed viral etiology treated with antibiotics increased by 86 percent for children and 115 percent for adults between 1981 through 1992 (Guillemot *et al.*, 1997).

One may postulate that treating URIs of likely viral etiology with antibiotics may reduce rates of secondary bacterial infections; however, this view is unproven. Studies looking at the efficacy of antibiotic therapy for specific manifestations of URIs, such as bronchitis, sinusitis, and pharyngitis, consistently have failed to demonstrate a beneficial treatment effect. Fahey and colleagues, for instance, analyzed eight randomized, placebo-controlled trials, with greater than 300 patients in each arm, looking at the role of antibiotics versus placebo for acute cough-related illnesses in adults. They found no significant differences in proportion of patients with resolution of cough or improvement in symptoms by 7–11 days (Fahey *et al.*, 1998).

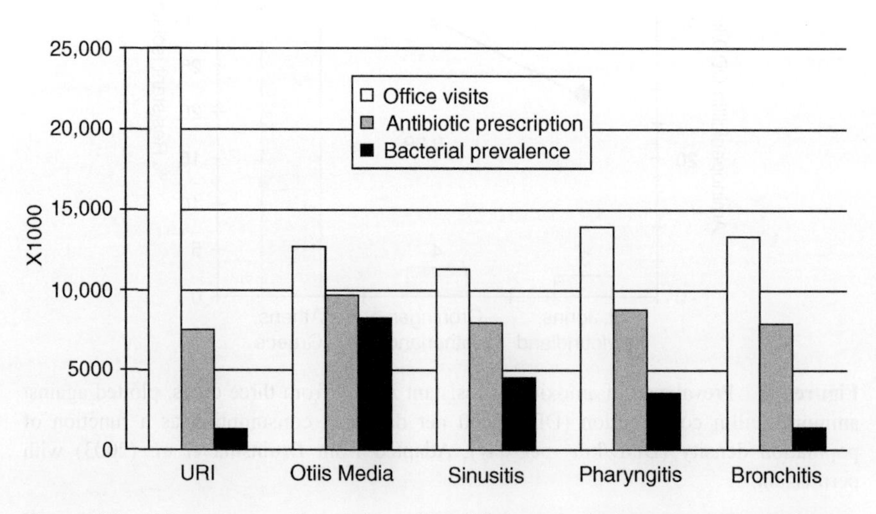


**Figure 9.9** Primary care office visits and antibiotic prescriptions for acute respiratory illnesses in the United States. Reproduced from Gonzalez *et al.* (2001), with permission.

The ecologic impact of inappropriate antibiotic prescribing for URIs may be illustrated most readily by examining the association between antibiotic use and resistance for a common, community-acquired bacterial isolate such as *S. pneumoniae*. This link has been demonstrated for both asymptomatic carriage and invasive pneumococcal disease, with patient- and population-level analyses. To evaluate this issue, general practitioners in Canberra, Australia, took nasal swab cultures and detailed antibiotic-use histories from the first 15 children, aged six years and under, who entered their practice on four separate occasions between 1997 and 1999. During the four sampling periods they collected a total of 1502 swabs, 631 (42 percent) of which grew *S. pneumoniae*; 14 percent were penicillin-resistant *S. pneumoniae* (PRSP). Utilizing a multivariate regression model, the authors identified beta-lactam use in the two months before swab collection as a key risk for carriage of PRSP. Children who had received both penicillin and a cephalosporin in the preceding two months were at even greater risk. Notably, any antibiotic use upto six months before specimen collection conferred an increased risk for PRSP carriage (Nasrin *et al.*, 2002; Figure 9.10). A similar study correlated nasal carriage rates of PRSP for children under seven years of age in Iceland with individual, as well as regional, antibiotic use. Children from the region with the highest overall antibiotic consumption (23 vs 9.6 DDD per 1000 children per day in the lowest use region) had an odds ratio of 20.3 for PRSP carriage. Additionally, children who had received a beta-lactam antibiotic in the previous year had a 6.75 odds ratio for PRSP carriage (Vilhjalmur *et al.*, 1996).

In addition to promoting PRSP carriage, outpatient antibiotic use has also been linked to invasive PRSP disease. In a retrospective cohort study, researchers reviewed records from 374 patients with invasive pneumococcal disease from

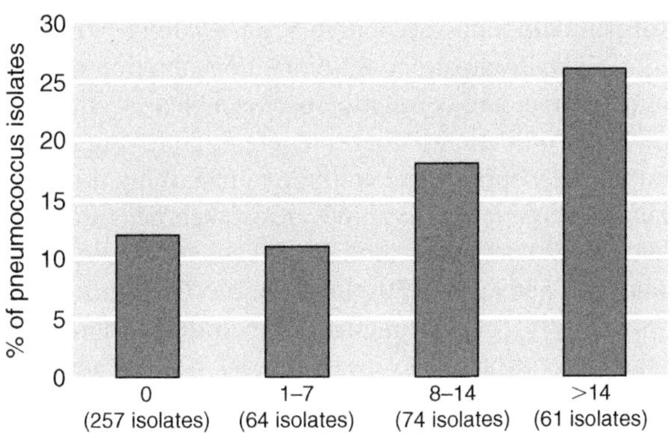


**Figure 9.10** Penicillin resistance in pneumococcus isolates from children, and beta-lactam use in the six months before swab collection. Taken from Nasrin *et al.* (2002).

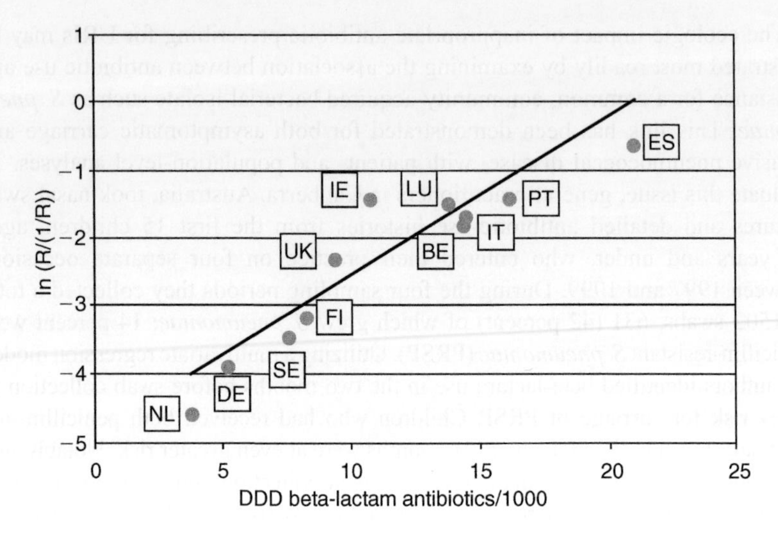


**Figure 9.11** The log odds of resistance to penicillin among invasive isolates of *Streptococcus neumoniae* (ln (R/{1/R})) is regressed against outpatient sales of beta-lactam antibiotics in 11 European countries; resistance data are from 1998–99 and antibiotic sales data are from 1997. DDD, defined daily dose; BE, Belgium; DE, Germany; FL, Finland; EI, Ireland; IT, Italy; LU, Luxembourg; NL, Netherlands; PT, Portugal; ES, Spain; SE, Sweden; UK, United Kingdom. From Bronzwaer *et al.* (2002).

five Spanish hospitals over a five-year period; 24 percent of the episodes were due to PRSP. Beta-lactam exposure in the prior three months was the only risk factor for penicillin resistance identified by multivariate analysis (Nava *et al.*, 1994). These data have been corroborated on a national level. EARSS collected national rates of penicillin non-susceptible *S. pneumoniae* (PNSP) from patients with invasive disease and outpatient antibiotic consumption data from 11 countries. They found a convincing correlation between beta-lactam consumption and national rates of PNSP infection (Bronzwaer *et al.*, 2002; Figure 9.11).

The recognition that inappropriate antibiotic prescribing for URIs has contributed to the global emergence of antibiotic resistance begs the simple question: what drives physicians to prescribe antibacterial agents for diseases of likely viral etiology? Although 97 percent of 350-plus Georgia (USA) practitioners surveyed about their antibiotic-prescribing practices agreed that antibiotic overuse represents a major factor contributing to resistance, 42 percent admitted to prescribing antibiotics for the common cold (Watson *et al.*, 1999). In a survey designed to elicit information about prescribing practices of nearly 500 Massachusetts primary care providers, 62 percent identified diagnostic uncertainty as a major reason for prescribing antibiotics for URIs (APUA, 1999). In the face of diagnostic uncertainty, physicians may feel obligated to err on the side of benefiting the patient by treating a possible bacterial infection, at the expense of the more distant and community-level problem of antibiotic resistance.

Patient or parent expectations also heavily influence physicians' antibiotic prescribing practices; 59 percent of 500 Massachusetts physicians surveyed indicated that patient requests increased their antibiotic prescribing (APUA, 1999). In a questionnaire-based study of over 600 pediatricians, 40 percent of respondents indicated that, on 10 or more occasions in the month before the survey, parents had requested an antibiotic when the physician did not feel it was indicated; 48 percent of respondents reported that parents always or most of the time pressured them to prescribe antibiotics; and one-third reported that they frequently complied with these parental requests. In this study, parental pressure, more than concerns about legal liability or practice efficiency, contributed most strongly to inappropriate antibiotic use (Bauchner *et al.*, 1999).

In a review of prescribing patterns in 15 practices, researchers found that 60 percent of the adult patients expected an antibiotic for a URI. Patients were more likely to expect an antibiotic if they felt that they had benefited previously from antibiotics for similar symptoms. In the same study, physicians believed that 62 percent of their patients expected antibiotics, and prescribed antibiotics to these patients more frequently (Dosh *et al.*, 2000). In a phone survey of over 5000 patients from nine countries, 11 percent admitted to exaggerating URI symptoms to get antibiotics from their physicians (Perchere, 2001). In a US survey, 32 percent of 12,000-plus people believed that taking antibiotics for colds prevented more serious illness, and 48 percent expected a prescription for antibiotics if they felt ill enough to seek medical attention (Eng *et al.*, 2003).

In addition to patient expectations, economic factors may also figure largely into a physician's decision to prescribe antibiotics for URIs. Pediatricians may prescribe antibiotics in an attempt to limit working parents' need for return visits (Pichichero, 1999) – but in fact an antibiotic prescription may "legitimize" the office visit for URI symptoms, thereby increasing the likelihood of an office visit and antibiotic expectations for the child's next cold (Watson *et al.*, 1999).

The details surrounding a physician's remuneration also may affect prescribing practices. Though all physicians in Canada receive payment from the single-payer government health system, some are salaried and some are paid on a fee-for-service basis. A study of antibiotic prescribing found that both fee-for-service payment and greater volume of patients strongly correlated with higher antibiotic prescription rates (Hutchinson and Foley, 1999). Additionally, in a managed-care setting that awards physicians based on patient satisfaction surveys, physicians may prescribe out of fear of dissatisfying their patients (Schwartz *et al.*, 1998).

Some of the factors discussed above may be operating on a national level, leading to dramatic disparities in prescribing practices in the neighboring countries of France and Germany. In 1998, 7 percent and 53 percent of the *S. pneumoniae* strains from Germany and France, respectively, were PNSP. This difference parallels the 8 percent versus 48 percent outpatient visits for colds that culminated in antibiotic prescriptions in Germany compared with France. Factors that may drive these differences include increased pressure to prescribe among French patients and increased utilization of day care by the French, whose households are more likely to have two working parents. Macro-economic forces also may drive the disparity in prescribing practices. Retail drug prices in France are among the lowest in all of Europe – 32 percent less than US prices – whereas German prices are 24 percent higher than US prices (56 percent higher than French prices). Low drug prices likely have contributed to France's ranking highest in *per capita* outpatient antibiotic consumption in Europe. Evidence also suggests that low drug prices in France encourage use of newer, broad-spectrum agents, whereas Germans more often rely on less expensive, older, generic antibiotics (Harbarth *et al.*, 2002b).

Antibiotics sold over the counter (OTC), without a physician's prescription, also account for a significant portion of antibiotic misuse, especially in developing countries, whose inhabitants may have limited access to physician-guided medical care. In addition to limited access to physicians, self-medication with antibiotics may be motivated by the desire to save money, the perceived need for urgent treatment of a suspected bacterial illness, or the desire to maintain privacy regarding potentially embarrassing symptoms. Varying legal and cultural forces shape pharmacy prescribing practices in different countries, including willingness to give advice on antibiotic use or to refill old prescriptions without a physician's approval (Radyowijati and Haak, 2002).

In many places, antibiotics may be sold without any physician advice. Of 49 Greek pharmacists presented with a fictionalized acute rhinosinusitis patient, 86 percent offered antibiotics, most often broad spectrum (Conotpoulos-Ioannidis *et al.*, 2001). A similar study that presented fictional scenarios to 100 pharmacies in Katmandu, Nepal, found that all retail pharmacists engaged in diagnostic and therapeutic behavior beyond the scope of their training; 97 percent suggested inappropriate antibiotics for diarrhea, while many (56 percent) failed to suggest oral rehydration therapy and only 3 percent recommended seeking a physician's care (Wachter *et al.*, 1999).

## Agricultural antibiotic use and abuse

The agricultural use of antibiotics contributes significantly – perhaps more than use of antibiotics to treat human disease – to selection pressures driving the emergence of antibiotic resistance. As farming has become more commercial, with larger numbers of animals being housed in fewer, more densely populated farms, the role of "agri-antibiotics" has grown. Farmers and veterinarians use antibiotics for three primary reasons: to treat sick animals, to halt the dissemination of infection, and to promote growth. While antibiotics are used at pharmacologic doses to treat or prevent infection in livestock, the common growth promotion-oriented practice uses sub-therapeutic doses, such as less than 200 grams per tonne of feed, for extended periods, with the unintended, but very efficient, selection of resistant bacterial mutants (McEwen and Fedorka-Cray, 2002).

Although the US agricultural industry has not published precise data on the quantity of antibiotics in use, several estimates have been made. An agricultural industry-sponsored group, the Animal Health Institute, approximated that in 1998, 17.8 million pounds of agri-antimicrobials were used, 17 percent of which, they estimate, was non-therapeutic use for growth promotion (McEwen and Fedorka-Cray, 2002). An independent organization, the Union of Concerned Scientists, calculated that annually, in the US, humans utilize 3 million pounds of antibiotics, while the agricultural industry non-therapeutic use is 24.6 million pounds – or 8 times the quantity used to treat human disease (Mellon *et al.*, 2001).

Despite regulations designed to strictly control the quantity of both the residual antibiotics found in food animals at the time of slaughter and spillage of animals' bacteria-laden intestinal contents during slaughter, and processing, antibiotic-resistant bacteria derived from food animals colonize humans and cause human disease. For instance, genetic sequencing allowed the tracing of an outbreak of antibiotic-resistant *Salmonella* serotype Newport in California from infected patients back to a fast-food hamburger chain, to a meat-processing plant, and finally to the farm where the source-cattle were raised (Swartz, 2002).

Concerns about the ecologic impact of agri-antibiotics are global. Before 1997, European Union (EU) farmers extensively used the glycopeptide antibiotic avoparcin as a growth promoter. Resistance to avoparcin confers cross-resistance to vancomycin, and the agricultural use of avoparcin has been linked to the emergence of VRE in Europe. While glycopeptides seldom had been used medically in Europe in the early 1990s, the presence of significant percentages of VRE carriers in the community suggested a problem. For instance, the former East German Government strictly limited vancomycin use, yet 12 percent of healthy patients demonstrated intestinal colonization with VRE (Bates, 1997).

While European physicians may not have used glycopeptide antibiotics in the early 1990s, the same cannot be said for the European agricultural industry. In Denmark, in 1993, physicians prescribed 24 kg of vancomycin, while farmers used 24,000 kg of avoparcin (Aarestrup, 1995). Not surprisingly, Danish researchers found that chicken farms using avoparcin had a 55 times greater chance of harboring VRE-colonized broilers compared to farms that did not utilize avoparcin (Bates, 1997). Research examining VRE colonization rates among Dutch turkey farmers who used avoparcin provides a circumstantial link between food animal VRE carriage and human colonization. The researchers found fecal samples colonized with VRE from 50 percent of turkeys, 39 percent of turkey farmers, 20 percent of turkey slaughterers, and 14 percent of area residents (van den Bogaard *et al.*, 1997).

Agricultural quinolone use has been linked to increasing rates of quinolone-resistant *Campylobacter jejuni*, but, unlike VRE colonization in healthy hosts, *C. jejuni* commonly causes clinical disease. The US first licensed agricultural fluoroquinolone use in 1995. To assess the effects, the Minnesota Department of Public Health studied quinolone resistance rates among human *Campylobacter* isolates. They found an increase in the state-wide proportion of quinolone-resistant *C. jejuni*, from 1.3 percent in 1992 to 10.2 percent in 1998. In 1997, they cultured quinolone-resistant *C. jejuni* from 14 percent of chicken parts sampled from 91 retail markets in the state. Using restriction enzyme-based sub-typing, the researchers showed the presence of identical quinolone-resistant *C. jejuni* isolates from both chickens and humans (Smith *et al.*, 1999). The National Antimicrobial Resistance Monitoring System for enteric bacteria (NARMS) corroborated these findings. In 1989–1990 they documented no cases of quinolone-resistant *Campylobacter*, while by 2001 19 percent of *Campylobacter* isolates demonstrated quinolone resistance. NARMS surveillance found that of 180 chicken products tested from three states, 10 percent harbored quinolone-resistant *Campylobacter* (Gupta *et al.*, 2004).

## Nosocomial infections

### Scope and magnitude

The threat that antibiotic resistance poses to human health seems even more dire when considered in the context of nosocomial infections. Ever since the 1840s, when Ignaz Semmelweis first demonstrated that hand hygiene by health-care workers could reduce rates of puerperal fever, there has been an awareness of both the importance of infection control and the impact of lapses in its application. Five percent of hospital patients, and 5–35 percent of patients in intensive care units (ICUs), acquire an infection during their stay (CDC, 1992; Roberts *et al.*, 2003).

The CDC estimate that greater than 2 million people contract nosocomial infections annually in the US (CDC, 1992). Similarly striking statistics have been cited in Europe. As part of the European Prevalence of Infection in Intensive Care (EPIC) study, researchers conducted a one-day point-prevalence study of ICU-acquired infections, drawing case-reports from over 1400 ICUs from 17 countries. They discovered that 21 percent of the ICU patients had acquired infections during the course of their ICU stay; pneumonia, urinary tract infection (UTI), and bloodstream infection (BSI) were most commonly reported (Vincent *et al.*, 1995). Although the crude rates of nosocomial infection have remained stable at 5–6 infections per 100 admissions between 1975 and 1995, progressively shorter average hospital stays over that time translate into higher incidence-densities of health-care associated infections (HAIs), increasing from 7 to 10 HAIs per 1000 patient-days in 1975 versus1995 (Weinstein, 1998).

In 1970, concern regarding HAIs prompted the CDC to establish the National Nosocomial Infection Surveillance (NNIS) System (now a part of the National Healthcare Safety Network – NHSN), a collection of nearly 300 hospitals that provided monthly data related to HAI rates. Most recently, the NNIS System reported average device-day incidence-densities for HAIs in US ICUs between January 2002 and June 2004. They documented 4.9 UTIs per 1000 urinary catheter days, 4.9 BSIs per 1000 central-line days, and 7.5 cases of ventilator-associated pneumonia (VAP) per 1000 ventilator days (NNIS System, 2004).

Many groups have demonstrated the clinical and economic impact of nosocomial infections. In the EPIC study, researchers showed that hospital-acquired BSIs, pneumonias, and clinical sepsis each acted independently to increase mortality (Vincent *et al.*, 1995). Others have estimated that HAIs confer rates of attributable mortality ranging from 4 percent for UTIs with secondary bacteremia to 47 percent for primary BSIs (Eggimann and Pittet, 2001). In addition to increasing mortality, HAIs increase cost. In a retrospective cohort study involving 139 non-infected patients and 25 patients with HAIs, researchers from Chicago's Cook County Hospital found that, after controlling for the confounding factors of severity of illness and ICU stay, an HAI diagnosis added over $15,000 in excess costs per associated hospital stay (Roberts *et al.*, 2003). Using this cost estimate and the annual approximated incidence of 2 million cases, HAIs may cost $30 billion in the US each year.

Due to the selection pressure created by frequent antibiotic use in hospital environments, antibiotic-resistant organisms frequently cause nosocomial infections. From 1988 to 1994, the percent of hospitalized patients who received antibiotics increased from 32 percent to 53 percent (Pestotnik *et al.*, 1996). The most recent NNIS System update reflects the consequences of increased use. NNIS reported antibiotic resistance rates for HAIs in ICU patients for 2003 and compared these with resistance rates for the previous five years, showing, for example, increasing methicillin resistance in staphylococcus and increasing resistance to third-generation cephalosporins in *Klebsiella* (NNIS System, 2004; Figure 9.12). The Surveillance and Control of Pathogens of Epidemiological Importance (SCOPE) study reported antibacterial resistance rates from over 24,000 nosocomial BSIs from 49 US hospitals from 1995 through 2002. They documented an increase in rates of MRSA, from 22 percent in 1995 to 51 percent in 2002. Ceftazidime resistance for *Pseudomonas* increased from 12 percent in 1995 to 29 percent in 2003 (Wisplinghoff *et al.*, 2004).

Increases in fluoroquinolone use may be particularly problematic. Looking at over 35,000 Gram-negative ICU isolates collected between 1994 and 2000, Neuhauser and colleagues correlated the 10 percent decline in *Pseudomonas* susceptibility to ciprofloxacin (86 percent in 1994 to 76 percent in 2000) with increased national fluoroquinolone use (Neuhauser *et al.*, 2003). More recent data have shown that rates of *Clostridium difficile*-associated diarrhea (CDAD) have increased 26 percent in the US between 2000 and 2001. Characterization of the latest outbreak strains of *C. difficile* revealed a predominant clone which produces higher quantities of cytotoxins A and B, secretes an additional "binary" toxin, may cause more severe colitis, and demonstrates markedly higher rates of

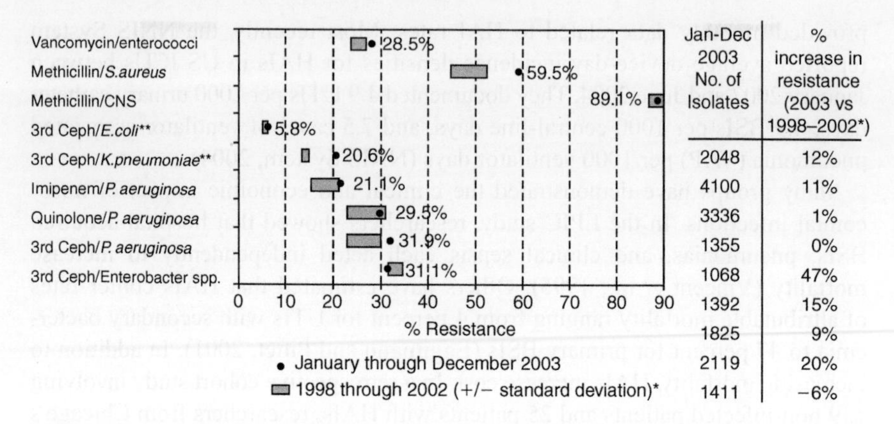


**Figure 9.12** Selected antimicrobial-resistant pathogens associated with nosocomial infections in ICU patients, comparison of resistance rates from January through December 2003 with 1998 through 2002. Percent increase in resistance rate of current year (2003) compared with mean rate of resistance over previous five years (1998–2002). From NNIS System (2004).

quinolone resistance than do historic *C. difficile* strains (McDonald *et al.*, 2005). Multivariable regression analysis from a recent case–control study of CDAD from Maryland's Veteran's Affairs medical system showed that previous fluoroquinolone use conferred the greatest risk of developing CDAD compared with use of other antibiotics (odds ratio 12.7 for quinolones, 2.2 for clindamycin, 0.4 for cephalosporins; McClusker *et al.*, 2003).

## Social dynamics driving the nosocomial infection epidemic

Just as hospitals require a diverse array of forces to act in concert in order to function properly, health-care institution-related problems often stem from a complex combination of factors. Here we discuss three key factors that affect rates of HAIs: the increasing intensity of care in modern hospitals, the aging of the population, and lapses in adherence to basic infection control practices.

### *Increased intensity of care*

Despite the decrease in average length of hospital stays, from 7.8 days in 1970 to 4.8 days in 2004, several lines of evidence suggest that hospitals now provide more intensive and invasive care. The National Hospital Discharge Survey (NHDS), conducted annually by the National Center for Health Statistics (NCHS), estimated that in 2003 nearly 44 million procedures were performed

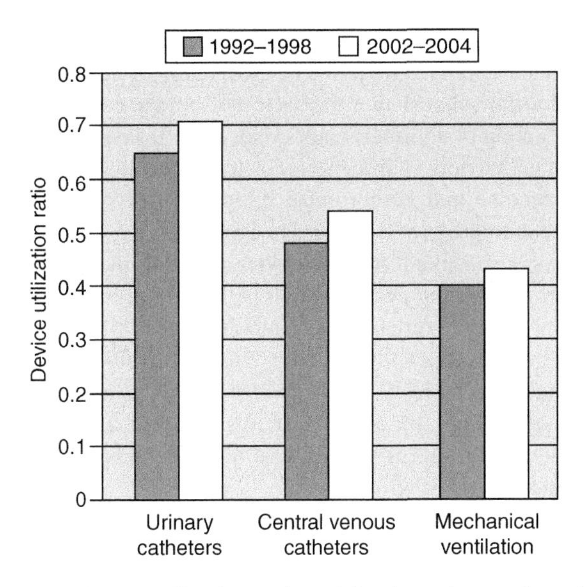


**Figure 9.13** Mean device utilization ratios, defined as the number of device days divided by the number of patient days, for US ICUs: comparison of two time periods. Adapted from NNIS System, 1998, 2004.

on inpatients during their hospital stays – 3 million more annual inpatient procedures than 10 years ago (Graves, 1995; DeFrances *et al.*, 2005). Many of these procedures require a high degree of invasiveness. In 1967, when surgeons performed the first coronary artery bypass graft, few would have guessed that nearly 600,000 such procedures would be performed in the US in 1996 (NCHS, 1996). In addition, many of these invasive procedures include the use of prosthetic devices. The American Academy of Orthopedic Surgeon reported that the annual number of total hip arthroplasties increased from 117,000 in 1991 to 220,000 in 2003, while knee replacements increased even more dramatically, from 160,000 in 1991 to 418,000 in 2003 (AAOS, 2005). Similarly, rates of prosthetic heart-valve replacements have increased since the procedure's 1960 inception; surgeons now perform almost 60,000 such procedures annually (Swartz, 1994).

Along with increases in invasive surgical procedures has come increased use of invasive devices to support sicker patients. By monitoring device-utilization ratios (defined as the number of device days divided by the number of patient days) for a given ICU, the NNIS System has shown progressive increases in the use of urinary catheterization, central venous catheterization (CVC), and mechanical ventilation in US ICUs (NNIS System 1998, 2004; Figure 9.13). Hemodialysis patients – 60,000 in 1980 and 450,000 in 2003 – represent another growing population that requires the use of invasive devices, in the form of long-term vascular access (USRDS, 2005).

Invasive devices increase risks for nosocomial infections in several ways. Foreign bodies such as prosthetic joints or CVCs provide a surface on which bacteria may form biofilms that shelter the bacteria from the patient's natural defense systems and the action of antimicrobials. Also, many invasive devices circumvent the body's physical barriers to infection. For instance, endotracheal tubes inhibit mucocilliary clearance and compromise the gag reflex, which makes patients more susceptible to micro-aspiration and pneumonia.

Several studies have linked invasive-device use with increased risk for HAIs. By prospectively monitoring nearly 17,000 patient-days, including patients from five separate ICUs, French researchers found that use of either mechanical ventilation or CVCs increased patients' risk for HAI three-fold (Legras *et al.*, 1998). In a separate study, pediatric ICU researchers examined risk factors for HAI in 945 ICU admissions. They found that patients who had a one-point increase in their device-utilization ratio (device days divided by patient days) were twice as likely to develop an HAI (Singh-Naz *et al.*, 1996).

## *Aging and nosocomial infections*

The aging of the population represents another factor driving the nosocomial infection epidemic. While in 1900 only 1 percent of the world's population (15 million people) was greater than 65 years of age, by 1992 this proportion had grown to 6 percent (342 million people). By 2050, it's estimated that 2.5 billion world inhabitants will be older than 65 (Strausbaugh, 2001). Several factors may increase elderly patients' risks of acquiring HAIs. Their immune function and natural defenses may be decreased. Older patients have less robust T-lymphocyte proliferation, leading to decreased antibody production and cell-mediated immune function with increasing age. Conditions more prevalent in the elderly, such as diabetes, malignancy, vascular disease, and dementia, also may decrease barriers to infection. Functional incapacity due to aging may necessitate use of invasive devices, such as urinary catheters or feeding tubes, which further bypass the body's natural defenses. Incapacity and immobility may also lead to skin breakdown and the risk of infected pressure sores (Strausbaugh, 2001).

Old-age specific risk factors have translated into increased rates of HAIs for the elderly. NNIS data from 1986 to 1990 showed that persons greater than 65 years of age acquired 54 percent of all nosocomial infections (Emori *et al.*, 1991). While decade-specific HAI rates were 10 per 1000 hospital days for patients up to the fifth decade of life, patients greater than 70 had more than 100 HAIs per 1000 hospital days (Gross *et al.*, 1983). HAIs in the elderly also may affect outcomes more dramatically. For instance, comparing younger patients with *S. aureus* surgical site infection (SSI) with elderly *S. aureus* SSI patients, researchers found that elderly patients had a three-fold greater chance of death, longer hospital stays (13 vs 9 days), and higher hospital costs ($85,658 vs $45,767) (McGarry *et al.*, 2004).

Living arrangements of elderly patients also may place them at greater risk for HAIs. Ninety percent of long-term care facility (LTCF) residents are greater than 65 years old (Capitano and Nicolau, 2003). The LTCF setting itself may contribute to elderly patients' increased risk for HAIs. The closed institutional setting increases LTCF residents' exposure to bacteria via frequent contact with staff and other residents, and in facilities with antiquated air-ventilation systems, exposure to airborne respiratory pathogens may be exacerbated (Yoshikawa, 2000).

LTCF residents acquire 2–4 million HAIs annually (Garibaldi, 1999), representing a rate of 4–8 infections per 1000 patient days (Capitano and Nicolau, 2003). The transfer into LTCFs of patients colonized or infected by resistant organisms during stays in acute-care hospitals, and selection pressure created by frequent LTCF use of antibiotics, lead to high rates of antimicrobial resistance. One survey of residents from 25 LTCFs found that 38 percent were receiving antibiotics; bacterial isolates from those residents were resistant to the prescribed antibiotic in 65 percent of cases (Capitano and Nicolau, 2003). The combination of at-risk, elderly residents and frequent antibiotic use led a prominent antibiotic resistance researcher to label LTCFs as antibiotic-resistance "factories" (Levy and Marshall, 2004).

## *Lapses in infection control: the dirt on hand-washing*

Although guidelines have repeatedly stressed the importance of infection control practices such as hand hygiene, lapses in basic infection control continue to drive HAI rates. A review of 12 studies examining health-care workers' hand hygiene practices found rates of compliance with hand-washing recommendations ranging from 16 to 80 percent, with an average rate of 40 percent compliance (Boyce, 2001). A study of neonatal ICU personnel documented the presence of Gram-negative bacilli on the hands of 75 percent (Goldmann *et al.*, 1978), and poor hand hygiene has been implicated directly as a cause of nosocomial infection. By comparing matched controls with 12 case-patients during an ICU VRE outbreak, Boyce and colleagues demonstrated that physical proximity to a case-patient and exposure to nurses who cared for a case-patient represented the primary risk factors for infection, implying that unwashed HCW hands moved VRE from patient to patient (Boyce *et al.*, 1994).

Numerous possible barriers to compliance with infection control recommendations have been cited, ranging from lofty excuses, such as man's yearning for liberty, to more mundane reasons, such as time constraints (Farr, 2000). A large survey of hand hygiene practices at a university teaching hospital found poor compliance more likely in clinical settings with greater intensity of care, as defined by number of opportunities for hand hygiene per hour. These researchers found that compliance decreased in progressive 5 percent decrements as hand-cleansing opportunities increased by 10 opportunities per hour (Pittet *et al.*, 2000). Although use of alcohol-based hand rubs greatly eases time constraints, adherence to recommendations is still embarrassingly low (Weinstein, 2001). It follows that increased workload related to staffing shortages could further reduce adherence to infection control guidelines. In an outbreak of catheter-related bloodstream infections in a surgical ICU, Fridkin and colleagues identified patient-to-nurse ratio as a major, independent risk factor for infection. Patients cared for by nurses with a 2 : 1 patient-to-nurse ratio had a relative risk for infection nearly 62 times that of patients cared for by nurses with a 1 : 1 ratio (Fridkin *et al.*, 1996).

## Reasons for hope: control and prevention

As antibiotic resistance and nosocomial infection have become leading global threats to human health, many individuals and institutions have dedicated themselves to limiting these threats. We note how three key components – antibiotic stewardship, improved infection control, and regulatory oversight – may help reduce rates of antibiotic resistance and HAIs.

### Antibiotic stewardship

Merriam-Webster's online dictionary defines stewardship as "the careful and responsible management of something entrusted to one's care." As evidence linking antibiotic misuse and resistance has accumulated, a primary focus of antibiotic stewardship has been to limit inappropriate antibiotic prescriptions in the inpatient setting. Several stewardship strategies, which intercede at various points in the prescription process, have been reviewed recently (Figure 9.14; MacDougall and Polk, 2005).

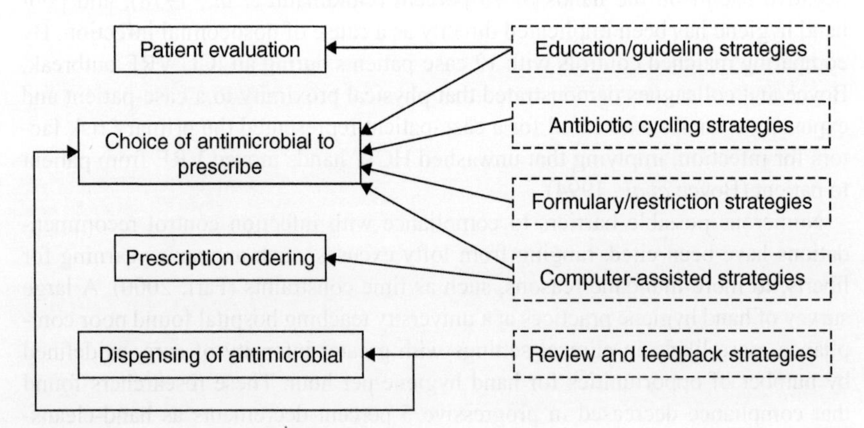


**Figure 9.14** Antimicrobial prescribing process and antimicrobial stewardship strategies. From MacDougall and Polk (2005).

Formulary restrictions have been used successfully to reduce rates of antibiotic resistance. Spurred on by a desire to reduce hospital antimicrobial expenditures, as well as a need to halt a resistant *Acinetobacter* bloodstream infection outbreak, Ben Taub County Hospital in Houston initiated a formulary restriction program in 1994. They required infectious disease approval before prescribing amikacin, ceftazidime, ciprofloxacin, fluconazole, ofloxacin, ticarcillin/clavulanate, or piperacillin/tazobactam. By encouraging selection of less expensive antibiotics, without reducing overall quantity of antibiotic prescribing, the restriction led to a $430,000 (32 percent) decrease in expenditures for parenteral antimicrobial agents during a six-month period. The restrictions also led to increased rates of antimicrobial susceptibility – for example, rates of susceptibility for *E. coli* to ticarcillin/clavulanate increased from 77 to 97 percent for ICU patients. Although some have raised concerns that antibiotic restrictions may worsen clinical outcomes by hindering a clinician's ability to immediately prescribe broad-spectrum antibiotics, this study showed equivalent or better patient outcomes during the restriction period; survival rates for bacteremic patients did not change, nor did time to administration of appropriate antibiotic therapy (White *et al.*, 1997).

Antibiotic restrictions limited to even a single unit can make a difference. In an attempt to decrease rates of quinolone-resistant *Pseudomonas aeruginosa*, ICU staff from a French hospital limited use of fluoroquinolones during a six-month period. Quinolone use decreased from 330 to 80 DDD per 1000 patient-days, and rates of quinolone resistance among *Pseudomonas* isolates decreased from 71 to 52 percent. Although unexpected, MRSA rates also fell from 30 percent to 18 percent (Aubert *et al.*, 2005).

Physician computer order-entry systems can bring automation to antibiotic stewardship. Computer systems may prompt prescribing physicians with regard to national guidelines, local resistance patterns, drug interactions, and indications for use (Weinstein, 2001). A group from Utah described antibiotic prescribing outcomes during use of an advanced physician order-entry system that linked directly with patients' microbiology records to make prescribing recommendations. During a one-year intervention, ICU physicians used the anti-infectives order-entry system for all patients. The system led to fewer orders for drugs to which the patients had reported allergies, improved drug dosing, lessened antibiotic susceptibility – drug choice mismatches (from 206 before to 12 during intervention), and reduced anti-infective costs, total hospital costs, and length of hospital stay (Evans *et al.*, 1998).

Health systems also have successfully instituted outpatient antibiotic stewardship programs to reduce inappropriate prescribing for URIs. Educational efforts directed at rural Utah physicians successfully reduced prescribing for URIs by 15 percent, with a 56 percent decrease in antibiotic prescriptions for acute bronchitis (Rubin *et al.*, 2005). A randomized controlled trial of the effects of a multi-faceted educational intervention aimed at reducing antibiotic prescriptions for URIs provided educational materials to patients, and pharmaceutical industry-style academic detailing and feedback of practice prescribing profiles for clinicians. During the intervention period there was a substantial reduction in rate of antibiotic treatment of acute bronchitis, from 74 percent to 48 (Gonzales *et al.*, 1999).

## Improvements in infection control

In an attempt to reduce rates of antibiotic resistance and nosocomial infections, multifaceted interventions also have been directed at bolstering infection control practices. In 1996, officials from the Siouxland region of Iowa, Nebraska, and South Dakota noted increased rates of VRE. A taskforce requested CDC assistance to institute comprehensive surveillance and isolation precautions. Under CDC guidance, they enacted active screening, via perianal swabs, for VRE at the majority of the 32 Siouxland acute and long-term care facilities. By 1999, 92 percent of the facilities actively screened for VRE and 88 percent of the facilities had implemented strict barrier isolation precautions and environmental disinfection policies for VRE-infected or -colonized patients. The overall prevalence of VRE decreased from 2.2 percent in 1997 to 0.5 percent in 1999 (Ostrowsky *et al.*, 2001).

Multifaceted interventions also have been aimed at improving hand hygiene. Between 1994 and 1997, Pittet and colleagues carried out a campaign to improve compliance with hand disinfection recommendations. Their interventions included educational posters in strategic areas, wide distribution of personal bottles of alcohol hand-rub, and bedside-mounted alcohol hand-rub dispensers. During 1994–1997, this sustained effort improved hand hygiene compliance from 48 to 66 percent. There also was a decreased prevalence of HAIs, from 17 to 10 percent, and a decreased rate of MRSA transmission, from 2.16 to 0.93 episodes per 10,000 patient-days (Pittet *et al.*, 2000).

Importantly, the intervention by Pittet and colleagues attacked the problem of poor hand hygiene on two fronts. While they attempted to change behavior through educational interventions, they also used the "technologic advance" of alcohol-based hand-rubs to simplify hand hygiene. By reducing skin irritation and the time needed for hand antisepsis, alcohol-based hand-rubs address key barriers to hand hygiene compliance. This study demonstrates how both the assiduous application of low technology (e.g. educational posters) along with technologic innovation (e.g. alcohol-based hand-gels) can play key roles in advancing infection control and limiting HAIs.

## Role of regulatory oversight

As reports of antibiotic resistance and HAIs have multiplied, the world's healthcare community has mobilized resources to respond to these problems. Large-scale surveillance projects, such as NNIS in the US and EARSS in Europe, have provided vital information about the scope of the problem. These alarming data have captured the attention of regulatory bodies. Laws designed to help reduce antibiotic misuse and limit nosocomial infections, thereby improving patient safety, have been enacted or are in the process of being enacted in several jurisdictions.

The strong correlation between increased rates of VRE colonization in Europe and the agricultural use of avoparcin has stimulated action. The agricultural use of avoparcin has been banned, first in Denmark in 1995, then by Germany in 1996, and by all European Union nations in 1997. The beneficial effects of this government-mandated ban have been dramatic. In Germany, for instance, the rate of poultry products testing positive for VRE has decreased from near 100 percent in 1994 to 25 percent in 1997. Correspondingly, human rates of VRE colonization in Germany have decreased from 12 to 3 percent (Swartz, 2002).

While a 1985 CDC report on nosocomial infection control showed that hospitals with four key infection control components – an effective hospital epidemiologist, one infection control practitioner for every 250 beds, active surveillance mechanisms, and ongoing control efforts – could reduce nosocomial infection rates by one-third, not until recently has the government taken a more aggressive approach to regulating hospital safety (Haley *et al.*, 1985). Growing concern regarding nosocomial infections has prompted several US states to consider legislation that would require hospitals to publish their HAI rates. As of July 2005, nosocomial infection "report card" laws have been passed in seven states, and are being considered in an additional 37 (Weinstein *et al.*, 2005; APIC, 2006). Hospital-specific HAI rates depend not only on controllable variables, such as hand hygiene compliance, but also on more intractable factors, such as the severity of illness of a hospital's patient population. Careful consideration of these confounding factors, along with emphasis on process measures (such as monitoring appropriate peri-operative antibiotic prophylaxis rates, and selected outcome measures) may facilitate the equitable application of nosocomial infection report card laws. Although legislating patient safety will prove to be complex, patients should be encouraged that hospitals are addressing this dire problem.

## Conclusion

Alexander Fleming, the discoverer of penicillin and a father of modern anti-infective therapy, once said, "A good gulp of hot whiskey at bedtime – it's not very scientific, but it helps" (Creative Quotations, 2006). As the problem of antibiotic resistance grows, we find ourselves on the brink of a new era, a post-antibiotic era during which a shot of ampicillin may be less helpful than Fleming's suggested shot of whiskey. We have reviewed the global scope, clinical consequences, and economic impact of antibiotic resistance and nosocomial infections. We have also highlighted how human behaviors, such as the misuse of antibiotics for viral disorders and the overuse of antibiotics in agriculture, create selection pressures that contribute to these epidemics. Although new drug discovery, continued vaccine research, and technologic advances in infection control may help to alleviate these problems, ultimately only changes in human behavior, whether mandated by government oversight or driven by individual conscience, will stem the tide of antibiotic resistance and nosocomial infection.

## References

Aarestrup, F.M. (1995). Occurrence of glycopeptides resistance among *Enterococcus faecium* isolates from conventional and ecological poultry farms. *Microbiological Drug Resistance* **1**, 255–257.

Alliance for the Prudent Use of Antibiotics (APUA) (1999). *Massachusetts Physician Survey*. Available at: http://www.tufts.edu/med/apua/Research/physicianSurvey1-01/physicianSurvey.htm (accessed 10 November 2005).

American Academy of Orthopaedic Surgeons (AAOS) (2005). *Arthroplasty and Total Joint Replacement Procedures: 1991 to 2003*. Available at: http://www.aaos.org/wordthml/research/stats/arthroplasty_all.htm (accessed 26 November 2005).

Association for Professionals in Infection Control and Epidemiology (APIC) (2006). *Mandatory Reporting of Infection Rates, Where does Your State Stand?* Available at: http://www.apic.org/Content/NavigationMenu/GovernmentAdvocacy/MandatoryReporting/state_legislation/state_legislation.htm (accessed 10 January 2006).

Aubert, G., Carricajo, A., Vautrin, A.C. *et al.* (2005). Impact of restricting fluoroquinolone prescription on bacterial resistance in an intensive care unit. *Journal of Hospital Infection* **59**, 83–89.

Bates, J. (1997). Epidemiology of vancomycin-resistant enterococci in the community and the relevance of farm animals to human infection. *Journal of Hospital Infection* **37**, 89–101.

Bauchner, H., Pelton, S.I. and Klein, J.O. (1999). Parents, physicians, and antibiotic use. *Pediatrics* **103**, 395–401.

Bonten, M.J., Hayden, M.K., Mathan, C. *et al.* (1996). Epidemiology of colonization of patients and environment with vancomycin-resistant enterococci. *Lancet* **348**, 1615–1619.

Bonten, .M.J., Slaughter, S., Ambergen, A.W. *et al.* (1998). The role of "colonization pressure" in spread of vancomycin-resistant enterococci. *Archives of Internal Medicine* **158**, 1127–1132.

Bonten, M.J., Bergmans, D.C., Speijher, H. and Stobberingh, E.E. (1999). Characteristics of polyclonal endemicity of *Pseudomonas aeruginosa* colonization in intensive care units. *American Journal of Respiratory and Critical Care Medicine* **160**, 1212–1219.

Boyce, J.M. (2001). Consequences of inaction: importance of infection control practices. *Clinical Infectious Diseases* **33**(Suppl. 3), S133–137.

Boyce, J.M., Opal, S.M., Chow, J.W. *et al.* (1994). Outbreak of multidrug-resistant *Enterococcus faecium* with transferable VanB class vancomycin resistance. *Journal of Clinical Microbiology* **32**, 1148–1153.

Bronzwaer, S.L., Carrs, O., Buchholz, U. *et al.* (2002). A European study on the relationship between antimicrobial use and antimicrobial resistance. *Emerging Infectious Diseases* **8**, 278–282.

Bruinsma, N., Hutchinson, J.M., van den Bogaard, A.E. *et al.* (2003). Influence of population density on antibiotic resistance. *Journal of Antimicrobial Chemotherapy* **51**, 385–390.

Capitano, B. and Nicolau, D.P. (2003). Evolving epidemiology and cost of resistance to antimicrobial agents in long-term care facilities. *Journal of American Medical Directors Association* **4**(Suppl), S90–98.

Carmeli, Y., Samore, M.H. and Huskins, C. (1999). The association between antecedent vancomycin treatment and hospital-acquired vancomycin-resistant enterococci. *Archives of Internal Medicine* **159**, 2461–2468.

Carmeli,Y., Eliopoulos, G.M. and Samore, M.H. (2002). Antecedent treatment with different antibiotic agents as risk factors for vancomycin-resistant *Enterococcus. Emerging Infectious Diseases* **8**, 802–807.

Centers for Disease Control and Prevention (1992). Public health focus: surveillance, prevention, and control of nosocomial infections. *Morbidity Mortality Weekly Report* **41**, 783–787.

Centers for Disease Control and Prevention (2004). Brief report: vancomycin-resistant *Stapylococcus* aureus – New York, 2004. *Morbidity and Mortality Weekly Report* **53**, 322–323.

Chang, S., Sievert, D.M., Hageman, J.C. *et al.* (2003). Infection with vancomycin–resistant *Staphylococcus aureus* containing the *vanA* resistance gene. *New England Journal of Medicine* **348**, 1342–1347.

Cohen, M.L. (1992). Epidemiology of drug resistance: implications for a post-antimicrobial era. *Science* **258**, 1050–1055.

Contopoulos-Ioannidis, D.G., Koliofoti, I.D., Koutroumpa, I.C. *et al.* (2001). Pathways for inappropriate dispensing of antibiotics for rhinosinusitis: a randomized trial. *Clinical Infectious Diseases* **22**, 76–82.

Creative Quotations (2006). http://www.creativequotations.com/one/2409.htm (accessed 13 January 2006).

DeFrances, C.J., Hall, M.J. and Podgomik, M.N. (2005). 2003 National Hospital Discharge Survey. *Advance Data from Vital Health Statistics* **359**, 1–20.

Deplano, A., Denis, O., Poirel, L. *et al.* (2005). Molecular characterization of an epidemic clone of panantibiotic-resistant *Pseudomonas aeruginosa. Journal of Clinical Microbiology* **43**, 1198–1204.

Deresinski, S. (2005). Methicillin-resistant *Staphylococcus aureus*: an evolutionary, epidemiologic and therapeutic odyssey. *Clinical Infectious Diseases* **40**, 562–573.

DiazGranados, C.A., Zimmer, S.M., Klein, M. and Jernigan, J.A. (2005). Comparison of mortality associated with vancomycin-resistant and vancomycin-susceptible enterococcal bloodstream infections: a meta-analysis. *Clinical Infectious Diseases* **41**, 327–333.

Diekema, D.J., Pfaller, M.A., Jones, R.N. *et al.* (1999). Survey of bloodstream infections due to Gram-negative bacilli: frequency of occurrence and antimicrobial susceptibility of isolates collected in the United States, Canada and Latin America for the SENTRY Antimicrobial Surveillance Program, 1997. *Clinical Infectious Diseases* **29**, 595–607.

Dosh, S.A., Hickner, J.M., Mainous, A.G. and Ebell, M.H. (2000). Predictors of antibiotic prescribing for non-specific upper respiratory infections, acute bronchitis, and acute sinusitis: and UPRnet study. *Journal of Family Practice* **49**, 407–414.

EARSS Management Team (2004). *EARSS Annual Report, 2003. European Antimicrobial Resistance Surveillance System.* Available at: http://www.earss.rivm.nl/index.html (accessed 31 October 2005).

Eggimann, P. and Pittet, D. (2001). Infection control in the ICU. *Chest* **130**, 2059–2093.

Emori, T.G., Banerjee, S.N., Culver, D.H. *et al.* (1991). Nosocomial infections in elderly patients in the United States, 1986–1990. *American Journal of Medicine* **91**(Suppl. 3b), S289–293.

Ena, J., Dick, R.W., Jones, R.N. and Wenzel, R.P. (1993). The epidemiology of intravenous vancomycin usage in a university hospital. A 10-year study. *Journal of the American Medical Association* **269**, 598–602.

Eng, J.V., Marcus, R., Hadler, J.L. *et al.* (2003). Consumer attitudes and use of antibiotics. *Emerging Infectious Diseases* **9**, 1128–1134.

Evans, R.S., Pestotnik, S.L., Classen, D.C. *et al.* (1998). A computer-assisted management program for antibiotics and other antiinfective agents. *New England Journal of Medicine* **338**, 232–238.

Fahey, T., Stocks, N. and Thomas, T. (1998). Quantitative systemic review of randomized controlled trials comparing antibiotic with placebo for acute cough in adults. *British Medical Journal* **316**, 906–910.

Farr, B.M. (2000). Reasons for non-compliance with infection control guidelines. *Infection Control and Hospital Epidemiology* **21**, 411–416.

Fauci, A.S. (2001). Infectious diseases: considerations for the 21st century. *Clinical Infectious Diseases* **32**, 675–685.

Feikin, D.R., Schuchat, A., Kolczak, A.M. *et al.* (2000). Mortality from invasive pneumococcal pneumonia in the era of antibiotic resistance, 1995–1997. *American Journal of Public Health* **90**, 223–229.

Fendrick, A.M., Monto, A.S., Nightengale, B. and Sarnes, M. (2003). The economic burden of non-influenza-related viral respiratory tract infection in the United States. *Archives of Internal Medicine* **163**, 487–494.

Fridkin, S.K., Pear, S.M., Williamson, T.H. *et al.* (1996). The role of understaffing in central venous catheter-associated bloodstream infections. *Infection Control and Hospital Epidemiology* **17**, 150–158.

Fridkin, S.K., Edwards, J.R., Courval, J.M. *et al.* (2001). The effects of vancomycin and third-generation cephalosporins on prevalence of vancomycin-resistant enterococci in 126 US adult intensive care units. *Annals of Internal Medicine* **135**, 175–183.

Garibaldi, R.A. (1999). Residential care and the elderly: the burden of infection. *Journal of Hospital Infection* **43**(Suppl.), S9–18.

Goldmann, D.A., Leclair, J. and Macone, A. (1978). Bacterial colonization of neonates admitted to an intensive care environment. *Journal of Pediatrics* **93**, 288–293.

Goni-Urriza, M., Capdepuy, M., Arpin, C. *et al.* (2000). Impact of an urban effluent on antibiotic resistance of riverine *Enterobacteriaceae* and *Aeromonas* spp. *Applied Environmental Microbiology* **66**, 125–132.

Gonzales, R., Steiner, J.F., Lum, A. and Parrett, P.H. (1999). Decreasing antibiotic use in ambulatory practice: impact of a multidimensional intervention on the treatment of uncomplicated acute bronchitis in adults. *Journal of the American Medical Association* **281**, 1512–1519.

Gonzales, R., Malone, D.C., Maselli, J.H. and Sande, M. (2001). Excessive antibiotic use for acute respiratory infections in the United States. *Clinical Infectious Diseases* **33**, 757–762.

Graves, E.J. (1995). Detailed diagnoses and procedures: National Hospital Discharge Survey, 1993. *Vital Health Statistics* **13**(122).

Gross, P.A., Papuano, C., Adrignola, A. and Shaw, B. (1983). Nosocomial infections: decade-specific risk. *Infection Control* **4**, 145–147.

Guillemot, D., Maison, P., Carbon, C. *et al.* (1997). Trends in antimicrobial drug use in the community – France, 1981–1992. *Journal of Infectious Diseases* **177**, 492–497.

Gupta, A., Nelson, J.M., Barrett, T.J. *et al.* (2004). Antimicrobial resistance among *Campylobacter* strains, United States, 1997–2001. *Emerging Infectious Diseases* **10**, 1102–1109.

Haley, R.W., Culver, D.H., White, J. *et al.* (1985). The efficacy of infection surveillance and control programs in preventing nosocomial infections in US hospitals. *American Journal of Epidemiology* **121**, 182–205.

Harbarth, S., Cosgrove, S. and Carmeli, Y. (2002a). Effects of antibiotics on nosocomial epidemiology of vancomycin-resistant enterococci. *Antimicrobial Agents and Chemotherapy* **46**, 1619–1628.

Harbarth, S., Albrich, W. and Brun-Buisson, C. (2002b). Outpatient antibiotic use and prevalence of antibiotic-resistant pneumococci in France and Germany: a sociocultural perspective. *Emerging Infectious Diseases* **8**, 1460–1467.

Hayden, M.K., Rezai, K., Hayes, R.A. *et al.* (2005). Development of daptomycin resistance *in vivo* in methicillin-resistant *Staphylococcus aureus*. *Journal of Clinical Microbiology* **43**, 5285–5287.

Howard, D.H. and Scott, R.D. (2005). The economic burden of drug resistance. *Clinical Infectious Diseases* **41**(Suppl. 4), S283–286.

Howard, D.H., Scott, R.D., Packard, R. and Jones, D. (2003). The global impact of drug resistance. *Clinical Infectious Diseases* **36**(Suppl, 1), S4–10.

Hutchinson, J.M. and Foley, R.N. (1999). Method of physician remuneration and rates of antibiotic prescription. *Canadian Medical Association Journal* **160**, 1013–1017.

Jacobs, M.R. (2003). Worldwide trends in antimicrobial resistance among common respiratory tract pathogens in children. *Pediatric Infectious Diseases Journal* **22**, S109–119.

Jacobs, M.R., Felmingham, D., Appelbaum, P.C. Gruneberg, R.N. and the Alexander Project Group (2003). The Alexander Project 1998–2000: susceptibility of pathogens isolated from community-acquired respiratory tract infection to commonly used antimicrobial agents. *Journal of Antimicrobial Chemotherapy* **52**, 229–246.

Kollef, M.H., Sherman, G., Ward, S. and Fraser, V.J. (1999). Inadequate antimicrobial treatment of infections: a risk factor for hospital mortality among critically ill patients. *Chest* **115**, 462–474.

Lautenbach, E., Patel, J.B., Bilker, W.B. *et al.* (2001). Extended-spectrum beta-lactamase-producing *Escherichia coli* and *Klebsiella pneumoniae*: risk factors for infection and impact of resistance on outcomes. *Clinical Infectious Diseases* **32**, 1162–1171.

Legras, A., Malvy, D., Quinioux, A.I. *et al.* (1998). Nosocomial infections: prospective survey of incidence in five French intensive care units. *Intensive Care Medicine* **24**, 1040–1046.

Levy, S. and Marshall, B. (2004). Antibacterial resistance worldwide: causes, challenges and responses. *Nature Medicine* **10**(Suppl.), S122–129.

Livermore, D.M. (2003). Bacterial resistance: origins, epidemiology, and impact. *Clinical Infectious Diseases* **36**(Suppl. 1), S11–23.

Low, D.E. (2005). Changing trends in antimicrobial-resistant pneumococci: It's not all bad news. *Clinical Infectious Diseases* **41**(Suppl. 4), S228–233.

Low, D.E., Keller, N., Barth, A. and Jones, R.N. (2001). Clinical prevalence, antimicrobial susceptibility, and geographic resistance patterns of enterococci: results from the SENTRY Antimicrobial Surveillance Program, 1997–1999. *Clinical Infectious Diseases* **32**(Suppl. 2), S133–145.

MacDougall, C. and Polk, R.E. (2005). Antimicrobial stewardship programs in health care systems. *Clinical Microbiology Reviews* **18**, 638–656.

Malbruny, B., Canu, A., Bozdogan, B. *et al.* (2002). Resistance to quinupristin-dalfopristin due to mutation of L22 ribosomal protein in *Staphylococcal aureus*. *Antimicrobial Agents and Chemotherapy* **46**, 2200–2207.

Manikal, V.M., Landman, D., Saurina, G. *et al.* (2000). Endemic carbapenem-resistant *Acinetobacter* species in Brooklyn, New York: citywide prevalence, interinstitutional spread, and relation to antibiotic usage. *Clinical Infectious Disease* **31**, 101–106.

McClusker, M.E., Harris, A.D., Perencevich, E. and Roghmann, M.C. (2003). Fluroquinolone use and *Clostridium difficile*-associated diarrhea. *Emerging Infectious Diseases* **9**, 730–733.

McDonald, L.C., Kilgore, G.E., Thompson, A. *et al.* (2005). An epidemic, toxin gene–variant strain of *Clostridium difficile*. *New England Journal of Medicine* **353**, 2433–2441.

McEwen, S.A. and Fedorka-Cray, P.J. (2002). Antimicrobial use and resistance in animals. *Clinical Infectious Diseases* **34**(Suppl. 3), S93–106.

McGarry, S.A., Engemann, J.J., Schmader, K. *et al.* (2004). Surgical site infection due to *Staphylococcus aureus* in the elderly: mortality, duration of hospitalization and cost. *Infection Control and Hospital Epidemiology* **25**, 461–467.

Mellon, M., Benbrook, C. and Benbrook, K. (2001). *Hogging It: Estimates of Antimicrobial Abuse in Livestock*. Cambridge: UCS Publications.

Merrer, J., Santoli, F., Appere de Vecchi, C. *et al.* (2000). "Colonization pressure" and risk of acquisition of methicillin-resistant *Staphylococcus aureus* in a medical intensive care unit. *Infection Control and Hospital Epidemiology* **21**, 718–723.

Mosdell, D.M., Morris, D.M. and Voltura, A. (1991). Antibiotic treatment for surgical peritonitis. *Annals of Surgery*, **214**, 543–549.

Mutnick, A.H., Rhomberg, P.R., Sader, H.S. and Jones, R.N. (2004). Antimicrobial usage and resistance trend relationships from the MYSTIC Programme in North America (1991–2001). *Journal of Antimicrobial Chemotherapy* **53**, 290–296.

Nasrin, D., Collignon, P.J., Roberts, L. *et al.* (2002). Effect of beta-lactam antibiotic use in children on pneumococcal resistance to penicillin: prospective cohort study. *British Medical Journal* **340**, 1–4.

National Center for Health Statistics (NCHS) (1996). *Number of Ambulatory and Inpatient Procedures by Procedure Category and Location: United States, 1996.* Available at: http://www.cdc.gov/nchs/data/hdasd/13_139t9.pdf (accessed 26 November 2005).

Nava, J.M., Bella, F., Garau, J. *et al.* (1994). Predictive factors for invasive disease due to penicillin-resistant *Streptococcus pneumoniae*: a population-based study. *Clinical Infectious Diseases* **19**, 884–890.

Neuhauser, M.M., Weinstein, R.A., Rydman, R. *et al.* (2003). Antibiotic resistance among Gram-negative bacilli in US intensive care units. *Journal of the American Medical Association* **289**, 885–888.

NNIS System (1998). *National Nosocomial Infection Surveillance System Report, Data Summary from October 1986–April 1998, issued June 1998*. Available at: http://www.cdc.gov/ncidod/hip/NNIS/sar98net.PDF (accessed 26 November 2005).

NNIS System (2004). National Nosocomial Infections Surveillance (NNIS) System Report, data summary from January 1992 through June 2004, issued October 2004. *American Journal of Infection Control* **32**, 470–485.

Ochoa, C., Eiros, J.M., Inglada, L. *et al.*, Spanish Study Group on Antibiotic Treatments (2000). Assessment of antibiotic prescription in acute respiratory infections in adults. *Journal of Infection* **41**, 73–83.

Ostrowsky, B.E., Trick, W.E., Sohn, A.H. *et al.* (2001). Control of vancomycin-resistant enterococcus in health care facilities in a region. *New England Journal of Medicine* **344**, 1427–1433.

Paris, M.M., Ramilo, O. and McCracken, G.H. (1995). Management of meningitis caused by penicillin-resistant *Streptococcus pneumoniae*. *Antimicrobial Agents and Chemotherapy* **39**, 2171–2175.

Paterson, D.L., Rossi, F., Baquero, F. *et al.* (2005). *In vitro* susceptibilities of aerobic and facultative Gram-negative bacilli from patients with intra-abdominal infections worldwide: the 2003 Study for Monitoring Antimicrobial Resistance Trends (SMART). *Journal of Antimicrobial Chemotherapy* **55**, 965–973.

Peeters, M.J. and Sarria, J.C. (2005). Clinical characteristics of Linezolid-resistant *Staphylococcus aureus* infections. *American Journal of Medical Sciences* **330**, 102–104.

Perchere, J.C. (2001). Patients' interviews and misuse of antibiotics. *Clinical Infectious Diseases* **33**(Suppl. 3), S170–173.

Pestotnik, S.L., Classen, D.C., Evans, R.S. and Burke, J.P. (1996). Implementing antibiotic practice guidelines through computer-assisted decision support: clinical and financial outcomes. *Annals of Internal Medicine* **124**, 884–890.

Phelps, D.E. (1989). Bug/drug resistance. *Medical Care* **27**, 194–203.

Pichichero, M.E. (1999). Understanding antibiotic overuse for respiratory tract infections in children. *Pediatrics* **104**, 1384–1388.

Pinner, R.W., Teutsch, S.M., Simonsen, L. *et al.* (1996). Trends in infectious diseases mortality in the United States. *Journal of the American Medical Association* **275**, 189–193.

Pittet, D. (2001). Compliance with hand disinfection and its impact on hospital-acquired infections. *Journal of Hospital Infection* **48**(Suppl.), S40–46.

Pittet, D., Hugonnet, S., Harbath, S. *et al.* (2000). Effectiveness of a hospital-wide program aimed at improving compliance with hand hygiene. *Lancet* **356**, 1307–1312.

Radyowijati, A. and Haak, H. (2002). Determinants of antimicrobial use in the developing world. *Child Health Research Project Special Report* **4**(1), 1–36. Available at: http://www.childhealthresearch.org/doc/AMR_vol4.pdf (accessed 14 November 2005).

Roberts, R.R., Scott, R.D., Cordell, R. *et al.* (2003). The use of economic modeling to determine the hospital costs associated with nosocomial infections. *Clinical Infectious Diseases* **36**, 1424–1432.

Rubin, M.A., Bateman, K., Alder, S. *et al.* (2005). A multifaceted intervention to improve antimicrobial prescribing for upper respiratory tract infections in a small rural community. *Clinical Infectious Diseases* **40**, 546–553.

Schwartz, B., Mainous, A.G. and Marcy, S.M. (1998). Why do physicians prescribe antibiotics for children with upper respiratory tract infections. *Journal of the American Medical Association* **279**, 881–882.

Seppala, H., Nissinen, A., Jarvinen, H. *et al.* (1992). Resistance to erythromycin in Group A Streptococci. *New England Journal of Medicine* **326**, 292–297.

Seppala, H., Klaukka, T., Vuopio-Varkila, J. *et al.* (1997). The effects of changes in the consumption of macrolide antibiotics on erythromycin resistance in Group A Streptococci in Finland. *New England Journal of Medicine* **337**, 441–446.

Singh-Naz, N., Sprague, B.M., Patel, K.M. and Pollack, M.M. (1996). Risk factors for nosocomial infection in critically ill children: a prospective cohort study. *Critical Care Medicine* **24**, 875–878.

Smith, K.E., Besser, J.M., Hedberg, C.W. *et al.* (1999). Quinolone-resistant *Campylobacter jejuni* infections in Minnesota, 1992–1998. *New England Journal of Medicine* **340**, 1525–1532.

Strausbaugh, L.J. (2001). Emerging health care-associated infections in the geriatric population. *Emerging Infectious Diseases* **7**, 268–271.

Stroser, V., Peterson, L.R., Postelnick, M. and Noskin, G.A. (1998). *Enterococcus faecium* bacteremia: does vancomycin resistance make a difference? *Archives of Internal Medicine* **158**, 522–527.

Swartz, M.N. (1994). Hospital-acquired infections: diseases with increasingly limited therapies. *Proceedings of the National Academy of Science* **91**, 2420–2427.

Swartz, M.N. (2002). Human diseases caused by foodborne pathogens of animal origin. *Clinical Infectious Diseases* **34**(Suppl. 3), S111–122.

United States Renal Data System (USRDS) (2005). *2005 Annual Data Report, Reference Section D: Treatment Modalities*. Available at: http://www.usrds.org/2005/ref/D.pdf (accessed 26 November 2005).

US Congress, Office of Technology Assessment (1995). *Impacts of Antibiotic-Resistant Bacteria, OTA-H-629*. Princeton: Princeton University. Available at: http://www.wws.princeton.edu/ota/ns20/year_f.html (accessed 31 October 2005).

van den Bogaard, A.E., Jensen, L.B. and Stobberingh, E.E. (1997). Vancomycin-resistant enterococci in turkeys and farmers. *New England Journal of Medicine* **337**, 1558–1559.

Vilhjalmur, A.A., Kristinsson, K.G., Sigurdsson, J.A. *et al.* (1996). Do antimicrobials increase the carriage rate to penicillin resistant pneumococci in children? Cross sectional prevalence study. *British Medical Journal* **313**, 387–391.

Vincent, J.L., Bihari, D.J., Suter, P.M. *et al.* (1995). The prevalence of nosocomial infection in the intensive care units in Europe: results of the European Prevalence of Infection in Intensive Care (EPIC) Study. *Journal of the American Medical Association* **274**, 639–644.

Wachter, D.A., Joshi, M.P. and Rimal, B. (1999). Antibiotic dispensing by drug retailers in Katmandu, Nepal. *Tropical Medicine and International Health* **4**, 782–788.

Watson, R.L., Dowwell, S.F., Jayaraman, M. *et al.* (1999). Antimicrobial use for pediatric upper respiratory infections: reported practice, actual practice and parent beliefs. *Pediatrics* **104**, 1251–1257.

Weinstein, R.A. (1998). Nosocomial infection update. *Emerging Infectious Diseases* **4**, 416–420.

Weinstein, R.A. (2001). Controlling antimicrobial resistance in hospitals: infection control and the use of antibiotics. *Emerging Infectious Diseases* **7**, 188–192.

Weinstein, R.A., Siegel, J.D. and Brennan, P.J. (2005). Infection control report cards – securing patient safety. *New England Journal of Medicine* **353**, 225–227.

White, A.C., Atmar, R.L., Wilson, J. *et al.* (1997). Effects of requiring prior authorization for selected antimicrobials: expenditures, susceptibilities, and clinical outcomes. *Clinical Infectious Diseases* **25**, 230–239.

Wisplinghoff, H., Bischoff, T., Tallent, S.M. *et al.* (2004). Nosocomial bloodstream infections in US hospitals: analysis of 24,179 cases from a prospective nationwide surveillance study. *Clinical Infectious Diseases* **39**, 309–317.

Yoshikawa, T.T. (2000). Epidemiology and unique aspects of aging and infectious diseases. *Clinical Infectious Diseases* **30**, 931–933.

# Vaccines and immunization 10

**Heather J. Lynch and Edgar K. Marcuse**

## Introduction

Vaccination is widely regarded as one of the great public health achievements of the twentieth century. The deadly scourge of smallpox was eradicated (CDC, 1997). Polio remains in only a handful of countries worldwide, and measles, mumps, rubella, diphtheria, tetanus, and *Haemophilus influenzae* type b are all in decline (CDC, 1999a, 2006; WHO, 2005a). New technologies and novel approaches to vaccine development have the potential to expand greatly the number of diseases that can be prevented or ameliorated by vaccines in the twenty-first century. Two new rotavirus vaccines have recently been developed that have the potential to reduce the morbidity and mortality of this ubiquitous infection worldwide. Two new Human Papilloma Virus (HPV) vaccines that prevent infection by the serotypes responsible for most cervical cancers will reduce the incidence of the most common cancer of women throughout much of the world. Promising results in the development of a malaria vaccine have led to recent major new investments to accelerate its development and distribution (PATH, 2005). Recent advances in the science of tuberculosis detection and efforts to develop more effective vaccines against the disease hold out hope that an old enemy may once again be forced to retreat (Hampton, 2004).

The success of widespread immunization has had major effects on human societies. Reductions in disease prevalence as a result of vaccination have contributed to increases in child survival, decreases in morbidity and long-term disability as well as their associated societal and health-care costs, and, in turn, increases in economic productivity. Since their earliest days, vaccination efforts have also raised, and continue to raise, important issues with which all societies must grapple – issues such as individual rights versus public good; risks and benefits of intervention versus doing nothing; equity and social justice in ensuring benefits accrue to all, not just an elite few; priority-setting and decision-making in the utilization and distribution of limited resources; and the political, social, financial, and institutional realities of creating and carrying out public health policies, and developing and disseminating new technologies.

In this chapter, we briefly the review the history of vaccines and public health immunization programs and then present the complex fabric of the modern vaccine enterprise and its effects on society, with a focus on the social, political, and economic dimensions that influence disease prevention today.

## A brief history of vaccines and public health immunization programs

### Development of vaccines

The roots of modern immunization can be traced back hundreds of years to early attempts to derive protection from disease through exposure to the putative cause. Buddhists in seventh-century India drank snake venom to protect against the deadly effects of a snakebite in what may have been an attempt to induce immunity. Variolation – the introduction of dried pus from smallpox pustules of a mild case into the skin of an unaffected person – as a regular practice dates back at least to sixteenth-century India, although it appears to have developed in Central Asia as early as the second century and then spread to China, Turkey, and eventually to Europe in the eighteenth century (Plotkin and Orenstein, 2004).

Edward Jenner's work on smallpox ushered in the modern vaccine era. Although his accomplishments built on the work of others – most notably that of an English cattle breeder named Benjamin Justy – and on earlier observations that cowpox infection provided immunity against smallpox, Jenner is considered the father of modern immunization because he took a systematic, scientific approach to vaccination and the understanding of immunity, and presented his findings for peer review to the community of physicians and scientists. The results of his experiments demonstrating the effectiveness of systematic cowpox inoculation in the prevention of smallpox were published in 1798 in his work entitled *Variolae Vaccinae* (Plotkin and Orenstein, 2004).

Louis Pasteur built on the work of Jenner to lay the foundation of modern vaccinology through his research on attenuation and the application of this principle to the development of vaccines against rabies, cholera, and anthrax. His work was also seminal because he demonstrated that standardized, reproducible vaccines could be produced in large quantities (Plotkin and Orenstein, 2004). The Institut Pasteur, established in 1887 to further Pasteur's work on rabies vaccine and the study of infectious disease, remains a leader in the fields of microbiology, immunology, and molecular biology (Institut Pasteur, 2006). Its academic research model and its collaboration with industry and the public sector to move discoveries from the bench to the bedside are enduring legacies that reflect the complex fabric of interconnected partnerships that is the modern vaccine enterprise.

This early history of immunization was also characterized by another important and enduring factor – public controversy. Two noteworthy early immunization advocates were, in England, Lady Montagu, and in the United States, Thomas Jefferson. Lady Mary Wortley Montagu, the wife of the British Ambassador to Constantinople, is credited with introducing variolation to England upon her return in 1721. She was an enthusiastic and persuasive advocate of the practice as a result of her own experience of smallpox, which had left her scarred (Glynn and Glynn, 2004; Plotkin and Orenstein, 2004). Her efforts resulted in variolation of the British army, which likely played a role in their defeat of the Americans at the battle of Quebec – an outcome that assured that Canada remained within the British Empire (Glynn and Glynn, 2004). Thomas Jefferson, a renaissance man whose passions extended to science and the prevention of disease, collaborated with Benjamin Waterhouse to research and promote smallpox vaccination. Waterhouse, a professor at Harvard Medical School who had trained in Europe, is credited with bringing smallpox vaccination to the United States. When unsuccessful in convincing President John Adams to support his smallpox immunization efforts, Waterhouse turned to Vice-President Jefferson for assistance. Jefferson, understanding the importance and the immense potential of smallpox vaccination, not only supported Waterhouse's efforts; he also conducted surveillance and immunizations himself, introducing smallpox vaccination to Virginia, Washington DC, and Philadelphia. He continued to promote immunization as President, and summarized the importance of these efforts in an 1806 letter to Edward Jenner in which he wrote: "{f}uture nations will know by history only that the loathsome smallpox has existed and by you has been extirpated" (Leavell, 1977).

## Vaccination and public health programs

In the late 1800s in the United States, efforts to combat infectious diseases such as cholera, malaria, diphtheria, smallpox, and tuberculosis were occurring in the midst of scientific advances that created new tools to combat these infectious diseases; this led to the establishment of more comprehensive public health programs and the development of municipal departments of health. The New York City (NYC) Department of Health, established in 1866 amidst fears of a new cholera epidemic in New York, was a leader in this arena. It opened a bacteriological laboratory in 1892 that became the first municipal laboratory in the world to provide for routine diagnosis of disease. This application of the new knowledge emerging in Europe at the turn of the nineteenth century regarding infectious diseases and their causative agents, as well as the emerging science of prevention, proved vital to the NYC Health Department's early successes in stemming the tide of epidemics and reducing deaths from common deadly infectious agents. A cornerstone of these efforts was combating diphtheria through the widespread distribution of diphtheria antitoxin (produced through the inoculation of horses in its bacteriology laboratory) (Junod, 2002); which, pivotally, included providing it free to the poor. The latter marked an important milestone in the history of vaccines and the development of public health disease prevention programs. As a result of the Department's widespread distribution of diphtheria antitoxin, the numbers of deaths due to diphtheria in New York City were cut in half by 1900. During this time period, the NYC Health Department went on to establish other key public health programs that led to further success. These included developing mandatory case-reporting requirements, conducting public education campaigns to enlist the public's help in reducing the spread of infectious diseases, creating a system of inspections of water, food and milk, and establishing a citywide baby-care program aimed at reducing the high child mortality rate (Jones, 2005).

During the same period, vaccine research and development was expanding worldwide, as evidenced by the sometimes competing work of Pasteur in France; Daniel Elmer Salmon and Theobald Smith on heat-killed vaccines for the US Department Agriculture; Henri Toussaint on anthrax; Richard Pfeiffer and Wilhelm Kolle in Germany and Almroth Wright in England on typhoid vaccines; Alexandre Yersin, Albert Calmette and Amédée Borelle on a plague vaccine for animals; and Waldemar Haffkine in India on a vaccine against human plague (Plotkin and Orenstein, 2004). Along with the immunobiologic discoveries that led to vaccine development, these research efforts also led to the development of epidemiologic methods to evaluate the safety and efficacy of new vaccines. Haffkine is credited with attempting to conduct the first controlled field trials with his cholera vaccine trials in the 1890s in India. However, the birth of the modern rigorous, controlled, large-scale field trial is considered to be the testing of the multivalent pneumococcal vaccine among 7730 army recruits during World War II. Ten years later, the field trial of the Salk polio vaccine involved the vaccination of more than 650,000 US children (Smith, 1990; Levine, 2004; Plotkin and Orenstein, 2004). The development of regulation to improve the safety, quality, and standardization of vaccines accompanied the advances in vaccine development at the turn of the twentieth century. The Biologics Control Act was passed by the US Congress in 1902 in response to the deaths of 13 children in St Louis and 9 children in New Jersey who received diphtheria immunizations contaminated with tetanus toxin (Junod, 2002; Milstien, 2004). The Act charged the Laboratory of Hygiene of the Marine Health Service (renamed the Hygienic Laboratory of the Public Health and Marine Hospital Service, and subsequently renamed the National Institute of Health in 1930) with regulating biologics. This work led to the development of vaccine standards and licensing requirements, including requiring that vaccines must be shown to be pure, safe, and efficacious; these provisions were formalized as part of the Public Health Service Act of 1944. Similar regulations were developed by the Food and Drug Administration (FDA) for drugs, which were codified in the Federal Food, Drug and Cosmetic Act of 1938. In 1941, Good Manufacturing Practice (GMP) was formally defined to ensure the safety and purity of medications during the manufacture process (Milstien, 2004). Following the Cutter Incident in 1955, in which incompletely inactivated polio vaccine caused polio disease and deaths, stricter standards for vaccine testing and manufacture were put into place and the Centers for Disease Control (CDC) developed a system for monitoring adverse events (Plotkin and Orenstein, 2004; Offit, 2005a). In 1972, the parallel regulation of biologics and drugs was brought together so that the provisions of the Food and Drug Control Act applied to vaccines as well. Since that time, the FDA has been responsible for the oversight and licensing of biological products including vaccines (Milstien, 2004).

## Scientific advances, vaccine research, and modern vaccine development

Prior to 1900, five human vaccines – against smallpox, rabies, typhoid, cholera, and plague – had been developed. Seven more were developed in the early twentieth century as a result of important discoveries about toxoid-producing bacteria that led to the development of "toxoid" vaccines. Further work in attenuation led to the creation of the Bacillus Calmette-Guerin (BCG) strain used in the tuberculosis vaccine, and E.W. Goodpasture's introduction of a technique for growing viruses in fertile hens' eggs made possible the development of influenza and yellow fever vaccines and then, by extension of the technique following the discovery of rickettsiae, of a vaccine against typhus (Plotkin and Orenstein, 2004).

After World War II, more than two dozen new vaccines were developed. Referred to as the "golden age" of vaccine development, the second half of the twentieth century was a period of great achievement in the field of immunization. Key to this success was the introduction of cell-culture techniques that made it possible to grow and propagate human viruses in a laboratory with relative ease, for which John Enders, Thomas Weller, and Frederick Robbins received the Nobel prize in 1954 (Smith, 1990; Plotkin and Orenstein, 2004). Other important advances included the development of vaccines to bacterial proteins and polysaccharides, the creation of conjugate vaccines, and, more recently, advances in genomics that have led to the development of recombinant vaccines (Wilson and Marcuse, 2001; Plotkin and Orenstein, 2004).

This age was golden in other ways as well. It was also a period of unparalleled national leadership. Franklin D. Roosevelt (FDR) was himself the victim of what would become, thanks in large part to his efforts, a vaccine-preventable disease. Having contracted paralytic polio in 1921, FDR became a key advocate and supporter of research to help victims of polio and to find a vaccine to prevent polio from claiming more lives. Seeing the continued devastation of polio epidemics, the then President Roosevelt established the National Foundation for Infantile Paralysis and enlisted the nation's help in the campaign against polio. FDR asked people to send dimes to the White House, and the effort became the "March of Dimes." Funds raised by the Foundation were used to support research to develop a polio vaccine. These efforts paid off. Jonas Salk, whose research was supported by the Foundation, successfully developed the first polio vaccine and demonstrated its efficacy in national field trials in 1954. Eight years later, the oral polio vaccine developed by Albert Sabin, also with support from the Foundation, was licensed. In 1979 the Foundation changed its name to the March of Dimes, as it is known today (March of Dimes, 2006).

Industry leadership also played an important role in basic and applied research and product development. A key innovator was Maurice Hilleman, a researcher and leader at Merck for more than 27 years, who led the development of numerous vaccines, including rubella, varicella, hepatitis A, hepatitis B, and combination vaccines for measles, mumps, and rubella (Merck, 2005).

## Twentieth-century multinational public health initiatives

The development of effective vaccines led to interest in worldwide disease control and the possibility that some diseases might be eliminated altogether. The first such effort was the smallpox eradication campaign, which began in 1966 as a joint proposal of the United States and the Soviet Union with the approval of the World Health Assembly (Barquet and Domingo, 1997). Fourteen years later, this effort came to fruition when the 33rd World Health Assembly declared the world free of smallpox in 1980, confirming Jefferson's prediction that Jenner's legacy would be the eradication of smallpox. This success set a precedent and inspired the search for other disease candidates with which similar success could be achieved (Foege, 1998; Henderson, 1999). Among these were poliomyelitis and measles.

In 1988, the World Health Assembly resolved to eradicate polio as an extension of the Expanded Programme on Immunization (EPI). Since then, the number of countries with endemic polio has declined from 125 to 6, with worldwide cases now less than 1000 (CDC, 2004a). Efforts are ongoing to maintain this success and reach the goal of eradication by sustaining polio immunization of young children, identifying new cases of polio through Acute Flaccid Paralysis surveillance, and focusing campaigns in areas with endemic disease with the goal of eliminating reservoirs of polio (Global Polio Eradication Initiative, 2006).

In contrast with eradication, which involves removing the disease altogether, elimination involves the containment of the disease to levels that no longer pose a public health threat. The goal to eliminate measles from the Western Hemisphere by the year 2000 was set in 1994. This was achieved in 2002 through efforts led by the Pan American Health Organization (PAHO), which included a major focus on increasing vaccine coverage in high-risk areas, conducting nationwide supplemental immunization activities to maintain high rates of measles vaccination in the population, and surveillance activities to identify, investigate, and respond to new measles cases in a timely manner (CDC, 2004b).

## Expanding vaccine coverage: equity and the promise of prevention

### Reaching populations at risk

#### *Developed nations*

The greatest benefits of vaccine development and widespread immunization have been realized in developed nations. In the 200 years since Edward Jenner, vaccines have been developed against more than 25 infectious agents (CDC, 1999b). National, regional, and local efforts to promote immunization have led to high rates of vaccine coverage, marked reductions in mortality and morbidity from vaccine-preventable disease, and expansion in the number of recommended vaccines (CDC, 1999b, 2006; Plotkin and Orenstein, 2004; see also Figures 10.1, 10.2).

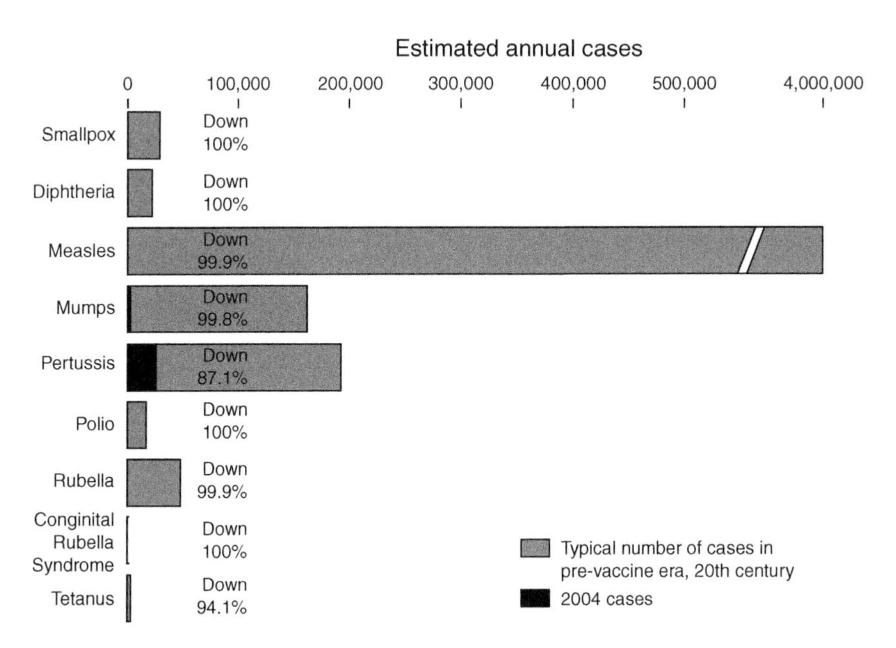


**Figure 10.1** Comparison of twentieth-century estimated annual and 2004 reported cases of vaccine-preventable diseases (pre-1990 vaccines). Source: US Centers for Disease Control.

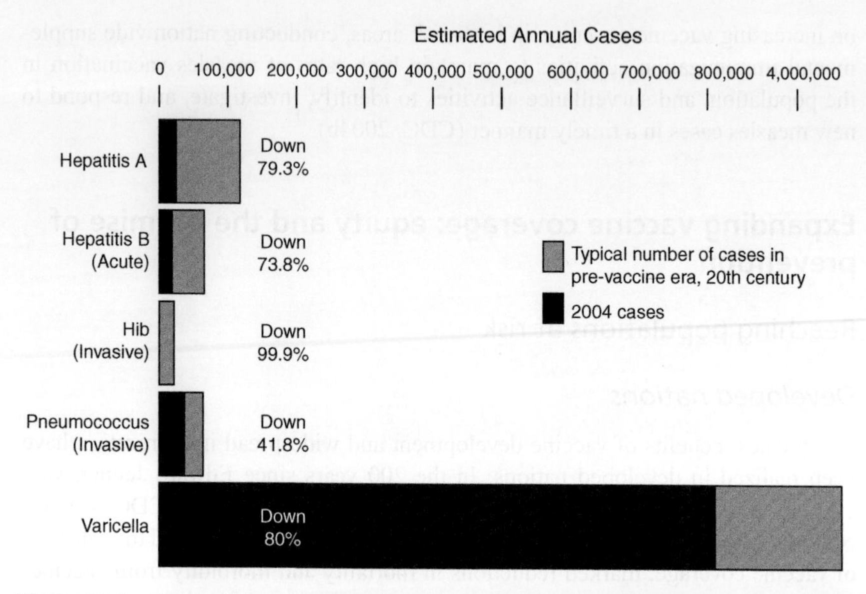


**Figure 10.2** Comparison of pre-vaccine era estimated annual cases and 2004 estimated cases of vaccine-preventable diseases (post-1990 vaccines). Source: US Centers for Disease Control.

The 2007 US Recommended Childhood and Adolescent Immunization Schedule includes vaccines against 15 infectious agents (Figure 10.3).

Findings from the 2003 US National Immunization Survey demonstrated high rates of vaccine coverage, with immunization rates of 80–90 percent or higher for children aged 19–35 months for most of these vaccines (Orenstein *et al.*, 2005; Figures 10.4, 10.5).

Vital to this success has been increasing access to vaccines and immunization services through the Vaccines for Children Program, which provides vaccines to children covered by Medicaid and to insured children. Ongoing surveillance of vaccine-preventable diseases, the safety and effectiveness of vaccines, and their use in the community is essential to sustain this success.

### *Developing nations*

In the developing world, large-scale efforts to prevent disease through widespread vaccination and immunization began in 1974, when success in smallpox eradication led the World Health Organization (WHO) to establish the Expanded Programme on Immunization (EPI) (Levine, 2004; Plotkin and Orenstein, 2004). By 1970 vaccines had been developed against numerous infectious agents, including diphtheria, pertussis, tuberculosis, tetanus, polio, measles, mumps, and

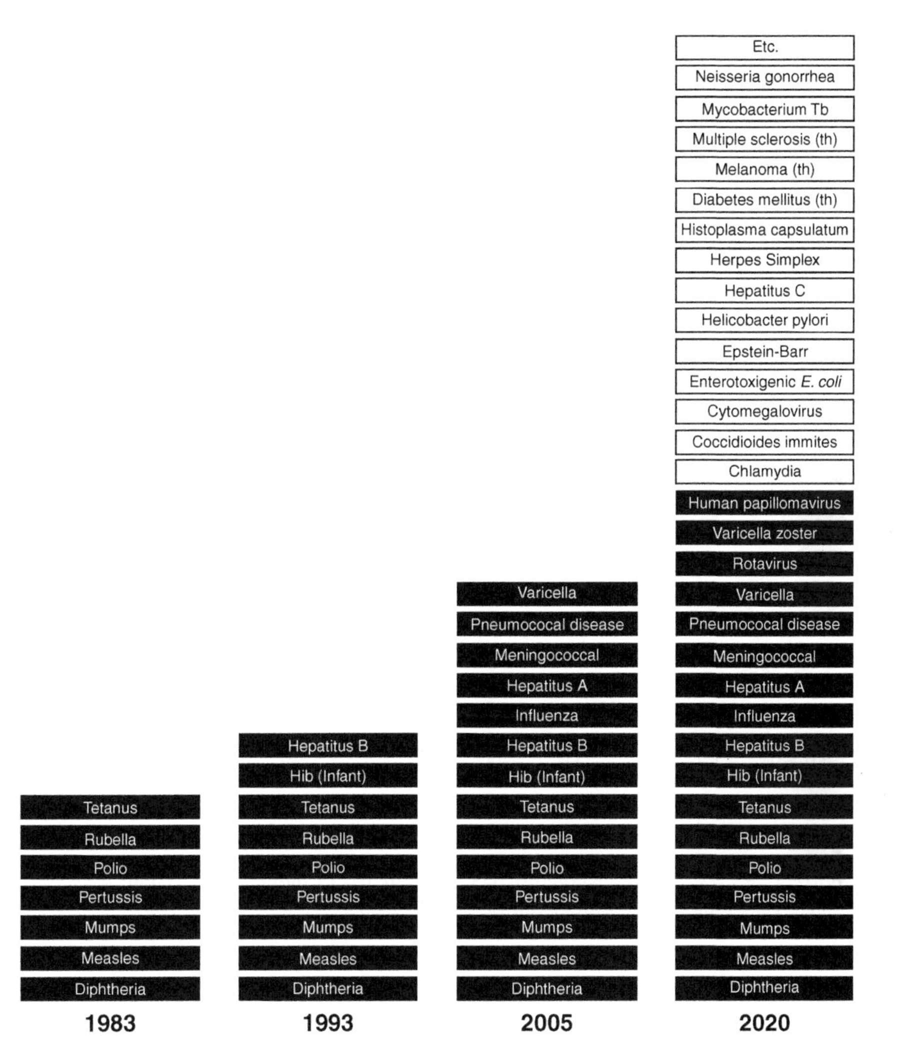


**Figure 10.3** Vaccine-preventable diseases – yesterday, today and tomorrow. Source: US Centers for Disease Control.

rubella. Despite this success, only 5 percent of children worldwide had access to these vaccines (see Figure 10.6).

The goal of EPI at its inception – so named because it sought to “expand” both the scope and reach of immunization services worldwide – was to make six key vaccines available to all children worldwide by 1990 (Keja *et al.*, 1988; Plotkin and Orenstein, 2004). Through the promotion of routine immunization services that led to the development of national immunization programs in most countries,

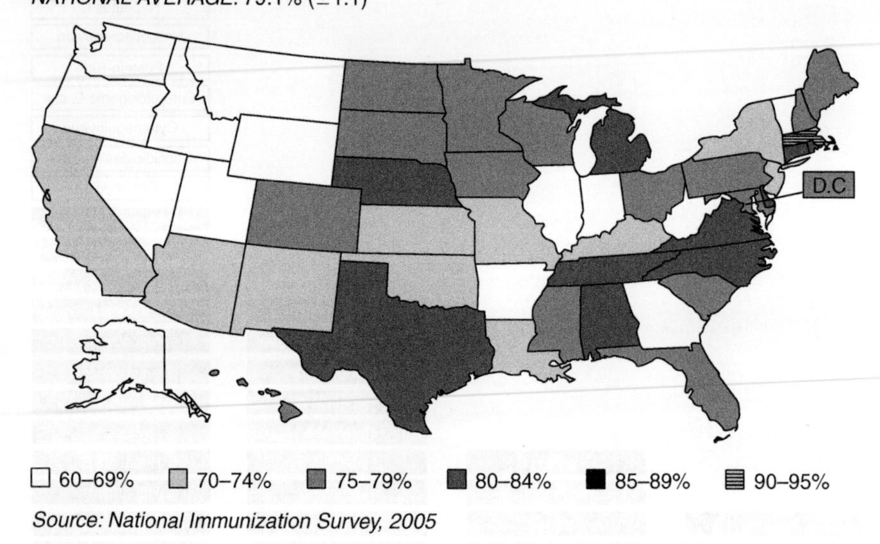


**Figure 10.4** Estimated coverage of US children 19–35 months of age with 4:3:1:3:3. Source: US Centers for Disease Control.

EPI achieved remarkable success in the 20 years following its inception. By 1994, the number of children worldwide receiving immunization services had risen to 80 percent (Levine and Levine, 1997; Bland and Clements, 1998). Success, however, was not uniform, and the momentum to maintain and expand these efforts was not consistent through the 1990s. Achievement of regional polio eradication and measles elimination during this period, though, led to renewed interest in and support for immunization programs. As a result, the Global Alliance for Vaccines and Immunization (GAVI) was launched in 2000 (Plotkin and Orenstein, 2004). GAVI is a collaboration of governments in both developed and developing countries; international organizations including the WHO, UNICEF, and the World Bank; the Bill and Melinda Gates Foundation; other non-governmental organizations; vaccine manufacturers in both developed and developing countries; and public health and research institutions. The primary goal of GAVI is to improve

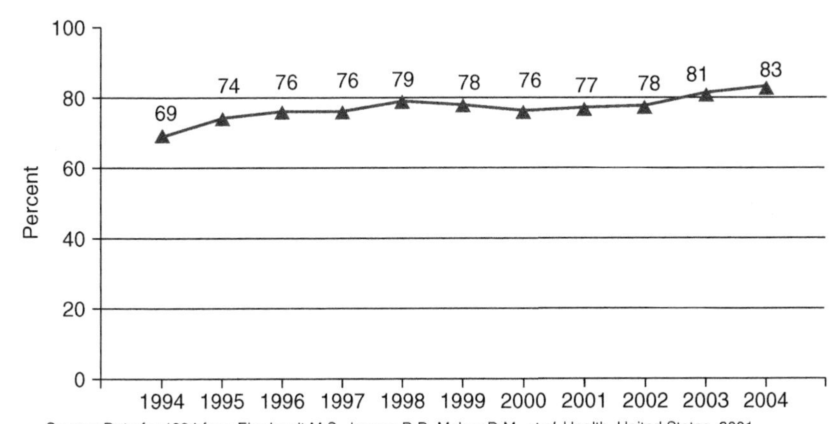
100
80
60
40
20
0
Percent
69
74
76
76
79
78
76
77
78
81
83
1994 1995 1996 1997 1998 1999 2000 2001 2002 2003 2004
Source: Data for 1994 from Eberhardt M.S., Ingram D.D. Makuc D.M. et al. Health, United States, 2001, with Urban and Rural Healthbook Hyattsville, Maryland National Center for Health Statisitic 2001: Table 73. Data for 1995-2001 from National Center for Health Statistics. (2003). Health United States, 2003 with Chartbook on trends in the Health of American National Center for Health Statisitcs Table 71 Data for 2002 from National International Program (2003). Immunization Coverage in the US. Results from National Immunization Survey Centers for Disease Control and prevention. Data for 2003 National Immunization Program (2004). Immunization Coverage in the US results from National Immunization Survey.
Centers for Disease Control and Prevention. Available online at http://www.edu.gowhipcoverage default htm.
Data for 2004 Centers for Disease control and Prevention, National Immunization Program (2005).
NIS data, tables, Jan-Dec 04. Available online at http://www.edu.gowhipcoveragedefault htm.

**Figure 10.5** Percentage of children aged 19–35 months receiving the combined series of vaccination (4:3:1:3), 1994–2004. Source: Child Trends Data Bank (2006), US Centers for Disease Control.

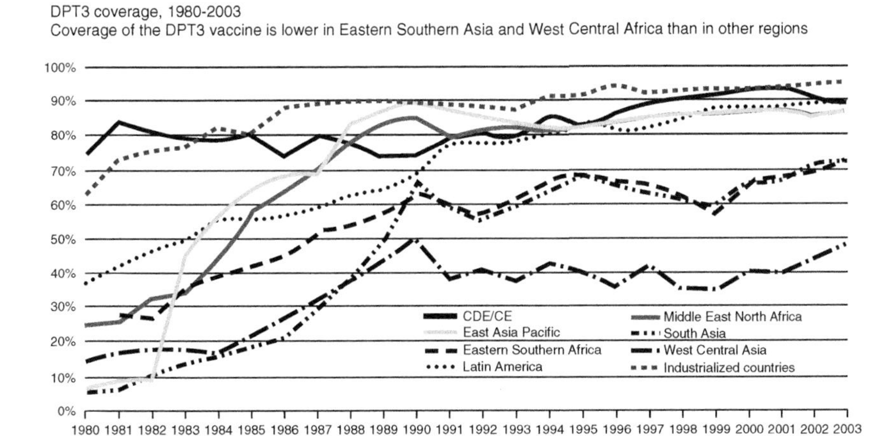
DPT3 coverage, 1980-2003
Coverage of the DPT3 vaccine is lower in Eastern Southern Asia and West Central Africa than in other regions
100%
90%
80%
70%
60%
50%
40%
30%
20%
10%
0%
CDE/CE
East Asia Pacific
Eastern Southern Africa
Latin America
Middle East North Africa
South Asia
West Central Asia
Industrialized countries
1980 1981 1982 1983 1984 1985 1986 1987 1988 1989 1990 1991 1992 1993 1994 1995 1996 1997 1998 1999 2000 2001 2002 2003

**Figure 10.6** Worldwide DTP3 coverage 1980–2003. Source: UNICEF (2005).

access to immunization for children in low-income countries, and to reduce the time it takes from the development and distribution of new vaccines in industrialized countries to their introduction and widespread use in developing nations. In the five years since GAVI's formation it has raised almost US$3.3 billion, and it estimates that its support has prevented more than 1.7 million premature deaths (GAVI Fact Sheet, 2006). As of 2003, all WHO regions realized gains in DPT3 coverage (percent of children receiving three doses of the combined diphtheria, pertussis, tetanus vaccine), with seven out of eight regions achieving more than 70 percent coverage (UNICEF, 2005).

## Going global: politics and economics of vaccine development and distribution

### Vaccine development at the turn of the twenty-first century

Just as the end of the nineteenth century marked the start of a new era of prevention through the invention and development of vaccines, so too did the end of the twentieth century. Advances and new discoveries in recombinant and conjugate technologies have made possible the development of vaccines against diseases such as hepatitis B, and *Haemophilus influenzae* type b, and the creation of additional vaccines against pathogens such as pneumococcus and meningococcus which could provide protection in infancy and perhaps durable, potentially lifelong, immunity (Plotkin and Orenstein, 2004).

Two second-generation rotavirus vaccines are the product of unprecedented multinational field trials that involved well over 100,000 children. Such large-scale trials were required to have the statistical power to detect an association with intussusception and thereby demonstrate the safety of these vaccines.

The development of vaccines which protect against infection with Human Papilloma Virus (HPV) types 16 and 18 can prevent cervical cancers associated with chronic infection by these serotypes. These vaccines have the potential to reduce the most common cause of cancer mortality among women worldwide. However, to realize the enormous promise of these new vaccines, two formidable challenges must be overcome: adolescent girls must be immunized prior to the onset of sexual activity, and this vaccine must be made widely available throughout the world.

These advances were the product of more complex and expensive vaccine development, which raised new challenges in the financing, distribution, and availability of vaccines; these will be discussed in more detail below. With more than 400 new vaccine projects currently underway around the world, the twenty-first century also looks to be a time of great achievement (Levine, 2004).

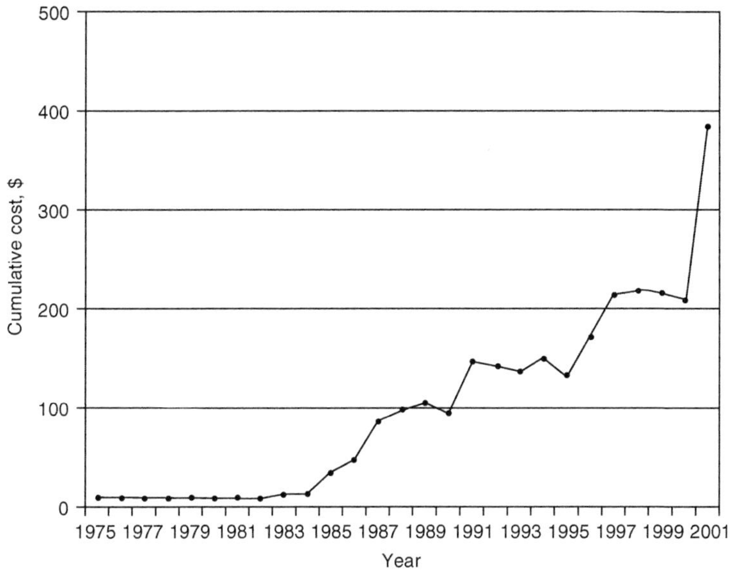


**Figure 10.7** Cumulative cost of recommended vaccine series per child at public-sector prices, 1975–2001. Source: Davis *et al.* (2002).

The making of vaccines has historically been a unique public–private partnership involving an interdependent collaboration of government, industry, and academia in vaccine research, design, and development (NVAC, 1997). Since the latter part of the twentieth century, the costs of vaccine development and manufacture have risen sharply. The use of more complex and expensive technologies to develop vaccines, the increase in regulatory and licensing requirements, and the proliferation of legal issues relevant to vaccine manufacture, distribution, and use have all contributed to this rise in cost (see Figure 10.7).

These costs, the relatively smaller return on investment realized by vaccines versus other pharmaceuticals, and the misalignment of need/disease burden versus potential lucrative markets, has resulted in a diminution in the number of vaccine manufacturers in the developed world, and a more tenuous vaccine supply.

## Financing the vaccine enterprise

Funding for vaccine research and development comes primarily from taxes/government sources, profits from sales of products, and venture capital. Compared with the worldwide pharmaceutical market, the vaccine market is very small and much less profitable. Revenues from worldwide vaccine sales in 2000 were approximately US$6 billion; this combined revenue was less than that from a single pharmaceutical blockbuster such as Lipitor (a cholesterol-lowering agent) (Offitt, 2005b). Furthermore, vaccines are often more difficult and more time-consuming to produce and have a much more limited potential for consumption, since an individual will be immunized with a specific vaccine only a handful of times at the most (Offitt, 2005b).

In addition, the dynamics of supply and demand are highly regulated and constrained in the vaccine market. In the United States, more than 50 percent of childhood immunizations purchased are bought by a single purchaser – the US Federal Government – principally through the Vaccines for Children Program of the Centers for Disease Control and Prevention (CDC). In Latin America the Pan American Health Organization (PAHO) manages vaccine purchases, and in the remainder of the developing world the United Nations Children's Fund (UNICEF) is the main purchaser of vaccines, accounting for 40 percent of the global volume of basic pediatric vaccines (Danzon *et al.*, 2005). Supply of vaccines is also highly constrained. In the US, there are only four major companies that manufacture vaccines (NVAC, 1997; Levine, 2004). As a result, most vaccines licensed in the US are made by only one or two manufacturers, leaving the vaccine supply at constant risk for disruption and shortages (Danzon and Pereira, 2005). In the developing world, the vaccine market has seen a divergence from the developed world in the vaccines produced, and a proliferation of local vaccine manufacturers to supply these vaccines; most of the vaccines now supplied to UNICEF come from developing country manufacturers, not the multinational corporations (Plotkin and Orenstein, 2004).

Also, the sectors of the vaccine market that are most profitable often differ from those where there is the greatest opportunity to reduce morbidity and mortality. Whereas the trends in developed nations are towards developing vaccines for the more profitable adolescent and senior adult markets, the greatest infectious disease burdens remain among children in the developing world (Levine, 2004). As a result, efforts are underway to develop innovative, sustainable financing mechanisms to enable all children worldwide not only to benefit from existing vaccines, but also to have access to new vaccines as they are developed, and to put in place incentives so that the vaccines needed to fight diseases predominantly affecting children in low-income countries will continue to be developed. The Vaccine Fund, established under the auspices of GAVI, was created to help support routine immunizations, the introduction of new vaccines, and safe delivery mechanisms for vaccines in the poorest countries worldwide (Levine, 2004; Batson, 2005). With the goal of ensuring the sustainability of these efforts, a requirement of this funding is the creation of a five-year national immunization plan that includes identifying long-term financing mechanisms beyond the period of the funding (Plotkin and Orenstein, 2004).

Among the solutions currently proposed to encourage innovation in immunization programs and facilitate the introduction of new vaccines to the developing world are the creation of advance-purchase agreements for low-income countries. These contracts would be made between sponsoring donors and vaccine manufacturers to create a guaranteed market for a new vaccine, thus establishing a "pull" mechanism to provide incentive to industry to develop needed immunizations by demonstrating demand in advance of vaccine development (Berndt and Hurvitz, 2005; Lieu *et al.*, 2005). Other solutions include "push" mechanisms, such as direct financing of vaccine research and development by donor agencies or governments, tax credits for research, and harmonization of regulations to create an environment favorable to multinational vaccine development (Lieu *et al.*, 2005).

One example of such a novel approach to vaccine financing is the partnership developing a vaccine against malaria. When a corporate restructure resulted in the manufacturer withdrawing funds for this vaccine, the Gates Foundation stepped in to fund continued research. When early trials yielded promising results, this catalyzed international collaboration to raise capital from multiple governments via the bond market. Finally, an advance purchase contract allowed for a purchase price of $15–25 dollars for the three-dose series for the first 200 million doses, to recoup research and development costs, followed by a price drop to $1.00–1.50 for subsequent doses (*Wall Street Journal*, 2005).

## Political challenges

In this global economy, important barriers to vaccine development and production include complex, sometimes conflicting regulatory environments that vary from one region or country to another, increasing the importance of the establishment and protection of intellectual property rights associated with vaccine discovery and development, and the burden of the legal defense of vaccines (Offit, 2005b). Harmonization of regulations between nations could help to address vaccine shortages by expanding the number of vaccine suppliers. For example, US vaccine shortages might be mitigated by facilitating US licensing of vaccines that are licensed and in widespread use in other countries. By standardizing vaccine production and licensing requirements, harmonization could also reduce costs and barriers associated with the introduction of new vaccines. The FDA committee established in response to the 2002 Institute of Medicine report on vaccine financing reported in the last year that progress has been made toward streamlining manufacturing requirements and harmonizing US and European regulations (Coleman *et al.*, 2005).

Intellectual property rights can also impede new vaccine development and production. Whereas patent protection is vital to encourage innovation, intellectual property rights protection can also be a barrier to vaccine development because collaboration and the sharing of knowledge from previous discoveries can lead to new avenues of inquiry. Finding the right balance between intellectual property rights protection and the need to share knowledge in the process of discovery remains a major challenge in the twenty-first century (Levine, 2004; Plotkin, 2005).

## Impact of vaccination on society

### Vaccine success

The last case of smallpox that was reported in the United States was in 1949, wild-type polio virus was eliminated from the Western Hemisphere by 1991, and cases of *Haemophilus influenzae* type b invasive disease have declined dramatically since the vaccine's introduction in the United States in 1987 (CDC, 1999b). Since 1998, measles outbreaks in the United States have all resulted from introduction by susceptible international travelers who entered the US incubating the infection. As noted above, the world was declared free of smallpox in 1980, polio eradication was achieved in the Americas, Western Pacific, and European regions of the WHO by 2002, and measles has recently been eliminated in the Americas (CDC, 1999b; Plotkin and Orenstein, 2004). In the US, the incidence of most diseases for which vaccines are available has been reduced by more than 85 percent (Orenstein *et al.*, 2005), demonstrating that immunization is cost-effective and enormously beneficial to both individual and public health (Lieu *et al.*, 2005; Orenstein *et al.*, 2005). Immunization is widely regarded as a public health triumph, with vaccine coverage often serving as a metric for judging public health programs.

Immunization results in direct cost savings within the health-care sector by reducing both the acute morbidity and disability due to vaccine-preventable disease, such as lameness following paralytic polio, or hearing loss following Haemophilus b meningitis. The benefits of widespread immunization extend beyond the direct effects of reducing effects on all sectors of society (Table 10.1).

Gains in life-years and health status result in healthier, more productive individuals, able to contribute to their nation's economy. Curtailing outbreaks of vaccine-preventable diseases avoids costly losses in both tourism and trade (Ehreth, 2003). Increases in child survival by preventing deaths from vaccine-preventable disease can lead to greater utilization of family planning, and concomitant reductions in family sizes (Shearley, 1999). When a disease such as smallpox can be eradicated the long-term economic benefits are enormous, because costly programs can be eliminated and scarce public health resources redirected.

### New threats to the public health

Unfortunately, the gains achieved by immunization programs may be offset by losses from infectious diseases that are not vaccine-preventable and non-infectious

**Table 10.1 Direct and indirect savings from vaccination**

| | Comparative savings | Direct or indirect savings (US$) |
|---|---|---|
| *Disease* | | |
| Smallpox[a,b] | NA | 300 million in direct costs per year |
| Polio[b,c] | NA | 13.6 billion in total savings worldwide by 2040<br>700 million in US between 1991 and 2000[d] |
| Measles[e] | One case of measles is 23 times the cost of vaccinating one child against measles[b] | 10 per disability-adjusted life-year (DALY) |
| Cholera | NA | 770 million lost in seafood export Peru, 1991 |
| Malaria | NA | 100 billion GDP lost annually in sub-Saharan Africa because of malaria[f] |
| MMR[g] | For every US$1 spent on MMR vaccine, more than US$21 is saved in direct medical care costs | 100 million in direct medical costs from 1989–91 measles outbreak |
| DTaP[g] | For every US$1 spent on DTaP vaccine, more than US$24 is saved | 23.6 billion in direct and indirect costs without DTP vaccines |
| Hib[g] | For every US$1 spent on Hib vaccine, more than US$2 is saved | 5 billion in direct costs and 12 billion in indirect costs incurred in US[h] |
| *Other public health problems* | | |
| Plague | NA | 1.7 billion lost tourist income and trade |
| AIDS | NA | 14 billion annual treatment cost in the US |
| Drug resistance | NA | 4 billion annual treatment cost in the US |

NA, not available; MMR, measles-mumps-rubella; DTaP, diphtheria-tetanus-acellular pertussis; Hib, *H. influenzae* type b.

[a]Based on eradication of smallpox in 1977

[b]CDC, Immunization Services Division, Health Services Research and Evaluation Branch, 1999

[c]Based on eradication of polio by 2005

[d]http:Clinton1.nara.gov/White_House/EOP/OSTP/CISET/html/iintro.html

[e]Canadian Institute of Health Information, http://www.cihi.com/Programme%20Information/Crosscutting%20Programmes/imm96p.pdf

[f]http://www.who.int/inf-pr-2000/en/pr2000-28.html

[g]Basic principles of immunization cited in *Why is Immunization Important Today? Module 1: Basic Principles of Immunization*; available at http://healthsoftonline.com/tip/matern111.htm (accessed 23 June 2001)

[h]http://www.ostp.gov/CISET/html/iintro.html table1

[i]WHO (1999). *WHO Infectious Diseases Report: Removing Obstacles to Healthy Development*; available at http://www.who.int.infectious-disease-report/pages/graph24.html

Source: Ehreth (2003).

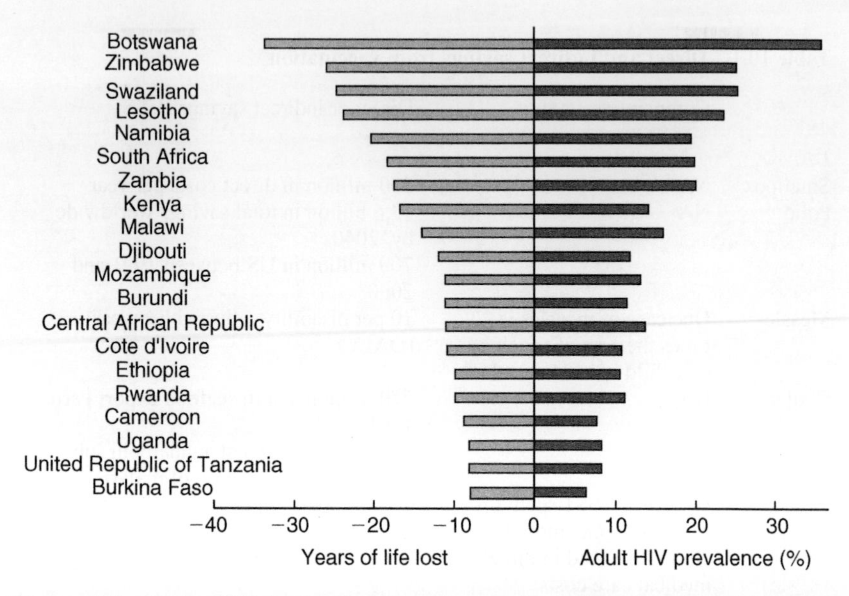


**Figure 10.8** Years of life lost to AIDS in 20 countries with highest adult HIV prevalence, 2000–2005. Source: World Health Organization (2005b).

diseases, or by increased costs associated with chronic illness care encountered when life expectancy is increased. How gains in one area can be overshadowed by profound losses in another is illustrated by the tragic effects of the HIV epidemic on life expectancy in sub Saharan Africa (Figure 10.8).

## Vaccine safety, risks, and the roots of mistrust

Despite their obvious success, vaccines have never been universally embraced as a public good. From the earliest days of vaccination to the present time, there have been concerns about an individual's right to refuse to be immunized and about vaccine safety, and some pivotal events have led to public outcry and contributed to a weakening of public confidence in immunization (Figure 10.9).

To lessen the credibility of those who opposed mandatory immunization because of concerns about the safety of smallpox vaccine, political cartoons from Jenner's day attempted to ridicule their fears by portraying early vaccinees sprouting the heads of cows (Wilson and Marcuse, 2001).

Pasteur's first inoculations of humans with rabies vaccine were met with alarm and dismay. Skeptical about the concept of attenuation, many questioned intentionally giving a potentially deadly agent to a healthy person, and considered the vaccine-associated deaths to be murder (Plotkin and Orenstein, 2004). Both real and perceived concerns about vaccine safety remain important issues today, eroding trust in public health and scientific entities promoting the use of vaccines.

**Figure 10.9** The cow pock or the wonderful effect of the new inoculation (1802). Source: *Nature Reviews*, a United Nations AIDS Wall Chart.

The long road to the development of vaccines has not been an unbroken series of successes; unanticipated severe adverse events and terrible accidents have occurred as well. Among these calamities were the Mulkowal disaster in 1902 in India, in which 19 people died of tetanus contracted from contaminated whole-cell plague vaccine; the Lubeck disaster in Germany, in which 72 infants died after receiving BCG vaccine inadvertently infected with a virulent strain of *Mycobacterium tuberculosis*; and the Cutter Incident in 1955, in which two lots of the Salk vaccine given to 120,000 children were discovered to be incompletely inactivated due to a manufacturing error after 60 of the children and 89 of their family members contracted polio (Levine, 2004). As a result of these and other vaccine tragedies, the need for greater quality control, testing, and surveillance was recognized, leading to the development of regulations to increase the safety of vaccines (Junod, 2002; Plotkin and Orenstein, 2004).

Vaccines are held to a higher safety standard than other medical interventions because, unlike most interventions that are aimed at the sick and the treatment of diseases, immunizations are given to the well to prevent disease. Therefore, the tolerance of side effects that sicken or disable a healthy person is understandably very low. Today, widespread immunization has reduced vaccine-preventable disease rates to such low levels that concern about the perceived risks of vaccination can weigh more heavily in individuals' vaccination decisions than concern for risks associated with acquiring the disease (Wilson and Marcuse, 2001). Therefore, a 1 in 10,000, or 1 in 100,000 chance of a serious side effect may result in a vaccine being considered unacceptable for a healthy child. The rhesus-human reassortant rotavirus vaccine was pulled from the US market by the manufacturer after the initial analysis of post-licensing surveillance reported that intussusception occurred in approximately 1 in 10,000 children receiving the vaccine. Because the vaccine was unacceptable in the US, it was deemed infeasible to attempt to use it in developing countries – this, despite the fact that more than 600,000 to 800,000 children die annually in low-income countries as a result of diarrheal disease caused by rotavirus (McPhillips and Marcuse, 2001).

While concern for documented vaccine-associated adverse events may contribute to vaccine hesitancy, concern for well-publicized but scientifically unsubstantiated claims of harms associated with vaccines has played a much greater role in modern vaccine hesitancy. Prominent among the current vaccine controversies today are concern about putative links between the development of autism and MMR (measles, mumps, rubella) vaccine, or between autism and vaccines containing the long-used ethyl mercury-based preservative, thimerosal. Widespread media coverage of subsequently discredited studies purporting to demonstrate a link between autism and MMR increased vaccine hesitancy, leading to reduced immunization rates in Great Britain, where the studies were conducted and first published, and prompting Congressional hearings in the United States (Wilson and Marcuse, 2001). Rigorous, systematic reviews of the evidence by scientific panels in multiple countries have failed to substantiate these claims. The US Institute of Medicine's Immunization Safety Committee's Review of Vaccines and Autism concluded (IOM, 2004):

> The body of epidemiological evidence favors rejection of a causal relationship between the MMR vaccine and autism. The committee also concludes that the body of epidemiological evidence favors rejection of a causal relationship between thimerosal-containing vaccines and autism. The committee further finds that potential biological mechanisms for vaccine-induced autism that have been generated to date are theoretical only. Given the lack of direct evidence for a biological mechanism and the fact that all well-designed epidemiological studies provide evidence of no association between thimerosal and autism, the committee recommends that cost-benefit assessments regarding the use of thimerosal-containing versus thimerosal-free vaccines and other biological or pharmaceutical products, whether in the United States or other countries, should not include autism as a potential risk.

These conclusions, however, have not served to fully reassure the American public about the safety of thimerosal as a vaccine preservative. In 2005, 22 states considered bills to ban the use of thimerosal-containing vaccines in children and pregnant women, with legislation being enacted in four of those states (Delaware, Illinois, Missouri, and New York) (AAP, 2005).

Reports of the associations between MMR and autism, and thimerosal and autism, were rapidly disseminated worldwide by the media and the Internet. Today the flow of scientific information often circumvents the process of peer review and its attendant scientific scrutiny, leading to widespread public awareness of preliminary findings or even hypotheses before they have been replicated or substantiated. Media attention to case reports of adverse events following immunization contributes to public misunderstanding of the difference between a temporal and a causal association. The standards for proof in the legal system differ from those in science, resulting in findings against vaccines that are not supported by the scientific evidence but nonetheless exert a powerful influence in the court of public opinion and in legislatures. When these perceptions result in a substantial proportion of the community becoming hesitant to follow a recommendation to vaccinate, the risk to the public health may be great as decline in immunization coverage results in vaccine-preventable disease outbreaks and deaths (Levine and Levine, 1997; Kane, 1998). Public health authorities must balance the protection of the public health and the protection of individual liberty, including the right to refuse vaccination (Feudtner and Marcuse, 2001).

## The future of vaccination and immunization

### Public education and effective communication

The paradox of immunization success requires ongoing efforts to educate the public about once-common deadly infectious diseases that few people today have ever seen, the role of immunizations in preventing these diseases, and the need for continued vaccination so that these diseases do not become commonplace once again (Kane, 1998). The challenge for public health authorities and policy-makers today is to communicate information convincingly and accurately, not only about the risks and benefits of vaccines, but also about the risks of the diseases that they prevent. As long as the public hears more about adverse events associated with vaccines and perceived, unfounded fears of harms caused by vaccines than it does about the deadly toll of these infectious diseases and the remarkable achievements in immunization over the last 200 years, declines in vaccine coverage are likely to increase, and the risk for turning back the clock on the gains achieved by widespread immunization in controlling deadly infectious diseases will also increase (Kane, 1998; Wilson and Marcuse, 2001).

### Sustaining success

Ensuring that the benefits of immunizations seen over the last 200 years will extend to and expand in the future requires not only the continued production and effective distribution of existing vaccines and the development of new vaccines, but also sustaining local and global immunization consensus. Understanding the public's knowledge about vaccine-preventable diseases and their perception of vaccines, involving the public in a dialogue about values and priorities in balancing and articulating public health and personal perspectives, and effectively communicating in today's media environment are all essential to sustaining consensus (Levine and Levine, 1997; Wilson and Marcuse, 2001). Future immunization success will also depend on assuring a stable supply of vaccines, continued mechanisms for funding and promoting vaccine development, assuring universal, global access to vaccines, and investing in the infrastructure required to achieve this.

## Conclusion

Immunization has fundamentally altered the global infectious disease ecology. Powerful pathogens, once so prevalent that they toppled empires and laid waste to communities and cultures, are now only a distant memory. Smallpox today only exists in laboratories, and its danger now comes from its perceived threat if employed as a biological weapon. Polio persists in only a few corners of the world just 50 years after development of a vaccine. The indigenous transmission of measles has been drastically reduced throughout the Western Hemisphere. Congenital rubella has been eliminated from the United States and, in much of the world, entire organizations and institutions no longer exist as a result of successful disease elimination through immunization. In developed nations, orthopedic and rehabilitation hospitals have evolved into acute care children's hospitals or have gone out of business altogether, and charitable institutions (such as the March of Dimes) that were once solely dedicated to fighting infectious diseases like polio have now branched out into other efforts, such as the prevention of birth defects. For many today, deadly infectious diseases are a historic relic without a personal memory or current face. However, the emergence and spread of HIV and its resulting devastation in the latter part of the last century, and the recent emergence of a virulent new strain of avian influenza, coupled with the recognition that the deadly 1918 influenza pandemic was due to an avian influenza strain that adapted to human-to-human transmission, has made clear mankind's continued vulnerability to epidemic infectious disease.

Vaccines have contributed to our modern prosperity and have changed our perceptions of the world and ourselves. The success of immunization undergirds our twenty-first century notions, born of affluence, that health is a right, that technology will conquer all, and, mythically, that all risk can be eliminated. The tragic irony of this evolution is that these notions now threaten to reverse the gains achieved through widespread immunization. The rise in vaccine hesitancy, the high cost of technology, regulation, and legal defense, and the focus of public discourse on rare or unsubstantiated harms from vaccines without a concomitant understanding of the severity of the infectious diseases prevented by vaccines, all have the potential to contribute to a decline in the development, availability, and use of vaccines, and the potential recrudescence of diseases now controlled.

The challenge for the twenty-first century is to build on past success toward a future in which new vaccines are developed, vaccine safety and efficacy remain paramount, relevant old vaccines that are safe and effective remain in widespread use, and their benefits are accrued by all – and to do so without having to re-learn lessons of the past through the return of old infectious disease foes.

## References


American Academy of Pediatrics (2005). *2005 State Legislation Report*. Available at: http://www.aap.org/advocacy/statelegrpt.pdf search=%22state %20legislation %20thimerosal %20map %22 (accessed 24 August 2006).

Barquet, N. and Domingo, P. (1997). Smallpox: the triumph over the most terrible ministers of death. *Annals of Internal Medicine* **127**(8), 635–642.

Batson, A. (2005). The problems and promise of vaccine markets in developing countries. *Health Affairs* **24**(3), 691–693.

Berndt, E.R. and Hurvitz, J.A. (2005). Vaccine advance-purchase agreements for low-income countries: practical issues. *Health Affairs* **24**(3), 653–665.

Bland, J. and Clements, J. (1998). Protecting the world's children: the story of the WHO's immunization programme. *World Health Forum* **19**, 162–173.

Centers for Disease Control and Prevention (1997). Smallpox surveillance – worldwide. *Morbidity and Mortality Weekly Report* **46**(42), 990–994.

Centers for Disease Control and Prevention. (1999a). Ten great public health achievements – United States, 1900–1999. *Morbidity and Mortality Weekly Report* **48**(12), 241–243.

Centers for Disease Control and Prevention (1999b). Achievements in public health, 1900–1999. Impact of vaccines universally recommended for children – United States, 1990–1998. *Morbidity and Mortality Weekly Report* **48**(12), 243–248.

Centers for Disease Control and Prevention (2004a). Brief report: global polio eradication initiative strategic plan, 2004. *Morbidity and Mortality Weekly Report* **53**(05), 107–108.

Centers for Disease Control and Prevention (2004b). Progress toward measles elimination – region of the Americas, 2002–2003. *Morbidity and Mortality Weekly Report* **53**(14), 304–306.

Centers for Disease Control and Prevention (2006). *A Global Commitment to Lifelong Protection through Immunization: National Immunization Program Annual Report*. Available at: http://www.cdc.gov/nip/webutil/about/annual-rpts/ar2006/00-06ar-entire-book.pdf (accessed 17 August 2006).

Child Trends Data Bank (2006). Available at: http://www.childtrendsdatabank.org/indicators/17Immunization.cfm trends (accessed 24 August 2006).

Coleman, M.S., Sangrujee, N., Zhou, F. and Chu, S. (2005). Factors affecting US manufacturers' decisions to produce vaccines. *Health Affairs* **24**(3), 635–642.

Danzon, P. and Pereira, N.S. (2005). Why sole-supplier vaccine markets may be here to stay. *Health Affairs* **24**(3), 694–700.

Danzon, P.M., Pereira, N.S. and Tejwani, S.S. (2005). Vaccine supply: a cross-national perspective. *Health Affairs* **24**(3), 706–717.

Davis, M.M., Zimmerman, J.L., Wheeler, J.R.C. and Freed, G.L. (2002). Childhood vaccine purchase costs in the public sector: past trends, future expectations. *American Journal of Public Health* **92**(12), 1982–1987.

Ehreth, J. (2003). The value of vaccination: a global perspective. *Vaccine* **21**, 4105–4117.
Fedson, D.S. (2005). Preparing for pandemic vaccination: an international policy agenda for vaccine development. *Journal of Public Health Policy* **26**, 4–29.
Feudtner, C. and Marcuse, E.K. (2001). Ethics and immunization policy: promoting dialogue to sustain consensus. *Pediatrics* **107**(5), 1158–1164.
Foege, W.H. (1998). Confronting emerging infections: lessons from the smallpox eradication campaign. *Emerging Infectious Diseases* **4**(3), 412–413.
Global Alliance for Vaccines and Immunization (2006). *The GAVI Alliance Fact Sheet*. Available at: http://www.vaccinealliance.org/resources/TVF.Insert.Gavi.1.06.pdf (accessed 1 March 2006).
Global Polio Eradication Initiative (2006). *The History*. Available at: http://www.polioeradication.org/history.asp (accessed 7 March 2006).
Glynn, I. and Glynn, J. (2004). *The Life and Death of Smallpox*. New York: Cambridge University Press.
Hampton, T. (2004). Funding advances invigorate TB fight. *Journal of the American Medical Association* **291**(21), 2529–2530.
Henderson, D.A. (1999). Eradication: lessons from the past. *Morbidity and Mortality Weekly Report* **48**(SU01), 16–22.
Institute of Medicine (2004). *Immunization Safety Review: Vaccines and Autism*. New York: IOM.
Institut Pasteur (2006). *The History of the Pasteur Institute*. Available at: http://www.pasteur.fr/pasteur/histoire/histoireUS/Histoire.html (accessed 1 March 2006).
Jones, M.M. (2005). *Protecting Public Health in New York City: 200 Years of Leadership*. New York: Bureau of Communications, Department of Health and Mental Hygiene. Available at: http://www.nyc.gov/html/doh/downloads/pdf/bicentennial/historical-booklet.pdf (accessed 1 March 2006).
Junod, S.W. (2002). *Biologics Centennial: 100 Years of Biologic Regulation. Update*. Rockville: FDA, Food and Drug Law Institute. Available at: http://www.fda.gov/oc/history/makinghistory/100yearsofbiologics.html (accessed 1 March 2006).
Kane, M.A. (1998). Commentary: public perception and the safety of immunization. *Vaccine* **16**, S73–75.
Keja, K., Chanc, C., Hayden, G. and Henderson, R.H. (1988). Expanded programme on immunization. *World Health Stat Q* **41**(2), 59–63.
Leavell, B.S. (1977). Thomas Jefferson and smallpox vaccination. *Transactions of the American Clinical and Climatological Association* **138**, 119–127.
Levine, M.M. (2004). *New Generation Vaccines*, 3rd edn. New York: Marcel Dekker.
Levine, M.M. and Levine, O.S. (1997). Influence of disease burden, public perception, and other factors on new vaccine development, implementation, and continued use. *Lancet* **350**, 1386–1392.
Lieu, T.A., McGuire, T.G. and Hinman, A.R. (2005). Overcoming economic barriers to the optimal use of vaccines. *Health Affairs* **24**(3), 666–679.
March of Dimes (2006). *The March of Dimes Story*. Available at: http://www.marchofdimes.com/aboutus/789_821.asp (accessed 1 March 2006).
McPhillips, H. and Marcuse, E.K. (2001). Vaccine safety. *Current Problems in Pediatrics* **31**, 95–121.
Merck (2005). *Merck Vaccines Play a Critical Role in Eliminating Health Threat from Rubella in the United States*. Corporate News. Available at: http://www.merck.com/newsroom/press_release/corporate/2005_0321.html (accessed 16 January 2006).
Milstien, J.B. (2004). Regulation of vaccines: strengthening the science base. *Journal of Public Health Policy* **25**(2), 173–189.

National Vaccine Advisory Committee (NVAC) (1997). United States vaccine research: a delicate fabric of public and private collaboration. *Pediatrics* **100**, 1015–1020.
Offit, P.A. (2005a). *The Cutter Incident: How America's First Polio Vaccine led to the Growing Vaccine Crisis*. New Haven: Yale University Press.
Offit, P.A. (2005b). Why are pharmaceutical companies gradually abandoning vaccines? *Health Affairs* **24**(3), 622–629.
Orenstein, W.A., Douglas, R.G., Rodewald, L.E. and Hinman, A.R. (2005). Immunizations in the United States: success, structure, and stress. *Health Affairs* **24**(3), 599–610.
Plotkin, S.A. (2005). Why certain vaccines have been delayed or not developed at all. *Health Affairs* **24**(3), 631–634.
Plotkin, S.A. and Orenstein, W.A. (2004). *Vaccines*, (4th edn). Philadelphia: Elsevier, Inc.
Program for Appropriate Technology in Health (PATH) (2005). *New Gates Funding will Enable MVI and GSK Biologicals to Complete Development of World's Most Advanced Malaria Vaccine Candidate*. Available at: http://www.path.org/news/pr-051027-malaria-vaccine-candidate.php (accessed 29 March 2006).
Shearley, A.J. (1999). The societal value of vaccination in developing countries. *Vaccine* **17**(Suppl. 3), S109–112.
Smith, J.S. (1990). *Patenting the Sun: Polio and the Salk Vaccine*. New York: William Morrow and Company, Inc.
Stern, I.S. and Markel, H. (2005). The history of vaccines and immunization: familiar patterns, new challenges. *Health Affairs* **24**(3), 611–621.
United Nations Children's Fund (UNICEF) (2005). *Progress for Children: A Report Card on Immunization, No. 3*. Available at: http://www.unicef.org/progressforchildren/2005n2/ (accessed 17 August 2006).
*Wall Street Journal*, New York, 26 April 2005.A1. *Vaccine "Mosquirix"* (available at: http://www.proquest.umi.com).
Wilson, C.B. and Marcuse, E.K. (2001). Vaccine safety – vaccine benefits: science and the public's perception. *Nature Reviews Immunology* **1**, 160–165.
World Health Organization (2005a). Immunization against disease of public health importance. *WHO Fact Sheet No. 288*. Available at: http://www.who.int/mediacentre/factsheets/fs288/en/index.html (accessed 16 January 2006).
World Health Organization (2005b). Wall Chart: *Years of Life Lost to AIDS, Twenty Countries with the Highest Adult HIV Prevalence, 2000–2005*. Department for Economic and Social Affairs/Population Division. Available at: http://www.un.org/esa/population/publications/aidswallchart/chart2.jpg (accessed 28 August 2006).

# Infectious diseases in the context of war, civil strife and social dislocation

# 11

**Ronald Waldman**

The spring of 1994 was a hellish time in the tiny central African country of Rwanda. For four months, following the death in April in a plane crash of the country's President, civil war reigned. The conflict pitted the majority Hutu ethnic group against its rival, the Tutsi. Although violence had erupted between the two on several occasions in the preceding decades, this time the Hutu intentions were frankly genocidal. In an atmosphere of unrelenting propaganda, emotions were stirred to the point that political leaders, military personnel, police, religious authorities, and Rwandans from all walks of life both precipitated and participated in murderous acts that resulted in the deaths of 800,000 people (Powers, 2002).

Amazingly, in late June the Rwanda Patriotic Front, a Tutsi militia trained in Uganda and supported by the government of that country, having crossed the border to Rwanda, penetrated the countryside and drove to and captured the capital Kigali, overthrowing the Hutu authorities and seizing both political and military power. Hutus, in well-founded fear of reprisals and retribution, took flight, and many were able to cross the border to Tanzania in the east, and to then-Zaire (now the Democratic Republic of Congo) to the west. An estimated 500,000–800,000 crossed the narrow bridge from the Rwandan town of Gisenyi, a picturesque community on the northern shore of Lake Kivu, to the equally beautiful and, at that time, calm city of Goma, a city that in many ways had closer ties to Rwanda than it did to Kinshasa, the distant capital of Zaire.

The scene in Goma that July, however, was far from idyllic. Relief workers arriving there to provide humanitarian assistance to the massive numbers of poorly housed, underfed, confused, and traumatized refugees during the second week of the month might as well have been entering a house of indescribable horrors. Their first sight, along the two-kilometer stretch of road leading from the small, one-runway airport to the center of town where the United Nations relief agencies were in the process of setting up their offices, was one of dead bodies stacked four and five high along both sides of the road. During the next three weeks, up to 45,000 people – close to 10 percent of the refugee population – died, and the carnage did not stop there (Goma Epidemiology Group, 2005).

What caused these deaths? During the preceding months, the months of genocide, the answer was clear: people died because of man's unspeakable cruelty to man. But in the aftermath of the genocide, when the killing by machete, by bullet, by beating, and by fires set to buildings in which large numbers of people had been forcibly assembled, had stopped, what could have been responsible for carnage of the magnitude seen in Goma?

From the biomedical perspective, the answer is also simple. The refugees were the victims of one of the most virulent epidemics of cholera ever recorded. The Central African Region, especially the area around Lake Kivu, had been a frequent location of cholera epidemics. In fact, a delegation from the World Health Organization had visited Rwanda and other countries in the area, just a few months before the genocide took place, holding consultations with national government officials on how to prepare and respond to cholera cases that were expected to occur during the summer, according to their usual seasonality.

The events that led to the situation in Goma that July, though, rapidly surpassed the capacity of local and national officials to cope. And, although literally thousands of humanitarian workers arrived in the days and weeks following the chaotic settlement of the refugees, the epidemic continued unabated.

The source of the outbreak has never been definitively established, but most experts feel that heavy contamination of the waters of Lake Kivu (from which the refugees obtained most of their water supplies), combined with the abominable sanitary conditions of the area – just as graves could not be dug, neither could latrines – and the progressively debilitated state of the population were responsible. The failure of the humanitarian community to respond effectively, providing neither adequate quantities of safe water nor effective treatment, was also an important factor.

Cholera is a disease that can, but need not, be fatal. The World Health Organization has stated that case-fatality rates can be kept to less than 1 percent. Deaths from cholera are due to the dehydration that results from the rapid loss of copious amounts of fluid and electrolytes due to the breakdown of homeostatic mechanisms in cells that have been poisoned by the toxin of the bacteria, *Vibrio cholerae*. The illness, however, is self-limiting in time, and the provision of replacement fluid and electrolytes until the cells of the intestinal lining regenerate is sufficient to carry a patient through it. Antibiotics may help reduce the duration of illness, but they are given primarily as a public health measure – reducing the bacterial load in the bowel of an infected individual renders that person less infectious to others.

Accordingly, the treatment of choice for cholera, as for other forms of acute, watery diarrhea, is oral rehydration salts (a formulation of sodium, potassium, chloride, a base compound, and glucose) dissolved in water in concentrations that take maximum advantage of active transport mechanisms that exist in the cells of the intestinal wall and that remain intact in the presence of most diarrheal diseases, including cholera. Oral administration of this formula is as effective as the administration of intravenous fluids and can be done at the household level in most cases, diminishing the need for clinics and hospitals, which are required only for those patients who present with the most severe cases of dehydration and those who are unable to drink rapidly enough to replace their losses.

A strong case can be made for the real cause of death of those 45,000 people in Goma in July 1994 being not cholera, a potential lethal but treatable infection, but the circumstances in which the epidemic occurred. Massive numbers of frightened refugees, gathered in a relatively small area (most refugees were concentrated in one of four refugee camps along the two principal roads that led from Goma toward the east and toward the north) with inadequate sources of unsafe water, non-existent sanitation, a heavily compromised food supply, and sub-standard shelter, were particularly vulnerable to the high fatality rates that have been documented. In sum, the massive number of excess deaths that occurred in Goma can more honestly be attributed not to cholera, but to war.

In today's world, similar circumstances are found only in what have been termed "complex emergencies." Public health experts have proposed various definitions of a complex emergency, but none is completely satisfactory. Some of these include civil strife or war as a defining characteristic, and add to it population displacement and elevated levels of mortality (Burkholder and Toole, 1995). However, if one considers situations that prompted large-scale humanitarian responses, such as the Balkan conflicts of the 1990s and the Timor Leste crisis that followed the referendum for independence from Indonesia in that tiny new country, it might reasonably be concluded that conflict is a more constant feature than an excess number of deaths – which was not a prominent feature of either of those two situations, despite the destruction of most elements of the health system in both settings, and accompanying reductions in the ability of the population to access health-care providers or to utilize health services of any nature.

Another definition focuses more on the impact of a situation on the potential responders than on the affected population. It defines a complex emergency as "a humanitarian crisis in a country, region or society where there is total or considerable breakdown of authority resulting from internal or external conflict and which requires an international response that goes beyond the mandate or capacity of any single agency and/or the ongoing United Nations country programs" (CDC, 2007). Although no definition is sufficiently precise, few would disagree with the statement that complex emergencies take an enormous toll on the societies they affect, and that they require a response that is all too frequently inadequate.

## Mortality and morbidity patterns in modern conflicts

The nature of conflict has changed considerably during the course of the past hundred years. No longer do young men don uniforms and fight each other far from major population centers. Today's battles do not take their toll only on soldiers. Instead, in our world, 90 percent of the victims of conflict are civilians, whether specifically targeted or innocent bystanders caught up in battles through no intention of their own (see Table 11.1). It is in civilian groups that most excess morbidity and mortality occurs, and childhood (those under five years of age) mortality rates, specifically, are generally found to be two to three times higher than those in the general population. A recent survey in the Democratic Republic of Congo (DRC), including its eastern regions, to which peace has never returned since the days of the Rwandan genocide more than a decade ago, is an excellent demonstration of the consequences of war. In this study, mortality rates are measured in the entire population, and in the under five-year-old population. Deaths are attributed not to the diseases that might be their proximate cause, but rather to the presence or absence of instances of violence in the health zones in which they occurred (Coghlan *et al.*, 2006).

Unfortunately, violence is and always has been part and parcel of human existence. According to one source, in early 2007 there were 41 separate violent conflicts occurring, some of which had begun as long ago as the 1960s (Globalsecurity.org, 2007). Accompanying the violence, the provision of humanitarian assistance to the victims of conflict has an equally long history. The Red Cross movement was founded after Henri Dunant, a Swiss merchant, witnessed the suffering of wounded soldiers at the Battle of Solferino. Recently, one commentator has compared the relationship between humanitarian agencies and war to that of coral divers and water – difficult environmental conditions are the fundamental nature of the business (Slim, 2004).

It is important to point out, however, that few deaths in the Democratic Republic of Congo study cited above were due to trauma resulting from violence. Although the direct consequences of war can be substantial, with many civilians killed – as they were in Rwanda and in the Balkans during the 1990s,

**Table 11.1 Crude mortality and under-five mortality in conflict zones**

| | Crude mortality rate, deaths per 10,000 per day (95% CI) | Under-five mortality rate (95% CI) |
|---|---|---|
| Health zones reporting violence | 3.1 (2.6–3.4) | 6.4 (5.7–7.2) |
| Health zones not reporting violence | 1.7 (1.5–1.9) | 3.1 (2.7–3.5) |

Source: Coghlan *et al.* (2006).

in Mozambique during the 1980s, and in Iraq in the first decade of the twenty-first century – it is more common, as in the case in Democratic Republic of Congo, for the vast majority of deaths to be from other causes. From a biomedical standpoint, most deaths in the DRC study were due to those conditions that are responsible for most deaths in the developing world, including infectious diseases such as pneumonia, diarrhea, malaria, and measles, compounded by malnutrition. In this part of the world, at this time, some may have been due to AIDS. Most remarkable, nevertheless, is the disparity between health zones in which violence was occurring and those in which it was not. Although it has always been intuitively known that a population's health must suffer in times of conflict, this study provides direct, quantitative proof.

Although it is not necessarily a prominent feature of the war in DRC, one of the more constant features of conflict, and one that makes an important contribution to increased morbidity and mortality, is the displacement of the affected civilian population. People, in their attempt to flee from violence and the perceived risk of danger to themselves and their families, abandon their homes, their belongings, and their land to seek refuge in nearby areas. Sometimes, members of the community – adult males, for example – will stay behind in an attempt to protect their economic interests, leading to a demographically skewed distribution of the population.

Forced migration results in two distinct groups of people. Under international law, a "refugee" is defined as someone who, "owing to well-founded fear of being persecuted for reasons of race, religion, nationality, membership of a particular social group, or political opinion, is outside the country of his nationality and is unable, or owing to such fear, is unwilling to avail himself of the protection of that country" (United Nations, 1950). According to the Office of the United Nations High Commissioner for Refugees (UNHCR), the organization explicitly charged with protecting and providing assistance to refugees, there were 8.4 million refugees in the world at the start of 2006 – a decline of 12 percent from the previous year. In fact, during the first half of the first decade of the twenty-first century, the number of refugees has fallen by one-third and has reached its lowest level since 1980 (UNCHR, 2006).

Although this trend may seem at first to be quite encouraging, it should not be taken as such. It is true that many refugees have been able to return to their countries of origin (but not necessarily to their land holdings) due to at least temporary improvements in the security situation. Such is the case in Afghanistan, to which almost three-quarters of a million refugees who had been living in Pakistan returned following the establishment of an elected government, and in Liberia, to which approximately 70,000 refugees returned from neighboring countries following the cessation of hostilities. On the other hand, the nature of today's conflicts is such that many people who would want to become refugees are prevented from doing so. For example, in the wake of the first Gulf War, in March 1991, several hundred thousand – perhaps as many as a million – Kurds in northern Iraq, justifiably fearing retribution from the Saddam Hussein regime for their support of its opponents, fled into the Kandil Mountains between Iraq and Turkey, hoping to find refuge in southern Turkey. However, because of longstanding animosity between the government of Turkey and secessionist Kurds in the east, the border between the two countries was effectively closed by either barbed wire or the presence of land mines. Similarly, Haitians fleeing from increasing disorder and lawlessness in their home country, a situation that has resulted in widespread abuses of human rights including killings, arbitrary arrests, human trafficking and sexual violence, have been prevented by the Coast Guard from seeking refuge in the United States and have been returned, without review of their situation, to Haiti.

As a result of these kinds of practices, the number of "internally displaced persons" (IDPs) – people who meet all of the criteria for being refugees but who have not crossed or been able to cross an internationally recognized border – has increased by 22 percent to approximately 23.7 million (UNHCR, 2006). The plight of IDPs is serious. They are proffered no protection under international law, and in fact their welfare is considered to be the responsibility of the government of the country of which they are citizens – the same governments whose abuses, in many cases, they are seeking to flee. For UNHCR, both refugees and IDPs are "people of concern," but the ability to provide assistance to the former is clearly far easier than for the latter, because of not only legal complexities, but also operational difficulties. Even the number of IDPs is difficult to determine, as many are "absorbed" by relatives or friends, imposing additional hardships even on those who are not displaced.

In addition to causing the displacement of millions of people, conflict also takes an important toll on both agrarian and industrial economies. Although there is some dispute as to the magnitude of the impact of conflict on national economies, some estimate that a 15-year civil war would reduce gross domestic product by as much as 30 percent (Collier, 1999; Imai and Weinstein, 2000). Local economies may be even more devastated, with more immediate consequences for the health of the population – particularly through food shortages resulting in a high prevalence of under-nutrition. In rural areas, where a relatively high proportion of food is derived from subsistence farming, farmers may be physically unable to plant as much as they might in the absence of conflict. Where even small-scale commercial food production is a way of life, farmers may be obstructed from bringing their produce to market. Resulting scarcities can be responsible for higher food prices, which, combined with the loss of jobs and currency inflation, leave the general population with less food with which to feed families. More direct effects of conflict, such as the destruction of irrigation systems and theft of produce by soldiers, have also been observed (Macrae and Zwi, 1994). Where conflict is exacerbated by drought, pest infestation, or other detrimental factors, famine can result.

For those city dwellers living in conflict, the food situation can become particularly problematic. Without the ability to revert to subsistence agricultural methods, to find alternative food sources, or to adopt "coping" mechanisms such as consuming seed stock, those whose nutrition depends entirely on a healthy marketplace can rapidly become deprived. For example, in 1992, during the Balkan war, residents of Sarajevo, accustomed to receiving 270 tonnes of food per day, were forced to survive on a total of 216 tonnes – an amount that translates to barely three-quarters of the minimum daily caloric requirement per person (Toole *et al.*, 1993).

Another serious threat to the health of civilians caught up in conflict is the destruction of utilities and other elements of the physical infrastructure, whether intentional, accidental, or through an inability to carry out essential maintenance functions. As above, the burden may fall mostly and most frequently on those living in an urban environment, but even in rural areas the disruption of irrigation systems and of local water supplies, for example, have taken an important toll. In one setting, soldiers intentionally destroyed hand pumps in rural parts of southern Sudan (Dodge, 1990). In some cities caught up in warfare, water supplies have been particularly compromised. In 1993, for example, residents of Sarajevo had only 5 liters of water per day – well below the minimum *per capita* requirement of 15 liters per capita per day suggested by the Sphere Project (Sphere Project, 2004). Not only was the amount of water severely restricted, but attempts to access water also put people at serious physical risk. Particularly notorious was the so-called "sniper's alley," where civilians were shot at from the surrounding hillsides while fetching water from outdoor fountains after their indoor taps ran dry. Similar situations prevailed in other urban areas of Bosnia as well, resulting in outbreaks of hepatitis A, watery diarrhea, and bacillary dysentery.

In the same vein, disruptions of the power supply can also have serious consequences on health. Clinic and hospital services are severely curtailed, and surgical procedures become more risky if they can be done at all. Drugs, vaccines, blood, and other products that require refrigeration are all likely to perish. In colder climates, the lack of electricity and/or fuel for heating not only increases caloric requirements, but also has an impact on rates of acute respiratory ailments and on exposure.

Finally, violent conflict has a major effect on the availability of health services, and on the ability of a population to use those services that might remain available. In many countries, of course, especially in sub Saharan Africa, health services are severely limited even in the absence of war, but even the most basic primary health care is usually denied to needy populations in a violent environment. At times clinics, even in the most peripheral areas, along with schools, are intentionally targeted by factions intent on disrupting all public functions. The actions of Renamo, in Mozambique, and of the Sandinistas, in Nicaragua, provide good examples of these destructive strategies (Cliff and Noormahomed, 1988; Garfield and Williams, 1992). In Afghanistan today, utilization rates of available health services are distinctly lower in areas in which personal safety has been compromised by Taliban threats to those utilizing government services than in those areas in which the government has been able to re-establish health care and where people can travel safely from their homes to health facilities.

The safety of health workers is also an important issue in times of conflict. Physicians, nurses, and other health professionals may be better able to flee dangerous areas than the rest of the population, should they choose to do so, thereby compromising the delivery of health services. Even more tragically, warring groups may target health workers, and they or their family members may be kidnapped and/or killed. Even health workers engaged by humanitarian agencies to provide relief to war-beleaguered populations are at increased risk. During the 1990s and 2000s, employees of the International Committee of the Red Cross have been killed in Afghanistan, Chechnya, Burundi, DRC, and Sierra Leone, among others. Health professionals working for the International Rescue Committee, Action Against Hunger, Médecins Sans Frontières, and CARE have been killed in Pakistan, Sri Lanka, Afghanistan, and Iraq, respectively, and many other humanitarian organizations have suffered losses of life in other countries, in what has become a particularly disturbing trend that has consequences not only for the health of the populations these individuals and organizations were serving, but also on the way and the extent to which humanitarian relief programs in general will be implemented in the future.

If, as discussed above, the biomedical causes of death in complex emergencies are the same as in developing countries that are not experiencing conflict, and only the circumstances are different, it stands to reason that, in emergencies, deaths from these common diseases occur at far higher rates. In the late 1970s, epidemiologists began to measure crude and age-specific mortality rates in emergency settings and to compare them to estimated baseline rates for developing countries. For example, in the so-called "death camps" of Thailand to which Cambodian refugees fled in 1979 to escape the genocide perpetrated by the Khmer Rouge, the crude mortality rate was 9.1 per 10,000 per day during the first week of resettlement. The incorporation of epidemiological evidence into the health planning process enabled those responsible for providing assistance to establish health-care priorities. Doing so contributed to lowering mortality to 0.71 per 10,000 per day by the fifth week of the relief effort (Glass *et al.*, 1980). A similarly rapid decline in crude mortality was documented among Kurdish refugees to the Turkey/Iraq border in the wake of the first Gulf War, in 1991 (CDC, 1991).

In Africa, crude mortality rates were measured in Somalia in the early 1980s, after the Ogaden War, combined with drought, resulted in the flight of more than 500,000 ethnic Somalis from Ethiopia into about 30 refugee camps, and in camps in eastern Sudan, to which Ethiopian nationals caught up in a complex web of circumstances that included drought, inequitable land reform practices, and violent conflict found themselves forced to flee. In each case, high mortality rates were recorded in the early stages of the emergency, and in each case these rates declined relatively rapidly toward baseline levels during the first 6–12 months of the international relief effort (CDC, 1992). Occasionally, less good outcomes have been documented. In the Hartisheik refugee camp in Ethiopia, in 1988–89, crude mortality rates actually rose during the first nine months that humanitarian assistance was being delivered (Toole and Bhatia, 1992).

The application of epidemiological methods and measurements has not been free of problems, though. While two-stage cluster sample surveys have become the method of choice for determining crude mortality rates in a population, along with age-specific (usually under five-year-old mortality) rates, these can be difficult to conduct for a variety of technical and operational reasons. For example, all of the results presented above come from surveys conducted in camp settings. In these settings, the affected population can usually be reasonably well enumerated and located – the selection of clusters is relatively simple, and access to samples selected is practical and safe. Increasingly frequently, however, the population in need of assistance is scattered, difficult (if not impossible) to count, and sometimes very difficult to reach. In these conditions, especially if violence is raging all around, the "epidemiological space" is contracted, and obtaining useful information to guide a relief effort is seriously compromised. Such has been the case, in recent years, in the eastern DRC, in Darfur (Sudan), and in Iraq.

In addition to the difficulties encountered in ascertaining and following trends in mortality, which include problems in determining both the number of deaths and the size of the target population, many humanitarian actors lack the proper training and experience to conduct and interpret epidemiological surveys. A review of more than 20 mortality surveys conducted in Somalia in the early 1990s by a wide variety of non-governmental organizations revealed gross inconsistencies in all aspects of sample design, conduct, analysis, and use of the data (Boss *et al.*, 1994). More recently, an evaluation of two-stage cluster sample surveys used to determine the nutritional status of children in Ethiopia also found serious methodological and analytical problems with the vast majority of surveys conducted by the humanitarian community responding to a perceived food shortage problem there (Spiegel *et al.*, 2004).

The problems with understanding and interpreting data from mortality surveys conducted in emergencies in general, and in conflict settings in particular, are perhaps best illustrated by the controversy surrounding estimates of civilian mortality in Iraq since the start of the war in 2003. Skepticism regarding the results of the survey, which demonstrated levels of mortality far higher than those reported from other, non-epidemiologically valid sources, focused on the two-stage cluster sampling method, which is perfectly suited for the purpose of determining mortality and has been repeatedly proven, even in conflict settings, to give accurate and reliable results (Roberts *et al.*, 2004; Burnham and Roberts, 2006; Burnham *et al.*, 2006). Because of this, the determination of crude and age-specific mortality rates using population-based studies has appropriately become an essential first step in determining the magnitude of an emergency, in designing appropriate humanitarian interventions, and in evaluating the degree to which they are being successfully implemented. Guidelines for interpreting the results of mortality surveys have recently been published (Checchi and Roberts, 2005).

The most common specific diseases that cause excess mortality are closely linked to the circumstances – the impact of a violent environment can be shown, in most cases, to contribute to higher rates of morbidity and of disease-specific mortality. The connection between conflict and food scarcity has been discussed above. The natural consequence of food shortages in a population is malnutrition, and indeed some of the highest rates of malnutrition, as documented by population-based anthropometric surveys, have been documented in populations affected by war. In Somalia, in 1992, following the death of the dictator Siad Barre, inter-clan violence disrupted the food supply throughout the country. At the start of the relief effort, the prevalence of malnutrition in internally displaced children living in camps and aged less than five years was as high as 75 percent. In south Sudan, when food shortages were compounded by a violent environment in 1994, malnutrition was measured at nearly 40 percent in the under-five population. In both the DRC in 1996 and Angola in 1995, malnutrition prevalence hovered around 30 percent (Toole *et al.*, 2001).

Interestingly, although the prevalence of malnutrition is certainly made worse by the presence of conflict, there are instances when it has actually increased during the relief effort. In Iraq in 1991, for example, malnutrition in children aged 12–23 months increased from 5 percent to 13 percent during several months following an epidemic of cholera in the general population, and while the population was receiving considerable external humanitarian assistance. In Hartisheik refugee camp in Ethiopia, cited above, malnutrition prevalence increased from less than 10 percent to almost 25 percent in 1988, with a concomitant rise in mortality, for the simple reason that the relief effort provided less than an adequate daily ration of food to the refugees.

Micronutrient deficiencies have also been documented frequently during complex emergencies. Scurvy, a disease that has become quite rare under normal circumstances, was documented on several occasions in refugee camps in Somalia and Sudan during the 1980s. In every case refugees were receiving external assistance, but relief foods did not contain an adequate source of ascorbic acid (Desenclos *et al.*, 1989). An outbreak of almost 20,000 cases of pellagra occurred among refugees in camps in Malawi fleeing from Renamo-instigated violence in Mozambique in 1990. Again, the outbreak was traced to a lack of niacin in the ration provided by the international relief community (Malfait *et al.*, 1993). Pellagra also occurred in war-torn Angola, among internally displaced persons, in 2000 (Baquet *et al.*, 2000).

There is a clearly established relationship between nutritional status and mortality from communicable diseases, especially in children, even though severe malnutrition may not be the immediate cause of death (Pelletier *et al.*, 1995). High case-fatality rates from measles, traditionally one of the most important causes of death in refugees and internally displaced persons, are probably due to a combination of malnutrition and vitamin A deficiency. Although it has now become an almost reflexive early element of humanitarian response, measles vaccination, which has been available as a safe and effective public health tool since the 1960s, was not routinely implemented until the 1980s. Epidemiologists had identified measles as the leading cause of death in Somalia in the late 1970s, and again in Wad Kowli and other refugee camps in Sudan in 1985. In the latter instance, measles was responsible for more than 50 percent of all mortality in both children and adults. By the end of the 1980s, due to the widespread adoption of a policy making mass vaccination one of the earliest interventions of relief efforts, measles had almost disappeared as a cause of death in complex emergencies to which vaccine could be supplied in adequate quantities. Exceptions occurred when refugees and/or IDPs were "absorbed" into local communities, beyond the reach of the international relief agencies, and were dependent for vaccination on weak national immunization programs.

As mentioned above, diarrhea has always been one of the most common causes of mortality, and its incidence and case-fatality rates are also exacerbated by conflict, forced migration, and malnutrition. Diarrhea has been responsible for between 25 and 85 percent of mortality, and is always among the leading causes of clinic visits in complex emergencies. Diarrhea in emergencies commonly occurs in three forms: acute watery diarrhea, usually of viral origin, and the two epidemic forms – cholera and dysentery – due to *Shigella dysenteriae* type 1. The cholera outbreak in Goma in July 1994, described in the opening section of this chapter, was atypical in its virulence, but cholera is a common occurrence in situations where there is a breakdown of public health infrastructure, where the population is dependent on primitive and insufficient water supplies prone to contamination, and where sanitation is rudimentary.

Acute respiratory infections, most commonly pneumonia, are also frequently recorded as being among the most common causes of mortality in emergencies, although documentation is not consistent. No studies of pneumonia have been conducted in emergency settings, and knowledge of its epidemiology in these circumstances is extremely limited. Pneumonia is, however, the leading cause of death in non-emergency settings (Black *et al.*, 2003). Also, it is intuitive that overcrowding and inadequate shelter and clothing are likely to increase both the incidence and mortality from respiratory infections (Connolly *et al.*, 2004).

Malaria, especially in Africa, is another leading cause of mortality. Forced migration has been a major risk factor for malaria, and the movement of people with low levels of immunity from areas of low endemicity to hyper-endemic areas has occurred on different occasions. Increased exposure to the elements, from inadequate shelter and clothing, increase the number of potentially infective mosquito bites. Overcrowding also increases the number of bites per person. All of these risk factors are compounded by growing resistance to chloroquine, which renders the effective treatment of malaria far more expensive than it had been. The provision of artemisinin combination therapy, the currently recommended treatment for *Plasmodium falciparum* in most parts of the world, puts a serious strain on traditional levels of funding for humanitarian assistance in the health sector.

The control of outbreaks of meningococcal meningitis, as well as that of other common infectious diseases such as tuberculosis and HIV/AIDS, and other locally endemic conditions, is rendered more difficult in areas where conflict occurs. As discussed above, all health services in a conflict zone are likely to be seriously disrupted and, even if they continue to be offered, the ability of the population to use them is greatly restricted.

## Return to normalcy

If, as discussed above, the presence of conflict were in itself a major risk factor for increased mortality, it would be reasonable to assume that the restoration of peace would result in lower mortality rates. Although this is usually the case, the situation is not as simple as might seem, because the destruction and disruptions caused by violence cannot be quickly undone. Instead, a major program of reconstruction of health facilities, rehabilitation of health systems, and restoration of health services is required before morbidity and mortality rates can be expected to return to those of the ante-bellum period.

One characteristic of civil and interstate violence that has not been mentioned is the gross violation of human rights that occurs when states are either unwilling or unable to discharge their responsibilities toward their citizens. Genocide and what has become known as "ethnic cleansing" are extreme examples of what can befall civilians caught up in violent settings, but even in situations of lesser conflict, instances of sexual violence and of torture have been increasingly documented (Inter-Agency Standing Committee, 2005).

Health is an integral component of human rights law, and the Universal Declaration of Human Rights states that "{e}veryone has the right to a standard of living adequate for the health and well-being of himself and his family, including food, clothing, housing and medical care…" The re-establishment of a legitimate government in a country that has been torn apart by violence should therefore be accompanied by the reinstatement of this "right" (United Nations, 1948). Although there are many processes by which health services can be made available in post-war environments, there is a growing tendency in recent years for fledgling governments to define and to offer a limited number of essential health interventions – a Basic Package of Health Services – to as many of its citizens as possible. Such an approach has been adopted in Cambodia, Afghanistan, the Democratic Republic of Congo, south Sudan, and Liberia, to name a few. These Basic Packages seek to address those health problems that are the greatest contributors to mortality during both wartime and times of peace; they typically consist of essential maternal and newborn services, interventions aimed at reducing childhood mortality, nutrition activities, and the control of communicable diseases. Depending on national priorities and on available resources, programs offering services in the areas of mental health and physical disability may also be included (Ministry of Health, 2003).

Although the development of a Basic Package of Health Services is a potentially important activity in the immediate post-conflict period for the rapid reduction of morbidity and mortality, and certainly an important statement of intention on the part of a new government trying to re-establish a legitimate and effective relationship with its citizenry, its implementation can be grossly impeded by the lack of financial and human resources. For these reasons, the role of non-governmental organizations may remain critical even when the humanitarian emergency has subsided. Donors may be reluctant to invest heavily in a new state authority, and may prefer to "hedge" their financial bets by continuing to fund non-governmental organizations, working at the grassroots level, to provide social services (including health care) to the population. Governments themselves, recognizing that their health personnel have become extremely limited during the war years – because they have not been paid, because more attractive jobs became available in the private sector, or because of emigration – may find that reliance on non-government organizations, both international and local, to implement health services in accordance with government policies, as spelled out in a Basic Package of Health Services, allows them to reach a large proportion of the population relatively quickly – especially in more peripheral areas, where civil servants may be reluctant to work.

One mechanism for the delivery of essential services to post-conflict societies that is becoming increasingly popular is performance-based contracting (Waldman and Hanif, 2002; Soeters *et al.*, 2006). Although there are many variations on the theme, the essential feature of this mechanism is that private-sector health personnel, usually non-government organizations, are reimbursed on the basis of how well they implement government policies; in some schemes they are given substantial financial bonuses for reaching pre-determined targets. Although there is little documentation attesting to the effectiveness of these schemes in improving the health status of target populations, early indications suggest that both access to health services and utilization of those services can be substantially increased in a relatively short time. In addition, although there is no evidence to suggest that this is in fact the case, the provision of health services to communities that had been deprived of them for years could have a stabilizing effect on shaky peace accords.

## Conclusion

Returning to the account of the cholera epidemic in Goma with which we began, it is important to emphasize that the circumstances in which that drama unfolded were not the only determinant of the unconscionably high mortality that was experienced. As discussed, cholera is a disease for which case-fatality ratios should not exceed 2–3 percent. The fact is that, in addition to the virulence of the outbreak, the physically challenging environment, and the traumatized population that had fled to Goma out of fear of retribution for the genocidal actions in which many of them had been at least peripherally engaged, fewer refugees might have died if relief workers had been more experienced in treating cholera specifically, and in working in complex emergencies more generally (Goma Epidemiology Group, 1995). Few health workers forge a career in humanitarian assistance, and many, despite the best of intentions and the generosity of spirit that they bring to strange and difficult situations, are unfamiliar with the basic approaches to health-care delivery in complex emergencies that have been forged since the debacle in Goma (Sphere Project, 2004). Recently there have been calls for further professionalization of health care in emergency settings, and in the future a medical specialization in this field may be developed.

Still, humanitarian workers are never responsible for the conflict and the accompanying violence that are the root causes of increased morbidity and mortality in emergency settings. Their job is to do what they can, in environments that are in many ways the "emergency rooms of international health," to provide the equivalent of first aid and to try to limit the impact of communicable diseases on the population, not to prevent them from claiming an excessive toll. The real cause of what are sometimes extraordinarily high rates of morbidity and mortality from common conditions like pneumonia, diarrhea, malaria, and measles in complex emergencies are not the bacteria, viruses, or parasites that might be identified in a laboratory, or even the fact that many of those engaged in providing assistance may be unfamiliar with the best ways with which to deal with these conditions. Instead, the underlying cause is armed conflict, which creates an environment of unspeakable violence that leads to forced migration, difficult access to available health services, and a besieged and beleaguered population that it is hard to reach and difficult to serve.

## References

Baquet, S., Wuillaume, F., van Egmond, K. and Ibanez, F. (2000). Pellagra outbreak in Kuito, Angola. *Lancet* **355**(9217), 1829–1830.

Black, R.E., Morris, S.S. and Bryce, J. (2003) Where and why are 10 million children dying every year? *Lancet* **361**, 2226–2234.

Boss, L.P., Toole, M.J. and Yip, R. (1994). Assessments of mortality, morbidity, and nutritional status in Somalia during the 1991–1992 famine. *Journal of the American Medical Association* **272**, 371–376.

Burkholder, B.T. and Toole, M.J. (1995). Evolution of complex disasters. *Lancet* **5**(346), 1012–15.

Burnham, G. and Roberts, L. (2006). A debate over Iraqi death estimates. *Science* **314**(5803), 1241.

Burnham, G., Lafta, R., Doocy, S. and Roberts, L. (2006). Mortality after the 2003 invasion of Iraq: a cross-sectional cluster sample survey. *Lancet* **368**(9545), 1421–1428.

Centers for Disease Control and Prevention (1991). Public health consequences of acute displacement of Iraqi citizens: March–May 1991. *Morbidity and Mortality Weekly Report* **40**, 443–446.

Centers for Disease Control and Prevention (1992). Famine-affected, refugee, and displaced populations: recommendations for Public Health Issues. *Morbidity and Mortality Weekly Report* **41**, RR–13.

Centers for Disease Control and Prevention (2007). Available at: http://www.cdc.gov/nceh/ierh/FAQ.htm (accessed 28 January 2007).

Checchi, F. and Roberts, L. (2005). Interpreting and using mortality data in humanitarian emergencies: a primer for non-epidemiologists. *HPN Network Paper 52*. London: Overseas Development Institute.

Cliff, J. and Noormahomed, A.R. (1988). Health as a target: South Africa's destabilization of Mozambique. *Social Science and Medicine* **27**, 717–722.

Coghlan, B., Brennan, R.J., Ngoy, P. *et al.* (2006). Mortality in the Democratic Republic of Congo: a nationwide survey. *Lancet* **367**(9504), 44–51.

Collier, P. (1999). On the economic consequences of civil war. *Oxford Economic Papers* **51**(1), 168–183.

Connolly, M.A., Gayer, M., Ryan, M.J. *et al.* (2004). Communicable diseases in complex emergencies: impact and challenges. *Lancet* **364**, 1974–1983.

Desenclos, J.C., Berry, A.M., Padt, R. *et al.* (1989). Epidemiological patterns of scurvy among Ethiopian refugees. *Bulletin of the World Health Organization* **67**(3), 309–316.

Dodge, C.P. (1990). Health implications of war in Uganda and Sudan. *Social Science and Medicine* **31**, 691–698.

Garfield, R. and Williams, G. (1992). *Health Care in Nicaragua: Primary Care under Changing Regimes*. New York: Oxford University Press.

Glass, R.I., Cates, W. Jr, Nieburg, P. *et al.* (1980). Rapid assessment of health status and preventive-medicine needs of newly arrived Kampuchean refugees, Sa Kaeo, Thailand. *Lancet* **1**(8173), 868–872.

Globalsecurity.org (2007). Available at: http://www.globalsecurity.org/military/world/war/index.html (accessed 21 February 2007).

Goma Epidemiology Group (1995). Public health impact of Rwandan refugee crisis: what happened in Goma, Zaire, in July, 1994? *Lancet* **11**(345), 339–344.

Imai, K. and Weinstein, J.M. (2000). Measuring the economic impact of civil war. Working Papers, Center for International Development, Harvard University.

Inter-Agency Standing Committee (2005). *Action to Address Gender-based Violence in Emergencies: IASC Statement of Commitment, 13 January 2005*. Available at: ochaonline.un.org/GetBin.asp?DocID=1083 (accessed 4 March 2007).

Macrae, J. and Zwi, A. (eds) (1994). *War and Hunger: Re-thinking International Responses to Complex Emergencies*. London: Zed Books.

Malfait, P., Moren, A., Dillon, J.C. *et al.* (1993). An outbreak of pellagra related to changes in dietary niacin among Mozambican refugees in Malawi. *International Journal of Epidemiology* **22**(3), 504–511.

Ministry of Health, Transitional Islamic Government of Afghanistan (2003). *A Basic Package of Health Services for Afghanistan*. Available at: unpan1.un.org/intradoc/groups/public/documents/APCITY/UNPAN018852.pdf (accessed 4 March 2007).

Pelletier, D., Frongillo, E.A. Jr, Schroeder, D.G. and Habicht, J.P. (1995). The effects of malnutrition on child mortality in developing countries. *Bulletin of the World Health Organization* **73**, 443–448.

Powers, S. (2002). *A Problem from Hell: America and the Age of Genocide*. New York: Basic Books.

Roberts, L., Lafta, R., Garfield, R. *et al.* (2004). Mortality before and after the 2003 invasion of Iraq: cluster sample survey. *Lancet* **364**(9448), 1857–1864.

Slim, H. (2004). *A Call to Alms: Humanitarian Action and the Art of War*. Geneva: Centre for Humanitarian Dialogue.

Soeters, R., Habineza, C. and Peerenboom, P.B. (2006). Performance-based financing and changing the district health system: experience from Rwanda. *Bulletin of the World Health Organization* **84**(11), 841–920.

Sphere Project (2004). *Humanitarian Charter and Minimum Standards in Disaster Response*. Oxford: Oxfam Publishing.

Spiegel, P.B., Salama, P., Maloney, S. and van der Veen, A. (2004). Quality of malnutrition assessment surveys conducted during famine in Ethiopia. *Journal of the American Medical Association* **292**(5), 613–618.

Toole, M.J. and Bhatia, R. (1992). A case study of Somali refugees in Hartisheik A camp, eastern Ethiopia: health and nutrition profile, July 1988–June 1989. *Journal of Refugee Studies* **5**, 313–326.

Toole, M.J., Galson, S. and Brady, W. (1993). Are war and public health compatible? Report from Bosnia-Herzogovina. *Lancet* **341**, 1193–1196.

Toole, M.J., Waldman, R.J. and Zwi, A.B. (2001). Complex humanitarian emergencies. In: M.H. Merson, R.E. Black and A.J. Mills (eds), *International Public Health – Diseases, Programs, Systems, and Policies*. Gaithersburg: Aspen Publishers, p. 464, Table 9.4.

Toole, M.J., Waldman, R.J. and Zwi, A.B. (2006). Complex emergencies. In: M.H. Merson, R.E. Black and A.J. Mills (eds), *International Public Health – Diseases, Programs, Systems, and Policies*, 2nd edn. Sudbury: Jones and Bartlett Publishers, p. 467, Table 10.4.

United Nations (1948). *Universal Declaration of Human Rights*. Available at: http://www.un.org/Overview/rights.html (accessed 5 March 2007).

United Nations (1950). Convention relating to the status of refugees. United Nations Conference of Plenipotentiaries on the Status of Refugees and Stateless Persons, convened under General Assembly resolution 429(V), 14 December.

UNHCR (2006). *Refugees by Numbers*. Geneva, Switzerland.

Waldman, R. and Hanif, H. (2002). *The Public Health System in Afghanistan*. Kabul: Afghanistan Research and Evaluation Unit.

# Bioterrorism 12

**Andrew W. Artenstein and Troy Martin**

Bioterrorism, broadly defined as the deliberate and malicious deployment of microbial agents or their toxins as weapons in a non-combat setting, represents perhaps the most overt example of human behavior impacting epidemic infectious diseases (Artenstein, 2004a). While most of the microbial threat agents of potential use in bioterrorism occur naturally in various ecological niches throughout the world, they are rare and sporadic causes of human disease in developed countries and urban environments. It is human behavior within the context of the extant geopolitical milieu that transforms these naturally-occurring organisms into potential weapons of mass terror.

There is historical precedent for the use of biological agents against both military and civilian populations. It is postulated that the fifth plague visited upon Pharaoh in approximately 1450 BC, "murrained carcasses… pestilence," signified cutaneous anthrax (Plaut, 1981). In the fourteenth century, Tartar invaders probably introduced the Black Death to Caffa by catapulting plague-infected corpses into the besieged Crimean city for that explicit purpose (Wheelis, 2002). British forces in mid-eighteenth century colonial America, under the command of Lord Jeffrey Amherst, distributed blankets and clothing used by smallpox victims to Native American tribes in an attempt to affect the balance of power during the French and Indian wars (Christopher *et al.*, 1997); it remains unclear whether these fomites resulted in contact transmission of smallpox to naïve hosts or whether the Native Americans were infected by direct contact with infected colonists.

The use of biological (and chemical) agents as weapons of war has been well documented (Christopher *et al.*, 1997). The German biological warfare program during World War I included covert infections of Allied livestock with anthrax and glanders. The Japanese army began conducting experiments on the effects of bacterial agents of biowarfare on Chinese prisoners in occupied Manchuria in 1932 at their infamous Unit 731; thousands of individuals were killed as a result of these experiments, which continued until 1945 (Harris, 1994). The United

States began its own offensive biological weapons program in 1942 and, during its 28-year official existence, weaponized and stockpiled lethal biological agents, such as anthrax, as well as incapacitating agents, such as the etiologic agent of Q fever (Christopher *et al.*, 1997).

The US program was ended by two presidential executive orders in 1969 and 1970; stockpiled weapons were destroyed by mandate from 1971–1973 (Christopher *et al.*, 1997). However, small quantities of pathogens were stored at Fort Detrick, Maryland, for biodefense research purposes. In 1972, under the auspices of the United Nations, the Convention on the Prohibition of the Development, Production, and Stockpiling of Bacteriological (Biological) and Toxin Weapons and on Their Destruction (BWC) was ratified with more than a hundred signatory nations, including the US and the Soviet Union (Christopher *et al.*, 1997).

Although a party to the BWC, the government of the Soviet Union apparently continued to weaponize biological agents at least through the mid-1990s (Alibek, 1999). There is direct evidence that the Soviets deployed weaponized ricin, a biological toxin, to carry out covert assassinations during the 1970s (Christopher *et al.*, 1997). Additionally, the corroborated statements of multiple high-level government defectors confirm decades of persistent Soviet violation of the BWC. Perhaps the most egregious example of these violations arose from the revelation, years after the event occurred, that an epidemic of inhalational anthrax in Sverdlovsk in 1979, responsible for the deaths of at least 66 people, resulted from the accidental release of weaponized spores from a biological weapons plant (Guillemin, 1999).

Other more recent examples of bioterrorism, though not necessarily resulting in attacks causing morbidity or mortality, may serve as harbingers of future events. Saddam Hussein's regime in Iraq developed and deployed anthrax- and botulinum-laden warheads in the years leading up to the Gulf War (Zilinskas, 1997); the reasons that these weapons were never used in an actual attack probably had more to do with the implicit threat of overwhelming US retaliation and Iraqi technological deficiencies rather than the regime's reluctance to violate any moral principles. Biological agents have also been used to forward political ideologies: in 1984 a religious cult, intent on influencing voter turnout during a local election, contaminated restaurant salad bars in The Dalles, Oregon, with *Salmonella*, resulting in over 750 cases of gastroenteritis among patrons, employees, and their contacts (Torok *et al.*, 1997). This event, coupled with revelations that the Japanese cult Aum Shinrikyo had attempted, multiple times, to release weaponized anthrax before their successful release of sarin nerve agent in the Tokyo subway system in 1995 (Olson, 1999), provides compelling evidence of the terrorist potential of these agents.

The catastrophic events of 11 September 2001 clearly ushered in a new era of global terrorism. The massive, simultaneous, and dramatic attacks on unarmed citizens in New York and Washington illustrate, convincingly, the mounting boldness of terrorists and their willingness to commit "unthinkable" acts. The anthrax attacks that followed 9/11 in the US killed 5 and sickened 17 additional people, and served to underscore the shifting sands of terrorism (Jernigan *et al.*, 2001). While these bioterror attacks have never been directly linked to the events of 9/11, their temporal connection reinforces the persistent global threat posed by bioterrorism.

## Social determinants of bioterrorism: the concept of "risk"

Any discussion of bioterrorism, and certainly one that involves mitigation strategies, hinges on the concept of "risk." "Risk" refers to the likelihood that exposure to a hazard will lead to a negative consequence; therefore, it is essential to understand both the threat and the potential range of consequences associated with bioterrorism in order to accurately assess risk in this regard (Ropeik and Gray, 2002). When applied to cause-specific mortality, risk can be viewed in a purely statistical sense: the risk of dying from cardiovascular disease in the US in the year 2000 was approximately 1 in 400, while the risk of succumbing in a lightning strike was approximately 1 in 4.5 million (Table 12.1). In the arenas of human biology and medicine, however, risk assessment is based on the complex interplay of genetics, environmental factors, and chance. Risk as it relates to bioterrorism is difficult to quantify; while the probability of exposure to a biologic attack is statistically low, it is not zero, and the consequences are potentially catastrophic. This, coupled with the fact that the likelihood of actual

**Table 12.1 US mortality risk analysis for selected public health concerns**

| | |
|---|---|
| Heart disease | 1 in 397 |
| Cancer | 1 in 511 |
| Stroke | 1 in 1699 |
| Alzheimer's | 1 in 5.752 |
| Motor vehicle accident | 1 in 6745 |
| Homicide | 1 in 15,440 |
| Drowning | 1 in 64,031 |
| Fire | 1 in 82,977 |
| Bicycle accident | 1 in 376,165 |
| Lightning strike | 1 in 4,478,159 |
| Bioterrorism (anthrax) | 1 in 56,424,800 |

Source: Artenstein, 2006; Harvard Center for Risk Analysis, http://www.hcra.harvard.edu/ © 2004 CBEP.

hazard exposure is dependent on the whims of terrorists (and therefore an unpredictable variable), renders accurate risk assessment impossible.

## Geopolitics and the "psyche" of terrorists

The opening act of the modern era of terrorism dates from 1972, when the Palestinian terrorist group Black September murdered 11 members of the Israeli Olympic team and a German police officer in Munich (Post, 2005). As the event played out, the enormous amplifying effect of international media coverage was recognized and serves as a legacy for today's terrorists. Three types of terror organizations have since been recognized: social-revolutionary groups, nationalist-separatist groups, and, more recently, religious fundamentalists. Social-revolutionary groups, as exemplified by the Red Brigades in Italy and the Red Army Faction in Germany, are leftist groups with strong ties to Communist parties seeking to overthrow the extant capitalist economic and social order. With the collapse of Communism in Europe and the end of the Cold War, their activity has dramatically declined.

Nationalist-separatist terrorism organizations are one of the two types seen commonly today. These groups are fighting to establish a new political order or state based on ethnic identity. They are influenced by the struggle of earlier generations to gain independence from a perceived oppressive regime, and their acts of terrorism are focused on this regime or its allies. Examples of such organizations include the Provisional Irish Republican Army of Northern Ireland (PIRA), the Basque separatist group Euskadi ta Askatasuna (ETA), the secular Palestinian organization al-Fatah, and the Liberation Tigers of Tamil Eelam (LTTE) in Sri Lanka (Arena and Arrigo, 2005).

The most influential type of organization with regard to US foreign policy appears to be religious fundamentalist terrorists, illustrated by organizations such as Al-Qaeda and Islamic Jihad. The goal of these organizations is broader in scope, in that they want not only to change the local political situation but also to expel representatives of the secular world from their lands. In the case of Islamic fundamentalist terrorists, their aim is to rid their society of Western influences. Islam is considered to be a comprehensive system that guides all aspects of life and gives meaning and direction to social, legal, and political systems (Wiechman, *et al.*, 1994). Western global influence, particularly the development of Western-style secularism in Islamic countries, is seen as an affront to the existence of Islamic societies.

Western colonization was responsible for the early development of Islamic fundamentalism. This movement is in large part traced to Hasan al-Banna, who came to believe that the nineteenth- and twentieth-century colonialism in the Middle East and the subsequent diffusion of secular ideas and Western values in the region had served to erode the fabric of Islam (Abu-Amr, 1993; Mitchell,

1993). As a response he formed the Egyptian Muslim Brotherhood in 1928, dedicated to re-establishing an Islamic state in which the tenets of the religion and the precepts of Islamic *shariah* (holy law) were both firmly established (Moussalli, 1998). As the movement evolved, the scope broadened to the establishment of a fundamentalist Islamic world-view (Davidson, 1998). In this way, a movement that began as predominantly Arab and anti-colonial has now become pan-Islamic, drawing from the entire, diverse, global Muslim population, in excess of 1 billion people. Al-Qaeda has emerged as the leading proponent of this new ideology, with its focus on both the local and worldwide struggle for influence.

Unique to these groups is the decision-making role of a pre-eminent leader who is generally thought to have insight to God's will. Actions sanctioned by the leader are thereby endowed with sacred significance, thus explaining the fervor with which group members kill innocent non-believers. Because their motivation is to expel Western secular values and create a pure Islamic state, they believe their actions are sanctioned by the *Koran* and not constrained by Western morals; they are willing to do the "unthinkable," including the use of biological agents as weapons (Post *et al*., 2003). Their enemies are anyone who is opposed to their worldview. While the primary goal is to attack symbolic targets that reflect the secular decadence of Western life and attract media attention, many attacks are conducted against smaller secondary targets, due to greater accessibility. A novel aspect of these terrorist's tactics is the focus on killing as many innocent people as possible. As illustrated by the 1998 US embassy bombings in Tanzania and Kenya, an additional objective appears to be the influencing of geopolitics through the induction of widespread fear in communities throughout the world where Western interests are present (Borum and Gelles, 2005).

## Threat assessment: bioterrorism in the overall terrorism context

Biologic agents are considered to be "weapons of mass destruction" (WMD) because, as with nuclear and chemical weapons, their use may result in potential mortality on a massive scale. Bioterrorism occupies a unique niche among WMD (Table 12.2); exposure to biologic agents entails a clinical incubation period of days to weeks during which time recognition of an attack is problematic (assuming a covert attack), detection of a specific agent is difficult, and infection may disseminate widely among a population. This is in contradistinction to other forms of WMD, where recognition of an actual "event" occurs with the deployment, allowing mitigation strategies to begin nearly immediately. The difference is important, as specific therapeutic or prophylactic measures may be available in bioterrorism, as opposed to simply general decontamination and supportive measures used in other arenas of terrorism response.

**Table 12.2 A comparison of weapons of mass terror**

| | Conventional | Biological | Chemical | Nuclear |
|---|---|---|---|---|
| Area involved | Limited | Moderately large | Moderate | Large |
| Rapid detection | Easy | Difficult | Moderate | Easy |
| Clinical incubation | Immediate | Days to weeks | Minutes to hours | Varies with dose |
| Medical Rx | Limited | Effective v some | Limited | Limited |
| Cost | High | Low | Low | Very high |
| Terror potential | High | Very high | Very high | Very high |



Multiple factors likely contribute to the attractiveness of biological agents as tools of terrorism: they are relatively inexpensive, available, and the technology for their production is generally accessible (Bhalla and Warheit, 2004); these agents can be aerosolized, deployed in occult fashion, and cause lethal or disabling disease in exposed individuals, and their impact may be amplified by long distance travel of aerosols (depending on extant environmental conditions) and the potential for person-to-person transmission. Perhaps of greatest utility from a terrorist perspective is that the specter of bioterrorism provokes fear and anxiety – "terror" – that is disproportionate to that seen with other forms of WMD. For this reason, it may be more fitting to consider these agents as "weapons of mass terror."

Vulnerability to bioterrorism is invariable in nations adhering to democratic principles. This is largely because of the freedom of movement and the access to public institutions that is afforded in such societies; terrorists intent on committing malicious acts can exploit these liberties.

The Center for Nonproliferation Studies has identified at least 11 nations with either known (e.g. former Soviet states, pre-war Iraq) or probable (e.g. Iran, China, North Korea) offensive bioweapons programs (Center for Nonproliferation Studies at the Monterey Institute of International Studies, 2002). Much of the underlying technical expertise for such programs may have derived from freelance scientists who became available for hire after the dissolution of the Soviet Union (Alibek, 1999). Because the technology needed for bioterrorism is "dual use," in that it can serve legitimate functions such as vaccine or pharmaceutical production, rogue states may either resist or be insulated from international scrutiny, as occurred in pre-Gulf War Iraq (Zilinskas, 1997).

The aims of bioterrorism are similar to those of other forms of terrorism: morbidity and mortality among civilians, disruption of social fabrics through panic and fear, and exhaustion or diversion of resources (Artenstein, 2006). However, a successful outcome from a terrorist's standpoint may be achieved without furthering all of these aims. The anthrax attacks in the United States in 2001 disrupted society and diverted scarce resources from other critical public health activities despite a limited number of casualties.

To be used in large-scale events, bioterrorism agents must undergo complex processes of production, chemical modification, and weaponization. Thus, state sponsorship or the support of organizations with significant resources and infrastructure would likely be necessary for the execution of substantial or multifocal attacks. However, recent revelations suggest the availability of bioweapons on the global black market (Miller *et al.*, 2001), and their relative simplicity, availability, and portability may make them preferable to expensive conventional or nuclear weapons for small, well-resourced terrorist organizations or even isolated terrorist cells. As demonstrated by the US anthrax attacks in 2001, bioterrorism attacks can be successful using only low-technology delivery methods such as the postal system.

The prospect of bioterrorism has been fueled by progress in the fields of molecular biology and biotechnology. While these advances have led to the possibility of new vaccines, medications, diagnostics, and genetic therapies to alleviate human disease, they have also introduced the potential to modify biological agents for malicious intent (Franz and Zajtchuk, 2002). Therefore the dissemination of information on developments in molecular biology, considered to be necessary to advance science, can serve the unintended dual purpose of providing terrorists with a virtual blueprint for developing genetically altered "designer" bioweapons, including hybrid organisms or drug-resistant mutants (Rappert, 2003). For example, publication of the genetic sequence of the 1918 influenza virus (Taubenberger *et al.*, 2005), and its reconstruction from viral RNA using reverse genetics (Tumpey *et al.*, 2005), while important to understand the pathogenesis of pandemic influenza, has raised concerns that terrorists might recreate the virus. Finally, our advancing knowledge of human genetics may have a dark side: the potential opportunity for terrorists to engineer biological agents targeted against our genomic vulnerabilities (Petro *et al.*, 2003).

## Epidemiologic principles as applied to bioterrorism

A World Health Organization (WHO) model based on the hypothetical effects engendered by the intentional release of 50 kilograms of aerosolized anthrax spores upwind from a population center of 500,000, a moderate-sized city, estimated that the agent would disseminate in excess of 20 kilometers downwind and that between 84,000 and 210,000 people would be killed or injured by the event, depending on whether the area was in a developed or developing country (WHO, 1970). The complete WHO theoretical analysis showed that casualty estimates depend on the properties of specific pathogens, the environmental setting, and the host population.

Numerous attributes contribute to the selection of a pathogen as a biologic weapon: availability of seed material; ease of cultivation; feasibility of large-scale production; capacity for aerosolization; stability of the product in storage, as a weapon, and in the environment (biologic entities differ in their physical properties); technology for dissemination; cost; and clinical virulence (Artenstein, 2004a). The latter refers to the consistency with which a biological agent causes high mortality, morbidity, and social disruption, and its intrinsic transmission characteristics. The Centers for Disease Control and Prevention (CDC) have prioritized biologic agent threats based on the aforementioned characteristics (CDC, 2000); the major purpose of this classification is to direct and focus public health preparedness strategies (Table 12.3). Category A agents, considered the highest priority, are those associated with high mortality and the greatest potential for major impact on the public health. Additionally, category A agents have been demonstrated to be capable of wide dissemination or person-to-person transmission. Category B agents are moderately high priority concerns. They may be considered "incapacitating" agents because of their potential for moderately high morbidity, but relatively low mortality. Most of the category A and B agents were experimentally weaponized and tested by the former Soviet Union, and are thus of proven feasibility (Alibek, 1999). Category C agents include emerging threats and pathogens that may be available for development into bioweapons in the future. As previously discussed, the potential exploitation of scientific progress by terrorists should prompt innovative thinking as it pertains to risk assessment and public health response. It is critical to be cognizant of future novel threats based upon engineered emergent or re-emergent pathogens (Madsen and Darling, 2006). Towards this end, the current authors have added a miscellaneous grouping of potential threat agents to the extant CDC categories (Table 12.3).

By definition, bioterrorism is insidious; absent of advance warning or specific intelligence information, clinical illness will be manifest before the circumstances of a release event are known. For this reason, health-care providers are likely to be the first responders to this form of terrorism, as symptomatic individuals present for medical attention. This contrasts with the more familiar scenarios in which police, firefighters, paramedics, and other emergency services personnel – traditional first responders – are deployed to the scene of a conventional attack or natural disaster. Physicians and other health-care workers must therefore maintain a high index of suspicion of bioterrorism and recognize suggestive epidemiologic clues and clinical features in order to enhance early recognition, optimize the initial management of casualties, and minimize the amplifying effect on the population (Artenstein *et al.*, 2002a).

Early recognition is hampered for multiple reasons. As discussed above, it is likely that the circumstances of any event will only be known in retrospect; therefore it may prove problematic immediately to discern the extent of exposure. Terrorists have an unlimited number of targets in most open, democratic

**Table 12.3 Agents of concern for use in bioterrorism**

| | |
|---|---|
| **Highest priority (Category A)** | |
| Microbe or toxin | Disease |
| *Bacillus anthracis* | Anthrax |
| Variola virus | Smallpox |
| *Yersinia pestis* | Plague |
| *Clostridium botulinum* | Botulism |
| *Fracisella tularensis* | Tularemia |
| Filoviruses | Ebola hemorrhagic fevers, Marburg disease |
| Arenaviruses | Lassa fever, South American hemorrhagic fevers |
| Bunyaviruses | Rift Valley fever, Congo-Crimean hemorrhagic fevers |
| **Moderately high priority (Category B)** | |
| *Coxiella burnetti* | Q fever |
| *Brucella* spp. | Brucellosis |
| *Burkholderia mallei* | Glanders |
| Alphaviruses | Viral encephalitides |
| Ricin | Ricin intoxication |
| *Staphylococcus aureas* enterotoxin B | Staphylococcal toxin illness |
| *Salmonella* spp., *Shigella dysenteriae, Escherichia coli* O157:H7, *Vibrio cholerae, Cryptosporidium parvum* | Food- and water-borne gastroenteritis |
| **Category C** | |
| Hantavirus | Viral hemorrhagic fevers |
| Flaviviruses | Yellow fever |
| *Mycobacterium tuberculosis* | Multi-drug resistant tuberculosis |
| **Miscellaneous** | |
| Genetically engineered vaccine- and/or antimicrobial-resistant category A or B agents | |
| HIV-1 | |
| Adenoviruses | |
| Influenza | |
| Rotaviruses | |
| Hybrid pathogens (e.g. smallpox–plague, smallpox–ebola) | |

Source: Artenstein (2003), reproduced with permission.

societies; it is unrealistic to expect that, without detailed intelligence data, all of these can be secured at all times. Government institutions, historic landmarks, or large social events may be predictable targets, but there are other, less predictable possibilities. US Department of State data reveal that businesses and other economic interests were the main targets of global terrorism during the period from 1996 to 2001 (US Department of State, 2002). Metropolitan areas are considered vulnerable, but, owing to the expansion of suburbs, commuters, and the clinical latency period between exposure and symptoms inherent with biologic agents, casualties of bioterrorism are likely to present for medical attention in diverse locations and at varying times after common exposures. A covert bioterrorism attack in New York City on a Wednesday morning may result in clinically ill persons presenting for medical attention over the ensuing weekend to a variety of emergency departments, urgent care centers, and physician offices within a 60-mile (~100-km) commuter radius. Additional cases may be seen hundreds or thousands of miles away at both national and international locations as infected, mobile individuals make use of modern modes of transportation during the clinical incubation period. This adds layers of complexity to an already complicated setting, and illustrates the critical importance of surveillance and real-time communication in this setting.

Further hindering the early recognition of bioterrorism is that the initial symptoms of many of the high priority agents may be non-diagnostic. In the absence of a known exposure, many symptomatic persons may either not seek medical attention early, or if they do they may be misdiagnosed as having a flu-like or other benign illness. Once beyond the early stages many illnesses related to bioterrorism progress rapidly, and treatment may be less successful. Because most of the diseases caused by agents of bioterrorism are rarely (if ever) seen in clinical practice, physicians are likely to be inexperienced with their clinical characteristics; physicians were only able to correctly diagnose diseases due to category A agents 47 percent of the time in one multicenter study (Cosgrove *et al.*, 2005). Additionally, these agents will by definition have been manipulated in a laboratory, and those affected may not present with the classic clinical features seen in naturally occurring infection. This was dramatically illustrated by some of the inhalational anthrax cases in the United States (Jernigan *et al.*, 2001).

Early recognition of bioterrorism is facilitated by the recognition of epidemiologic and clinical clues. Clustered presentations of patients with common symptoms and signs may suggest a common exposure source, and should prompt expeditious notification of local public health authorities. Aside from capturing the low-probability event of bioterrorism, this approach will also lead to enhanced recognition of outbreaks of naturally occurring disease, or those due to emerging pathogens. The recognition of a single case of a rare or non-endemic infection, in the absence of an appropriate travel history or other potential natural exposure, should raise the suspicion of bioterrorism and should prompt notification of public health authorities. Finally, unusual patterns of disease, such as an acute, fulminant febrile illness in an otherwise healthy young individual, or concurrent illness in human and animal populations, should raise suspicions of bioterrorism or another novel, emerging infection. Since multifocal attacks are expected, attention must be paid to effective, ongoing communication between public health jurisdictions to ensure that a single unusual case is not viewed in a vacuum, as it may not represent an isolated event. An effective response to bioterrorism requires coordination of the medical system at all levels, from the community physician to the tertiary care center, with public health, emergency management, and law enforcement contributions.

## Threat agents

CDC category agents are those biologic threat agents thought to be of major public health concern. Extensive coverage of these and other pathogens of concern in bioterrorism can be found elsewhere (Sidell *et al.*, 1997). Data concerning clinical incubation periods, transmission characteristics, and infection control procedures for selected agents of bioterrorism are provided in Table 12.4. Syndromic differential diagnoses for select clinical presentations are detailed in Table 12.5.

**Table 12.4 Infection control issues for selected agents of bioterrorism**

| Disease | Incubation period | Person-to-person transmission | Infection control practices |
|---|---|---|---|
| Inhalation of anthrax | 2–43* days | No | Standard |
| Botulism | 12–72 hours | No | Standard |
| Primary pneumonic plague | 1–6 days | Yes | Droplet |
| Smallpox | 7–17 days | Yes | Contact and air-borne |
| Tularemia | 1–14 days | No | Standard |
| Viral hemorrhagic fevers | 2–21 days | Yes | Contact and air-borne |
| Viral encephalitides | 1–14 days | No | Standard |
| Q fever | 2–14 days | No | Standard |
| Brucelloses | 5–60 days | No | Standard |
| Glanders | 10–14 days | No | Standard |

*Based on limited data from human outbreaks; experimental animal data support clinical latency periods of up to 100 days
Source: Artenstein (2003), reproduced with permission.

**Table 12.5 Syndromic differential diagnoses of selected bioterrorism agents**

| Clinical presentation | Disease | Differential diagnosis |
|---|---|---|
| Non-specific flu-like symptoms with nausea, emesis; cough with or without chest discomfort, without coryza or rhinorrhea, leading to abrupt onset of respiratory distress with or without shock; mental status changes, with chest radiograph abnormalities (wide mediastinum, infiltrates, pleural effusions) | Inhalational anthrax | Bacterial mediastinitis, tularemia, Q fever, psittacosis, Legionnaires' disease, influenza, *Pneumocystis carinii* pneumonia, ruptured aortic aneurysm, superior vena cava syndrome, histoplasmosis, coccidioidomycosis, sarcoidosis |
| Pruritic, painless papule, leading to vesicle(s), leading to ulcer, leading to edematous black eschar with or without massive local edema and regional adenopathy and fever, evolving over 3–7 days | Cutaneous anthrax | Recluse spider bite, plague, staphyloccal lesion, atypical Lyme disease, orf, glanders, tularemia, rat-bite fever, ecthyma gangrenosum, rickettsial pox, atypical mycobacteria, diphtheria |
| Rapidly progressive respiratory illness with cough, fever, rigors, dyspnea, chest pain, hemoptysis; possible gastrointestinal symptoms; lung consolidation with or without shock | Primary pneumonic plague | Severe community-acquired bacterial or viral pneumonia, inhalational anthrax, inhalational tularemia, pulmonary infarct, pulmonary hemorrhage |
| Sepsis, disseminated intravascular coagulation, purpura, acral gangrene | Septicemic plague | Meningococcemia, Gram-negative, streptococcal, pneumococcal or staphylococcal bacteremia with shock; overwhelming postsplenectomy sepsis, acute leukemia, Rocky Mountain spotted fever, hemorrhagic smallpox, hemorrhagic varicella (in immunocompromised patients) |

(*Continued*)

**Table 12.5 (Continued)**

| Clinical presentation | Disease | Differential diagnosis |
| --- | --- | --- |
| Fever, malaise, prostration, headache, myalgias followed by development of synchronous, progressive papular leading to vesicular and then pustular rash on face, mucous membranes (extremities more than the trunk); rash may become generalized, with a hemorrhagic component and system toxicity | Smallpox | Varicella, drug eruption, Stevens-Johnson syndrome, measles, secondary syphilis, erythema multiforme, severe acne, meningococcemia, monkeypox, generalized vaccinia, insect bites, Coxsackie virus infection, vaccine reaction |
| Non-specific flu-like illness with pleuropneumonitis; bronchiolitis with or without hilar lymphadenopathy; variable progression to respiratory failure | Inhalational tularemia | Inhalational anthrax, pneumonic plague, influenza, mycoplasma pneumonia, Legionnaires' disease, Q fever, bacterial pneumonia |
| Acute onset of afebrile, symmetric, descending flaccid paralysis that begins in bulbar muscles; dilated pupils, diplopia or blurred vision; dysphagia; dysarthria; ptosis; dry mucous membranes leading to airway obstruction with respiratory muscle paralysis; clear sensorium and absence of sensory changes | Botulism | Myasthenia gravis, brain-stem cerebrovascular accident, polio, Guillain-Barrè syndrome variant, tick paralysis, chemical intoxication |
| Acute-onset fevers, malaise, prostration, myalgias, headache, gastrointestinal symptoms, mucosal hemorrhage, altered vascular permeability, disseminated intravascular coagulation, hypotension leading to shock with or without hepatitis and neurologic findings | Viral hemorrhagic fever | Malaria, meningococcemia, leptospirosis, rickettsial infection, typhoid fever, borrelioses, fulminant hepatitis, hemorrhagic smallpox, acute leukemia, thrombotic thrombocytopenic purpura, hemolytic uremic syndrome, systemic lupus erythematosus |

Source: Artenstein (2003), reproduced with permission.

## Anthrax

Anthrax results from infection with *Bacillus anthracis*, a Gram-positive, spore-forming, rod-shaped organism that exists in its host as a vegetative bacillus and in the environment as a spore. In nature, anthrax is a zoonotic disease of herbivores that is ubiquitous in the soil of many geographic regions; sporadic human disease results from environmental or occupational contact with endospore-contaminated animal products (Dixon *et al.*, 1999). The cutaneous form of anthrax is the most common presentation of naturally-occurring disease; gastrointestinal and inhalational forms are exceedingly rare. Cutaneous anthrax occurred regularly in the first half of the twentieth century in association with contaminated hides and wools used in the garment industry, but it is uncommonly seen in current-day industrialized countries due to importation restrictions. The last known case of naturally occurring inhalational anthrax in the US occurred in 1976 (Suffin *et al.*, 1978).

It is ironic that, as American policies and regulations alleviated the risk of industrial outbreaks of anthrax, latter-day governmental policies may have shifted that risk to bioterror-related outbreaks. What was once an occupational disease of slaughterhouse workers, ranchers, and mill workers has now become an occupational hazard of politicians, journalists, postal workers, and the public at large (Witkowski and Parish, 2002). Prevailing wisdom had previously held that a large-scale bioterrorism attack with anthrax would employ aerosolized endospores and result in outbreaks of inhalational disease. The attacks in the US in 2001 illustrate the difficulties in predicting modes and outcomes in bioterrorism: the attacks were on a relatively small scale, and while endospores were used, the delivery method – envelopes – resulted in a significant proportion of cutaneous cases (Inglesby *et al.*, 2002). However, as with the Sverdlovsk outbreak in 1979, the serious morbidity and mortality in the US attacks were related to inhalational disease. Thus, it still seems warranted to plan for larger-scale events with aerosolized agents.

The clinical presentations and differential diagnoses of cutaneous and inhalational anthrax are described in Table 12.5. The lesion of cutaneous anthrax may be similar in appearance to other lesions, including cutaneous forms of other agents of bioterrorism such as tularemia or glanders; however, it may be distinguished by epidemiologic as well as certain clinical features. Unless secondarily infected, anthrax is traditionally a painless lesion and associated with significant local edema. The bite of *Loxosceles reclusa*, the brown recluse spider, shares many of the local and systemic features of anthrax, but is typically painful from the outset and lacks such significant edema (Freedman *et al.*, 2002). Cutaneous anthrax is associated with systemic disease and its attendant mortality in up to 20 percent of untreated cases, although with appropriate antimicrobial therapy mortality is less than 1 percent (Inglesby *et al.*, 1999).

Once the inhaled endospores reach the terminal alveoli of the lungs, generally requiring particle sizes of 1–5 μm, they are phagocytosed by macrophages and transported to regional lymph nodes, where they germinate into vegetative bacteria and subsequently disseminate hematogenously (Dixon *et al.*, 1999). Spores may remain latent for extended periods of time in the host, up to 100 days in experimental animal exposures (Henderson *et al.*, 1956). This correlates with the potential for prolonged clinical incubation periods after exposure to endospores; cases of inhalational anthrax occurred up to 43 days after exposure in the Sverdlovsk experience (Meselson *et al.*, 1994). The calculated incubation period based on the known dates of exposure in 6 of the 11 cases of inhalational anthrax from 2001 ranged from 4 to 6 days (Jernigan *et al.*, 2001), and from 1 to 10 days for the cutaneous cases (Bell *et al.*, 2002). Studies in non-human primates suggest the incubation period is influenced by exposure inoculum (Dixon *et al.*, 1999; Inglesby *et al.*, 2002).

Prior to the US anthrax attacks in October 2001, most of the clinical data concerning inhalational anthrax derived from Sverdlovsk – the largest previous outbreak recorded. Although there is much overlap among the clinical manifestations noted in both outbreaks, more detailed data are available from the recent US experience. Mailed letters containing anthrax spores in late September and early October of 2001 from a still-unidentified terrorist source(s) resulted in 22 cases of bioterrorism-associated anthrax (Lucey, 2005). Of these, 11 were of the cutaneous form; 11 were confirmed persons with inhalational anthrax, 5 (45 percent) of whom died. Although this contrasts with a case-fatality rate of greater than 85 percent reported from Sverdlovsk, the reliability of reported data from this outbreak is questionable (Inglesby *et al.*, 2002).

Patients almost uniformly presented an average of 3.3 days after symptom onset, with fevers, chills, malaise, myalgias, non-productive cough, chest discomfort, dyspnea, nausea or vomiting, tachycardia, peripheral neutrophilia, and liver enzyme elevations (Jernigan *et al.*, 2001; Barakat *et al.*, 2002). Many of these findings are non-diagnostic and overlap considerably with those of influenza and other common viral respiratory tract infections, rendering clinical diagnosis problematic in the absence of a known outbreak. Recently compiled data suggest that shortness of breath, mental status abnormalities, nausea, and vomiting are significantly more common in anthrax, whereas rhinorrhea and sore throat are uncommonly seen in anthrax, but noted in the majority of viral respiratory infections (CDC, 2001; Hupert *et al.*, 2003).

Other common clinical manifestations of inhalational anthrax, as informed by the 2001 outbreak, include abdominal pain, headache, mental status abnormalities, and hypoxemia. Abnormalities on chest radiography appear to be universally present, although these may only be identified retrospectively in some cases (Jernigan *et al.*, 2001). Pleural effusions appear to be the most common abnormality; infiltrates, consolidation, and/or mediastinal adenopathy/widening are also noted in the majority. The latter is thought to be an early indicator of disease, but computed tomography was more sensitive than chest radiography for this finding.

Clinical manifestations of inhalational anthrax generally evolve to a fulminant septic picture with progressive respiratory failure and shock. *B. anthracis* is routinely isolated in blood cultures if obtained before the initiation of antimicrobials.

Pleural fluid is typically hemorrhagic; bacteria can either be isolated in culture or documented by antigen-specific immunohistochemical stains of this material in the majority of patients (Jernigan *et al.*, 2001). The average time from hospitalization until death was three days (range 1–5 days) in the US series, consistent with other reports of the clinical virulence of this infection. Autopsy data typically reveal hemorrhagic mediastinal lymphadenitis and disseminated metastatic infection. Pathology data from the Sverdlovsk outbreak confirm meningeal involvement, typically hemorrhagic meningitis, in 50 percent of disseminated cases (Abramova *et al.*, 1993). Meningitis was the presenting manifestation in the index anthrax case in 2001 (Bush *et al.*, 2001).

The diagnosis of inhalational anthrax should be entertained in the setting of a consistent clinical presentation in the context of a known exposure, a possible exposure, or epidemiologic factors suggesting bioterrorism (e.g. clustered cases of a rapidly progressive illness). The diagnosis should also be considered in a single individual with a consistent or suggestive clinical illness in the absence of another etiology. The early recognition and treatment of inhalational anthrax is likely to be associated with a survival advantage (Jernigan *et al.*, 2001); however, patients appear to evolve rapidly to a late stage of infection in which survival appears unlikely (Lucey, 2005). Therefore, prompt empiric antimicrobial therapy should be initiated if infection is clinically suspected.

Combination parenteral therapy is appropriate in the ill person for a number of reasons – to cover the possibility of antimicrobial resistance, to target specific bacterial functions (e.g. the theoretical effect of clindamycin on toxin production), to ensure adequate drug penetration into the central nervous system, and perhaps to favorably affect survival (Jernigan *et al.*, 2001; Lucey, 2005). In the future, it is likely that novel therapies such as toxin inhibitors or receptor antagonists will be available, in combination with antimicrobials, to treat anthrax (Artenstein *et al.*, 2004; Opal *et al.*, 2005; see also Friedlander, 2001). Detailed therapeutic and postexposure prophylaxis recommendations for adults, children, and special groups have been recently reviewed elsewhere (Inglesby *et al.*, 2002; Lucey, 2005). Anthrax vaccine adsorbed has been proved to be effective in preventing cutaneous anthrax in human clinical trials, and in preventing inhalational disease after aerosol challenge in non-human primates (Friedlander *et al.*, 1999). The current vaccine has generally been found to be safe, but requires six doses over 18 months with the need for frequent boosting. Its availability is currently limited, although it is hoped that second-generation anthrax vaccines, currently in clinical trials, will prove effective.

## Smallpox

The last known naturally acquired case of smallpox occurred in Somalia in 1977; the disease was officially declared eradicated in 1980, the culmination of a 12-year intensive campaign undertaken by the WHO (Fenner *et al.*, 1988). At that time all laboratories involved in the global eradication effort were asked to voluntarily destroy or relocate their variola virus stocks to the CDC and the State Research Center of Virology and Biotechnology in Russia (Rotz *et al.*, 2005). However, because of concerns that variola virus stocks may have either been removed from, or sequestered outside of, their officially designated repositories, smallpox is today considered to be a potential bioterror threat. It is a cruel irony of the modern world that perhaps mankind's greatest triumph over nature – the eradication of smallpox – could be undone, volitionally and maliciously, by man.

Multiple features make smallpox an attractive biologic weapon and ensure that its reintroduction into human populations would be a global public health catastrophe: it is stable in aerosol form with a low infective dose; case fatality rates are historically high, approaching 30 percent; secondary attack rates among unvaccinated close contacts are 37 percent to 88 percent and are amplified; and much of the world's population is susceptible, as routine civilian vaccination was terminated more than three decades ago, vaccine-induced immunity wanes over time, and there is no virus circulating in the environment to provide low-level booster exposures (Breman and Henderson, 2002). Additionally, vaccine supplies are currently limited, although this problem has begun to be addressed, and there are currently no antiviral therapies of proven effectiveness against this pathogen.

After an incubation period of 7–17 days (average 10–12 days), the patient experiences the acute onset of a prostrating prodrome of fever, rigors, headache, and backache that may last 2–3 days. This is followed by a centrifugally distributed eruption that generalizes as it evolves through macular, papular, vesicular, and pustular stages in synchronous fashion over approximately eight days, with umbilication in the latter stages (Fenner *et al.*, 1988). Enanthema in the oropharynx typically precedes the exanthem by a day or two. The rash typically involves the palms and soles early in the course of the disease. The pustules begin crusting during the second week of the eruption; separation of scabs is usually complete by the end of the third week. The differential diagnosis of smallpox is delineated in Table 12.5. Historically, varicella and drug reactions have posed the most problematic differential diagnostic dilemmas (Breman and Henderson, 2002).

Smallpox is transmitted person-to-person by respiratory droplet nuclei and, less commonly, by contact with lesions or contaminated fomites. Historically, therefore, most transmission has resulted from prolonged face-to-face contact, such as within families or health-care settings. Air-borne transmission by fine-particle aerosols has, under certain conditions, been documented (Wehrle *et al.*, 1970). The virus is communicable from the onset of the enanthema until all of the scabs have separated, although transmissibility is thought to peak during the first week of the rash due to high titers of replicating virus in the oropharynx (Henderson *et al.*, 1999). Thus, hospitalized cases are placed in negative-pressure rooms with contact and air-borne precautions; cases that do not require hospital-level care should remain isolated at home to avoid infecting others.

The suspicion of a single smallpox case should prompt immediate notification of local public health authorities and infection-control specialists. Containment of smallpox is predicated on the "ring vaccination" strategy, which was successfully deployed in the WHO global eradication campaign and mandates the identification and vaccination of all directly exposed persons, including close contacts, health-care workers, and laboratory personnel. Vaccination, if deployed within four days of infection during the early incubation period, can significantly attenuate or prevent disease and may reduce secondary transmission (Henderson *et al.*, 1999). Because variola virus does not exist in nature, and legitimate stocks were confined to the two sites in the US and Russia, the occurrence of even a single case of smallpox outside of an accidental laboratory exposure would be tantamount to bioterrorism. An epidemiologic investigation would be necessary to ascertain the perimeter of the initial release, so that tracing of initially exposed persons could be accomplished.

## Botulism

Botulism, an acute neurologic disease resulting from intoxication with *Clostridium botulinum*, occurs sporadically and in focal outbreaks throughout the world, related to wound contamination by the bacterium or ingestion of food-borne toxin (Bleck, 2005). Aerosol forms of the toxin, while a rare mode of acquisition in nature, have been weaponized for use in bioterrorism (Zilinskas, 1997). Botulinum toxin is considered to be the most toxic molecule known; it is lethal to humans in minute quantities. It acts by blocking the release of the neurotransmitter acetylcholine from presynaptic vesicles, thereby inhibiting muscle contraction (Arnon *et al.*, 2001). Botulism therefore possesses a number of attributes of concern: it is lethal in small quantities; it has been successfully weaponized in the past; and its deployment by terrorists could paralyze a health-care system.

Botulism presents clinically as an acute, afebrile, symmetric, descending, flaccid paralysis. The disease manifests initially in the bulbar musculature, and is unassociated with mental status or sensory changes. Fatigue, dizziness, dysphagia, dysarthria, diplopia, dry mouth, dyspnea, ptosis, ophthalmoparesis, tongue weakness, and facial muscle paresis are early findings seen in more than 75 percent of cases (Arnon *et al.*, 2001). Progressive muscular involvement leading to respiratory failure may ensue. The clinical presentations of food-borne and inhalational botulism are indistinguishable in experimental animals (Arnon *et al.*, 2001).

The diagnosis of botulism is largely based on epidemiologic and clinical features and the exclusion of other possibilities (see Table 12.5). Clinicians should recognize that any single case of botulism could be the result of sporadic food-borne exposure, the sentinel case of a larger-scale "natural" outbreak, or a bioterrorism attack. A large number of epidemiologically unrelated, multifocal cases should be clues to an intentional release of the agent, either in food or water supplies or as an aerosol.

The mortality from food-borne botulism has declined from 60 percent to 6 percent over the last four decades, representing progress in supportive care and mechanical ventilation more than specific therapies (Arnon *et al.*, 2001). The prolonged need for ventilatory support would rapidly deplete the availability of limited resources, such as ventilators, in the event of a large-scale bioterrorism event involving botulism. Treatment with an equine antitoxin may ameliorate disease if given early, but this is available only in very limited supply from the CDC.

## Plague

Plague, the disease caused by the Gram-negative pathogen *Yersinia pestis*, presents in a variety of clinical forms in naturally acquired disease. Pandemic plague has significantly impacted world history; its impact may have been so great in the Middle Ages as to have led to genetic selection within Europeans, thus possibly affecting the course of future epidemic diseases such as HIV through changes in one of the viral co-receptors (Galvani and Slatkin, 2003). Plague is endemic in parts of Southeast Asia, Africa, and the western United States, with nearly all of the 13 annual US cases occurring in four states of the desert southwest (CDC, 1996).

The allure of plague as an agent of bioterrorism is related to a number of factors: it can be mass produced and disseminated as an aerosol, as successfully accomplished experimentally by both the US (Christopher *et al.*, 1997) and the Soviet (Alibek, 1999) bioweapons programs in the past; the pneumonic form of the disease is communicable from person-to-person and associated with a high mortality rate if untreated; drug-resistant mutants occur in nature (Galimand *et al.*, 1997); and an effective vaccine is not widely available. Perhaps the greatest appeal to terrorists is the stigma attached to plague, largely based on its historical track record of social and economic devastation. While the outbreak in Surat, India, in 1994 resulted in only 52 deaths, hundreds of thousands fled the city and mass chaos followed in its wake (Ramalingaswami, 2001).

Aerosolized preparations of the agent, the expected vehicle in bioterrorism, would be predicted to result in cases of primary pneumonic plague outside of usual endemic areas. As was the case with the anthrax attacks in 2001, however, additional forms of the disease, such as bubonic and septicemic plague, might also be expected to occur. Primary pneumonic plague classically presents as an acute, febrile, pneumonic illness with prominent respiratory and systemic symptoms; gastrointestinal symptoms, purulent sputum production or hemoptysis occur variably (Artenstein and Lucey, 2000). Chest roentgenogram typically shows patchy, bilateral, multilobar infiltrates or consolidations. In the absence of appropriate treatment, there may be rapid progression to respiratory failure, vascular collapse, purpuric skin lesions, necrotic digits, and death. The differential diagnosis, as noted in Table 12.5, is largely that of rapidly progressive pneumonia. The diagnosis may be suggested by the characteristic small Gram-negative coccobacillary forms in stained sputum specimens with bipolar uptake – the "safety pin" appearance – of Giemsa or Wright stain (Inglesby *et al.*, 2000). Culture confirmation is necessary to confirm the diagnosis; the microbiology laboratory should be notified in advance if plague is suspected, because special techniques and precautions must be employed to avoid inadvertent exposures.

Treatment recommendations for plague have been reviewed elsewhere (Inglesby *et al.*, 2000). Pneumonic plague can be transmitted from person-to-person by respiratory droplet nuclei, thus placing close contacts, other patients, and health-care workers at risk. Prompt recognition and treatment, appropriate deployment of postexposure prophylaxis, and early institution of droplet precautions will interrupt secondary transmission of plague.

## Tularemia

*Francisella tularensis*, the causative agent of tularemia, is another small Gram-negative coccobacillus that would likely cause a primary pneumonic presentation if delivered as an aerosol agent of bioterrorism. The agent is associated with a high attack rate due to its virulence: as few as 10 organisms can cause a pneumonic infection (Dennis *et al.*, 2001). Inhalational tularemia presents with the abrupt onset of a febrile, systemic illness with prominent upper respiratory symptoms, pleuritic chest pain, and the variable development of pneumonia, hilar adenopathy, and progression to respiratory failure and death in excess of 30 percent of those who do not receive appropriate therapy (Dennis *et al.*, 2001). The diagnosis is generally based on clinical features after other infectious etiologies are ruled out. Laboratory personnel should be notified in advance if tularemia is suspected, because the organism can be very infectious under culture conditions.

## Viral hemorrhagic fevers

The agents of viral hemorrhagic fevers (VHF) are members of four distinct families of ribonucleic acid viruses that cause clinical syndromes with overlapping features: fever, malaise, headache, myalgias, prostration, mucosal hemorrhage, and other signs of increased vascular permeability and circulatory dysregulation, leading to shock and multiorgan system failure in advanced cases (Borio *et al.*, 2002). Specific agents of VHF may also be associated with specific target organ effects. These pathogens include the agents of Ebola, Marburg, Lassa fever, Rift Valley fever, and Congo-Crimean hemorrhagic fever.

Hemorrhagic fever viruses have been viewed as emerging infections in nature due to their sporadic occurrence in focal outbreaks throughout the world; they are thought to be the results of human intrusion into a viral ecologic niche. They are, however, potential weapons of bioterrorism because they are highly infectious in aerosol form, are transmissible in health-care settings, cause high morbidity and mortality, and are purported to have been successfully weaponized (Alibek, 1999). Additionally, VHF frequently produce dramatic clinical pictures that have received worldwide attention, thus fulfilling another terrorist goal – to induce maximum fear and panic in the civilian population.

Blood and other body fluids from infected patients are extremely infectious, and person-to-person air-borne transmission may occur; therefore, strict contact and air-borne precautions should be instituted in these cases (Borio *et al.*, 2002). Treatment is largely supportive, and includes the early use of vasopressors as needed. Ribavirin is effective against some forms of VHF, but not those caused by Ebola and Marburg viruses. Nonetheless, this drug should be initiated empirically in patients presenting with a syndrome consistent with VHF until the etiology is confirmed.

## Genetically modified weapons

While modern-day terrorists may be constrained by the physics of aerosols, dispersion clouds and pulmonary alveolar dimensions, advances in molecular genetics and biotechnology have afforded them the possibility of manipulating the genetic composition of biologic organisms in order to enhance their threat potential. Theoretically, at least, such science could result in a Cold War-style "arms race" between bioterrorist states and biodefense organizations, since the molecular technology would serve dual purposes (Fraser, 2004).

The application of genomic science to bioterrorism may include the insertion of select genes for heightened infectivity, virulence, enhanced aerosol stability or antibiotic resistance into the agent's genome; it may also involve modification of the sequences recognized by detection devices or the host immune response (Petro *et al.*, 2003). One example is the concept of a multi-drug resistant anthrax strain created by the insertion of plasmids carrying multiple antibiotic-resistant genes. There is evidence that the Soviets had some success in developing such variants (Alibek, 1999); the STI-1 strain was engineered with plasmid-based resistance to penicillin, rifampicine, tetracycline, chloramphenicol, macrolides, and lincomycin (Stepanov *et al.*, 1996). This strain was purportedly developed as a live bacterial vaccine for prophylaxis and treatment purposes in a bioterrorism setting, thus illustrating a dramatic example of dual-use technology.

More ominous is the concept of genetic or genomic warfare, in which biological threat agents are tailored in the laboratory to attack populations of specific genetic backgrounds. The initial draft of the human genome has identified many essential genes that may be targets for future pharmaceuticals; conceivably, specific genomic segments may also be exploited as targets for custom-designed biological threat agents (Black, 2003). Potential weapons may include infectious agents, toxins, or small molecules targeted to subjects who display select genetic profiles. There is evidence that the Iraqi Government was working on weaponizing the camelpox virus prior to 1990 specifically for use as a possible "ethnic weapon," as it is most toxic to populations reared in areas without camels and therefore immunologically inexperienced with this organism (Zilinskas, 1997).

## Bioterrorism and the public health response: problem areas

The response to bioterrorism is unique among weapons of mass destruction, because it necessitates the consequence management that is common to all disasters as well as the application of basic principles of infectious disease: disease surveillance, infection control, antimicrobial therapy and prophylaxis, and vaccine prevention. For these reasons, and factors related to the epidemiology of bioterrorism (see above), physicians and other clinicians are the likely first responders to bioterrorism and are expected to be reliable sources of information for their patients, colleagues, and public health authorities (Artenstein *et al.*, 2002b).

There remain a number of potential pitfalls regarding bioterrorism that must be identified and managed to optimize the public health. As discussed above, emergencies involving conventional threats, natural disasters, or even chemical attacks, have immediate consequences; assessments of casualties can begin as can containment and mitigation strategies. In bioterrorism, the clinical latency period between exposure to an agent and the manifestation of signs and symptoms is in the order of days to weeks with most of the CDC category A, B, or C agents, other than pre-formed, pathogen-derived toxins. For this reason, early diagnoses of the first cases are likely to prove problematic; heightened clinical vigilance is required to recognize presentations of diseases that are rarely seen in clinical practice (Artenstein, 2003). The fear of the unknown, exacerbated by the "stealth" property of biologic attacks, may result in a panicked society and paralyzed economy.

Even after the initial victims have been diagnosed, communications among hospitals and other health-care institutions on local, regional, national, and global levels will be essential to define the epidemiology and possibly to identify exposure sources. Given the extent and ease of transit within our world, clinical presentations from a point-source, unifocal biologic attack could occur in widely disparate geographic locations. Additionally, it is likely that a terrorist attack would be multifocal, thus further confounding efforts to delineate extents and sources of exposure. A classic epidemiologic approach using case definitions, case identification, surveillance, and real-time communications is necessary whether the event is a malicious attack, emergent from nature, or of unknown source (Artenstein *et al.*, 2002b).

Other potential pitfalls reside in the arena of diagnostic techniques, treatment, and prevention of disease related to biologic agents. Although an active area of research, the development of field-ready, highly predictive, rapid screening tests for agents of bioterrorism has not, as yet, progressed to the point at which such assays are approved by the US Food and Drug Administration and available for deployment. Treatment and prevention issues – such as the absence of effective treatments for many forms of viral hemorrhagic fevers, shortages in the availability of multivalent anti-toxin for botulism, projected shortages in the availability of mechanical ventilators to manage a large-scale attack using botulism, lack of human data regarding the use of antiviral agents in smallpox, and the unfavorable toxicity profiles of currently available smallpox vaccines – remain unresolved but active areas of research. The fact that modern molecular biologic techniques have been used to produce genetically altered pathogens with "designer" phenotypes, such as antimicrobial or vaccine resistance, adds additional layers of complexity to an already complex problem. Finally, as has been vividly illustrated during the recent epidemic of severe acute respiratory syndrome (Svoboda *et al.*, 2004) and had been well recognized when epidemic smallpox occurred with regularity (Breman and Henderson, 2002), transmission of infection within hospitals is common. Health-care workers, our first line of defense against an attack using biologic agents, remain at significant occupational risk.

## Vaccines

Perhaps the most effective approach towards mitigation of the bioterrorist threat is the development of an effective, scaleable, technologically advanced vaccine platform that can not only respond to likely threat agents, but also has the flexibility to respond to novel and re-emergent pathogens. Vaccines will need to be designed for imminent threats and post-exposure settings; products targeted against the likeliest threats will need to be stockpiled for rapid, practical deployment. A brief review of the current state of vaccines for select Category A agents follows; smallpox vaccine is discussed on p. 342. More comprehensive reviews of biodefense vaccines can be found elsewhere (Cieslak *et al.*, 2000; Ales and Katial, 2004).

Anthrax vaccine adsorbed (AVA) is currently the only licensed anthrax vaccine in the US, and consists of a cell-free filtrate derived from a non-encapsulated, attenuated strain of *B. anthracis* developed in the 1950s (Friedlander *et al.*, 1999). It is licensed and beneficial for adults in both the pre- and post-exposure settings. It has a complicated dosing schedule and requires frequent boosting

(Nass, 1999). While generally safe, the vaccine is difficult to produce and is associated with significant local reactions. Recent work has focused on the development of a recombinant subunit vaccine targeted against the anthrax protective antigen (PA); antibodies to PA inhibit binding to its cellular receptor and correlate with protection against anthrax (Friedlander, 2001). Purified PA and DNA plasmids that express PA *in vivo* are in clinical trials as next-generation anthrax vaccines.

Although antibodies to PA address the initiation of anthrax infection, antibodies against additional virulence factors such as the capsule or somatic antigens in the spore may be needed to induce sterilizing immunity (Brey, 2005). DNA vaccines provide an attractive new platform, as they are thought to be relatively safe and easy to produce. A plasmid DNA vaccine encoding genetically detoxified PA and lethal factor, the latter a major component of anthrax lethal toxin, has been effective in protecting animals from aerosolized spore challenge, and is currently undergoing human clinical trials (Hermanson *et al.*, 2004).

*Francisella tularensis* represents an example of an intracellular pathogen that requires the induction of a wide range of immune responses, in particular CD8+ T-cell activation, to achieve protective immunity. The existing vaccine, consisting of live attenuated strains of *F. tularensis*, has been used extensively in the former Soviet Union (Alibek, 1999). One strain, LVS, was produced by multiple passages of a fully virulent strain of *F. tularensis* subspecies *holarctica*, and was shown to protect against aerosol challenge in animal and human models of the disease (Saslaw *et al.*, 1961; Isherwood *et al.*, 2005). However, LVS vaccine licensure was recently revoked due to several problems: it affords incomplete protection against laboratory-acquired tularemia (Eigelsbach and Down, 1961; Saslaw, *et al.*, 1961; Burke, 1977); the genetic and immunological basis for its attenuation and immunogenicity remains unknown (Oyston *et al.*, 2004); and it provides suboptimal protection against aerosol challenge in animal and human studies (Eigelsbach *et al.*, 1961; Hornick and Eigelsbach, 1966). To date, an incomplete understanding of correlates of protection in tularemia has hindered development of an effective subunit vaccine. Completion of the genetic sequence for the infective strain *F. tularensis* SCHU S4 and the vaccine strain LVS may further the discovery of proteins which are likely to induce protective immunity (Larsson *et al.*, 2005; Oyston and Quarry, 2005; Twine *et al.*, 2005).

Whereas anthrax and tularemia efforts demonstrate models for bacterial vaccines, *Clostridium botulinum* is an example of a vaccine effort directed against an important toxin agent of bioterrorism. Antitoxin remains a scarce resource, and is only useful in certain clinical settings; vaccines are an important approach to mass prophylaxis against this toxin (Artenstein, 2003). Botulinum toxin is expressed by *C. botulinum* in seven structural forms designated toxins A–G (Arnon *et al.*, 2001). There is currently a pseudo-licensed US vaccine against serotypes A–E, developed in the early 1970s (Byrne and Smith, 2000). It consists of formalin-deactivated purified toxins (toxoids) combined to form a pentavalent vaccine. An individual monovalent vaccine against serotype F has been developed in the United Kingdom (Hatheway, 1976). For purposes of mass production, DNA-based vaccines are under development. The carboxyl half of botulinum toxin appears to be the best vaccine candidate, and several vaccine expression systems in yeast and viral vectors are currently being tested in animal models (Lee *et al.*, 2005; Middlebrook, 2005).

## Bioterrorism in special populations

The public health approach to bioterrorism must be broadened to include special populations, including not only children, pregnant women, and immunocompromised persons, but also other at-risk vulnerable populations – disabled persons, non-English speakers, the homeless, substance abusers, mentally ill persons, and those that are geographically or culturally isolated. A general approach to the management of biothreat infections requires an assessment of the risk of certain drugs or products in select populations versus the potential risk of the infection in question, accounting for extent of exposure and the agent involved. While specific recommendations for treatment and prophylaxis have been recently reviewed (Inglesby *et al.*, 1999, 2000; Dennis *et al.*, 2001), these only address the biological part of the issue. A larger and more complex problem is ensuring that risk communication regarding bioterrorism and other emergencies is appropriately formulated and delivered in a fashion that is accessible and understandable by at-risk populations in our communities, and that these people have access to the public health system in a manner that optimizes health and minimizes the transmission of contagion (McGough *et al.*, 2005).

## Psychosocial issues

An often overlooked but vitally important issue in bioterrorism is that of psychosocial sequelae. These often take the form of acute anxiety reactions and exacerbations of chronic psychiatric illness during the stress of the event, or posttraumatic stress disorder (PTSD) in its aftermath. Nearly half of the emergency department visits during the Gulf War missile attacks on Israel in 1991 were related to acute psychological illness or exacerbations of underlying psychopathology (Karsenty *et al.*, 1991). Data from recent acts of terrorism in the US suggest that PTSD may develop in as many as 35 percent of those affected by the events (Yehuda, 2002). In the early period after the 11 September 2001 attacks in New York, PTSD and depression were nearly twice as prevalent as in historical control subjects (Galea *et al.*, 2002). Although close proximity to the events and personal loss were directly correlated with PTSD and depression, respectively, there was a substantial burden of morbidity among those only indirectly involved. The psychological impact of these events and of persistent international concern over terrorism can be expected to be significant and sustained for society as a whole.

## Public health response as informed by recent events: anthrax attacks, 2001

During October and November of 2001, beginning just weeks after the events of 11 September, the US experienced a series of biological attacks using weaponized anthrax spores deployed in mailed letters. In total there were 22 confirmed or suspected cases, 11 of inhalational anthrax and 11 of the cutaneous form (Jernigan *et al.*, 2001); five of the inhalational cases were fatal. Although these attacks were small scale and employed a low-technology approach to anthrax delivery, their impact was substantial: two branches of the Federal Government were temporarily closed, postal operations were severely disrupted, thousands of potentially exposed persons received post-exposure prophylaxis, total mitigation costs approached US$3 billion, and scarce public health resources were diverted away from other concerns to manage the inordinate volume of false alarms that accompanied the actual exposures (Heyman, 2002). Not unexpectedly, a host of after-action reports and analyses have subsequently reviewed the salient features of the response to these significant acts of bioterrorism (Gursky *et al.*, 2003; Lucey, 2005).

A number of important lessons, at all levels, were learned from the response to the anthrax attacks of 2001. First and foremost, they exposed the numerous deficiencies in the national and local public health infrastructures, including laboratory and diagnostic capabilities. Second, the events revealed significant knowledge gaps in the scientific community regarding biological threat agents – for instance, the finding of secondary spore aerosolization with experimental routine office activities in the US Hart Senate Office Building during the decontamination phase suggests an additional risk from anthrax weapons (Weis *et al.*, 2002). Third, the attacks caused the public health community to question previously held assumptions regarding bioterrorism; the idea that such substantial social and economic disruption could result from such a small event represents a new potential paradigm for terrorists and planners alike (Artenstein, 2003).

Perhaps the most durable lesson resulting from the anthrax attacks revolves around the difficult yet important issue of communication. Local and federal authorities struggled with imparting appropriate crisis and risk communication, and at times were viewed as giving contradictory messages to the media and the public (Gursky *et al.*, 2003). This uncertainty, along with a rapidly evolving situation on the ground and heightened public anxieties (no doubt magnified in the immediate post-9/11 period), led to inconsistent statements and actions by public health authorities, which exacerbated the public's lack of confidence in its leaders. One such example was the recommended use of ciprofloxacin as post-exposure prophylaxis for the Senate Office workers, while the mostly African-American postal workers were given doxycycline. As both drugs are effective, a consistent message and recommendation would have gone a long way towards assuaging public concern. The overarching theme highlighted by the anthrax attacks of 2001 is that our public health response planning must be proactive, not reactive to future events.

## Smallpox vaccination program, 2003

The terrorist attacks on the World Trade Center and the Pentagon on 9/11, and the anthrax attacks that followed shortly thereafter, served to focus attention on the threat of bioterror. Smallpox, for reasons delineated above, is widely considered to be a high priority threat agent, and one for which an effective vaccine exists, although not in sufficient quantity for broad application. In late 2002 the Federal Government authorized the implementation of a smallpox vaccination program for over half-a-million operational military personnel, and a second program, presided over by the states, to vaccinate civilian "smallpox response teams" comprising health-care workers and other emergency response personnel (Faden *et al.*, 2003).

Routine use of smallpox vaccine ceased in 1972, at which point it was determined that the potential risk of vaccine-associated adverse events significantly outweighed the risk of smallpox (Fenner *et al.*, 1988). Because smallpox vaccine may be associated with a number of potentially life-threatening toxicities, and at the time the programs were implemented there was no clearly definable risk of a smallpox exposure, the decision to proceed with pre-event vaccinations provoked a vocal national debate, raising a number of statistical, scientific, political, legal, and ethical issues that informed public health response planning in general.

Many of the issues were controversial and complex: whether vaccination of health-care workers is preferable to mass vaccination of the public (Kaplan *et al.*, 2002; Bozzette *et al.*, 2003); and the risk-to-benefit ratio for health-care workers, and how to compensate them in the event of vaccine-related injury (Faden *et al.*, 2003). While these and other issues sparked useful discussion, the most significant development related to the smallpox vaccination program was probably the recognition of hitherto unrecognized toxicity information concerning the vaccine itself.

Smallpox vaccine has a well-described toxicity profile that includes such serious but predictable complications as post-vaccinial encephalitis, a rare, potentially fatal neurologic syndrome generally seen in young children; progressive vaccinia, frequently fatal and seen in immunocompromised hosts; generalized vaccinia; eczema vaccinatum, dissemination of the vaccine virus seen in hosts with eczema or atopic disease; and contact transmission of vaccinia to an unvaccinated host (Artenstein, 2004b). Historically, 1 per million vaccinees developed a fatal complication (Lane *et al.*, 1970). The vaccination program was designed to screen participants carefully in order to minimize the potential for serious sequelae (CDC, 2003); in general it was successful in this regard, with over 730,000 military personnel and 40,000 civilian volunteers vaccinated, and very low rates of predictable, serious adverse events noted (Grabenstein and Winkenwerder, 2003; Poland *et al.*, 2005).

Because of the questions concerning risk outweighing benefit in the pre-event setting, the vaccine program was subjected to more rigorous scrutiny than historical smallpox mass vaccination. As may be seen with such large, well-studied datasets, novel and unanticipated effects were observed. Ischemic cardiac events occurred in 24 military and 9 civilian vaccinees; in retrospect, most had underlying coronary artery disease, and the incidence of ischemia did not exceed the level expected in an age-matched, unvaccinated population (Poland *et al.*, 2005). On the other hand, 86 cases of myopericarditis in military vaccinees and 22 cases among civilians were recognized, leading an expert panel to conclude that smallpox vaccination is casually related to myopericarditis and increases the risk of this complication (Poland *et al.*, 2005). Thus, data gleaned from preparations for a potential bioterrorist attack using smallpox will inform future vaccine efforts for smallpox (Artenstein *et al.*, 2005) as well as other areas of biodefense.

## Pandemic influenza

The public health response to the potential for pandemic influenza, although not a bioterror threat *per se*, represents an opportunity to implement some of the lessons learned from other, recent experiences in biodefense. While it is not clear that the cause of the current avian influenza epidemic, H5N1, will be the next pandemic strain (Bartlett and Hayden, 2005), the possibility of a future influenza pandemic appears to be all but certain (Mermel, 2005). H5N1 has met two of three criteria for a pandemic: it represents a novel subtype of influenza to which the population is immunologically naïve, and it is capable of infecting humans (albeit in limited fashion to date) and causing potentially lethal disease (WHO, 2005). The remaining pandemic hurdle for the virus is the ability for efficient human-to-human transmission (Fauci, 2006).

Given this, and the state of scientific knowledge that is currently available, we are in a much more favorable circumstance than our predecessors were at the time of the 1918 influenza pandemic. We are in a position to couple the recent public health lessons related to bioterror threats with our expanding databases of genetics, molecular biology, and biotechnology, and apply all of this knowledge to the immense challenge posed by the threat of pandemic influenza.

## Challenges to global public health

The threat of bioterrorism will likely persist and continue to present challenges to global public health. Adding to the concern is the possibility that advances in biomedical research may be used for malicious purposes – a possibility that has recently resulted in the creation of the National Science Advisory Board for Biosecurity by the US Department of Health and Human Services, to counsel government agencies regarding the dissemination of results from "controversial experiments" (Steinbrook, 2005). While the overall risk of bioterrorism is probably low from a practical standpoint, the consequences are potentially quite high; thus it is essential that we continue to develop countermeasures and response plans. There is otherwise a tendency to move on in our thinking to "the next big thing" and to leave these threats incompletely addressed. This concept of "bioterrorism fatigue" can be quantified (Figure 12.1).

Bioterrorism represents the ecological niche that lies at the confluence of global geopolitics, sociology, biology, public health, and medicine. So too do

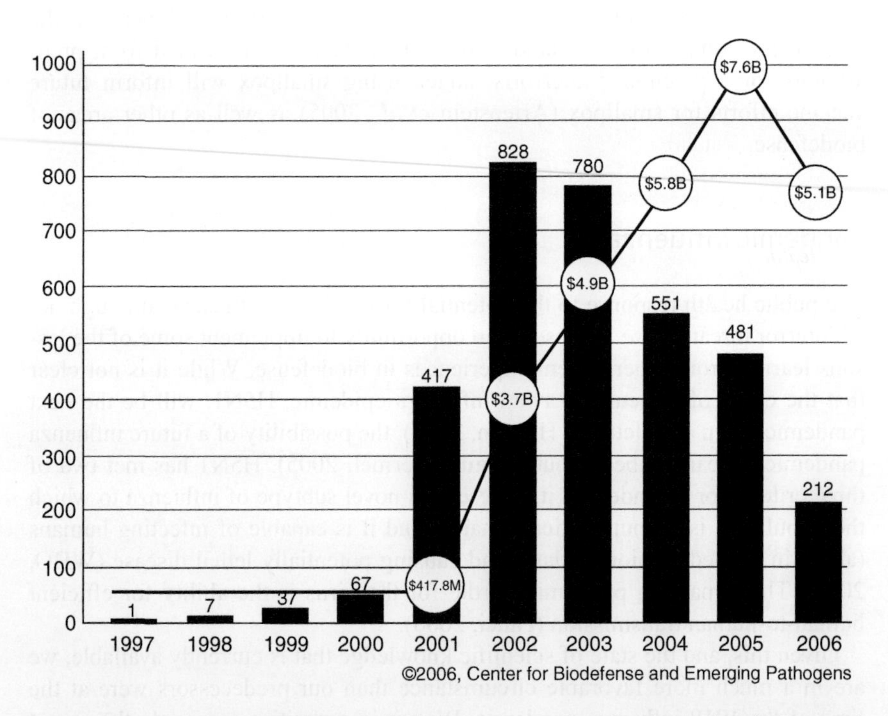


**Figure 12.1** "Bioterrorism fatigue": declining volume of publications using the keyword "bioterrorism" as referenced in the PubMed database from 1997 through 2006 (the latter year represents an annualized number) with superimposed federal bioterrorism funding from 2001 through 2006 (©2006, Center for Biodefense and Emerging Pathogens).

emerging infectious disease threats, such as pandemic influenza. Fortunately, the resources, human and economic, and technology that must be allocated to a cogent biodefense strategy are similar to those that are needed to combat naturally occurring disease threats (Artenstein, 2003; Relman, 2006). The duality of biodefense offers society the luxury of not having to choose between the two; it instead speaks to the need for a nimble and robust approach that can be adapted to changing circumstances.

## References


Abramova, F.A., Grinberg, L.M., Yampolskaya, O., *et al.* (1993). Pathology of inhalational anthrax in forty-two cases from the Sverdlovsk outbreak of 1979. *Proceedings of the National Academy of Sciences of the United States of America* **90**, 2291–2294.

Abu-Amr, Z. (1993). Hamas: A historical and political background. *Journal of Palestine Studies* **22**, 5–19.

Ales, N.C. and Katial, R.K. (2004). Vaccines against biologic agents: uses and developments. *Respiratory Care Clinics* **10**, 123–146.

Alibek, K. (1999). *Biohazard*. New York: Random House.

Arena, M.P. and Arrigo, B.A. (2005). Social psychology, terrorism, and identity: a preliminary re-examination of theory, culture, self, and society. *Behavioral Sciences and the Law* **23**, 485–506.

Arnon, S.S., Schechter, R., Inglesby, T.V. *et al.* (2001). Botulinum toxin as a biological weapon: medical and public health management. *Journal of the American Medical Association* **285**, 1059–1070.

Artenstein, A.W. (2003). Biodefense: medicine in the time of bioterrorism. *Medicine and Health/Rhode Island* **86**, 201–203.

Artenstein, A.W. (2004a). Bioterrorism and biodefense. In: J. Cohen, W.C. Powderly, S.F. Berkley *et al.* (eds), *Infectious Diseases*, 2nd edn. London: Mosby, pp. 99–107.

Artenstein, A.W. (2004b). Initial management of a suspected outbreak of smallpox. In: J. Cohen, W.C. Powderly, S.F. Berkley *et al.* (eds), *Infectious Diseases*, 2nd edn. London: Mosby, pp. 1022–1024.

Artenstein, A.W. (2006). Biologic attack. In: G.R. Ciottone, P.D. Anderson, auf der Heide, E. *et al.* (eds), *Disaster Medicine*. Philadelphia: Mosby, pp. 415–423.

Artenstein, A.W. and Lucey, D.R. (2000). Occupational plague. In: A.J. Couturier (ed.), *Occupational and Environmental Infectious Diseases*. Beverly: OEM Press, pp. 329–335.

Artenstein, A.W., Neill, M.A. and Opal, S.M. (2002a). Bioterrorism and physicians. *Annals of Internal Medicine* **137**, 626.

Artenstein, A.W., Neill, M.A. and Opal, S.M. (2002b). Bioterrorism and physicians. *Medicine and Health/Rhode Island* **85**, 74–77.

Artenstein, A.W., Opal, S.M., Cristofaro, P. *et al.* (2004). Chloroquine enhances survival in *Bacillus anthracis* intoxication. *Journal of Infectious Diseases* **190**, 1655–1660.

Artenstein, A.W., Johnson, C., Marbury, T.C. *et al.* (2005). A novel, cell culture-derived smallpox vaccine in vaccinia-naïve adults. *Vaccine* **23**, 2203–2209.

Barakat, L.A., Quentzel, H.L., Jernigan, J.A. *et al.* (2002). Fatal inhalational anthrax in a 94-year-old Connecticut woman. *Journal of the American Medical Association* **287**, 863–868.

Barlett, J.G. and Hayden, F.G. (2005) Influenza A (H5N1): will it be the next pandemic influenza? Are we ready? *Annals of Internal Medicine* **143**(6), 460–462.

Bell, D.M., Kozarsky, P.E. and Stephens, D.S. (2002). Conference summary: clinical issues in the prophylaxis, diagnosis, and treatment of anthrax. *Emerging Infectious Diseases* **8**, 222–223.

Bhalla, D.K. and Warheit, D.B. (2004). Biological agents with potential for misuse: a historical perspective and defensive measures. *Toxicology and Applied Pharmacology* **199**, 71–84.

Black, J.L. (2003). Genome projects and gene therapy: gateways to next generation biological weapons. *Military Medicine* **168**, 864–871.

Bleck, T.P. (2005). *Botulinum* toxin as a biological weapon. In: G.L. Mandell, J.E. Bennett and R. Dolin (eds.), *Principles and Practice of Infectious Diseases*, 6th edn. Philadelphia: Elsevier, pp. 3624–3625.

Borio, L., Inglesby, T., Peters, C.J., *et al.* (2002). Hemorrhagic fever viruses as biological weapons: medical and public health management. *Journal of the American Medical Association* **287**, 2391–2405.

Borum, R. and Gelles, M. (2005). Al-Qaeda's operational evolution: behavioral and organizational perspectives. *Behavioral Sciences and the Law* **23**, 467–483.

Bozzette, S.A., Boer, R., Bhatnagar, V. *et al.* (2003). A model for a smallpox-vaccination policy. *New England Journal of Medicine* **348**, 416–425.

Breman, J.G. and Henderson, D.A. (2002). Diagnosis and management of smallpox. *New England Journal of Medicine* **346**, 1300–1308.

Brey, R.N. (2005). Molecular basis for improved anthrax vaccines. *Advanced Drug Delivery Reviews* **57**(9), 1266–1292.

Burke, D.S. (1977). Immunization against tularemia: analysis of the effectiveness of live *Francisella tularensis* vaccine in prevention of laboratory acquired tularemia. *Journal of Infectious Diseases* **135**, 55–60.

Bush, L.M., Abrams, B.H., Beall, A. *et al.* (2001). Index case of fatal inhalational anthrax due to bioterrorism in the United States. *New England Journal of Medicine* **345**, 1607–1610.

Byrne, M.P. and Smith, L.A. (2000). Development of vaccines for prevention of botulism. *Biochemie* **82**, 955–966.

Centers for Disease Control and Prevention (1996). Prevention of plague: recommendations of the Advisory Committee on immunization practices (ACIP). *Mortality and Morbidity Weekly Report* **45**(RR-14), 1–15.

Centers for Disease Control and Prevention (2000). Biological and chemical terrorism: strategic plan for preparedness and response. *Mortality and Morbidity Weekly Report* **49**(RR-4), 1–14.

Centers for Disease Control and Prevention (2001). Considerations for distinguishing influenza-like illness from inhalational anthrax. *Mortality and Morbidity Weekly Report* **50**, 984–986.

Centers for Disease Control and Prevention (2003). Recommendations for using smallpox vaccine in a pre-event vaccination program. *Mortality and Morbidity Weekly Report* **52**(RR-7), 1–14.

Center for Nonproliferation Studies at the Monterey Institute of International Studies (2002). *Chemical and Biological Weapons: Possession and Programs Past and Present.* Available at: http://cns.miis.edu/research/cbw.possess.htm (accessed 3 April 2006).

Christopher, G.W., Cieslak, T.J., Pavlin, J.A. *et al.* (1997). Biological warfare: a historical perspective. *Journal of the American Medical Association* **278**, 412–417.

Cieslak, T.J., Christopher, G.W., Kortepeter, M.G. *et al.* (2000). Immunization against potential biological warfare agents. *Clinical Infectious Diseases* **30**, 843–850.

Cosgrove, S.E., Perl, T.M., Song, X. *et al.* (2005). Ability of physicians to diagnose and manage illness due to category A bioterrorism agents. *Archives of Internal Medicine* **165**, 2002–2006.

Davidson, L. (1998). *Islamic Fundamentalism*. Westport: Greenwood.

Dennis, D.T., Inglesby, T.V., Henderson, D.A. *et al.* (2001). Tularemia as a biological weapon: medical and public health management. *Journal of the American Medical Association* **285**, 2763–2773.

Dixon, T.C., Meselson, M., Guillemin, J. *et al.* (1999). Anthrax. *New England Journal of Medicine* **341**, 815–826.

Eigelsbach, H.T. and Down, C.M. (1961). Prophylactic effectiveness of live and killed tularemia vaccines. I. Production of vaccine and evaluation in the white mouse and guinea pig. *Journal of Immunology* **87**, 415–425.

Eigelsbach, H., Tulis, H., Overholt, E. *et al.* (1961). Aerogenic immunization of the monkey and guinea pig with live tularemia vaccine. *Proceedings of the Society of Experimental Biology and Medicine* **108**, 732–734.

Faden, R.R., Taylor, H.A. and Keirer, N.K. (2003). Consent and compensation: a social compact for smallpox vaccine policy in the event of an attack. *Clinical Infectious Diseases* **36**, 1547–1551.

Fauci, A.S. (2006). Emerging and re-emerging infectious diseases: influenza as a prototype of the host–pathogen balancing act. *Cell* **124**, 665–670.

Fenner, F., Henderson, D.A., Arita, I. *et al.* (1988). *Smallpox and its Eradication.* Geneva: World Health Organization.

Franz, D.R. and Zajtchuk, R. (2002). Biological terrorism: understanding the threat, preparation, and medical response. *Disease-a-Month* **48**(8), 490–564.

Fraser, C. (2004). A genomics-based approach to biodefence preparedness. *Nature Reviews* **5**, 23–33.

Freedman, A., Afonja, O., Chang, M.W. *et al.* (2002). Cutaneous anthrax associated with microangiopathic hemolytic anemia and coagulopathy in a 7-month-old infant. *Journal of the American Medical Association* **287**, 869–874.

Friedlander, A.M., Pittman, P.R. and Parker, G.W. (1999). Anthrax vaccine: evidence for safety and efficacy against inhalational anthrax. *Journal of the American Medical Association* **282**, 2104–2106.

Friedlander, A.M. (2001). Tackling anthrax. *Nature* **414**, 160–161.

Galea, S., Ahern, J., Resnick, H. *et al.* (2002). Psychological sequelae of the September 11 terrorist attacks in New York City. *New England Journal of Medicine* **346**, 982–987.

Galimand, M., Cuiyoule, A., Gerbaud, G. *et al.* (1997). Multidrug resistance in *Yersinia pestis* mediated by a transferable plasmid. *New England Journal of Medicine* **337**, 677–680.

Galvani, A.P. and Slatkin, M. (2003). Evaluating plague and smallpox as historical selective pressures for the CCR5-Delta 32 HIV-resistance allele. *Proceedings of the National Academy of Sciences* **100**(25), 15,276–15,279.

Gursky, E. Inglesby, T.V. and O'Toole, T. (2003). Anthrax 2001: observations on the medical and public health response. *Biosecurity and Bioterrorism: Biodefense Strategy, Practice, and Science* **1**(2), 97–110.

Grabenstein, J. and Winkenwerder, W. (2003). US military smallpox vaccination program experience. *Journal of the American Medical Association* **289**, 3278–3282.

Guillemin, J. (1999). *Anthrax: The Investgation of a Deadly Outbreak.* Berkeley: University of California Press.

Harris, S.H. (1994). *Factories of Death: Japanese Biological Warfare, 1932–45, and the American Cover-up*. New York: Routledge.

Hatheway, C.L. (1976). Toxoid of *Clostridium* outline: purification and immunization studies. *Applied Environmental Microbiology* **31**, 234–242.
Henderson, D.A., Inglesby, T.V., Bartlett, J.G. *et al.* (1999). Smallpox as a biological weapon: medical and public health management; working group on civilian biodefense. *Journal of the American Medical Association* **281**, 2127–2137.
Henderson, D.W., Peacock, S. and Belton, F.C. (1956). Observations on the prophylaxis of experimental pulmonary anthrax in the monkey. *Journal of Hygiene* **54**, 28–38.
Hermanson G., Whitlow V., Parker S. *et al.* (2004). A cationic lipid-formulated plasmid DNA vaccine confers sustained antibody-mediated protection against aerosolized anthrax spores. *Proceedings of the National Academy of Science* **101**, 13,601–13,606.
Heyman, D. (2002). *Lessons from the Anthrax Attacks*. Washington, DC: Center for Strategic and International Studies.
Hornick, R.B. and Eigelsbach, H.T. (1966). Aerogenic immunization of man with live tularemia vaccine. *Bacteriological Reviews* **30**, 532–538.
Hupert, N., Bearman, G., Mushlin, A. *et al.* (2003). Accuracy of screening for inhalational anthrax after a bioterrorist attack. *Annals of Internal Medicine* **139**, 337–345.
Inglesby, T.V., Henderson, D.A., Bartlett, J.G. *et al.* (1999). Anthrax as a biological weapon: medical and public health management. *Journal of the American Medical Association* **281**, 1735–1745.
Inglesby, T.V., Dennis, D.T., Henderson, D.A. *et al.* (2000). Plague as a biological weapon: medical and public health management. *Journal of the American Medical Association* **283**, 2281–2290.
Inglesby, T.V., O'Toole, T., Henderson, D.A. *et al.* (2002). Anthrax as a biological weapon, 2002: updated recommendations for management. *Journal of the American Medical Association* **287**, 2236–2252.
Isherwood, K.E., Titball, R.W., Davies, D.H. *et al.* (2005). Vaccination strategies for *Francisella tularensis*. *Advanced Drug Delivery Reviews* **57**, 1403–1414.
Jernigan, J., Stephens, D.S., Ashford, D.A. *et al.* (2001). Bioterrorism-related inhalational anthrax: the first 10 cases reported in the United States. *Emerging Infectious Diseases* **7**, 933–944.
Kaplan, E.H., Craft, D.L. and Wein, L.M. (2002). Emergency response to a smallpox attack: the case for mass vaccination. *Proceedings of the National Academy of Sciences* **99**, 10,935–10,940.
Karsenty, E., Shemer, J., Alshech, I. *et al.* (1991). Medical aspects of the Iraqi missile attacks on Israel. *Israeli Journal of Medical Science* **27**, 603–607.
Lane, J.M., Ruben, F.L., Neff, J.M. *et al.* (1970). Complications of smallpox vaccination, 1968: results of ten statewide surveys. *Journal of Infectious Diseases* **120**, 303–309.
Larsson, P., Oyston, P.C.F., Chain, P. *et al.* (2005) The complete genome sequence of *Francisella tularensis*, the causative agent of tularemia. *Nature Genetics* **37**, 153–159.
Lee, J.S., Hadjipanayis, A.G. and Parker, M.D. (2005). Viral vectors for use in the development of biodefense vaccines. *Advanced Drug Delivery Reviews* **57**, 1293–1314.
Lucey, D. (2005). Anthrax. In: G.L. Mandell, J.E. Bennett and R. Dolin (eds), *Principles and Practice of Infectious Diseases*, 6th edn. Philadelphia: Elsevier, pp. 3618–3624.
Madsen, J.M. and Darling, R.G. (2006). Future biologic and chemical weapons. In: G.R. Ciottone, P.D. Anderson, E. auf der Heide *et al.* (eds), *Disaster Medicine*. Philadelphia: Mosby, pp. 424–433.
McGough, M., Frank, L.L., Tipton, S. *et al.* (2005). Communicating the risks of bioterrorism and other emergencies in a diverse society: a case study of special populations in North Dakota. *Biosecurity and Bioterrorism: Biodefense Strategy, Practice, and Science* **3**(3), 235–245.

Mermel, L.A. (2005). Pandemic avian influenza. *Lancet Infectious Diseases* **5**(11), 666–667.
Meselson, M., Guillemin, J., Hugh-Jones, M. *et al.* (1994). The Sverdlovsk anthrax outbreak of 1979. *Science* **266**, 1202–1208.
Middlebrook, J.L. (2005). Production of vaccines against leading biowarfare toxins can utilize DNA scientific technology. *Advanced Drug Delivery Reviews* **57**, 1415–1423.
Miller, J., Engelberg, S. and Broad, W. (2001). *Germs: Biological Weapons and America's Secret War*. New York: Simon and Schuster.
Mitchell, R. (1993). *The Society of the Muslim Brothers*. New York: Oxford University Press.
Moussalli, A.S. (1998). Introduction to Islamic fundamentalism: realities, ideologies and international politics. In: A.S. Moussalli (ed.), *Islamic Fundamentalism: Myths and Realities*. Reading: Garnet, pp. 3–39.
Nass, M. (1999). Anthrax vaccine. Model of a response to the biological warfare threat. *Infectious Diseases Clinics of North America* **13**(1), 187–210.
Olson, K.B. (1999). Aum Shinrikyo: Once and future threat? *Emerging Infectious Diseases* **5**(4), 513–516.
Opal, S.M., Artenstein, A.W., Cristofaro, P.A. *et al.* (2005). Inter-alpha-inhibitor proteins are endogenous furin inhibitors and provide protection against experimental anthrax intoxication. *Infection and Immunity* **73**(8), 5101–5105.
Oyston, P.C.F. and Quarry, J.E. (2005). Tularemia vaccine: past, present and future. *Antonie van Leeuwenhoek* **87**, 277–281.
Oyston, P.C.F., Sjostedt, A. and Titbull, R.W. (2004). Tularemia: bioterrorism defense renews interest in *Francisella tularensis*. *Nature Reviews Microbiology* **2**, 967–978.
Petro, J.B, Plasse, T.R. and McNulty, J.A. (2003). Biotechnology: impact on biological warfare and biodefense. *Biosecurity and Bioterrorism: Biodefense Strategy, Practice, and Science* **1**(3), 161–168.
Plaut, W.G. (1981). *The Torah*. New York: Union of American Hebrew Congregations.
Poland, G.R., Grabenstein, J.D. and Neff, J.M. (2005). The US smallpox vaccination program: a review of a large modern era smallpox vaccination implementation program. *Vaccine* **23**, 2078–2081.
Post, J.M. (2005). The new face of terrorism: socio-cultural foundations of contemporary terrorism. *Behavioral Sciences and the Law* **23**, 451–465.
Post, J.M., Sprinzak, E. and Denny, L.M. (2003). The terrorists in their own words: interviews with 35 incarcerated Middle Eastern terrorists. *Terrorism and Political Violence* **15**, 171–184.
Ramalingaswami, V. (2001). Psychosocial effects of the 1994 plague outbreak in Saurat, India. *Military Medicine* **166**, 29–30.
Rappert, B. (2003). Biological weapons, genetics, and social analysis: emerging responses, emerging issues – II. *New Genetics and Society* **22**(3), 297–314.
Relman, D.A. (2006). Bioterrorism – preparing to fight the next war. *New England Journal of Medicine* **354**, 113–115.
Ropeik, D. and Gray, G. (2002). *Risk*. Boston: Houghton Mifflin Company.
Rotz, L.D., Cono, J. and Damon, I. (2005). Smallpox and bioterrorism. In: G.L. Mandell, J.E. Bennett and R. Dolin (eds), *Principles and Practice of Infectious Diseases*, 6th edn. Philadelphia: Elsevier, pp. 3612–3617.
Saslaw, S., Eigelsbach, H.T., Prior, J.A. *et al.* (1961). Tularemia vaccine study. II. Respiratory challenge. *Archives of Internal Medicine* **107**, 702–714.
Sidell, F.R., Takafuju, E.T. and Franz, D.R. (eds) (1997). *Textbook of Military Medicine Series. Part I, Warfare, Weaponry and the Casualty*. Washington, DC: Office of the Surgeon General, Department of the Army.

Steinbrook, R. (2005). Biomedical research and biosecurity. *New England Journal of Medicine* **353**, 2212–2214.
Stepanov, A.V., Marinin, L.I., Pomerantsev, A.P. *et al.* (1996). Development of novel vaccines against anthrax in man. *Journal of Biotechnology* **44**, 155–160.
Suffin, S.C., Carnes, W.H. and Kauffman, A.F. (1978). Inhalation anthrax in a home craftsman. *Human Pathology* **9**, 594–597.
Svoboda, T., Henry, B., Shulman, L. *et al.* (2004). Public health measures to control the spread of the severe acute respiratory syndrome during the outbreak in Toronto. *New England Journal of Medicine* **350**, 2352–2361.
Taubenberger, J.K., Reid, A.H., Lourens, R.M. *et al.* (2005). Characterization of the 1918 influenza virus polymerase genes. *Nature* **437**, 889–893.
Torok, T.J., Tauxe, R.V., Wise, R.P. *et al.* (1997). A large community outbreak of *Salmonellosis* caused by intentional contamination of restaurant salad bars. *Journal of the American Medical Association* **278**, 389–395.
Tumpey, T.M., Basler, C.F., Aguilar, P.V. *et al.* (2005). Characterization of the reconstructed 1918 Spanish influenza pandemic virus. *Science* **310**, 77–80.
Twine, S., Bystrom, M., Chen, W. *et al.* (2005). A mutant of *Francisella tularensis* Strain SCHU S4 lacking the ability to express a 58-kilodalton protein is attenuated for virulence and is an effective live vaccine. *Infection and Immunity* **73**, 8345–8352.
United States Department of State (2002). *Patterns of Global Terrorism 2001 Report.* Available at: http://www.state.gov/s/ct/rls/crt.2001/html/ (accessed 3 April 2006).
Wehrle, P.F., Posch, J., Richter, K.H. *et al.* (1970). An outbreak of smallpox in a German hospital and its significance with respect to other recent outbreaks in Europe. *Bulletin of the World Health Organization* **43**, 669–679.
Weis, C.P., Intrepido, A.J., Miller, A.K. *et al.* (2002). Secondary aerosolization of viable *Bacillus anthracis* spores in a contaminated US senate office. *Journal of the American Medical Association* **288**(22), 2853–2858.
Wheelis, M. (2002). Biological warfare at the 1346 siege of Caffa. *Emerging Infectious Diseases* **8**, 971–975.
Wiechman, D.J., Azarian, M. and Kendall, J.D. (1994) Islamic courts and corrections. *International Journal of Comparative and Applied Criminal Justice* **19**, 33–47.
Witkowski, J.A. and Parish, L.C. (2002). The story of anthrax from antiquity to the present: a biological weapon of nature and humans. *Clinics in Dermatology* **20**, 336–342.
World Health Organization (1970). *Health Aspects of Chemical and Biological Weapons: Report of a WHO Group of Consultants*. Geneva: World Health Organization, pp. 98–99.
World Health Organization (2005). Avian influenza a (H5N1) infection in humans. *New England Journal of Medicine* **353**(13), 1374–1385.
Yehuda, R. (2002). Post-traumatic stress disorder. *New England Journal of Medicine* **346**, 108–114.
Zilinskas, R.A. (1997). Iraq's biological weapons. The past as future? *Journal of the American Medical Association* **278**, 418–424.

# Infectious diseases associated with natural disasters

# 13

John G. Bartlett

Natural disasters include earthquakes, hurricanes, tornadoes, and floods. Some authorities also use the term "complex emergencies," which also includes emergencies that impact large populations through war, civil strife, famine, and other events leading to large population displacements with common humanitarian crises (Toole and Waldman, 1997; Connolly *et al.*, 2004). These are characterized by mass population movement with resettlement and crowding, often with limited infection control bringing risk of epidemics (see also Chapter 11). Some of the characteristic features of natural disasters are summarized in Table 13.1 (Sphere Health Services, 2007).

Factors that contribute to infectious disease risks with these complex emergencies include:

- massive population movement and temporary settlement
- breakdown of public health

**Table 13.1 Public health impact of some complex emergencies**

| Effect | Earthquakes | Winds | Floods |
|---|---|---|---|
| Deaths | Many | Few | Few |
| Injuries | Many | Moderate | Few |
| Infectious diseases | Few | Few | Variable |
| Food scarcity | Rare | Rare | Variable |
| Population displacements | Rare | Rare | Common |

Adapted from: Sphere-Health Services (www.sphereproject.org/content/view/114/84/lang,English).

- loss of health-care infrastructure
- poor sanitation
- food and water contamination
- loss of shelter
- crowding.

The result is that a common feature of these complex emergencies is outbreaks of infectious disease that may contribute substantially to the morbidity and mortality (Spiegel *et al.*, 2002; Salama *et al.*, 2004; Orellana, 2005; Waring and Brown, 2005; Wilder-Smith, 2005). The consequences of these events depend to a large extent on the type and location of the disaster, as well as the ability of the affected population to respond, and the skill and resources of those responsible for the response.

This chapter will deal with the infectious disease consequences of natural disasters, using the tsunami of 26 December 2004 as a prototype.

## The risk of post-natural disaster infectious diseases

On 26 December 2004 a tsunami struck South Asia, resulting in 230,000 deaths and displacement of 5 million persons (Ishii *et al.*, 2005; Lay *et al.*, 2005). This represents what has been regarded as the third greatest natural disaster in recorded history. The initial event caused death by trauma and drowning, but WHO emphasized the risk of subsequent infectious disease outbreaks that might even double the mortality (Moszynski, 2005).

The following is a summary of the infectious disease risks associated with natural disasters. It should be emphasized that natural disasters most commonly involve flooding, earthquakes, hurricanes, and tornadoes. A review of the published literature on epidemics associated with such disasters indicates that much of the morbidity and mortality (Salama *et al.*, 2004) is not a direct result of the natural disaster *per se*, but rather of the crowded conditions and disrupted services that result from the disaster (Wilder-Smith, 2005). Contributing factors to this association are listed in Table 13.2. This shows that most of the associated epidemics are diseases that characterize refugee settings.

### Vector-borne diseases

Water pools and disrupted vector control are the contributing factors to vector-borne diseases, and the most common consequences are diseases transmitted by *Aedes* mosquitoes in endemic areas. These include malaria, dengue, and scrub typhus.

Malaria is perhaps the greatest threat by precedent (Connolly *et al.*, 2004; Orellana, 2005; Wilder-Smith, 2005). An example is the malaria epidemic of

**Table 13.2 Most common infectious diseases in disaster camp settings**

| Condition | Comment |
|---|---|
| *Food-borne and water-borne*: primarily diarrheal disease, cholera, shigella, salmonella, hepatitis A, hepatitis E | Account for > 40% of deaths and > 80% children < 2 years |
| Contributing factors: polluted water and food. | |
| *Vector-borne*: region-specific – malaria, dengue, scrub typhus, Japanese B encephalitis | Contributing factors: inadequate vector control, crowding; flooded areas to promote mosquito breeding |
| Measles | Contributing factors: crowding, populations with low measles vaccination rates |
| Acute respiratory tract infections | Often most common cause of morbidity (Diaz and Achi, 1989) |
| Meningococcal meningitis | Most common in the "meningitis belt" of Africa |
| Tetanus | Contributing factors: wounding |

Adapted from Wilder-Smith (2005); Connelly *et al.*, 2004.

Burundi in 2001, which included 2.8 million cases in a country of 7 million (Connnolly *et al.*, 2004). This was a reflection of overcrowding, use of temporary shelters, inadequate access to health-care services, a lapse in vector control and high levels of chloroquine-resistant *P. falciparum*.

With regard to the tsunami, most of the affected areas are endemic for malaria, with an incidence prior to the disaster of about 1 per 1000 population (Briet *et al.*, 2005; Wilder-Smith, 2005). The malaria risk was not magnified by flooding from salt water, since this does not support the lifecycle of mosquitoes, but the salt water was turned brackish by monsoon rains in Indonesia and Sri Lanka (VanRooyen and Leaning, 2005). Dengue is also endemic in this area, particularly in Indonesia, where there was an ongoing epidemic prior to the disaster (Wilder-Smith, 2005).

## Food- and water-borne disease

Diarrheal diseases are probably the major cause of most epidemic disease associated with natural disasters (Hatch *et al.*, 1994; Goma Epidemiology Group, 1995; Toole and Waldman, 1997; Spiegel *et al.*, 2002; Connolly *et al.*, 2004;

Orellana, 2005; Waring and Brown, 2005; Wilder-Smith, 2005; Izadi *et al.*, 2006). In camps, these usually account for about 40 percent of deaths during the initial phase and over 80 percent of deaths for children less than two years of age (Connolly *et al.*, 2004). The contributing factors are primarily polluted water from fecal contamination, which may occur during transport and storage; the sharing of cooking utensils; a scarcity of hygiene products; and food contamination. An example is the largest recorded outbreak of diarrhea, which afflicted Rwandan refugees in Goma in 1994, when diarrhea was responsible for over 40,000 deaths; about 60 percent were due to cholera and 40 percent to shigellosis (Goma Epidemiology Group, 1995; see also Chapter 11). Contributing factors were that the most common antibiotic used for treatment, doxycycline, was inactive against the strain implicated, the rate of rehydration was too slow, and the health-care workers managing these cases simply lacked experience (Siddique *et al.*, 1995). Particularly important is fecal contamination of the water supply resulting in risks for cholera in endemic areas, but also typhoid fever, salmonellosis, shigellosis, hepatitis A, and hepatitis E.

## *Measles*

There is substantial risk of epidemics of measles in camps (Toole *et al.*, 1989; Toole, 1995; Connolly *et al.*, 2004; Wilder-Smith, 2005). The mortality rate in stable populations is about 1 percent, but may be as high as 33 percent in natural disasters (Toole, 1995). Transmission of the virus is promoted by crowding combined with poor national rates for vaccination. For the refugee camp in Sudan in 1985, this infection accounted for 53 percent of deaths (Connolly *et al.*, 2004). These rates decreased with subsequent disasters as a result of increased awareness and expanded immunization plans (Toole *et al.*, 1989; Toole, 1995). Nevertheless, there are many countries that still are far behind in national measles vaccination coverage. With natural disaster in such areas, immunization for measles should receive a high priority – in fact, this is probably second only in importance to the provision of adequate food (Toole *et al.*, 1989). Children aged six months to five years should be immunized on entry to a camp or settlement if possible. If supplies of vaccine are inadequate, the highest priority should be given to children who are considered undernourished, since they are the most vulnerable to lethal outcome.

## *Upper respiratory infections*

These are usually the most common cause of morbidity, but there is little information about the specific agents (Connolly *et al.*, 2004; Salama *et al.*, 2004; Wilder-Smith, 2005). The risk appears to reflect crowding and lack of hygiene. Serious consequences, including pneumonia and occasional mortality, are shown primarily in children. In 1993, 30 percent of deaths in children aged under five years in Kabul, Afghanistan, were attributed to URIs (Gessner, 1994). Some of these were presumably due to pneumonia, but, again, specific agents are usually not defined (Talley *et al.*, 2001).

### *Meningitis*

Large outbreaks of meningococcal meningitis have been reported, but nearly all reports have been limited to the "meningitis belt" of increased prevalence of meningitis that includes East, Southern, and Central Africa – Brundi, Rwanda and Tanzania (Santaniello-Newton and Hunter, 2000).

### *Tuberculosis*

Tuberculosis (TB) has caused outbreaks in displaced persons, presumably due to crowding or a frequently unrecognized case, especially in countries with limited diagnostic resources (Rutta *et al.*, 2001; Connolly *et al.*, 2004; Wilder-Smith, 2005). TB is endemic in most of the developing world and is particularly problematic in areas with high rates of HIV infection, where approximately half of those with tuberculosis have HIV and nearly half with TB have HIV infection. Also of concern is the MDR-TB and the more recently described extremely resistant form of tuberculosis (XDR-TB) (Gandhi, 2007), which presumably could lead to a devastating epidemic with crowding and undiagnosed cases. Nevertheless, TB has not been reported to be a major problem in most complex emergencies, and did not appear to be important in the tsunami case.

## Mental health

Traumatic events, such as tsunami, produce substantial psychological trauma to individuals and communities (Ursano *et al.*, 1995, 2006; Lopez-Ibor, 2006; CDC, 2007). This applies to a variety of events, including not only earthquakes and tsunami, but also hurricanes, tornadoes, pandemics, and bioterrorism. In general, there are several features in common when these sudden and often unexpected events afflict a community (CDC, 2007):

1. Everyone in the disaster area is affected
2. In general, people collaborate and cooperate during and after the disaster, but their effectiveness is usually diminished
3. There are concerns about mental health during the response and recovery
4. The stress and grief reactions noted in the disaster are regarded as normal responses to an abnormal situation
5. Mental health assistance is often more important than psychological support; the emphasis here is on activities such as distributing food, listening,

encouraging, reassuring and comforting, rather than intervention by mental health professionals
6. Disaster relief is often confusing, and may result in anger or a perception of helplessness in the response by governmental or non-profit agencies.

## Psychological implications of the 2004 tsunami

Estimates are highly variable and compounded by definitions. One WHO estimate was that 5–10 percent had serious psychological problems requiring professional assistance (UN, 2007). Another survey claimed that 40 percent of children had post-traumatic stress disorders. A different assessment was that mild or moderate mental disorders in the general population totaled about 10 percent, and this increased to 20 percent after a disaster. Severe mental problems such as psychosis and depression typically affect 2–3 percent of the population.

## Disaster psychiatry

The role of the psychiatrist generally changes in a disaster, at least in the initial response (Norwood, *et al.*, 2007). Most psychiatrists deal with treatment of persons with mental illness, but in a disaster the major initial effort is dealing with people who experience normal psychological and behavioral symptoms that would be expected after the disaster. The initial responses that indicate "normal" psychological and behavioral response include anger, sadness, fear, irritability, sleep disturbances, and increased use of alcohol, caffeine, and tobacco. There are some caveats that are important to remember in the psychiatric evaluation of patients with possible psychiatric complications from a disaster, including evaluation for the possibility of an authentic mental disorder due to head injury, exposure to toxins, illness such as a CNS infection, and dehydration. There also may be the loss of standard medications for the individual.

The major psychiatric diagnosis following a disaster is post-traumatic stress disorder (PTSD) (Neuner *et al.*, 2006; Thienkrua *et al.*, 2006). This is a result of a serious threat to life or injury to self or others that is followed by terror, helplessness or abnormal fear lasting over one month. It may also occur acutely and is then referred to "acute stress disorder" (ASD), described as symptoms similar to those of PTSV that occur within one month of the event and persist for at least two days with a maximum of four weeks. Other disaster-associated responses that are relatively common include adjustment disorders, substance abuse, major depression, generalized anxiety disorders, and complicated bereavement (van Griensven *et al.*, 2006).

The psychological impact of disasters on children may be quite different, and is age-associated. As with adults, the risk is greater in children with prior psychiatric difficulties, and the range of conditions include PTSD, depression, and separation anxiety. For pre-school children through grade 2, the common reactions are fear, confusion, sleep disturbance, separation anxiety, and regressive symptoms. Older children, aged 8–11 years, often have difficulty in concentrating, somatization, concerns for safety, and sleep disturbances. For adolescents the responses are more like those of adults, and are often characterized by increased risk-taking.

## Community response

Natural disasters are often associated with a sudden increase in the number of strangers in the community, including volunteers, representatives of the press, etc. There are also epidemics of rumors, and of course there is the confrontation with death and dying among families and friends. The community response is quite variable. Often there is anger that is usually directed toward accountability and the search for someone who is responsible, and also anger about inequities in the distribution of resources. It is important for psychiatrists to work with governmental agencies in developing a disaster psychiatric response plan. Some of the observations that should be included are as follows:

1. There should be minimal exposure to dead bodies or other images that are likely to be disturbing
2. Parents should be warned to limit children's exposure to television until news coverage of the event has passed
3. There should be respect for the need for family privacy, including limitation of exposure to media
4. The media should be viewed as an opportunity to educate the public and provide information that must be credible and helpful in terms of risk, resource utilization, and recommendations for necessities such as food, water, shelter, hygiene, health care, etc.

## Timelines

The phases of the disaster that are critical for planning, including psychiatric support, include the warning of threat, and the onset and tempo of its onset.

The warning of threat ranges from no advance notice (as in tsunami) to days to weeks for hurricanes. The actual onset and tempo is variable. The 2004 tsunami struck with enormous impact, resulting in 230,000 deaths in a few days with 1 million persons displaced, but three months later there were an additional 1300 deaths from one aftershock. The events of bioterrorism may linger over days and months, as they did with anthrax in the US in 2001.

## Recommendations for mental health intervention based on CDC guidance (CDC, 2007)

1. Potential risk groups that are regarded as particularly vulnerable include:

- selected age groups (infants, children and the elderly)
- cultural and ethnic groups (immigrants, those who do not speak English, undocumented aliens)
- "low visibility groups" (homeless, mobility-impaired, mentally challenged, mentally ill)
- persons in group facilities (hospitals, chronic care facilities, prisons)
- those who provide human services, including health-care and disaster relief work.

2. The needs of survivors and anticipated reactions include the following:

- concern for survival
- grief for personal losses, including loved ones and possessions
- fear and anxiety about personal safety and safety of loved ones
- sleep disturbances
- concerns about relocation and crowding
- the need to ventilate about events related to the disaster
- the need to be part of the recovery efforts.

3. Reactions that indicate the need for mental health referral include:

- disorientation (memory loss, disorientation in time or place, or inability to recall recent events)
- depression (including withdrawal, feeling of despair that is disproportionate to the event)
- anxiety (including obsessive fear of another disaster)
- mental illness (hallucinations, delusional thinking, etc.)
- inability to provide self-care (absence of eating, etc.)
- suicidal or homicidal thoughts
- abuse of alcohol or drugs
- domestic violence or child abuse.

4. Organizational approaches to reduce stress include the following:

- there must be effective management structure and leadership that provides a clear chain of command, and disaster orientation for all workers – shifts that are no longer than 12 hours per day, daily briefings with credibility, appropriate supplies, and communication tools such as cell phones, radios, etc.

- there must be a clearly defined goal and purpose
- roles must be defined
- there must be positive reinforcement with consideration for a buddy system for monitoring stress reactions
- stress management is necessary and should include assessment, rotation between low- and high-stress tasks, time breaks, education about signs and symptoms of stress, individual and clearly defined tours of duty.

## Lessons from refugee situations

A central feature of complex emergencies such as tsunami is large-scale displacement of people (Nieburg *et al.*, 2005). In many cases there is little prior notice for any planning. The international relief community has dealt with these issues for 30 years. The following are some of the lessons that are emphasized by Sphere, which represents over 100 humanitarian organizations that have responded to natural disasters and provided guidance on some of the priorities for the emergency phase of disaster relief.

1. *Public health issues must be a high priority.* This will include the need for medical skills, since some people will have injuries, many will have chronic illnesses (e.g. heart disease, chronic lung disease, seizure disorders, etc.), babies will be born, and some evacuees will need emergency mental health care. However, after the first 24–48 hours basic public health interventions are needed, and these include adequate quantities of food, safe water, general hygiene, and adequate shelter.
2. *There must be accurate information about refugees that will clarify the health status, outbreaks, and nutritional status of the population.* Particularly important will be surveillance of diseases, including many that are predicted on the basis of the type and location of the complex emergency. For example, in tsunami much of the region affected is endemic for malaria and dengue; all such displacements can anticipate outbreaks of diarrheal disease and respiratory infections, while measles is particularly common because it is so contagious with crowding in an inadequately vaccinated population; hepatitis E is endemic in the area; the extensive injuries brings the risk of tetanus. Also, in the 2004 tsunami, Indonesia was the country with the highest rate of Avian influenza, another concern. Thus, surveillance can be quite disease-targeted, but the need for personnel that are skilled in this process is emphasized.
3. *There must be appropriate attention paid to, and priority in managing, large groups of displaced persons.* This includes attention to survival necessities, such as food, water, sanitation, health care, etc. There is often the desire to provide services to large population aggregations because this seems most efficient, but experience has shown that smaller groupings are preferred.

Thus, although the large gatherings in the Superdome and the New Orleans Convention Center during Hurricane Katrina provided for efficient use of resources in the early phase of the disaster, it was not viewed as suitable past the early phase.

4. *There must be attention to the urgency of food and water requirements.* Calculated water needs are 15–20 liters/day per person, but this may be greater in hot weather. This means that for 20,000 people in the New Orleans Convention Center who were housed after Hurricane Katrina, the requirement would have been 300,000 liters (or 80,000 gallons) of water per day. Inadequate food and water is likely to precipitate human-made disasters.
5. *There must be adequate attention paid to informing the affected population.* It must be credible and useful.
6. *There must be awareness of the most vulnerable populations who are at greatest risk.* Many patients with serious chronic illnesses will require substantial medical care, and many will have separation from disease-dependent medicines.

## The tsunami of 26 December 2004

The Sumatra-Andaman earthquake was a great undersea earthquake that occurred at 07:58:53 (local time) on 26 December 2004, with an epicenter near the west coast of Sumatra, Indonesia (WHO, 2005a). The magnitude was graded 9.1–9.3 on the Richter Scale, and it was reported to be the longest earthquake on record, lasting between 500 and 600 seconds (8.3–10 minutes). This was large enough to cause the entire globe to vibrate approximately 1 centimeter. The original death toll was originally reported as over 275,000, but the updated UN figure is 229,866, including 186,983 dead and 42,883 missing. There were also approximately 2 million people displaced. The greatest toll was in Indonesia, particularly the Aceh Province (see Table 13.3).

**Table 13.3 Tsunami toll (four major countries)**

| Country | Deaths | Missing | Total | Displaced |
|---|---|---|---|---|
| Indonesia | 166,760 | 6222 | 172,982 | 452,845 |
| Sri Lanka | 30,920 | 6020 | 36,940 | 431,224 |
| India | 10,747 | 5553 | 16,300 | 647,556 |
| Thailand | 5373 | 3141 | 8514 | 8500 |

Source: WHO (2005b).

## Assessment of risks

Risk assessments for infectious diseases that represent important potential risks after tsunami are listed for five major affected countries in Table 13.4.

## Interventions

The following represents recommendations for response in natural disasters from the WHO Communicable Diseases Working Group on Emergencies (WHO, 2005a). The basic document was released just prior to the 2004 tsunami, and was then adapted after it. It represents the work plan that was largely followed in the response.

### *Emergency medical care*

The highest priority for the initial response is the provision of emergency medical and surgical care to persons who have been severely injured, and to provide psychosocial support to communities. Treatment follows standard protocols in health facilities, and uses the standard first-line drugs for the anticipated infectious disease complications including malaria, cholera, dysentery, typhoid fever,

**Table 13.4 Assessment of communicable disease threats with tsunami (26 December 2004)**

| | Indonesia | Sri Lanka | Thailand | India | Maldives |
|---|---|---|---|---|---|
| *Food and water* | | | | | |
| Cholera | + | + | + | + | + |
| Typhoid | + | + | + | + | + |
| Shigellosis | + | + | + | + | + |
| Hepatitis A | + | + | + | + | + |
| Hepatitis E | + | + | + | + | + |
| *Vector-borne* | | | | | |
| Malaria | + | + | + | + | – |
| Dengue | + | + | + | + | + |
| Scrub typhus | + | + | + | + | + |
| *Crowding* | | | | | |
| Measles | + | + | + | + | + |
| URI | + | + | + | + | + |
| Tuberculosis | + | + | + | + | + |

hepatitis, dengue, measles, meningitis, leptospirosis, and sexually transmitted infections. There is the concurrent emphasis on infection control guidelines.

## Water and sanitation

A high priority is given to safe drinking water to prevent the major risk of water-borne diseases. The major method to accomplish the goal is with chlorine, which is readily available, easily used, inexpensive, and highly active against all water-borne pathogens. It needs to be available for point-of-use in practical forms of free chlorine, as liquid sodium hypochlorite, sodium calcium hypochloride, and/or bleaching powder. The amount given should produce a concentration of active chlorine at 0.2–0.5 mg/l. This can be determined using a readily available test kit. The recommendation is for at least 20 liters of clean water per person per day. There also needs to be attention to adequate water containers, adequate cooking utensils, and clean latrines as additional methods to prevent water- and food-borne epidemics.

## Shelter and site planning

These need to be planned with sufficient space to prevent crowding; this is important for controlling outbreaks of measles, respiratory infections, diarrheal disease, and vector-borne diseases.

## Safe food

The importance of food in contributing to outbreaks of diarrheal and other forms of disease is often underrated. This needs to be included as a priority, along with safe water. Boiling will eliminate most microbial pathogens, but will not eliminate chemical risks. There needs to be targeted health education; this is discussed further below.

## Surveillance system

The recommendation is to establish the EWARN/Surveillance System, which focuses on communicable diseases of public health significance that are most likely in this setting. The reporting forms should be simple, with a standardized case definition. Data from laboratory resources need to be included. The system established should complement any existing surveillance structures. There needs to be the ability for real-time evaluations and prompt investigation of unusual events and rumors of outbreaks. This should be led by one agency that is identified as responsible for coordinating activities and linking all other agencies.

## *Immunization*

Health-care facilities should be used as appropriate sites for vaccinations. The following are considerations:

1. Vaccinations offered as part of national immunization programs should be made available to infants and other persons.
2. Measles immunization should be a high priority, along with supplementation with vitamin A. This should be provided for all children aged 6 months through 14 years, regardless of prior vaccination or disease history. The highest priority should be given to children aged 6 months to 4 years, and to those who are malnourished. It should also be noted that one case of suspected measles should prompt immediate implementation of measles control, and acceleration of the measles vaccination strategy.
3. Hepatitis A vaccine is not recommended for routine use, but may be used to control outbreaks.
4. Tetanus vaccination should be offered for persons who have open wounds, as routinely recommended.
5. Oral cholera vaccine should be given in the context of a cholera epidemic, but this must be accompanied by other strategies of disease prevention, including sanitation.
6. Typhoid vaccines are *not* recommended for routine use, but may be used to control typhoid outbreaks.

## *Vector control*

A high priority should be given to vector control in areas with malaria risks, and should include spraying of shelters with insecticide and/or the distribution of insecticide-treated mosquito nets. Other preventive measures are directed at reducing mosquito-breeding sites, by covering water-storage containers, eliminating pooled water, and efficient collecting and disposal of garbage.

## *Health education*

The following recommendations are part of the public health message to the public:

1. *Hygiene practice*: there should be at least 250 g of soap per person per month. Hands should be washed with soap, ashes, or lime before cooking and eating, and after latrine use. Hand-washing should be meticulous.
2. *Water use*: consumers should know that water can be heavily contaminated even if it looks clear. Drinking water may be made safe by boiling or adding drops of chlorine. Drinking water should be kept in a container with a small opening, holding no more than a 24-hour supply.

3. *Food safety*: food must be cooked until steaming hot and eaten immediately. Food may be considered safe if refrigerated or kept hot, but it must not be left at room temperature for more than two hours. Safe water (as defined above) must be used, vegetables cooked and fruit peeled – "Cook it, peel it or forget it."
4. *Water sources*: latrines must be separated from cooking areas and must be kept clean. Stagnant water should be removed and open wells covered. People must avoid washing themselves, their clothes, or cooking utensils with potentially contaminated water such as that from streams, rivers, or water holes.
5. *Early treatment*: with fever or diarrhea, it is important to get health care within 24 hours.

## Aceh Province, Indonesia

The Aceh Province of Indonesia was the area most severely affected by the tsunami of 26 December 2004 (CDC, 2005; Ministry of Health, Indonesia *et al.*, 2005). As of 22 March 2005, there were 126,602 known dead and 93,638 missing. In the public sector, 53 of 244 (23 percent) health facilities were destroyed or severely incapacitated; 42 of 481 (9 percent) health professionals died. This area was also the site of civil unrest for 39 years, which had some severe social and economic consequences.

At the time of the earthquake, Dr David Navarro, Head of the UN Crisis Operations, warned of a second wave of morbidity and mortality, based on anticipation of multiple harsh epidemics of communicable diseases. The major concerns were contaminated water supplies, overcrowded refugee camps, and destroyed sewer lines. It was also an area in which malaria and dengue were endemic, and the combination of the tsunami plus heavy rains that followed created great mosquito-breeding grounds.

### *The response*

For the acute phase of the emergency, the Aceh Provincial Health Office was augmented by staff from the Ministry of Health in Jakarta and with teams from the WHO. The latter included representatives from the Global Outbreak Alert and Response Network (GOARN) (Ministry of Health, Indonesia *et al.*, 2005). The major goal of the Provincial Health Office and GOARN was to establish an early warning and surveillance response network (EWARN) system for prompt detection of epidemic-prone diseases, to facilitate outbreak investigation, and to implement control measures.

The EWARN was rapidly established. The target population included displaced persons and residents. The major sources of information were health facilities and public health laboratories under the supervision of national and international governmental and non-governmental organizations. The system was developed for syndromic surveillance to identify acute diarrhea, bloody diarrhea, fever of unknown origin, dengue, jaundice, measles, meningitis, malaria, tetanus, and URIs. Data were collected on morbidity and mortality on a weekly basis for persons under or over the age of five years. The system was complemented by an alert system based on daily telephone calls, text messages, and e-mail reporting. Alerts were followed by verification and investigation. The results of these activities were presented and discussed at conferences held twice weekly, with the involvement of over 50 agencies.

## *Experience of this system*

The analysis of the initial, acute response with the system established in Aceh Province showed that, despite the need to establish emergency needs for over 500,000 displaced persons, there were no large outbreaks of communicable diseases (Ekdahl, 2005). There was a limited number of cases attributed to water-borne diseases, including cholera, shigellosis, typhoid fever, hepatitis A, and hepatitis E; some vector-borne diseases, including malaria and dengue; and conditions attributed to overcrowding, including URIs, influenza, and meningitis. Nevertheless, no major outbreaks were reported. The reasons for this are thought to be as follows:

1. Some have opined that communicable diseases are actually more uncommon following natural disasters than conventional wisdom would suggest.
2. The population served was well versed in the importance of hand-washing and of boiling drinking water prior to consumption. Also, this population was generally healthy, with relatively low levels of malnutrition and infant mortality.
3. The EWARN system was highly effective in rapid response to possible outbreaks.
4. There was an outbreak of measles which was somewhat expected due to low vaccine coverage prior to the event, but the outbreak was limited as a result of a rapid vaccination strategy.

Despite what appears to be a highly successful program, there were some concerns about the EWARN system, which included:

1. Inconsistent reporting by some agencies
2. Difficulty in obtaining denominator data due to the high mobility of displaced persons
3. Difficulty in reaching rural areas
4. Duplicate reporting due to health care at multiple sites.

Despite these problems, many concluded that the data generated by the EWARN system were actually more accurate than the health data obtained for the general population prior to the event, and the current plan is to integrate the EWARN system (including its data management component) within the routine surveillance used in this area prior to the tsunami. This would include monitoring for outbreaks of typhoid fever, hepatitis, cholera, measles, malaria, dengue, and shigellosis.

## *Epidemic-prone disease surveillance and results after the tsunami in Aceh, Indonesia*

The following report is based on that from the European Center for Disease Prevention and Control, Stockholm, Sweden (Ekdahl, 2005), and was prepared by the Ministry of Health, Indonesia, the World Health Organization assisted by the Global Outbreak Alert and Response Network Partners (GOARN), the Centers for Disease Control and Prevention, USA, and multiple other organizations.

> Aceh Province, Indonesia, has a population of 4.8 million persons, and represents the area most severely affected by the tsunami on 26 December 2004. This report is based on the system described above, which is designed to provide surveillance, including an early warning and response system (EWARN) to detect epidemic-prone diseases. This includes the capability to investigate outbreaks, confirm pathogens, identify modes of transmission, identify risk factors, and design outbreak management and control methods. The diseases of epidemic potential that were targeted included acute watery diarrhea, bloody diarrhea, dengue, fever of unknown origin (FUO), acute jaundice (as a marker of hepatitis A and E), measles, pyogenic meningitis (meningococcal meningitis), malaria, acute respiratory infections, and tetanus. Data were collected for morbidity and mortality, and were updated weekly.
>
> The data were separately reported for age-groups of less than five years, and for those of five years and older. The EWARN system was established to connect with health facilities (mobile clinics, field hospitals, permanent hospitals, etc.) and laboratories in affected districts. Reporting was augmented by daily telephone calls, text messages or e-mail reporting. Alerts were followed by verification, investigation, and response, as defined above, under the supervision of the Aceh Provincial Health Office and the WHO. The total was up to 123 reporting units from 10 districts.
>
> The height of the emergency phase was between weeks 4 and 10. The following report is from week 12, which included a cumulative total of 184,864 medical consultations. Of these, there were 40,706 consultations for epidemic-prone diseases, and a total of 11 deaths. The specific results for morbidity and mortality from the targeted epidemic-prone diseases are summarized in Table 13.5.
>
> Outbreak investigations following reported alerts were performed for bloody diarrhea (11), acute watery diarrhea (1), dengue (5), typhoid fever (3), acute jaundice (11), malaria (4), meningitis (4), encephalitis (1), measles (24), and scrub typhus (1). These investigations indicated some clusters of cases (tetanus, dengue, bloody diarrhea, typhoid fever, scrub typhus, hepatitis A, hepatitis E), "false alarms" (cholera, malaria, and encephalitis), and a single outbreak of measles.
>
> There was a cluster of tetanus cases with a total of 106 patients who were hospitalized during the month following the tsunami. The case–fatality ratio was 19 percent,

**Table 13.5 Morbidity and mortality due to infectious diseases for weeks 4–10, Aceh Province**

| Diseases | 0–4 years | | ⩾5 years | | Total | |
|---|---|---|---|---|---|---|
| | Morbidity | Mortality | Morbidity | Mortality | Morbidity | Mortality |
| Acute diarrhea | 3589 | 1 | 5894 | 0 | 9483 | 1 |
| Bloody diarrhea | 128 | 0 | 448 | 0 | 576 | 0 |
| Malaria | 77 | 0 | 562 | 1 | 639 | 1 |
| Fever > 38° (other) | 1533 | 2 | 3056 | 0 | 4589 | 2 |
| Measles | 70 | 0 | 75 | 0 | 145 | 0 |
| URI | 6599 | 3 | 18,613 | 3 | 25,212 | 6 |
| Acute jaundice | 4 | 0 | 45 | 0 | 49 | 0 |
| Meningitis | 1 | 0 | 12 | 1 | 13 | 1 |
| Total | 12,001 | 6 | 28,705 | 5 | 40,706 | 11 |

Reproduced from Ekdahl (2005), with permission.

with the peak onset of symptoms at 2–3 weeks after the tsunami. There was also an outbreak of 35 cases of measles in children aged 5 months–15 years, with the onset of rash in these cases occurring 2–4 weeks after the tsunami. The median age was 4 years. The response included a vaccination campaign targeting children aged 6 months–15 years, which was initiated in mid-January for all displaced persons in camps, and then subsequently extended to surrounding communities.

The analysis of this experience represents data for over 500,000 displaced persons who were housed in temporary shelters or moved to host families across the province. Despite substantial risk factors for epidemic-prone diseases, there were no large outbreaks. These results suggest several conclusions to explain the favorable outcome.

## Lessons from the tsunami

On 4–6 May 2005, the WHO sponsored a conference entitled "Health Aspects of the Tsunami Disaster in Asia" in Phuket, Thailand (WHO, 2005c). The conference involved 400 senior policy advisors and experts from Asia and other parts of the world, and included representatives from the UN, various governments, non-governmental organizations, and academic institutions, and representatives of countries and organizations that provided relief assistance after the tsunami.

The major objectives of the conference were to review the health-care experience, identify components that could have been done better, and tabulate lessons learned for future natural disasters. The following is a brief summary of the most important conclusions from this conference (www.who.int/hac/events/tsunamiconf/proceedings/en/print.html).

## Developing national capacity for disaster preparedness

1. There was a consensus that nations need to be better prepared for natural disasters, including having better capacity to address health-care needs. This should include funding to support national capacity for this activity.
2. The findings of the tsunami relief experience indicated that areas that had faced past disasters were better prepared. Another correlate with good preparedness was communities that had established disaster plans, had practiced drills, and responded rapidly. These also included national and international health agencies that had experience in this type of response. Rapid military deployment facilitated the response, especially in hard-to-reach areas.
3. Of concern, and possibly obvious, is the fact that the tsunami occurred in an area that had an inadequate infrastructure to respond well. This included the lack of systems for warning, response, and evacuation for health-care needs. Supplies and drugs on site were largely absent, obsolete, or in short supply. Health-care facilities were destroyed, although some could have survived had there been better standards of construction. There was undue concern for disposal of dead bodies under the misconception that these were epidemiologic hazards.
4. There was a strong impression that preparedness at the community and local level was particularly important. This planning must deal with displaced persons, surveillance systems, managing and preventing psychological problems, food shortages, water purification, and methods to prevent damage to health-care facilities.
5. There was consensus that there are some good models for cost-effective disaster preparedness that could be generally applied to virtually all areas. There was a need for dissemination of this information, which is generally available from the WHO/PAHO. This includes standards of care, and establishing partnerships between various organizations. There needs to be attention to methods to address the specific needs of women, children, the elderly, the disabled, and other vulnerable groups.

## Post-disaster needs assessment

1. The tsunami experience was considered uncoordinated due in part to competing and overlapping assignments by different agencies. The result was a badly traumatized population being subjected to repeated questioning by different groups.

2. Reliable and relevant information are critical to crisis management. These data must be collected, collated and analyzed in real time.
3. Baseline information was critical in identifying needs. For example, some communities had public health surveillance systems well in place, while others did not. The pre-event data, therefore, needed to be local. The WHO is currently developing the standard health assessment tools that will be used by the Red Cross, the Red Crescent movement, UN agencies, non-governmental organizations, and other groups that participate in the response.

## Neglected issues in health care

1. *Emergency rapid needs assessment.* This should focus on basic health needs, including needs assessment for food, water, sanitation, and shelter. There must be attention to an equitable distribution of these resources, and special attention to the vulnerable populations as defined above.
2. *Mental health.* Emphasis was made on efforts to "normalize the life of individuals, families and communities" in terms of livelihoods, schooling, and housing. Psychological support should not be restricted to medical practitioners, since only a minority of those affected actually requires this level of professional help. It was recommended that the WHO develop guidelines for the use of psychotropic medications in the setting of natural disasters.
3. *Management of mass casualties.* The areas affected by the tsunami appeared to be unable to manage large numbers of casualties due to the lack of standardized triage, of a pre-established network of health-care facilities for sharing the burden, and of appropriate plans for allocation based on expertise. The immediate health-care assistance to those who were injured was usually provided by someone else who was injured, indicating the need for public education in first aid techniques. Particularly important was the special role played by the National Red Cross and Red Crescent societies, and the collaborative interactions of these groups with local health authorities and the WHO.
4. *Forensic issues.* As noted, there were excessive human resources devoted to handling of dead bodies, under the erroneous impression that these generate disease. The appeal was to avoid rushing to bury or cremate rapidly, since this detracts from the requirements of the living and the need to respect cultural factors such as grieving, proper burial, etc. However, there was also emphasis on the need for a rapid method for body storage facilities.
5. *Benchmarks, standards and ethics.* Ethical standards have been established by the Sphere Project (WHO), which has provided standards for health care, including a document on "Standardized Monitoring and Assessment of Relief and Transitions Initiative." Nevertheless, there was an impression that, although standards are available, the tsunami experience highlighted the gap between what is desirable and what happens in actual practice. The greatest

concern was that different organizations used different measurements, methods, and interpretations, so that it was difficult to compare results, establish priorities or achieve consensus. This emphasizes the need for standardized assessment and reporting, and the request was that the WHO take leadership in such a guidance document. Another need was guidance in health-care standards, with a curriculum for training institutions.

### Coordinated disaster response

1. The local capacity for disaster management was overwhelmed by the magnitude and suddenness of the tsunami, and the subsequent global response was viewed as having been uncoordinated, congested, and competitive for scarce resources.
2. One of the major concerns was the administrative burden imposed on the health-care relief workers at the time that they were needed for response to the acute disaster.
3. It is acknowledged that coordination and supervision requires the consent and cooperation of all stakeholders, and the WHO, representing the UN, is generally seen by most governments as the authoritative source. It was also noted that many countries are now revamping their national and local coordination structures, based on the tsunami experience, with a disaster management center under the guidance of the WHO.

### Voluntary groups

1. Major responses and resources were provided by voluntary groups such as the Red Cross, Red Crescent, and non-governmental organizations. The major recommendation in reviewing the tsunami response was that the WHO should work more closely to ensure efficient and good coordination between groups.
2. It was also noted that many of the NGOs lack expertise and codes of conduct, and often have high turnover that can be disruptive. The emphasis here was on the need for clarity on the mutual roles, expectation, and accountability.

### Private sector

1. The tsunami experience showed an extraordinarily high level of national and international public sector involvement, with contributions in terms of professional skills, supplies, and funding. It was noted that these work best when there was a public–private partnership made before people or materials were sent. Best results were with pre-existing relationships or ongoing programs between private groups and UN agencies, the Red Cross, the Red Crescent movement, NGOs, or government agencies. There was emphasis on the need

for mutual investment, with an understanding of who and what is needed, and where and when, and under what management arrangements.

2. In the tsunami, private sector experts were successfully deployed, their roles were carefully defined, and support was provided by key representatives of local government and of the UN. Nevertheless, there was concern about some of the motives of the private sector, with emphasis on the need for neutrality and integrity in order to ensure trust, credibility, and efficiency.

## Donor funding

1. The tsunami was benefited by generous donations from multiple government sources. There was emphasis on the need to catalog reliable data on pledges, commitments, and disbursements.
2. It was noted that some of the contributions had earmarking that was not necessarily consistent with the needs. One of the requests was for a system that matches resources and needs.
3. A source of concern by donors was accountability, including inconsistent tracking and better reporting of how funds were used. Another criticism was the excessive use of resources for health care and not enough for public health interventions in peripheral areas.
4. There was concern that funding needed to be rapidly available and directed to high priorities, with sufficient flexibility to respond to needs. It was also noted that funding decisions should not promote already existing inequities between populations, special populations, and nations.

## Military cooperation

1. The military response was coordinated by a combined Support Force in Utapao, Thailand, with military from 30 nations plus a UN civil–military liaison.
2. It was emphasized that the interface between civil and military resources worked best if civilian authorities identified the priorities for the military.
3. The discussion of civil–military cooperation following the tsunami was the subject of substantial debate, based on concern for the ability of the military "to operate within accepted humanitarian principles … and ensure integrity of many humanitarian agencies." This focuses attention on the need to establish methods for collaboration that would be best achieved with memoranda of understanding.
4. Although there was the tension noted above, it was also acknowledged that the military response was highly effective in some areas, particularly in hard-to-reach locations and Aceh, and here the military response provided the first comprehensive needs assessment of 500,000 people.

### Media

1. It was acknowledged that media personnel play a critical role in providing information, and journalists are generally the only source of information. This is important for the local community as well, since it can provide critical health information.
2. It is important for the organizations engaged in the response to provide the journalists with the accurate information that they need. After the tsunami, many journalists felt that there was sparse information on population-based health risks. The main problem was that the emphasis was anecdotal, and disproportionately emphasized psychological trauma, diarrhea, malaria risks, etc. The assessment emphasized the needs between responsible agencies and media.
3. Emphasis was also placed on the need for journalists and broadcasters to be considered part of the response team, as key players in helping to develop a policy agenda by disseminating public health messages, identifying areas of greatest need, and correcting myths – which are so common.
4. It was also a recommendation that the WHO hire local journalists to collect better information.

## Recommendations for responding health-care professionals

The following recommendations are made for health-care professionals who travel to sites of natural disasters. These are based on the advice from the CDC as relevant to tsunami-affected areas (CDC, 2005), including that the pre-travel evaluation should include a complete medical history to determine medical fitness, psychological stability, and pregnancy. The following are components of the complete evaluation.

### Travel advisories

There should be a review of the US Department of State advisories against travel to selected areas, as provided at http://travel.state.gov. It should also be noted that some affected countries may restrict travel due to health or security reasons.

### Immunizations

- *Tetanus diphtheria*: the vaccine or boosters should be given to those not fully vaccinated or not vaccinated for the past five years (fully vaccinated is defined as three doses of a tetanus toxoid-containing vaccine with the last dose within five years).

- *Hepatitis A*: this should be given even if travel is imminent.
- *Hepatitis B*: the recommendation is for the three-dose series (months 0, 1 and 6), but it may be given in an accelerated schedule at days 0, 7 and 14.
- *Influenza*: this can be given if the vaccine is locally available using the inactivated formulation for injection; the alternative is the live virus vaccine that can be used for most healthy persons age 5–49 years who are not pregnant and not working directly with populations affected by tsunami. This decision may be largely based on the probability of influenza in the affected area (www.cdc.gov/nip/publications/acip-wist.htm).
- *Typhoid*: the injectable vaccine is given as a single dose, but protection does not begin for two weeks.
- *Polio*: a booster should be provided if there has been none since childhood.
- *Measles*: immunity is assumed if there is physician-diagnosed laboratory evidence of measles immunity, or proof of receipt of two doses of live measles vaccine after one year or birth before 1957. MMR vaccine may be given if there is concern about susceptibility, and this is one of the more important components of pre-travel immunizations.
- *Rabies*: this is a recognized risk, although rare. The problem is that pre-exposure immunization requires at least three weeks and incomplete pre-exposure provides minimal value. In the event of potential exposure, the traveler should receive post-exposure prophylaxis with Rabies Immune Globulin (RIG), plus five doses of vaccine. If this is not available on site, the exposed person should return to home or to the major city where it is available.
- *Japanese encephalitis*: this was a major concern in the 2004 tsunami due to risk in all affected countries. Full vaccination requires 2–4 weeks, but an abbreviated schedule is two doses separated by seven days; this results in seroconversion in 80 percent. One potential concern is that reactions to the vaccine can occur in some at up to one week after vaccination; these include generalized pruritis, angioedema, or even anaphylaxis. If travel is planned later than two weeks, this vaccine should be given (www.cdc.gov/travel/diseases/jenceph.htm). For those who need to travel earlier, the traveler should be given advice about prevention of mosquito bites using insect repellent, and should sleep in insecticide-treated bed nets treated with permethrin (www.cdc.gov/travel/bugs.htm).
- *Cholera*: this vaccine is recommended if there is cholera in the area, but availability is a potential problem.
- *Yellow fever*: there is no yellow fever risk in Asia, but this vaccine should be considered for those who travel to natural disasters in East Africa.
- *Malaria prophlaxis*: recommendations are variable depending on the country, but risk may be enhanced by migration, breakdown in mosquito control, and flooding. Specific recommendations vary by country (www.cdc.gov/travel/malariadrugs.htm). This site also provides a valuable list of side effects. The preferred agents are atovaquone/proguanil (A/P, malarone), doxycycline, or

mefloquine (lariam and generic). Doxycycline is often considered preferred because of the protection afforded against other common infectious diseases, and its established merit for preventing malaria. The traveler should be warned that unexplained fever that occurs within a year of returning should prompt consideration of malaria. There is also a recommendation for empiric treatment of malaria for patients in areas of risk who have compatible clinical conditions. The recommendations for presumptive self-treatment are available at www.cdc.gov/travel/diseases/malaria. Another resource is the CDC malaria hotline on 770-488-7788 and, for emergency consultation after hours, 770-488-7100, with the request to speak to the CDC malaria branch clinician.

## Other precautions

- *Protective footwear*: this is particularly important due to the potential for tetanus. Injuries should be evaluated on site by a health-care professional. Of particular importance is room cleansing with soap and water.
- *Food and water*: major concerns are enteric pathogens, including hepatitis A and hepatitis E. The general recommendation is to consume food that is thoroughly cooked, and to avoid salads and ice cubes. If bottled water or a trusted source of water is not available, water should be boiled or disinfected (www.cdc.gov/travel/foodwater.htm). The traveler is recommended to carry loperamide and an antibiotic for self-treatment of acute diarrhea; this could be ciprofloxacin or azithromycin. These should be taken until the symptoms resolve, which is usually within three days. Diarrhea associated with blood or high fever indicates the need for consultation with a health-care professional.
- *Mosquito bites*: the risks include Japanese encephalitis, dengue, and malaria. Prevention will include use of insect repellent containing deet, wearing long-sleeved shirts and long pants when out of doors, and, as noted above, sleeping under a bed-net, preferably one treated with permethrin.
- *Health kit*: the health kit for travel should include medicines that are routinely taken, an antidiarrheal (usually loperamaide), an antibiotic for self-treatment for acute bacterial illness (such as a fluoroquinolone or azithromycin as recommended for diarrhea), insect repellent, sunscreen, and antimalarial drugs. It should be emphasized that professional health care in areas of natural disasters may be sharply limited. Also important is consideration of a food supply (preferably canned or processed) and a method of water purification.

## References

Briet, O.J., Galappaththy, G.N., Konradsen, F. *et al.* (2005). Maps of the Sri Lanka malaria situation preceding the tsunami and key aspects to be considered in the emergency phase and beyond. *Malaria Journal* **4**, 8.

Centers for Disease Control and Prevention (CDC) (2005). Assessment of health-related needs after tsunami and earthquake – three districts, Aceh Province, Indonesia, July–August 2005. *Morbidity and Mortality Weekly* **55**(4), 93–97

Centers for Disease Control and Prevention (2006). Tsunami Disaster: Health Information for Humanitarian Workers. Available at: www.bt.cdc.gov/disasters/rsunami/humanitarian.asp (accessed 29 December 2006).

Centers for Disease Control and Prevention (2007). *Disaster Mental Health Primer: Key Principles, Issues and Questions*. Available at: www.bt.cdc.gov/mentalhealth/pdf/states (accessed 1 February 2007).

Connolly, M.A., Gayer, M., Ryan, M.J. *et al.* (2004). Communicable diseases in complex emergencies: impact and challenges. *Lancet* **364**, 1974–1983.

Diaz, T. and Achi, R. (1989). Infectious diseases in a Nicaraguan refugee camp in Costa Rica. *Tropical Doctor* **19**, 14–17.

Ekdahl, K. (2005). Epidemic-prone disease surveillance and response after the tsunami in Aceh, Indonesia. *Eurosurveillance Weekly Release*, **10**, issue 5.

Gandhi, N.R., Moll, A., Sturm, A.W. *et al.* (2007). Extensively drug-resistant tuberculosis as a cause of death in patients co-infected with tuberculosis and HIV in a rural area of South Africa. *Lancet* **368**, 1575–1580.

Gessner, B.D. (1994). Mortality rates, causes of death, and health status among displaced and resident populations of Kabul, Afghanistan. *Journal of the American Medical Association* **272**, 382–385.

Goma Epidemiology Group (1995). Public health impact of Rwandan refugee crisis: what happened in Goma. *Lancet* **345**, 339–344.

Hatch, D.L., Waldman, R.J., Lungu, G.W. and Piri, C. (1994). Epidemic cholera during refugee resettlement in Malawi. *International Journal of Epidemiology* **23**, 1292–1299.

Ishii, M., Shearer, P.M., Houston, H. and Vidale, J.E. (2005). Extent, duration and speed of the 2004 Sumatra-Andaman earthquake imaged by the Hi-Net array. *Nature* **435**, 933–936.

Izadi, S., Shakeri, H., Roham, P. and Sheikhzadeh, K. (2006). Cholera outbreak in southeast of Iran: routes of transmission in the situation of good primary health care services and poor individual hygienic practices. *Japanese Journal of Infectious Diseases* **59**, 174–178.

Lay, T., Kanamori, H., Ammon, C.J. *et al.* (2005). The great Sumatra-Andaman earthquake of 26 December 2004. *Science* **308**, 1127–1133.

Lopez-Ibor, J.J. (2006). Disasters and mental health: new challenges for the psychiatric profession. *World Journal of Biological Psychiatry* **7**, 171–182.

Ministry of Health, Indonesia; World Health Organization; Global Outbreak Alert and Response Network (GOARN) Partners; Centers for Disease Control and Prevention-USA; Epicentre-France; European Programme for Intervention Epidemiology Training (EPIET)-Sweden; Health Protection Agency-UK; Institut de Veille Sanitaire-France; Australian Biosecurity CRC at Curtin University-Australia; McFarlane Burnet Institute-Australia; Mailman School of Public Health, Columbia University-USA; University of Malaysia-Sarawak; UNICEF; Khalakdina A (2005). Epidemic-prone disease surveillance and response after the tsunami in Aceh, Indonesia. *Eurosurveillance* **10**, E050623.2.

Moszynski, P. (2005). Disease threatens millions in wake of tsunami. *British Medical Journal* **330**, 59.

Neuner, F., Schauer, E., Catani, C. *et al.* (2006). Post-tsunami stress: a study of post-traumatic stress disorder in children living in three severely affected regions in Sri Lanka. *Journal of Traumatic Stress* **19**, 339–347.

Nieburg, P., Waldman, R.J. and Krumm, D.M. (2005). Hurricane Katrina. Evacuated populations – lessons from foreign refugee crises. *New England Journal of Medicine* **353**, 1547–1549.

Norwood, A.E., Ursano, R.J. and Fullerton, C.S. (2007). *Disaster Psychiatry*. Available at: www.psych.org/disasterpsych/s1/principlespractice.cfm (accessed 15 February 2007).

Orellana, C. (2005). Tackling infectious disease in the tsunami's wake. *Lancet Infectious Diseases* **5**, 73.

Redwood-Campbell, L.J. and Riddez, L. (2006). Post-tsunami medical care: health problems encountered in the International Committee of the Red Cross Hospital in Banda Aceh, Indonesia. *Prehospital Disaster Medicine* **21**, S1–7.

Rutta, E., Kipingili, R., Lukonge, H. *et al.* (2001). Treatment outcome among Rwandan and Burundian refugees with sputum smear-positive tuberculosis in Ngara, Tanzania. *International Journal of Tuberculosis and Lung Disease* **5**, 628–632.

Salama, P., Spiegel, P., Talley, L. and Waldman, R. (2004). Lessons learned from complex emergencies over past decade. *Lancet* **364**, 1801–1813.

Santaniello-Newton, A. and Hunter, P.R. (2000). Management of an outbreak of meningococcal meningitis in a Sudanese refugee camp in Northern Uganda. *Epidemiology and Infection* **124**, 75–81.

Siddique, A.K., Salam, A., Islam, J.S. *et al.* (1995). Why treatment centres failed to prevent cholera deaths among Rwandan refugees in Goma, Zaire. *Lancet* **345**, 359–361.

Sphere Health Services (2007). Available at www.sphereproject.org/content/view/114/84/lang.English (accessed 1 March 2007).

Spiegel, P., Sheik, M., Gotway-Crawford, C. and Salama, P. (2002). Health programmes and policies associated with decreased mortality in displaced people in postemergency phase camps: a retrospective study. *Lancet* **360**, 1927–1934.

Talley, L., Spiegel, P.B. and Girgis, M. (2001). An investigation of increasing mortality among Congolese refugees in Lugufu camp. Tanzania, May–June. *Journal of Refugee Studies* **14**, 412–427.

Thienkrua, W., Cardozo, B.L., Chakkraband, M.L. *et al.* (2006). Symptoms of posttraumatic stress disorder and depression among children in tsunami-affected areas in southern Thailand. *Journal of the American Medical Association* **296**, 576–578.

Toole, M.J. (1995). Mass population displacement. A global public health challenge. *Infectious Disease Clinics of North America* **9**, 353–366.

Toole, M.J. and Waldman, R.J. (1997). The public health aspects of complex emergencies and refugee situations. *Annual Review of Public Health* **18**, 283–312.

Toole, M.J., Steketee, R.W., Waldman, R.J. and Nieburg, P. (1989). Measles prevention and control in emergency settings. *Bulletin of the World Health Organization* **67**, 381–388.

UN (2007). *UN Office of the Special Envoy for Tsunami Recovery: Psychosocial Effects*. Available at www.tsunamispecialenvoy.org/briefs/psychosocial.asp (accessed 1 February 2007).

Ursano, R.J., Rullerton, C.S. and Norwood, A.E. (1995). Psychiatric dimensions of disaster: patient care, community consultation, and preventive medicine. *Harvard Review of Psychiatry* **3**, 196–209.

Ursano, R.J., Cerise, F.P., Demartino, R. *et al.* (2006). The Impact of Disasters and Their Aftermath on Mental Health. *Primary Care Companion to the Journal of Clinical Psychiatry* **8**, 4–11.

van Griensven, F., Chakkraband, M.L., Thienkrua, W. *et al.* (2006). Mental health problems among adults in tsunami-affected areas in southern Thailand. *Journal of the American Medical Association* **296**, 537–548.

VanRooyen, M. and Leaning, J. (2005). After the tsunami – facing the public health challenges. *New England Journal of Medicine* **352**, 435–438.

Waring, S.C. and Brown, B.J. (2005). The threat of communicable diseases following natural disasters: a public health response. *Disaster Management Response* **3**, 41–47.

WHO (2005a). Communicable Diseases Working Group on Emergencies: Communicable Disease Risks and Interventions (Communicable Diseases Toolkit for Tsunami-affected Areas). Available at: www.who.int/hac/techguidance/pht/communicabledisease/CDtsunami_rusjubterebtuibs12012005.pdf (accessed 11 February 2007).

WHO (2005b). *Tsunami and Health Situation Report* **24**, 1–22.

WHO (2005c). *Conference on the Health Aspects of the Tsunami Disaster in Asia, Phuket, Thailand, May 4–6, 2005*. Available: at www.who.int/hac/events tsunamiconf/en/index.html (accessed 25 January 2007).

Wilder-Smith, A. (2005). Tsunami in South Asia: what is the risk of post-disaster infectious disease outbreaks? *Annals Academy of Medicine Singapore* **34**, 625–631.

# *Climate change and infectious diseases* 14

**Anthony J. McMichael and Rosalie E. Woodruff**

In recent decades there has been a well-recognized and widespread upturn in the rate of emergence, incidence, and spread of infectious diseases in all regions of the world. Many long-established infectious diseases have increased their geographic range and incidence. Antimicrobial resistance is becoming more prevalent. Of particular significance, over the past three decades there has been a succession of apparently new (mostly viral) infectious diseases. These include HIV/AIDS, Ebola virus, legionellosis, hepatitis C, hantavirus pulmonary syndrome, Nipah virus, Severe Acute Respiratory Syndrome (SARS), and, most recently, H5N1 avian influenza.

This changing tempo and pattern of infectious disease is, via diverse pathways, a consequence of the increases in intensity and scale of the human enterprise. The main influences arising from this unprecedented profile of changes in human demography, ecology and environmental impact are summarized in Box 14.1. They include increased physical mobility, extended trading, other forms of inter-population contact, changes in social relations (including changes in sexual networks and drug practices), newer commercial technologies and the scale of agriculture, intensified land clearance, biodiversity losses and other environmental disturbances, and global climate change (Weiss and McMichael, 2004).

Of particular background relevance to this chapter are the various large-scale environmental changes that human societies are now causing. Frequently these entail changes to complex natural systems and processes, which then exert influences on infectious disease patterns and risks. For example, deforestation and habitat fragmentation can facilitate the mobilization of microbes that were not previously agents of human infection, and thereby prompt the emergence of new infectious diseases in humans (see Chapter 5). Changes to ecosystems often alter the profile of species, and may thus disturb the natural constraints on vector species (e.g. mosquitoes and ticks) and intermediate host species

### Box 14.1 Modern-world factors that influence emerging and re-emerging infectious diseases

- Population growth and increasing density (and persistence of crowded peri-urban poverty)
- Urbanization: changes in mobility and in social and sexual relations
- Globalization (distance, speed, volume) of travel and trade
- Intensified livestock production methods (especially inter-species contacts)
- Live animal markets: longer, quicker supply lines
- Changes to ecosystems (deforestation, biodiversity loss, etc.)
- Global climate change
- Biomedical exchange of tissues: transfusion, transplants, hypodermic injection
- Misuse of antibiotics (in humans and livestock)

(e.g. mammals and birds) that are integral to the spread of various human infectious diseases.

Human-induced global climate change is the best known of these contemporary "global environmental changes." This past decade this has progressed, in the assessment of scientists, from being a likely future problem, to a definite future problem, to now a process that is actually underway (IPCC, 2001; Karl and Trenberth, 2003). Climate change will affect the patterns of many of the infectious diseases that are known to be sensitive to climatic conditions – particularly various of the insect-borne infections and the infections that are spread person-to-person via contaminated food and water.

## The links between climate variation and infectious diseases

The transmission of infectious disease agents is influenced by many factors, including social, economic, climatic, and ecological conditions (Weiss and McMichael, 2004). In situations where low temperature, low rainfall, or lack of vector habitat are significant limiting factors for vector-borne disease transmission, climatic changes may tip the ecological balance and trigger outbreaks. A change in transmission rate, range, or seasonality can also result from climate-related migration of reservoir host species or human populations (Hales *et al.*, 2000).

## Defining climate and weather – and their basic relationship to infectious disease transmission

The term *weather* refers to the state of the atmosphere in a particular place and time, with respect to wind, temperature, cloud, moisture, etc. Weather data are transformed into climate data when averaged over time. That is, *climate* refers to the weather that prevails in an area over a long period. Climate is what you expect; weather is what you get!

It is helpful to think of climatic conditions as setting the basic spatial and seasonal limits of infectious disease transmission – via the survival and replication of pathogens, the season of activity (pathogen and vector), the geographic range of vectors and hosts, and the behaviors of susceptible human populations. Within these climatic constraints, shorter-term weather can affect the timing and intensity of outbreaks (as with outbreaks of cholera and other diarrheal diseases in the wakes of Hurricane Mitch in 1998, and Hurricane Katrina in 2005).

The distribution of malaria illustrates the importance of climate in determining the geographic regions where the disease has the *potential* to become established. That is, climatic conditions set the geographic limits of the disease. Malaria is caused by four species of a protozoan parasite (*plasmodium*), transmitted between humans by the bite of infective female *Anopheles* mosquitoes. Of the four protozoa that infect humans, two of them predominate overwhelmingly: *Plasmodium vivax* and *P. falciparum*. Falciparum (which causes far greater mortality than vivax) requires warmer conditions, and predominates in most tropical and sub-tropical regions. Vivax malaria, with its capacity for overwintering dormancy, predominates in cooler, temperate zones. The many environmental factors that affect malaria incidence include altitude (Bodker *et al.*, 2003), topography (Balls *et al.*, 2004), land use, irrigation, and other environmental disturbance (Carlson *et al.*, 2004).

Meanwhile, it is important to stress that various other complex and interacting factors influence the occurrence and prevalence of infectious diseases within climatically suitable envelopes. The most upstream of these factors may well be poverty (Winch, 1998), as the availability of population wealth enables development and maintenance of the public health infrastructure necessary for detecting, tracking, and protecting against infectious diseases, as well as the treatments for managing them. Dengue infection provides a good illustration of how proximal geographic regions that share the same climate and weather, but have widely disparate economic resources, can exhibit significantly different rates of this disease (Reiter *et al.*, 2003; see also Chapter 4).

Historically, vivax malaria occurred widely within temperate zones (e.g. Europe and Scandinavia, North America, Australia), but has been effectively eradicated from those regions for the past half-century. Other social and demographic factors that affect malaria transmission include human population density and immunity;

housing location and condition (screens, air conditioning, piped water); and the use of bed nets, mosquito control programs, and medical treatments.

## Infectious disease transmission cycles

A further point should be made at this stage of the chapter. There is a great diversity of types and biological modes of infectious diseases. The simplest transmission cycle is one where a pathogen is transmitted (by "contagion") from an infected person to another susceptible person directly (e.g. via droplet secretion or sexual contact). Cycles of medium complexity are those where the pathogen is transmitted indirectly (through an intermediate plant, animal, or environmental factor such as water). Vector-borne pathogens – those that rely on a vector (such as a mosquito, fly, cockroach, tick or rodent) to infect humans – are a major subset of this second group.

Pathogens that are human-adapted are the *anthroponoses*. They circulate from human to human, either with (e.g. malaria) or without (e.g. cholera) the intervention of vector species. The *zoonoses* are pathogens that naturally infect (but do not necessarily affect) non-human animal species (the *reservoir host*), and which occasionally infect "bystander" humans. These too may have an intervening vector (a mosquito, as in West Nile Virus) or not (such as direct transmission, as in rabies from the bite of a dog). Some of these cycles are highly complex, with more than one animal reservoir needed in the transmission cycle (e.g. tick and deer, Lyme spirochete) or with numerous different species capable of acting as reservoirs (in the case of Ross River virus). Figure 14.1 summarizes the main types of transmission cycles for infectious agents.

## Climatic influences on pathogen, vector, host

All infectious organisms (bacteria, viruses, protozoa, helminths, and others) are thought likely to be affected by some aspect(s) of climatic conditions. The most affected stages of the pathogen's lifecycle are the free-living, intermediate, or within-vector stages – that is, the stages spent outside either the human host or the intermediate (reservoir) host. Many studies have documented how climatic variations influence the occurrence of a wide range of infectious diseases. Some of these studies have depended on field observations of natural variations; other have tested specific relationships in laboratory-experimental fashion – such as the relationship between temperature and malarial parasite maturation within the vector mosquito.

The following section describes the biology underpinning the climate and infectious disease relationship, and refers to epidemiological studies that have recognized or tested this.

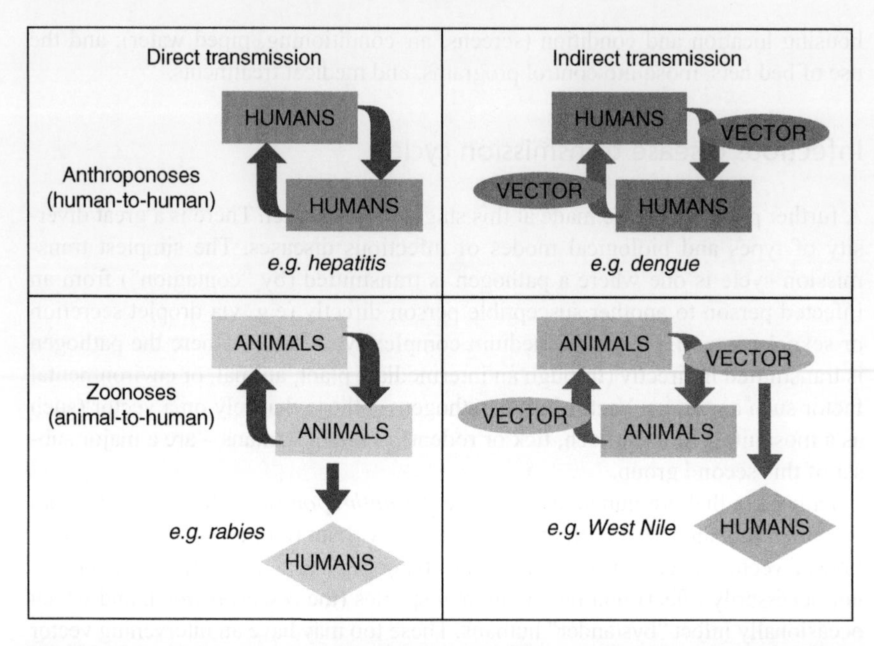


**Figure 14.1** Main types of transmission cycles for infectious agents. Based on Wilson (2001).

## *Pathogen*

Many vector-borne diseases are climate-limited because the pathogens cannot complete development before the vectors die. Laboratory and field studies have shown that the *extrinsic incubation period* (the time needed for a pathogen to replicate in the salivary gland of the mosquito to sufficiently high titers for infection to occur) becomes shorter as temperatures increase – thus enhancing transmission, because mosquitoes become infectious more quickly (Patz *et al.*, 1998). In cooler areas of the world, malaria is hindered by the lack of development of the malaria parasites, rather than by the presence of the vector. Various estimates of the parasite's temperature requirements indicate that *P. vivax* has a developmental threshold temperature of around 14–16ºC – lower than the 16–19ºC needed for *P. falciparum* (Martens *et al.*, 1999).

As some pathogens are dependent on warm temperatures for survival, others survive better in colder temperatures. Rotavirus infections (which cause diarrheal disease in children) occur at much higher rates in winter than in summer (Turcios *et al.*, 2006). Respiratory syncytial virus, the major contributor to lower respiratory tract infections in children, has been observed to survive longer under cold conditions (Hambling, 1964), and to become more infective (Rechsteiner and Winkler, 1969). In temperate climates, an increase in respiratory tract

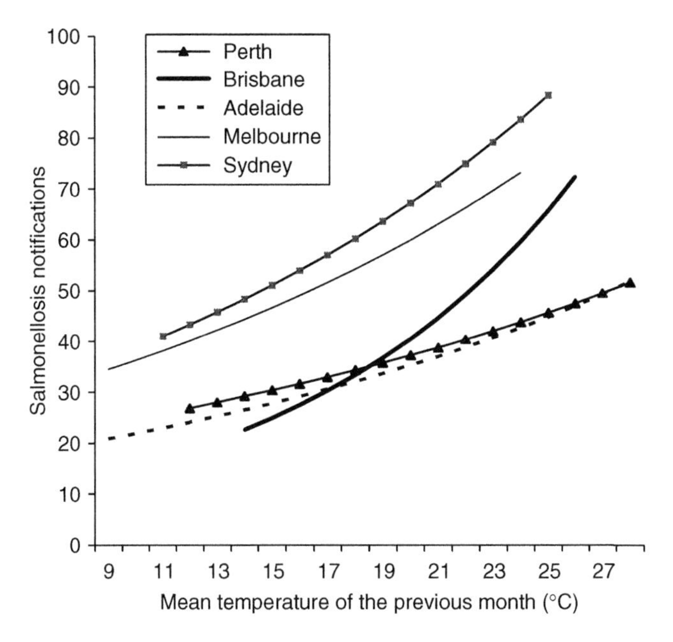


**Figure 14.2** Predicted increase in monthly notifications of food poisoning (salmonellosis) with increasing previous month's temperature for five Australian cities. Reprinted from d'Souza *et al.* (2004), with permission.

infections among infants during the cold season is commonly reported (Isaacs and Donn, 1993; Chan *et al.*, 2002; Al-Khatib *et al.*, 2003). Interestingly, however, respiratory syncytial virus has a completely different seasonal pattern in the tropical climates of most Asian, African, and South American countries, appearing during the rainy season (Chew *et al.*, 1998; Shek and Lee, 2003). This suggests that factors other than – or in combination with – temperature may be important in the transmission of this virus.

*Salmonella*, and other bacteria responsible for food poisoning, proliferate more rapidly at higher temperatures (Baird-Parker, 1990). The rate of multiplication of *Salmonella* species is directly related to temperature within the range 7.5–37°C (Baird-Parker, 1994). Studies have shown that an increase in notifications of (non-specific) food-poisoning in the United Kingdom (Bentham and Langford, 1995; Bentham and Langford, 2001) and of diarrheal diseases in Peru and Fiji (Checkley *et al.*, 2000; Singh *et al.*, 2001) has accompanied short-term increases in temperature. Linear associations have been found between temperature and notifications of salmonellosis in European countries (Kovats *et al.*, 2004) and Australia (D'Souza *et al.*, 2004; see Figure 14.2), and a weaker seasonal relationship exists for infection by *Campylobacter* (Kovats *et al.*, 2005; Nichols, 2005). As cooking destroys most food-borne pathogens, it is likely that inadequate storage and the spread from contaminated to non-contaminated food are risk factors for transmission in sporadic cases (Schmid *et al.*, 1996). Outdoor temperatures might also affect people's exposure to salmonella bacteria through seasonal changes in eating patterns (such as the consumption of foods with a higher risk of *Salmonella* contamination from buffets, barbecued foods, and salads, etc.) (Kovats *et al.*, 2004).

One of the world's most feared diarrheal diseases, cholera, is also sensitive to environmental temperature. There is now diverse evidence of the proliferative response of the cholera vibrio to warmer water in lakes, estuaries, and coastal waters (Long *et al.*, 2005; Wilcox and Colwell, 2005). It is likely that the combination of persistent poverty (with lack of sanitation), warmer temperatures, and population displacement and movement in poorer tropical and sub-tropical regions will exacerbate the occurrence of cholera in future.

Meningococcal meningitis in the Sahel region ('meningitis belt') of West Africa provides a tantalizing example of an epidemic infectious disease possibly related to climatic conditions. The fact that outbreaks occur approximately periodically may reflect cyclical fluctuations in climatic conditions (Sultan *et al.*, 2005). The person-to-person spread of meningococcal meningitis appears to be related to temperature, rainfall (especially aridity), and other environmental (especially winds and dust) conditions (Greenwood *et al.*, 1984; Haberberger *et al.*, 1990). Also, the infrequency of outbreaks in humid, forested, or coastal region areas may be because high continuous humidity impairs transmission (Molesworth *et al.*, 2003).

## Vector

Weather factors such as temperature, rainfall, and humidity are capable of assisting or interrupting the biology and population dynamics of vector mosquitoes (Reeves *et al.*, 1994), thereby influencing their abundance and distribution. Rainfall (or lack of it) plays a crucial role in the epidemiology of arboviral diseases, as it provides the medium for the aquatic stages of the mosquito lifecycle. Temperature impacts on mosquito productivity and on viral replication. Humidity affects mosquito survival, and hence the probability of transmission (Sellers, 1980; Reiter, 1988; Leake 1998).

Temperature directly affects the distribution and nutritional requirements of mosquitoes. Extreme temperatures will kill mosquito populations – for example, *Culex annulirostris* larvae die at temperatures below 10°C and above 40°C (Lee *et al.*, 1989). Consequently, mosquitoes are limited both in latitudinal and altitudinal range. High temperatures (up to a limit) reduce the period needed for larval development, meaning that more generations can fit into a given time period. *C. annulirostris* has an egg-to-adult time of 12–13 days at 25°C, and of only 9 days at 30°C (Kay and Aaskov 1989). A 10°C increase in temperature can reduce the development time of *Schistosoma mansoni*, a human pathogen in an intermediate host, by more than half – from 35 to 12 days (Harvell *et al.*, 2002).

Mosquitoes have a high surface area to mass ratio, and are susceptible to loss of body water if a rise in ambient temperature is not accompanied by a rise in humidity. This has particular implications for arid regions, and to a lesser extent for temperate ones. High humidity influences the survival of mosquitoes (Reeves *et al.*, 1994). As the proportion of old mosquitoes in the population increases, so does the risk of pathogen transmission (older female mosquitoes are more likely than younger ones to have had two or more blood meals). For example, Hales and others (2002) have shown that the single climatic variable, vapor pressure, could predict the current global distribution of dengue fever epidemics with 89 percent accuracy.

Water is essential for the breeding cycle of mosquitoes, given both the larval and pupal stages are aquatic. The effect of rainfall on mosquito breeding, however, is not always direct and positive. The pattern of precipitation is critical for mosquito survival. A moderate increase in rainfall can be beneficial (Lindsay and Mackenzie, 1997), while excessive increases can wash away the mosquito larvae or dormant eggs and interrupt the transmission cycle – a particular problem for species that prefer to breed in still water. The length of a rainfall event (i.e. the number of contiguous days with recorded rainfall), the number and duration of events, and the total amount of rain that falls are all factors that differently affect the breeding of mosquito vectors of Rift Valley fever in epizootic regions (Davies *et al.*, 1985). The timing of rainfall across the year (*seasonality*) is also decisive in determining whether disease outbreaks occur, and whether they are epidemic in proportion (Woodruff *et al.*, 2002).

Box 14.2 provides a summary of the sensitivity of infectious disease pathogens, vectors, and reservoir host species to changes in climatic conditions.

## Box 14.2 Infectious disease pathogens, vectors, and reservoir host species: sensitivity to changes in climatic conditions

- Changes in temperature, rainfall and humidity can affect the range, population density, and biological behaviors of various vector organisms (mosquitoes, ticks, water snails)
- Temperature affects the rate of bacterial proliferation (e.g. food-poisoning species); the proliferation of cholera vibrios increases in response to warmer coastal/estuarine waters, facilitating their subsequent dissemination in the aquatic food web
- The rate of maturation (incubation) of various viruses (e.g. dengue virus) and protozoa (e.g. malaria plasmodium) within mosquito and other vectors is typically very temperature-dependent

## Host

Climatic variation influences host growth and immunity levels. In arid and semi-arid regions (and in temperate regions in severe drought years), the main determinant of host population breeding is the level of food supply. For mammalian vertebrates, rainfall is the dominant factor governing the control of pasture biomass (Noy-Meir, 1973). Kangaroos, for example – the main host for Ross River (see Box 14.3) and Barmah Forest viruses – respond to changes in food supply by adjusting their rates of reproduction (Bayliss, 1987), and come into breeding condition almost immediately following rainfall (Strahan, 1991).

A well-known example of the indirect influence of meteorological factors on a complex infectious transmission cycle is of hantavirus pulmonary syndrome in the southwest USA. The first outbreak in 1993 was subsequently attributed to a

### Box 14.3 Complex transmission cycles: Ross River virus

The epidemiology of Ross River virus (RRv) disease illustrates the varying effect of climatic phenomena on a complex infectious disease transmission cycle. RRv disease is the most prevalent vector-borne disease in Australia, and has also been reported in several Pacific island countries. The virus causes a non-fatal epidemic polyarthritis, with arthritic symptoms that range from mild to severe and debilitating, and which can last several years in some people. There is no cure, and treatment is palliative.

The disease is caused by an alphavirus that can be transmitted by more than 30 different mosquito species. Mosquitoes such as *Ochlerotatus camptorhynchus* and *O. vigilax* breed in inter-tidal wetlands, and are the main vectors in coastal regions. *Culex annulirostris* is the main inland vector: it breeds in vegetated fresh water, and is common in both tropical and temperate regions that are subject to flooding or irrigation during summer. Other species, such as *O. notoscriptus*, are important vectors in semi-rural and urban areas. The main natural vertebrate hosts for the virus are marsupials (kangaroos and wallabies), with other animals also implicated (notably possums and horses).

The primary enzootic cycle of the virus is between the non-immune reservoir host and a mosquito vector. When immunity in the host population is low, and climatic conditions are suitable, massive amplification of the virus occurs in the host and mosquito populations. In this situation the abundance of mosquitoes typically results in transmission of the virus to humans, when infected mosquitoes are forced to seek blood meals outside their natural targets. Changes in climate strongly influence the replication of this virus (Kay and Aaskov, 1989), the breeding, abundance and survival of the mosquito vectors (Lee *et al.*, 1989), and the breeding cycle of the natural hosts (Caughley *et al.*, 1987). The epidemiology of the disease reflects this, with different seasonal and inter-annual patterns being observed between the broad climatic regions of the Australian continent (Russell, 1994; Tong and Hu, 2001; Woodruff, 2005). In the northern tropical region the disease is endemic, and cases occur in most months of every year. In the southern temperate and inland arid parts of the country the pattern is sporadic, with wide variation between years in the number of reported cases.

RRv disease is one of the few infectious diseases that can be predicted by climate-based early warning systems (WHO, 2004). Weather conditions at relatively coarse temporal and spatial resolutions have been used to predict epidemics with sufficient accuracy and advance notice for public health planning. In the dry temperate south-eastern part of the country, Woodruff and colleagues have shown that sustained winter/spring rainfall (i.e. the total number of rain days) and warm late spring temperatures, in conjunction with low rainfall in the spring of the preceding year, increased the risk of summertime RRv epidemics (Woodruff *et al.*, 2002). They speculated there are two linked mechanisms that lead to epidemic potential in this region. First, floodwater *Aedes* species maintain the natural cycle of the virus by transovarial transmission through embryonated eggs (Lindsay *et al.*, 1993). These mosquitoes "overwinter" as drought-resistant eggs in mud flats and creek beds (Marshall, 1979). After heavy winter rainfall, the females emerge and infect the vertebrate hosts (Lindsay *et al.*, 1993). Substantial rainfall from late winter enables early and prolific breeding of *Aedes* populations and an extended period of virus build-up, thus increasing the transmission potential. The period of prolonged heavy rainfall also acts to raise the water table across this flat region, reducing absorption and runoff. As a consequence, pools of water remain on the ground into summer (even if there is low summer rainfall). This provides breeding sites for the summer breeding *Culex* mosquitoes, which preferentially bite both humans and kangaroos – thus extending the infection beyond the natural cycle to humans.

The second mechanism relates to rainfall in the spring of the year before an epidemic, and to its role in host–virus population dynamics. High spring rainfall supports large numbers of mosquitoes. When this occurs, a greater proportion of kangaroo hosts become infected (the period of viraemia lasting about one week) and then immune for life. This results in a reduction in the pool of susceptible kangaroos in the following year, and consequently minimal viral amplification and a lower probability of human cases. Conversely, several years of low spring rainfall dramatically raises the proportion of susceptible kangaroos so that in subsequent years – if climatic conditions are suitable – the probability of a large outbreak becomes very high.

sequence of climatic and ecological changes that created optimum conditions for the proliferation and spread of the virus (Engelthaler *et al.*, 1999). Six preceding years of drought appear to have reduced the populations of natural predators – birds and snakes, in particular – of the white-footed mouse, which naturally harbors the virus. In early 1993, heavy rainfall resulted in an abundance of piñon nuts and grasshoppers upon which rodents feed. The resulting rapid expansion of the mouse population caused a huge increase in the amount of mouse-excreted virus entering the local environment, drying, and then blowing around in the wind. Human exposure increased greatly, and the apparently first-ever outbreak of hantavirus pulmonary syndrome occurred in North America.

## Climate-related human behavior and infectious disease occurrence

In addition to the direct influence of climate on pathogen proliferation and the lifecycles of vectors and intermediate hosts, climatic conditions can also affect infectious disease transmission indirectly by changing human physiology or behavior. For example, a child's susceptibility to respiratory syncytial virus infection may be increased if there is an alteration in the mucociliary activity of the respiratory epithelium or mucosa thickness. This has been hypothesized to occur in response to increasing dryness and cold (Chan *et al.*, 2002). Another example is cerebrospinal meningitis, which typically occurs world-wide in seasons and regions of low absolute humidity (Besancenot *et al.*, 1997). The natural habitat of the meningococcus is the throat, the lining of which is likely to deteriorate in susceptible people during periods of low humidity.

These relationships may, at least partly, be explained by seasonally-related change in patterns of children's play and person-to-person contact (see also Chapter 6). Children who spend more time indoors in cold weather may inadvertently increase their exposure to indoor air pollutants, such as tobacco smoke, smoke from wood fires, and nitrogen dioxide emitted from un-flued gas heaters, depending on the amount of ventilation, which could confound the association between infections and cold temperatures (Jones, 1998).

Prolonged close proximity to other infectious children can increase the opportunity for transmission of infectious agents (Sennerstam and Moberg, 2004). The well-known seasonality of influenza in elderly people, with epidemic outbreaks occurring in winter, has generally been attributed to indoor crowded conditions. The strong inverse association between cases of pneumococcal disease and temperature (with peaks in midwinter and troughs in midsummer) may relate to the coincident high concentration of circulating viruses in winter (such as respiratory syncytial virus, influenza virus, and adenovirus). Infection with such viruses predisposes to otitis media, which becomes suppurative and leads to pneumococcal bacteremia or meningitis (Kim *et al.*, 1996).

However, social behaviors, and, in particular, increased contact with other people, cannot explain all variations and outbreaks of infectious diseases. For example, several studies have found that the time-pattern of respiratory syncytial virus epidemics is not consistent with increased social contact among children during school time. Meanwhile, there is increasing evidence that infectious disease emergence, reactivation, and spread reflect the various large-scale environmental changes that are now arising in response to the burgeoning human pressures on the world's environment. Widespread deforestation, other land-use changes, water-damming and irrigation, biodiversity losses that occur because of those and other reasons, more intensive and extensive trading patterns, and increased human crowding (especially in peri-urban shanty towns and slums) are all likely to influence patterns of infectious disease occurrence. As humans encroach further into previously uncultivated environments, new contacts between wild fauna, insect vectors, and humans and their livestock increase the risk of cross-species infection.

The recent emergence of the Nipah virus, a highly virulent paramyxovirus, as a human-infecting pathogen illustrates the interplay between social, behavioral, and environmental influences in generating a new circumstance for infectious disease emergence – in this case, more specifically, the interplay between climatic conditions, deforestation, wild species disturbances, and intensive livestock production (Chua *et al.*, 2002; Chua, 2003; Weiss and McMichael, 2004). The first recorded Nipah virus outbreak followed the establishment of commercial piggeries in conjunction with fruit orchards located close to the tropical forest in northern Malaysia. The causal constellation underlying this emergent infectious disease remains uncertain, and almost certainly complex. The following are likely components. During the 1980s and 1990s, the forest habitat of the local fruit bats (*Pteropus* species – the natural host of the virus) had been reduced by deforestation for pulpwood and industrial plantation. Then, in 1997–1998, slash-and-burn deforestation escalated and resulted in the formation of a severe haze that blanketed much of the region. This was exacerbated by a drought and associated forest fires, driven by a severe El Niño event in the same year (see Box 14.4). Forest fruit yields declined, and hungry food-seeking bats encroached into cultivated fruit orchards. Pigs are thought to have been infected via the eating of shared fruits (and bat droppings), and this new mammalian host then infected the pig farmers. This zoonosis ultimately infected several hundred Malaysian rural workers, causing a fatal encephalitic disease in approximately half of them (Daszak *et al.*, 2000, 2006).

## Global climate change

There is now general agreement among climate scientists that the human-induced increase in greenhouse gas concentration in the lower atmosphere

## Box 14.4 El Niño and infectious diseases

The El Niño Southern Oscillation (ENSO) is the world's dominant source of year-to-year climatic variation. The oscillation originates in the Pacific Ocean region, where a natural quasi-periodic variation in atmospheric pressure between east and west Pacific affects the east–west directional flow of low-latitude ocean surface waters, and hence the flow of moist air in the lower atmosphere. Reflecting this process, when the oscillation "index" is low, the sea surface temperatures in the Pacific Ocean rise in the east and fall in the west. This El Niño event has wide-ranging consequences for weather in low-to-mid latitudes around the world. It is especially associated with droughts and floods. During these El Niño events, which occur approximately twice per decade, there is heavy rain on the west coast of South America (especially Peru) and reduced rainfall (often drought) in eastern Australia, parts of Southeast Asia, South Asia, the Horn of Africa, southern Africa, and Venezuela and its environs. During a La Niña, the opposite phase of the cycle, the climate pattern is typically reversed.

The ENSO phenomenon is important for the understanding of climatic influences on infectious diseases for two reasons – as recently discussed by Kovats and colleagues (2003). First, it provides a substantial contrast in temperature and rainfall between the two extremes of the cycle – a "natural experiment" that can reveal clearly how climatic variation affects infectious diseases. Second, since climate variability is anticipated to increase with climate change, the ENSO phenomenon may well intensify.

Several time series studies have examined ENSO in relation to dengue fever outbreaks in the Asia-Pacific region, where El Niño and La Niña events appear to have influenced the occurrence of dengue fever outbreaks (Hales *et al.*, 1996, 1999; Hopp and Foley, 2003). In South Asia and South America (Venezuela and Columbia), both phases of the ENSO cycle have been associated with malaria outbreaks (Bouma and van der Kaay, 1996; Bouma *et al.*, 1996, 1997; Bouma and Dye, 1997). Similarly, ENSO-related variations in climatic conditions in Australia have influenced outbreaks of Ross River virus disease (Maelzer *et al.*, 1999; Woodruff *et al.*, 2002; Tong *et al.*, 2004).

Those studies that incorporate multi-decadal time series data, entailing a long series of El Niño and La Niña events, are best able to reveal an association between ENSO and infectious diseases. Many of the observed associations have a plausible climatic explanation. In particular, the higher temperatures characteristic of El Niño events can affect both the vector species and the pathogen (in ways described in the main text). Tidal inundation is essential for salt-marsh mosquito breeding: sea levels and tide heights rise, and wetland areas are more frequently inundated in years when sea surface temperatures are regionally warmer (such as during a La Niña phase in Australia, or an El Niño phase on the west coast of South America). The effect of ENSO cycles on rainfall and subsequent disease is more complex. For example, in poor and highly crowded tropical and sub-tropical regions, heavy rainfall and flooding may result in outbreaks of diarrhea – whereas very high rainfall can also reduce mosquito populations by flushing larvae from their aquatic habitat.

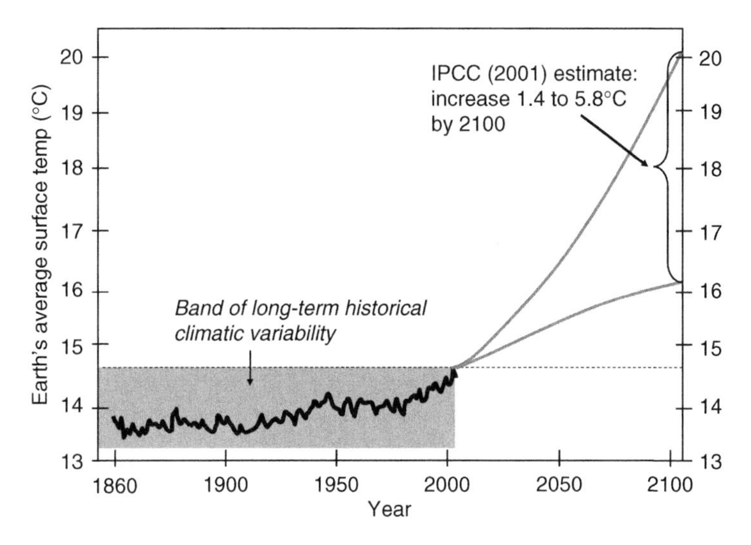


**Figure 14.3** Increasing average global temperatures.

(especially carbon dioxide and methane from industrial, transport, mining, and agricultural practices) will affect the global climate system. In particular, Earth's surface will warm and precipitation patterns will change. Recent scientific consensus is that this predicted process of climate change is now becoming evident, and that most of the warming that has occurred over the past half-century has been due to human actions (IPCC, 2001; see also Figure 14.3).

The approximately 0.5°C warming that has occurred in Earth's average surface temperature since the mid-1970s has evidently carried us above the upper range of the millennium-long statistical band of climatic variability. The UN's Intergovernmental Panel on Climate Change forecasts a rise of between 1.4 and 5.8°C this century, depending on future levels of greenhouse gas emissions, the sensitivity of the climate system to these higher concentrations of atmospheric greenhouse gases (IPCC, 2001), and other processes that could amplify or dampen the rate of change – such as ice-albedo and carbon-cycle feedbacks. At the time of writing, the scientific literature is beginning to point in the direction of more rapid change in climatic conditions, and a greater risk that the upper end of the Intergovernmental Panel's estimate will be reached or exceeded by 2100.

If, as predicted by mainstream climate science, Earth should warm by 2–3ºC this century, it would be an extraordinarily rapid event that would reverse the global cooling that has occurred over the past 20–30 million years. Many of the natural systems upon which human societies depend would be adversely affected. Further, if the process of climate change becomes non-linear, with critical thresholds being passed – such as the recent positive feedback from the thawing Siberian permafrost and its release of methane – warming could proceed faster than has been foreseen.

## Potential health impacts of climate change: pathways, and examples of infectious disease impacts

The various pathways by which climate change could affect patterns of human health, acting via its various manifestations in climatic conditions and weather patterns, are summarized in Figure 14.4. The range of risks to human health is extensive – and potentially catastrophic in some circumstances. The risk of infectious disease is affected mostly by climatic influences on the biology and behavior of pathogen, vector species, and intermediate or natural host species.

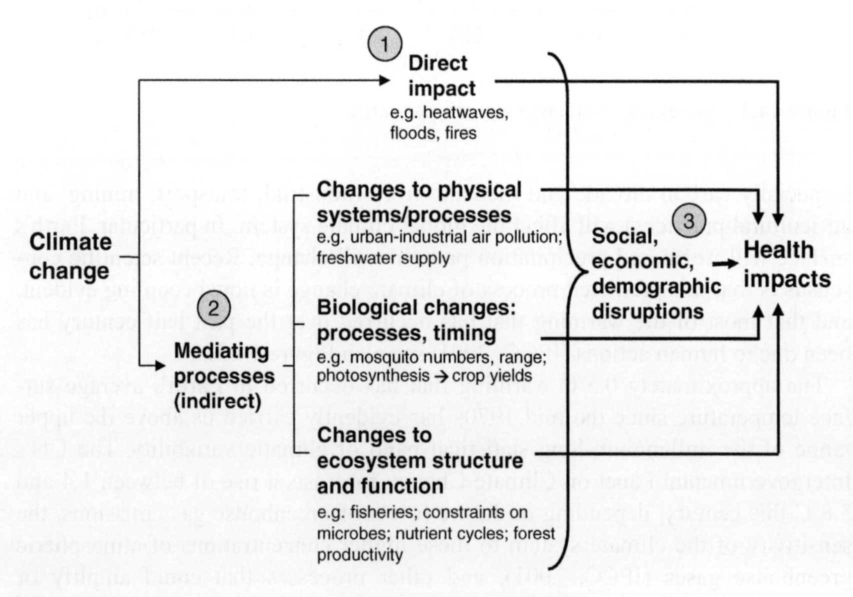


**Figure 14.4** Climate change and health: pathways, impacts.

The risks span readily understood risks to health and survival from extremes of ambient temperature and from other extreme weather events, disruptions to food production and availability, and changes in the range and activity of pathogens and vector organisms. Infectious disease risk may also be affected more indirectly via social disruption, poverty, and population displacement (environmental refugees) occurring in response to climate change.

All infectious diseases that have climate warming will affect host–pathogen interactions by:

- increasing pathogen development rates, transmission, and number of generations per year
- constraining over-wintering restrictions on pathogen lifecycles (Figure 14.5)
- modifying host susceptibility to infection.

Changes in these mechanisms could cause pathogen range expansions and host declines, or could release hosts from disease control by interfering with the precise conditions required by many parasites. Clearly, not all pathogens have

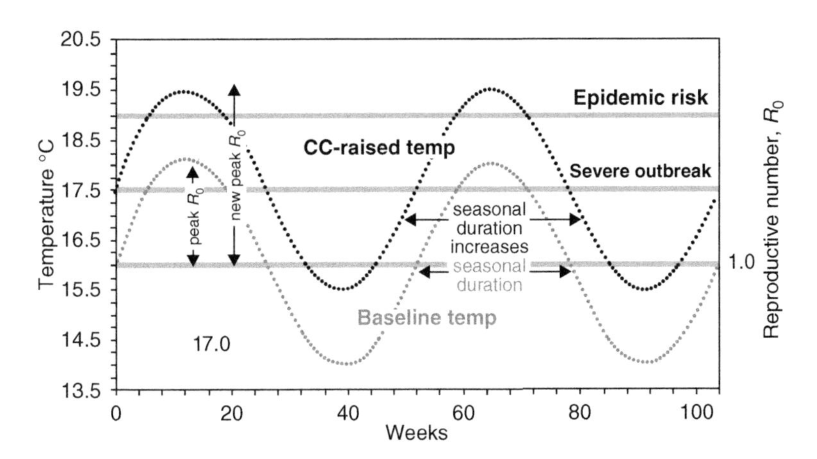


**Figure 14.5** The influence of an average 1.5°C rise in temperature on the growth rate (reproductive number, $R_0$) of a typical pathogen. If $R_0 > 1$, pathogen multiplies. When $R_0$ is more than 1, pathogen growth will increase. The lower dotted curve represents the average weekly temperature before climate change; the upper dotted curve represents average weekly temperature after a 1.5°C increase. The lower horizontal line corresponds to $R_0 = 1$ (below this temperature the pathogen population declines). The rate of pathogen growth increases above this temperature. We assume the risk of disease outbreaks becomes severe when temperature reaches the middle horizontal line, and epidemic at the upper horizontal line. Temperature increases also lead to an extension in the duration of the season when the pathogen is a problem. Reprinted from Harvell *et al.* (2002), with permission; ©2002 AAAS.

equal potential to control host populations or to be affected by warming. Climate warming is expected to disproportionately affect pathogens with complex lifecycles, or those that infect mosquitoes during one or more lifecycle phases (Harvell *et al.*, 2002).

Of climate-sensitive infectious diseases, vector-borne diseases are strong candidates for altered abundance and geographic range shifts, because rising temperatures will affect vector distribution, parasite development, and transmission rates (Kovats *et al.*, 2001). Climate change will affect the *potential* geographic range, seasonal transmission, and incidence of various vector-borne diseases. These would include malaria, dengue fever, and yellow fever (all mosquito-borne); various types of viral encephalitis; schistosomiasis (water-snails); leishmaniasis (found on the South America and Mediterranean coasts, and spread by sand-flies); Lyme disease (ticks); and onchocerciasis (West African "river blindness," spread by black flies).

## Modeling current and future influences of climate on infectious disease transmission

Various model-based estimates have been made of how climate change scenarios would affect the future transmissibility (both geographic range and seasonal incidence) of malaria (Lindsay and Birley, 1996; Rogers and Randolph, 2000; Tanser *et al.*, 2003; Kovats *et al.*, 2004; Thomas *et al.*, 2004; van Lieshout *et al.*, 2004). Two contrasting types of models have been used. One is based on known climate–disease relationships from laboratory and local studies. Such models comprise an integrated set of equations that express those relationships mathematically. They are referred to as "biological models". The other type, the "statistical" model, uses an empirical–statistical approach. A statistical equation is derived that expresses the currently observed relationship between the geographic distribution of the disease and local climatic conditions, and then applies that same equation to the specified future climate scenario.

The controversy about the relative roles of climatic conditions, which set spatial–seasonal limits for transmission, and the many non-climatic variables, which (separately or together) may even preclude occurrence of the infectious disease entirely within some locations, has regrettably bred some confusion within the scientific literature. To model how a given scenario of climate change would alter the receptive zone and season limits for some infectious disease is not to model where and when the disease *will* occur; it is to model where and when it *could* occur. We cannot, of course, *know* what the future pattern of infectious disease transmission will be, because we cannot know the future of vaccine technologies, vector controls, public health surveillance strategies, and antimicrobial resistance – nor, more generally, the future impacts of changes in levels of wealth, mobility, social organization, and public literacy about infectious diseases.

Nevertheless, in accord with classic experimental scientific practice, we can sensibly ask the following question: if all non-climatic factors were held constant over coming decades, how would a change in climate alter the potential geographic range and seasonality of infectious disease transmission? Indeed, with the increasing sophistication of both our knowledge base and our modeling capacities, we can incorporate plausible scenarios of how at least some of those non-climatic factors will change in future, and thus estimate the net impact of climate change on potential infectious disease transmission. There has been a second, cruder, confusion: some critics appear to presume that those who publish modeled estimates of future climate-induced changes in the potential transmission of a specified infectious disease therefore also assume that the current pattern of transmission reflects recent climate changes. This criticism has its basis more in politics than in logic, and will not be explored further.

Several modeling studies have projected limited geographical expansion of malaria transmissibility over the next few decades (Rogers and Randolph, 2000; Thomas *et al.*, 2004), while some estimate more extensive changes later this century (Martens *et al.*, 1999; Thomas *et al.*, 2004; van Lieshout *et al.*, 2004). Those studies that have modeled how climate change would affect seasonal changes in transmission project a substantial increase. One study, based on thorough documentation of current malaria occurrence in sub Saharan Africa, estimates that climate change by 2100 would cause a 16–28 percent increase in person-months of exposure to malaria (Tanser *et al.*, 2003).

Dengue fever, the world's most common mosquito-borne viral infectious disease, is also well known to be sensitive to climatic conditions. Various research groups have developed ways of modeling how future changes in climate would be likely to affect the geographic range and seasonality of this disease. As with malaria, it is well understood that many other non-climate factors influence, indeed can preclude, the occurrence of dengue – as is well illustrated by the huge differential in rates between Texas (very low rates) and adjoining Mexico (very high rates). Public health programs of monitoring, mosquito control, and rapid case detection and treatment are important. Holding constant such non-climate factors around the world as at present, statistical modeling indicates a substantial potential for increased geographic spread of dengue in warmer and wetter conditions over the coming century (Hales *et al.*, 2002). In a recent further development, the non-stationary temporal–spatial relationship between El Niño and the spread of dengue in Thailand has been modeled (Cazelles *et al.*, 2005). This study suggests that the El Niño event acts as a "pacemaker," resulting in a point-source "surface ripple" spread of the infectious outbreak.

A recent study in Canada has modeled the impact of projected climate change on the potential geographic extension of Lyme disease in that country, to 2080 (Ogden *et al.*, 2006). The disease, currently confined to the southern extremity of the country, would, approximately, become transmissible throughout much of the southern half of the country by the middle of this century.

There is, in all of this recent modeling of how climate change would affect infectious disease risks within a specified single country, a strong tendency towards an "inverse law." Those countries that have the professional and economic resources to carry out such research are generally countries at relatively low risk, and *vice versa*. However, a welcome recent development has been the advent of such studies for countries such as India and Zimbabwe. In the Zimbabwe study (Ebi *et al.*, 2005), plausible country-level climate change scenarios were generated and then applied to a mathematical model of how climatic parameters affect malaria transmissibility. The study showed that, with rising average and minimum daily temperatures accompanied by minimum necessary monthly rainfall, the future risk of malaria would progressively extend to higher altitudes. An important corollary here is that even if Zimbabwe were to become very wealthy and socially modernized, it would still cost much more than today to prevent the population's risk (exposure) from rising temperatures each morning.

*Legionella pneumophila* lives in the water of (evaporative) air-conditioning cooling towers, and is spread by aerosolized droplets. There is therefore the possibility of increased outbreaks of legionellosis with climate change, especially in developed countries that are becoming increasingly dependent on air-conditioning to cool both private and public buildings.

As noted earlier, weather disasters may also affect outbreaks of infectious diseases. One important manifestation of climate change is a change in climatic variability. Hence, regional patterns of extreme weather events are expected to alter as climate change proceeds. Following Hurricane Mitch in 1998, which directly killed 11,000 people in Central America, dramatic increases occurred in rates of cholera, malaria, dengue leptospirosis, and dengue fever – especially in Honduras, with estimates of 30,000 cholera cases, 30,000 malaria cases, and 1000 dengue cases (Epstein, 1999). In similar fashion, extreme flooding in Mozambique in early 2000 caused a surge in malaria cases three months later (see Figure 14.6).

This genre of modeling has, so far, usually not included various non-climate characteristics of the future world that would also affect infectious disease transmission probabilities, since many of those characteristics are not easily foreseen. If the pathogen were not locally present (e.g. because of efficient case surveillance and treatment) or if the vector species had been eliminated (e.g. by mosquito control programs), then the disease could not be transmitted. Future modeling will become more versatile if it can incorporate plausible scenarios (or, better, probabilistic projections) of these non-climatic contextual changes. Nevertheless, estimating how the intrinsic probability of infectious disease transmission would alter in response to climate change alone is itself informative – and, indeed, accords with classical experimental science. It serves to alert us to the range of future potential risks, and it focuses attention on areas that need more attention and research (see Box 14.5, page 398).

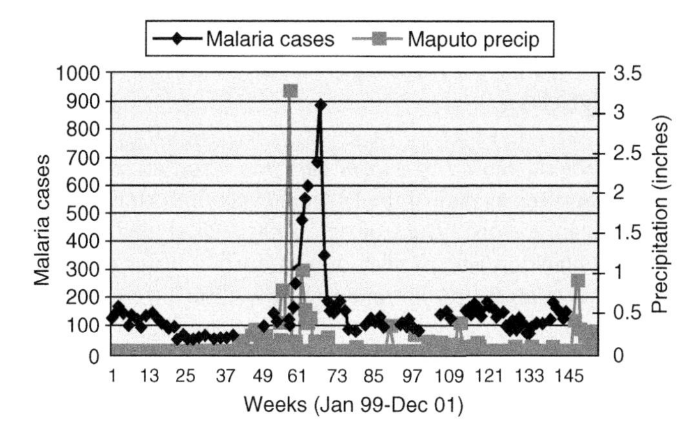


**Figure 14.6** Weekly cases of malaria (dark gray) and association with floods (pale gray) in Maputo, Mozambique, 2000. Reprinted from Milne (2005), with permission.

## Attributing observed change in disease occurrence to climate change

There are well-documented observations of various non-human systems, both physical and biotic, that have undergone changes that are reasonably attributable to associated regional warming over the past few decades. This includes melting of glaciers, shrinkage of sea-ice, changes in seasonal timing of bird nesting and plant flowering, and changes in timing and paths of insect migration. Attribution of climatic influence is much easier for these relatively simple systems, mostly lacking compelling alternative explanations. This, however, is not the case for patterns of infectious diseases in human populations – there are always several, sometimes many, plausible explanations for any observed change in pattern of occurrence. Hence, caution is needed in the interpretation of climate-associated changes in human infectious diseases. In general, no single report is conclusive.

Within a particular "climate envelope," many other social, economic, behavioral, and environmental factors also influence disease transmission. Against this complex and "noisy" background, it is unavoidably difficult to make a quantitative attribution to climate change of any observed change in the occurrence of some specified infectious disease. Vector-borne infectious diseases vary greatly in the complexity of their transmission modes, and hence some are much easier to study via modeling than are others. To date, the formal modeling of how climate change would affect vector-borne diseases has focused on malaria and dengue fever. Modeling future impacts is conceptually simpler for dengue than for malaria. The two main pathogen variants of malaria (falciparum and vivax) and its transmission relies on several dozen regionally dominant mosquito species, whereas dengue fever is transmitted principally by one mosquito vector, *Aedes aegypti*.

## Box 14.5 A tale of two vibrios – *cholerae* and *parahaemolyticus*

Outbreaks of cholera, caused by *Vibrio cholerae*, show an association with coastal water warming, as part of the El Niño cycle, in both Bangladesh and Peru. An American scientist, Rita Colwell (1996), has argued that the proliferation of phytoplankton and zooplankton in warmer water provides a biological culture medium for proliferation of this vibrio, whose natural home is in coastal waters, estuaries, and rivers. The proliferating vibrio then enters the aquatic food web, and reaches humans via fish that are caught and eaten.

Other more recent research (Long *et al.*, 2005) has found that, as water temperatures warm, the inhibition of proliferation of the cholera vibrio exerted by other bacterial species wanes and the vibrio becomes an increasingly dominant bacterium within that ecosystem. Support for one or both of these mechanisms is evident in the very strong correlation between the observed incidence of cholera in Matlab, near coastal Bangladesh, and the incidence predicted on the basis of sea-surface temperature and planktonic blooms during 2000–04 (Willcox and Colwell, 2005). These findings suggest that warming ocean waters will increase the risk of cholera outbreaks, particularly in vulnerable populations around the world.

Meanwhile, another temperature-sensitive vibrio inhabits some of the world's coastal waters. *Vibrio parahaemolyticus* is the main cause of seafood-associated food poisoning in the US. A major outbreak occurred on a cruise ship off northern Alaska in summer 2004, after passengers had eaten oysters (McLaughlin *et al.*, 2005). The record showed that mean coastal water temperatures had increased by 0.2ºC per year since 1997 – and, most interestingly, that 2004 was the only year in which the temperature exceeded the *critical temperature* of 15ºC throughout the July–August oyster harvest season. The authors concluded that: "Rising temperatures of ocean water seem to have contributed to one of the largest known outbreaks of *V. parahaemolyticus* in the US." They suggested that, with global warming, this elevated risk will persist in future.

Several recent reports are nevertheless suggestive of how climate change may affect the transmission of infectious agents. For example, tick-borne (viral) encephalitis (TBE) in Sweden appears to have increased in both its (northern) geographic range and, in Stockholm County, its annual incidence in response to a succession of warmer winters during the 1980s and 1990s (Lindgren 1998; Lindgren *et al.*, 2000). During that time there was evidence of an inter-annual correlation between winter–spring temperatures and the incidence of TBE in

Stockholm County. The geographic range of the ticks that transmit this disease extended northwards in Sweden during the 1980s and 1990s (Lindgren *et al.*, 2000). The range of the ticks has also increased in altitude in the Czech Republic (Danielova, 1975), in association with a recent warming trend (Zeman, 1997). However, these interpretations have been contested, including in relation to climatic influences on the complex seasonal dependence of the three life-stages of the tick (Randolph and Rogers, 2000).

There has been much interest in whether or not recent regional warming in parts of eastern and southern Africa has been the cause of increases in malaria incidence in the highlands, or whether human influences (such as habitat alteration or drug-resistant pathogen strains) were responsible. For the moment, the evidence remains equivocal. Several studies have noted an increase in highland malaria in recent decades (Loevinsohn, 1994; Lindblade *et al.*, 1999; Ndyomugyenyi and Magnussen, 2004), with some such increases occurring in association with local warming trends (Tulu, 1996; Bonora *et al.*, 2001). Two studies concluded that there had been no statistically significant trends in climate in those same regions (Hay *et al.*, 2002; Small *et al.*, 2003), although the medium-resolution climate data (New *et al.*, 1999) that were used in those two studies were subsequently deemed not well suited to research at this smaller geographical scale (Patz *et al.*, 2002). A recent re-analysis of the study that found no evidence of a temperature effect, updated to the present from 1950 to 2002 for four high-altitude sites in East Africa where malaria has become a serious public health problem, found evidence for a significant warming trend at all sites (Pascual *et al.*, 2006). It is most likely that the expansion of anti-malarial drug resistance and failed vector control programs, in addition to climate, are also important factors driving recent malaria expansions in these regions (Harvell *et al.*, 2002).

Recent studies in China indicate that the increase in incidence of schistosomiasis over the past decade may incorporate an influence of the warming trend. The critical "freeze line" limits the survival of the intermediate host (Oncomelania water snails) and hence the transmission of the parasite *Schistosomiasis japonica*. This has moved northwards, and now an additional 20 million people are at risk of schistosomiasis (Yang *et al.*, 2005).

Depending on the temperature preferences of pathogen and host, it is plausible to imagine that the season for proliferation of infections would expand or contract in future, with implications for the length of climate-sensitive infectious diseases. Donaldson (2006) and others have observed that the season associated with laboratory isolation of respiratory syncytial virus, and RSV-related emergency department admissions, now ends 3.1 and 2.5 weeks earlier, respectively, per 1°C increase in annual central England temperatures ($P = 0.002$ and 0.043, respectively). They conclude that climate change may be shortening the RSV season.

## Future research: recognizing and documenting complex patterns

There has been a rapid accrual of evidence of changes in physical and non-human biological systems, in association with regional warming trends. Taken together, these indicate a signal that global climate change is now beginning to change the conditions of the biophysical environment around us (Root *et al.*, 2003). This raises an obvious question. Is similar evidence emerging of changes in occurrence of infectious diseases that are plausibly due to recent or ongoing regional changes in climate? That is, can we develop formal methods of "pattern recognition"? Particular attention has recently been paid to reports of changes in malaria in eastern and southern Africa, tick-borne encephalitis in Sweden, and the temporal correlation of cholera outbreaks in Bangladesh with recently-intensifying El Niño events.

### Modeling future health risks under climate change scenarios

While epidemiologists have often projected from observed recent "exposures" and/or current disease trends to estimate future disease risks and burdens, they have much less experience in doing this in relation to the health risks of scenarios of future environmental conditions. Such scenarios usually entail plausible ranges of the underlying drivers (such as fossil fuel combustion as a major determinant of greenhouse gas emissions) rather than formal probability distributions. The scenarios also entail substantial uncertainties about both future societal trajectories and (climate) system responses to as-yet unexperienced (in human records) atmospheric composition. The former category of uncertainty can be better addressed by achieving a higher level of horizontal integration (of non-climate effect-modifiers) into the model. The latter category will require more empirical observation by climate scientists, both now and as the process unfolds (WHO, 2004).

Further, climate–health risk functions that extend into future decades, entailing higher climate-change exposures, may not be linear, and, anyway, the exposures may change in an unforeseen discontinuous fashion. Overall, then, this is a setting in which close interdisciplinary collaboration is needed, often across wide conceptual and content divides.

### Infectious disease outbreak forecasting systems

As the understanding of these complex causal influences of social and environmental conditions on infectious disease occurrence patterns improves, so the capacity to develop forecasting models will improve. This will be of public health importance – particularly in higher-risk regions of the world where prevention is usually very much better than "cure" (the latter often being unaffordable).

Vector control and public notification remain the only public health response to the majority of vector-borne diseases. Both measures require knowledge of an impending outbreak, and a suitable response time. Climate factors that drive the growth of vector populations and the replication of pathogens have the potential to be used as a proxy for early warning of the probability of an outbreak of disease. The objective is to detect conditions suitable for pathogen amplification in the natural cycle at the earliest possible time so that public health interventions can have the greatest opportunity for success. Climate forecasts will be helpful tools in the public health management of vector-borne diseases, if they can (i) improve the targeting and sensitivity of surveillance and increase the length of the response time, or (ii) reduce the cost of traditional surveillance activities. An interesting recent example comes from research carried out in southern Africa, showing how the empirically derived relationship between observed summer rainfall and subsequent annual malaria incidence can then be used successfully for forecasting malaria incidence in the coming year (Thomson *et al.*, 2006) – thereby providing 6–9 months advance warning for the pubic health and health-care facilities.

## Conclusion

The recent worldwide upturn in the occurrence of both new (emerging) and re-emerging or spreading infectious diseases highlights the importance of underlying environmental and social conditions as determinants of the generation, spread, and impact of infectious diseases in human populations. Human ecology, worldwide, is undergoing rapid transition. This encompasses urbanization, rising consumerism, changes in working conditions, population aging, marked increases in mobility, changes in culture and behavior, evolving health-care technologies, and other factors.

Global climate change is becoming a further, and major, large-scale influence on the pattern of infectious disease transmission. It is likely to become increasingly important over at least the next half-century, as the massive, high-inertial, and somewhat unpredictable process of climate change continues. As discussed in this chapter, the many ways in which climate change does and will influence infectious diseases are subject to a plethora of modifying (precluding, constraining, amplifying) influences by other factors and processes: constitutional characteristics of hosts, vectors and pathogens; the prevailing ambient conditions (topography, disease control programs, and others); and coexistent changes (local and global) in other social, economic, behavioral and environmental factors. This global anthropogenic process, climate change, along with other unprecedented global environmental changes, is beginning to destabilize and weaken the planet's life-support systems. Infectious diseases, unlike other diseases, depend on the biology and behavior – each often climate-sensitive – of two or more parties (pathogen, vector, intermediate host, human host). Hence, these diseases will be particularly susceptible to changes as the world's climate and its climate-sensitive geochemical and ecological systems undergo change over the coming decades.

## References

Al-Khatib, I., Ju'ba, A., Kamal, N. *et al.* (2003). Impact of housing conditions on the health of the people at al-Ama'ri refugee camp in the West Bank of Palestine. *International Journal of Environmental Health Research* **13**(4), 315–326.

Baird-Parker, A.C. (1990). Foodborne salmonellosis. *Lancet* **336**(8725), 1231–1235.

Baird-Parker, A.C. (1994). Fred Griffith Review Lecture. Foods and microbiological risks. *Microbiology* **140**, 687–695.

Balls, M.J., Bodker, R., Thomas, C.J. *et al.* (2004). Effect of topography on the risk of malaria infection in the Usambara Mountains, Tanzania. *Transactions of the Royal Society of Tropical Medicine and Hygiene* **98**(7), 400–408.

Bayliss, P. (1987). Kangaroo dynamics. In: G. Caughley, N. Shepherd and J. Short (eds), *Kangaroos: Their Ecology and Management in the Sheep Rangelands of Australia.* Cambridge: Cambridge University Press, pp. 119–134.

Bentham, G. and Langford, I.H. (1995). Climate change and the incidence of food poisoning in England and Wales. *International Journal of Biometeorology* **39**(1), 81–86.

Bentham, G. and Langford, I.H. (2001). Environmental temperatures and the incidence of food poisoning in England and Wales. *International Journal of Biometeorology* **45**(1), 22–26.

Besancenot, J.P., Boko, M. and Oke, P.C. (1997). Weather conditions and cerebrospinal meningitis in Benin (Gulf of Guinea, West Africa). *European Journal of Epidemiology* **13**(7), 807–815.

Bodker, R., Akida, J., Shayo, D. *et al.* (2003). Relationship between altitude and intensity of malaria transmission in the Usambara Mountains, Tanzania. *Journal of Medical Entomology* **40**(5), 706–717.

Bonora, S., de Rosa, F., Boffito, M. *et al.* (2001). Rising temperature and the malaria epidemic in Burundi. *Trends in Parasitology* **17**, 572–573.

Bouma, M.J. and Dye, C. (1997). Cycles of malaria associated with El Nino in Venezuela. *Journal of the American Medical Association* **278**(21), 1772–1774.

Bouma, M.J. and van der Kaay, H.J. (1996). The El Nino Southern Oscillation and the historic malaria epidemics on the Indian subcontinent and Sri Lanka: an early warning system for future epidemics? *Tropical Medicine & International Health* **1**(1), 86–96.

Bouma, M.J., Dye, C. and van der Kaay, H.J. (1996). Falciparum malaria and climate change in the northwest frontier province of Pakistan. *American Journal of Tropical Medicine & Hygiene* **55**(2), 131–137.

Bouma, M.J., Poveda, G., Rojas, W. *et al.* (1997). Predicting high-risk years for malaria in Colombia using parameters of El Nino Southern Oscillation. *Tropical Medicine & International Health* **2**(12), 1122–1127.

Carlson, J., Byrd, B. and Omlin, F. (2004). Field assessments in western Kenya link malaria vectors to environmentally disturbed habitats during the dry season. *BMC Public Health* **4**: doi:10.1186/1471-2458-4-33.

Caughley, G., Shepherd, N. and Short, J. (1987). *Kangaroos: Their Ecology and Management in the Sheep Rangelands of Australia*. Cambridge: Cambridge University Press.

Cazelles, B., Chavez, M., McMichael, A.J. *et al.* (2005). Non-stationary influence of El Nino on the synchronous dengue epidemics in Thailand. *Public Library of Science, Medicine* **2**(4), e106.

Chan, P.W., Chew, F.T., Tan, T.N. *et al.* (2002). Seasonal variation in respiratory syncytial virus chest infection in the tropics. *Pediatric Pulmonology* **34**(1), 47–51.

Checkley, W., Epstein, L.D., Gliman, R.H. *et al.* (2000). Effects of El Niño and ambient temperature on hospital admissions for diarrheal diseases in Peruvian children. *Lancet* **355**, 442–450.

Chew, F.T., Doraisingham, S., Ling, A.E. *et al.* (1998). Seasonal trends of viral respiratory tract infections in the tropics. *Epidemiology and Infection* **121**(1), 121–128.

Chua, K.B. (2003). Nipah virus outbreak in Malaysia. *Journal of Clinical Virology* **26**(3), 265–275.

Chua, K.B., Chua, B.H. and Wang, C.W. (2002). Anthropogenic deforestation, El Nino and the emergence of Nipah virus in Malaysia. *Malaysian Journal of Pathology* **24**(1), 15–21.

Colwell, R. (1996). Global climate and infectious disease: the cholera paradigm. *Science* **274**(5295), 2025–2031

Danielova, V. (1975). Overwintering of mosquito-borne viruses. *Medical Biology* **53**, 282–287.

Daszak, P., Cunningham, A.A. and Hyatt, A.D. (2000). Emerging infectious diseases of wildlife – threats to biodiversity and human health. *Science* **287**(5452), 443–449.

Daszak, P., Plowright, R., Epstein, J.H. *et al.* (2006). The emergence of Nipah and Hendra virus: pathogen dynamics across a wildlife-livestock-human continuum. In: S.K. Collinge and C. Ray (eds), *Disease Ecology: Community Structure and Pathogen Dynamics*. Oxford: Oxford University Press, 186–201.

Davies, F.G., Linthicum, K.J. and James, A.D. (1985). Rainfall and epizootic Rift Valley Fever. *Bulletin of the World Health Organization* **63**(5), 941–943.

Donaldson, G.C. (2006). Climate change and the end of the respiratory syncytial virus season. *Clinical Infectious Diseases* **42**(5), 677–679.

D'Souza, R.M., Becker, N.G., Hall, G. *et al.* (2004). Does ambient temperature affect food-borne disease? *Epidemiology* **15**(1), 86–92.

Ebi, K.L., Hartman, J., Chan, N. *et al.* (2005). Climate suitability for stable malaria transmission in Zimbabwe under different climate change scenarios. *Climatic Change* **73**, 375–393.

Engelthaler, D.M., Mosley, D.G., Cheek, J.E. *et al.* (1999). Climatic and environmental patterns associated with hantavirus pulmonary syndrome, Four Corners region, United States. *Emerging Infectious Diseases* **5**(1), 87–94.

Epstein, P.R. (1999). Climate and health. *Science* **285**, 347–348.

Greenwood, B.M., Blakebrough, I.S., Bradley, A.K. *et al.* (1984). Meningococcal disease and season in sub-Saharan Africa. *Lancet* **1**(8390), 1339–1342.

Haberberger, R.L., Fox, E., Asselin, P. *et al.* (1990). Is Djibouti too hot and too humid for meningococci? *Transactions of the Royal Society of Tropical Medicine and Hygiene* **84**, 588.

Hales, S., Weinstein, P. and Woodward, A. (1996). Dengue fever epidemics in the South Pacific: driven by El Niño Southern Oscillation? *Lancet* **348**, 1664–1665.

Hales, S., Weinstein, P., Souares, Y. *et al.* (1999). El Niño and the dynamics of vector-borne disease transmission. *Environmental Health Perspectives* **107**, 99–102.

Hales, S., Kovats, S. and Woodward, A. (2000). What El Niño can tell us about human health and global climate change. *Global Change and Human Health* **1**, 66–77.

Hales, S., de Wet, N., Maindonald, J. *et al.* (2002). Potential effect of population and climate changes on global distribution of dengue fever: an empirical model. *Lancet* **360**(9336), 830–834.

Hambling, M.H. (1964). Survival of the respiratory syncytial virus during storage under various conditions. *British Journal of Exploratory Pathology* **45**, 647–655.

Harvell, C.D., Mitchell, C.E., Ward, J.R. *et al.* (2002). Climate warming and disease risks for terrestrial and marine biota. *Science* **296**(21 June), 2158–2162.

Hay, S.I., Cox, J., Rogers, D.J. *et al.* (2002). Climate change and the resurgence of malaria in the East African highlands. *Nature* **415**(6874), 905–909.

Hopp, M. and Foley, J. (2003). Worldwide fluctuations in dengue fever cases related to climate variability. *Climate Research* **25**, 85–94.

IPCC (2001). *Climate Change 2001: The Scientific Basis. Contribution of Working Group I to the Third Assessment Report of the Intergovernmental Panel on Climate Change.* Cambridge: Cambridge University Press.

Isaacs, N. and Donn, M. (1993). Housing and health – seasonality in NZ mortality. *Australian and New Zealand Journal of Public Health* **17**, 68–70.

Jones, A.P. (1998). Asthma and domestic air quality. *Social Science and Medicine* **47**, 755–764.

Karl, T.R. and Trenberth, K.E. (2003). Modern global climate change. *Science* **302**(5651), 1719–1723.

Kay, B.H. and Aaskov, J.G. (1989). Ross River virus (epidemic polyarthritis). In: T.P. Monath (ed.), *The Arboviruses: Epidemiology and Ecology*, Vol. 4. Boca Raton: CRC Press, pp. 93–112.

Kim, P.E., Musher, D.M., Glezen, W.P. *et al.* (1996). Association of invasive pneumococcal disease with season, atmospheric conditions, air pollution, and the isolation of respiratory viruses. *Clinical Infectious Diseases* **22**(1), 100–106.

Kovats, R.S., Campbell-Lendrum, D.H., McMichael, A.J. *et al.* (2001). Early effects of climate change: do they include changes in vector-borne disease? *Philosophical Transactions of the Royal Society of London Series B – Biological Sciences* **356**(1411), 1057–1068.

Kovats, R.S., Bouma, M.J., Hajat, S. *et al.* (2003). El Niño and health. *Lancet* **362**, 1481–1482.

Kovats, R.S., Edwards, S.J., Hajat, S. *et al.* (2004). The effect of temperature on food poisoning: a time-series analysis of salmonellosis in ten European countries. *Epidemiology and Infection* **132**, 443–453.

Kovats, R.S., Edwards, S.J., Charron, D. *et al.* (2005). Climate variability and campylobacter infection: an international study. *International Journal of Biometeorology* **49**(4), 207–214.

Leake, C.J. (1998). Mosquito-borne arboviruses. In: S.K. Palmer, L. Soulsby and D.I.H. Simpson (eds), *Zoonoses*. Oxford: Oxford University Press, pp. 401–413.

Lee, D.J., Hicks, M.M., Debenham, M.L. *et al.* (1989). *The Culicidae of the Australasian Region*. Canberra: Australian Government Publishing Service.

Lindblade, K.A., Walker, E.D., Onapa, A.W. *et al.* (1999). Highland malaria in Uganda: prospective analysis of an epidemic associated with El Nino. *Transactions of the Royal Society of Tropical Medicine and Hygiene* **93**(5), 480–487.

Lindgren, E. (1998). Climate change, tick-borne encephalitis and vaccination needs in Sweden – a prediction model. *Ecological Modelling* **110**(1), 55–63.

Lindgren, E., Tälleklint, L. and Polfeldt, T. (2000). Impact of climatic change on the northern latitude limit and population density of the disease-transmitting European tick *Ixodes ricinus*. *Environmental Health Perspectives* **108**(2), 119–123.

Lindsay, M., Broom, A.K., Wright, A.E. *et al.* (1993). Ross River virus isolations from mosquitoes in arid regions of Western Australia: implications of vertical transmission

as a means of persistence of the virus. *American Journal of Tropical Medicine and Hygiene* **49**(6), 686–696.
Lindsay, M. and Mackenzie, J. (1997). Vector-borne viral diseases and climate change in the Australasian region: major concerns and the public health response. In: P. Curson, C. Guest and E. Jackson (eds), *Climate Change and Human Health in the Asia-Pacific Region*. Canberra: Australian Medical Association and Greenpeace International, pp. 47–62.
Lindsay, S.W. and Birley, M.H. (1996). Climate change and malaria transmission. *Annals of Tropical Medicine and Parasitology* **90**(6), 573–588.
Loevinsohn, M.E. (1994). Climatic warming and increased malaria incidence in Rwanda. *Lancet* **343**(8899), 714–718.
Long, R.A., Rowley, D.C., Zamora, E. *et al.*, (2005). Antagonistic interactions among marine bacteria impede the proliferation of *Vibrio cholerae*. *Applied and Environmental Microbiology* **71**, 8531–8536.
Maelzer, D., Hales, S., Weinstein, P. *et al.* (1999). El Niño and arboviral disease prediction. *Environmental Health Perspectives* **107**(10), 817–818.
Marshall, I.D. (ed.) (1979). *The Epidemiology of Murray Valley Encephalitis in Eastern Australia – Patterns of Arbovirus Activity and Strategies of Arbovirus Survival. Arbovirus Research in Australia*. Brisbane: CSIRO Division of Animal Health and Queensland Institute of Medical Research.
Martens, P., Kovats, R.S., Nijhof, S. *et al.* (1999). Climate change and future populations at risk of malaria. *Global Environmental Change* **9**, S89–107.
McLaughlin, J.B., DePaola, A., Bopp, C.A. *et al.* (2005). Outbreak of *Vibrio parahaemolyticus* gastroenteritis associated with Alaskan oysters. *New England Journal of Medicine* **353**(14), 1463–1470.
Milne, A. (2005). Flooding in Mozambique. In: P.R. Epstein and E. Mills (eds), *Climate Change Futures: Health, Ecological and Economic Dimensions*. Boston: Center for Health and the Global Environment, Harvard Medical School, p. 36.
Molesworth, A.M., Cuevas, L.E., Connor, S.J. *et al.* (2003). Environmental risk and meningitis epidemics in Africa. *Emerging Infectious Diseases* **9**(10), 1287–1293.
Ndyomugyenyi, R. and Magnussen, P. (2004). Trends in malaria – attributable morbidity and mortality among young children admitted to Ugandan hospitals, for the period 1990–2001. *Annals of Tropical Medicine and Parasitology* **98**, 315–327.
New, M., Hulme, M. and Jones, P. (1999). Representing twentieth-century space–time climate variability. Part I: Development of a 1961–90 mean monthly terrestrial climatology. *Journal of Climate* **12**(3), 829–856.
Nichols, G.L. (2005). Fly transmission of *Campylobacter*. *Emerging Infectious Diseases* **11**, 361–364.
Noy-Meir, I. (1973). Desert ecosystems: environments and producers. *Annual Review of Ecology and Systematics* **4**, 25–51.
Ogden, N.H., Maarouf, A., Barker, I.K. *et al.* (2006). Climate change and the potential for range expansion of the Lyme disease vector *Ixodes scapularis* in Canada. *International Journal of Parasitology* **36**(1), 63–70.
Pascual, M., Ahumada, J.A., Chaves, L.F. *et al.* (2006). Malaria resurgence in the East African highlands: temperature trends revisited. *Proceedings of the National Academy of Sciences of the USA* **103**(15), 5829–5834.
Patz, J.A., Strzepek, K., Lele, S. *et al.* (1998). Predicting key malaria transmission factors, biting and entomological inoculation rates, using modelled soil moisture in Kenya. *Tropical Medicine and International Health* **3**(10), 818–827.
Patz, J.A., Hulme, M., Rosenweig, C. *et al.* (2002). Climate change – regional warming and malaria resurgence. *Nature* **420**(6916), 627–628.

Randolph, S.E. and Rogers, D.J. (2000). Fragile transmission cycles of tick-borne encephalitis virus may be disrupted by predicted climate change. *Proceedings of the Royal Society of London Series B – Biological Sciences* **267**, 1741–1744.

Rechsteiner, J. and Winkler, K.C. (1969). Inactivation of respiratory syncytial virus in aerosol. *Journal of General Virology* **5**(3), 405–410.

Reeves, W.C., Hardy, J.L., Reisen, W. *et al.* (1994). Potential effect of global warming on mosquito-borne arboviruses. *Journal of Medical Entomology* **31**, 323–332.

Reiter, P. (1988). Weather, vector biology, and arboviral recrudescence. In: T.P. Monath (ed.), *The Arboviruses: Epidemiology and Ecology*, Vol. 1. Boca Raton: CRC Press, pp. 245–255.

Reiter, P., Lathrop, S., Bunning, M. *et al.* (2003). Texas lifestyle limits transmission of dengue virus. *Emerging Infectious Diseases* **9**(1), 86–89.

Rogers, D.J. and Randolph, S.E. (2000). The global spread of malaria in a future, warmer world. *Science* **289**, 1763–1766.

Root, T. L., Price, J.T., Hall, K.R. *et al.* (2003). Fingerprints of global warming on wild animals and plants. *Nature* **421**(6918), 57–60.

Russell, R.C. (1994). Ross River virus: disease trends and vector ecology in Australia. *Bulletin of the Society of Vector Ecology* **19**(1), 73–81.

Schmid, H., Burnens, A.P., Baumgartner, A. *et al.* (1996). Risk factors for sporadic salmonellosis in Switzerland. *European Journal of Clinical Micriobiology & Infectious Diseases* **15**, 725–732.

Sellers, R.F. (1980). Weather, host and vector – their interplay in the spread of insect-borne animal viral diseases. *Journal of Hygiene* **85**, 65–102.

Sennerstam, R.B. and Moberg, K. (2004). Relationship between illness-associated absence in day-care children and weather parameters. *Public Health* **118**(5), 349–353.

Shek, L.P. and Lee, B.W. (2003). Epidemiology and seasonality of respiratory tract virus infections in the tropics. *Paediatric Respiratory Reviews* **4**(2), 105–111.

Singh, R.B., Hales, S., de Wet, N. *et al.* (2001). The influence of climate variation and change on diarrheal disease in the Pacific Islands. *Environmental Health Perspectives* **109**(2), 155–159.

Small, J., Goetz, S.J., Hay, S.I. *et al.* (2003). Climatic suitability for malaria transmission in Africa, 1911–1995. *Proceedings of the National Academy of Sciences of the United States of America* **100**(26), 15,341–15,345.

Strahan, R. (ed.) (1991). *Complete Book of Australian Mammals*. Sydney: The Australian Museum and Cornstalk Publishing.

Sultan, B., Labadi, K., Guegan, J.F. *et al.* (2005). Climate drives the meningitis epidemics onset in west Africa. *Public Library of Science, Medicine* **2**(1), e6.

Tanser, F., Sharp, B.L. and le Sueur, D. (2003). Potential effect of climate change on malaria transmission in Africa. *Lancet* **362**, 1792–1798.

Thomas, C.J., Davies, G. and Dunn, C.E. (2004). Mixed picture for changes in stable malaria distribution with future climate in Africa. *Trends in Parasitology* **20**(5), 216–220.

Thomson, M., Doblas-Reyes, F.J., Mason, S.J. *et al.* (2006). Malaria early warnings based on seasonal climate forecasts from multi-model ensembles. *Nature* **439**, 576–579.

Tong, S. and Hu, W. (2001). Climate variation and incidence of Ross River virus in Cairns, Australia: a time-series analysis. *Environmental Health Perspectives* **109**(12), 1271–1273.

Tong, S.L., Hu, W.B. and McMichael, A.J. (2004). Climate variability and Ross River virus transmission in Townsville region, Australia, 1985–1996. *Tropical Medicine & International Health* **9**(2), 298–304.

Tulu, A.N. (1996). *Determinants of Malaria Transmission in the Highlands of Ethiopia. The Impact of Global Warming on Morbidity and Mortality ascribed to Malaria.* London: University of London.
Turcios, R.M., Curns, A.T., Holman, R.C. *et al.* (2006). Temporal and geographic trends of rotavirus activity in the United States, 1997–2004. *Pediatric Infectious Disease Journal* **25**(5), 451–454.
van Lieshout, M., Kovats, R.S., Livermore, M.T.J. *et al.* (2004). Climate change and malaria: analysis of the SRES climate and socio-economic scenarios. *Global Environmental Change – Human and Policy Dimensions* **14**(1), 87–99.
Weiss, R.A. and McMichael, A.J. (2004). Social and environmental risk factors in the emergence of infectious diseases. *Nature Medicine* **10**(12 Suppl.), S70–66.
WHO (2004). *Using Climate to Predict Infectious Disease Outbreaks: A Review.* Geneva: World Health Organization, p. 55.
Willcox, B.A. and Colwell, R. (2005). Emerging and re-emerging infectious diseases: biocomplexity as an interdisciplinary paradigm. *Ecosystem Health* **2**, 244–257.
Wilson, M.L. (2001). Ecology and infectious disease. In: J.L. Aron and J.A. Patz (eds), *Ecosytem Change and Public Health: A Global Perspective.* Baltimore: Johns Hopkins University Press, pp. 283–324.
Winch, P. (1998). Social and cultural responses to emerging vector-borne diseases. *Journal of Vector Ecology* **23**(1), 47–53.
Woodruff, R.E. (2005). Epidemic early warning systems: Ross River virus disease in Australia. In: K.L. Ebi, J. Smith and I. Burton (eds), *Integration of Public Health with Adaptation to Climate Change: Lessons Learned and New Directions.* London: Taylor & Francis Group, pp. 91–113.
Woodruff, R.E., Guest, C.S., Garner, M.G. *et al.* (2002). Predicting Ross River virus epidemics from regional weather data. *Epidemiology* **13**(4), 384–393.
Yang, G.J., Vounatsou, P., Zhou, X.N. *et al.* (2005). A potential impact of climate change and water resource development on the transmission of *Schistosoma japonicum* in China. *Parassitologia* **47**(1), 127–134.
Zeman, P. (1997). Objective assessment of risk maps of tick-borne encephalitis and Lyme borreliosis on spatial patterns of located cases. *International Journal of Epidemiology* **26**(5), 1121–1130.

# 15 Governance, human rights and infectious disease: theoretical, empirical and practical perspectives

**Jonathan Cohen and Joseph J. Amon**

The relationship between the health status of a population and the behavior of the government under which that population lives has been previously explored in war or civil crisis settings, and in the context of chronic health issues such as nutrition, famine, and child mortality (Sen, 1999; Dreze and Sen, 2002). These studies have found that government behavior (sometimes captured by the term "governance"), as measured by indicators such as accountability, stability, rule of law, respect for human rights, and the existence of an independent civil society, plays a significant role in health outcomes – a role independent of, and perhaps even superior to, host genetics, insect vectors, or individual behaviors. Famines stem not solely from bad weather or genetics, but also from the failure of governments to protect their populations from civil strife, or to equitably distribute food aid. Occupational illnesses, such as pneumoconiosis (or "black lung"), can be understood in terms of the risk behaviors and biological susceptibility of mineworkers, but also as a failure of governments to protect individuals in the workplace.

Despite the widespread acceptance of these findings, the influence of governance on infectious disease spread has received far less attention. As the chapters in this volume illustrate, there is an increasing appreciation that factors outside what is traditionally considered the health sphere can contribute to creating or controlling endemic and epidemic infectious disease. Some of these causes are socio-cultural, while others are the result of migrating populations, transportation infrastructure, and the workforce. There is increasing evidence, from diverse infectious diseases and using diverse research approaches, that suggests that government behavior has an influence on infectious disease spread and control as well.

We begin this chapter by exploring the fundamental obligation of governments to protect health, drawing on both historical state practice and explicit obligations under international human rights law. We note that this obligation extends not only to the provision of health care or to the control of infectious disease outbreaks, but also to the cessation of human rights violations that contribute to poor health or disease risk. We then review recent attempts to quantify the relationship between good (or bad) governance and public health, noting that such quantification is often limited by confounding factors and methodological limitations. By reference to three infectious diseases – HIV and AIDS, Guinea Worm, and SARS – we explore some of the causal mechanisms by which human rights abuses might fuel disease spread and/or constrain the ability of governments to arrest disease. We conclude by reflecting on the importance of developing a coherent framework for understanding the precise relationship between governance, human rights, and infectious disease.

## The obligation to protect health

The written records of ancient China, Egypt, India, and Peru document humankind's earliest efforts to provide safe water, sewage, and drainage systems so as to protect public health. In the absence of a clear scientific understanding of infectious diseases, these efforts were largely linked to religious beliefs. City states in ancient Greece created sanitation systems for the entire community, and medical care for the poor. By the thirteenth century, Italian cities had laws modeled after ancient Roman standards to prevent epidemic disease through maintaining clean water supplies, controlling refuse disposal, and monitoring migrants to the city that might be carrying infectious disease (Kiple, 1995). The Elizabethan Poor Laws in Britain in the early seventeenth century strengthened the responsibility of local authorities for health and welfare. The Public Health Act of 1848, passed to improve sanitation in England and Wales, was one of the great milestones in public health history (Fee and Brown, 2005). By the twentieth century, the authority of local, state, and national governments was extended from sanitation and indigent medical care to activities such as chlorinating and fluoridating community water supplies, conducting insect vector control, screening and inoculating against infectious diseases, partner notification and tracking of sexually transmitted diseases, and regulating food, drugs, and the blood supply. Each of these actions acknowledged the widening responsibility of government to maintain and promote public health, especially in the area of infectious disease control.

In more recent years there has been an explicit recognition of the obligation of governments to protect health as a matter of human rights. The Universal Declaration of Human Rights (1948) recognizes that:

> Everyone has the right to a standard of living adequate for the health and well-being of himself and of his family, including food, clothing, housing and medical care and necessary social services, and the right to security in the event of unemployment, sickness, disability, widowhood, old age or other lack of livelihood in circumstances beyond his control.

The UDHR further recognizes that all people are entitled to "realization … of the economic, social and cultural rights indispensable for his dignity and the free development of his personality." Among the social rights recognized in the subsequently ratified International Covenant on Economic, Social and Cultural Rights is the right to "the highest attainable standard of health" (International Covenant on Economic, Social and Cultural Rights, 1966). Under this provision, governments are obligated to respect, protect, and fulfill the "right to health" by taking positive actions that ensure access to high-quality health services, and by refraining from or preventing negative actions that interfere with health, such as denying health care to certain populations or censoring health-related information. The right to the highest attainable standard of health is also intimately linked to the enjoyment of a full range of civil and political rights, such as the right to information, equality, and due process under law. So, for example, a government ban on the reporting of a newly identified infectious disease may violate the right to information under the International Covenant on Civil and Political Rights (1966) and also infringe upon the right to health by preventing individuals from protecting themselves from illness. Violence, discrimination, and arbitrary actions by the state also can have both direct and indirect public health impacts – for example, when police officers arbitrarily detain outreach workers providing life-saving HIV-prevention services, this implicates not only due process rights but also the health of those who benefit from these services (Human Rights Watch, 2006a).

While not always defined in the language of human rights obligations, epidemiologists, national governments, and international agencies have increasingly recognized that since health is shaped by the broadest spectrum of social, cultural, and political factors, the analysis of health status and programs designed to respond to poor health must take all of these factors into account. Areas of public health, such as health promotion, that traditionally emphasized individual lifestyle choices and personal responsibility are now being reframed to recognize broader influences on individual behaviors and broader responsibility of government for promoting healthy social environments. A milestone of this shift was the 1986 World Health Organization's Ottawa Charter, which set out five key areas of action for influencing healthy behaviors: building healthy public policy, creating supportive environments, strengthening community action, developing personal skills, and reorienting health services (WHO, 1986; Breslow, 1999). Critics of "black box" epidemiology, who advocated for more contextualized and structural analyses to be incorporated into epidemiological studies, mirrored this movement (Susser and Susser, 1996a, 1996b). The field of social epidemiology, which looks at social factors that shape disease vulnerability, also recognizes the important role of political context and human rights violations in fueling infectious disease spread.

## Linking measures of governance to general health indicators

Both governance and health can be measured in many different ways, and one factor complicating efforts to measure the association between the two is the difficulty finding accurate, sensitive, and specific enough indicators for either variable. Broad measures of population health – such as life expectancy, and infant and maternal mortality – have been chosen by a number of authors, in part because these indicators capture (and average the effects of) multiple specific diseases and are broadly distributed across the population. To measure governance, two approaches have been used. Some authors have chosen the rankings of the organization Freedom House, which scores countries based upon political rights and civil liberties, electoral process, political pluralism and participation, functioning of government (including transparency and corruption), freedom of expression and belief; association; organizational rights, rule of law and personal autonomy; and individual rights (Freedom House, 2006). Other authors have used a set of governance indicators collected by the World Bank that includes measures of voice and accountability, political stability, government effectiveness, regulatory quality, rule of law, and corruption (Kaufmann *et al.*, 2003).

In a 2004 article in the *British Medical Journal*, Alvaro Franco and his colleagues plotted life expectancy and maternal and infant mortality in 170 countries against the "freedom index" produced by Freedom House (Figure 15.1). Controlling for determinants of health that include socio-economic and political measures such as wealth, equality, and the size of the public sector, the authors found a statistically significant relationship ($r$ = .XX, $P < .0X$) between freedom ratings and health indicators at all income levels. In their conclusion, they speculate that democracies produce better health outcomes because they "allow more space for social capital," like social networks and pressure groups, opportunities for empowerment, better access to information, and better recognition by government of people's needs (Franco *et al.*, 2004).

In a similar study, again using life expectancy, infant and maternal mortality as health indicators, Alvarez-Dardet and Franco-Giraldo (2006) analyzed data from 23 post-Communist countries during the last decade of the twentieth century. Again there was a significant correlation between the level of democratization

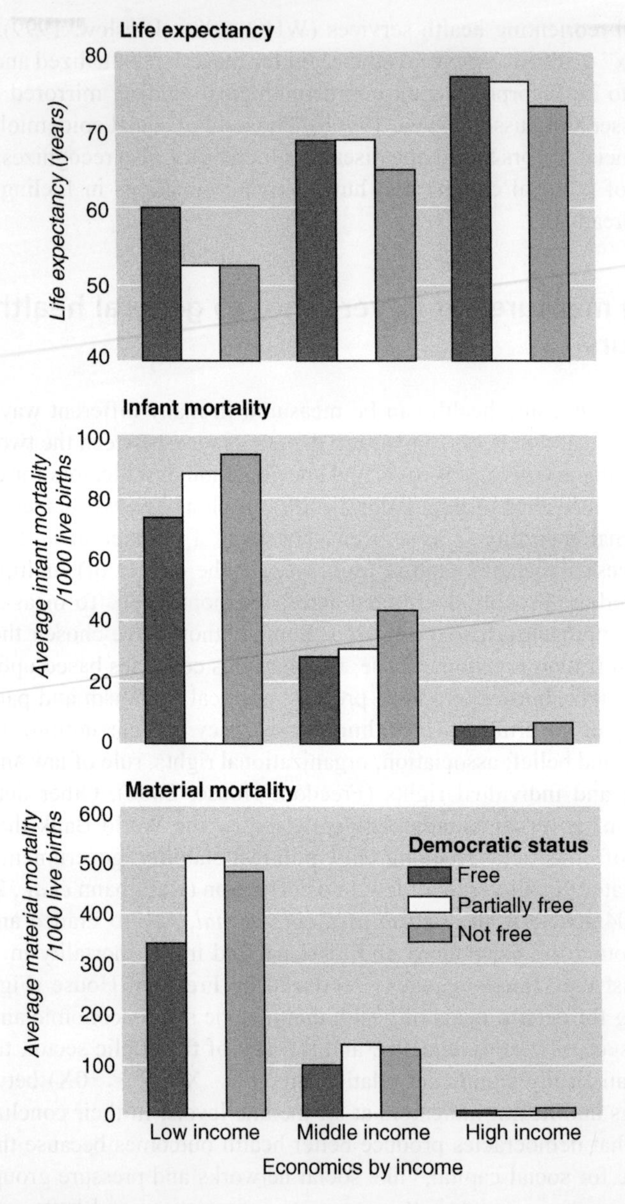


**Figure 15.1** Health indicators in 170 countries by classification of economies (World Bank) and democracy (Freedom House), 1998. Source: Franco *et al.* (2004).

and health ($r = 0.XX$, $P < 0.0X$), taking into account both wealth and the level of inequality.

In a third study, using governance indicators collated by the World Bank, Reidpath and Allotey (2006) plotted a composite governance indicator against infant mortality and healthy life expectancy (disability adjusted life expectancy) in 176 countries, and found significant correlation ($r = -0.68$ and $r = 0.72$, respectively, $P < 0.001$ for both). To control for *per capita* wealth the authors performed regression analyses, which identified both governance and wealth (measured as *per capita* gross domestic product (GDP)) as independently, and significantly, associated with life expectancy and with each other. The authors noted that these multiple correlations, as well as the correlation with another variable studied, the adequate supply of water, made it difficult to fully describe causality.

## Linking measures of governance to infectious disease indicators

The attempt to demonstrate quantitatively a link between governance factors and a more narrow measure of infectious disease risk in particular has been more challenging, in part because infectious disease risk is not spread evenly across populations, and comparisons between countries of the prevalence of a specific disease miss the over-burdening of some communities or subgroups within a country. In addition, governance influences different infectious disease risks in different ways, and presents distinct challenges to governments – compelling different types of government policies and approaches. For example, the risks posed by mosquito-borne diseases such as malaria or dengue fever present different possibilities for spread and control than, for example, hepatitis B or C, which are transmitted through sex or blood contact.

Nonetheless, the belief that there is a link between governance and specific infectious disease spread stems from an understanding that how governments handle infectious disease is determined not just by epidemiologic characteristics, but also by the overall political and social climate of the nation. Experience shows, for example, that the response to infectious disease epidemics will be strongly influenced to the detriment of the public health by the level of social opprobrium against the populations affected, by social discomfort with the means of transmission implicated (such as drug use or sex), and by fear and ignorance surrounding the disease or its means of transmission. These factors may make it important for governments to address infectious disease spread through working respectfully with at-risk populations, rather than adopting top-down approaches (such as criminalizing disease transmission, instituting mandatory testing, and quarantining people living with infectious diseases) that risk driving these populations even further to the margins of society where they cannot be reached with prevention services.

The theoretical basis for this observation is built upon the pioneering work of Jonathan Mann and Paul Farmer, who were among the first to describe the impact of human rights violations on health. Mann argued that pervasive human rights abuses perpetrated against socially marginalized groups (e.g. sexual violence, discrimination, police abuse) increased their risk of acquiring HIV, and that coercive public health responses such as quarantining and forced testing served to drive these groups further into hiding and fuel the epidemic (Mann, 1999). Informed by years of delivering HIV care in Haiti, Paul Farmer used the term "structural violence" to describe conditions of poverty, sexism, racism, and political violence that constrain individuals' ability to make informed and autonomous choices about their health (Farmer, 1999; Farmer, 2004).

Since 2001, the HIV/AIDS and Human Rights Program at Human Rights Watch has gathered thousands of testimonies from persons living with and at high risk of HIV, documenting the link between human rights abuses against them and their risk of HIV. These abuses have included rape, domestic violence, sex discrimination, and other abuses against women and girls; arbitrary arrest, beatings, torture, and the over-incarceration of injecting drug users, gay and bisexual men, sex workers, and other vulnerable groups; arbitrary detention of AIDS activists and outreach workers; and censorship of science-based HIV/AIDS information (Human Rights Watch, 2006b).

One of the first attempts to explain quantitatively the link between a specific infectious disease and governance factors was the study conducted by Menon-Johansson (2005) using World Bank governance indicators to examine HIV prevalence in 149 countries (see Table 15.1). The study found a significant negative correlation between HIV prevalence and all six governance dimensions ($r$ ranged from 0.12 to 0.20, and $P$ ranged from 0.03 to 0.001). The study, though statisti-

**Table 15.1 HIV prevalence correlations for each governance dimension and mean governance**

| Governance dimension | Correlation coefficient ($n = 149$) | $P$ value |
|---|---|---|
| Voice and accountability | −0.123 | 0.032 |
| Political stability and absence of violence | −0.164 | 0.004 |
| Government effectiveness | −0.204 | 0.000 |
| Regulatory quality | −0.157 | 0.006 |
| Rule of law | −0.194 | 0.001 |
| Corruption | −0.184 | 0.001 |
| Mean governance | −0.170 | 0.003 |

Source: Menon-Johansson *et al.* (2005); original publisher, Biomed Central.

cally significant, suggested that governance accounted for only a small percentage of the variance in HIV prevalence from one country to the next.

To address the relatively weak correlation found by Menon-Johansson (2005), Reidpath and Allotey (2006) re-analyzed the data using a single composite indicator from the six governance dimensions provided by the World Bank. The authors found a similar result ($r = 0.2$, $P < 0.05$), and concluded that analyzing structural measures such as governance versus single diseases was bound to show a weaker correlation than broader measures unless the diseases were ubiquitous. However, the authors say little about the limitations inherent in comparing across countries an infectious disease such as HIV that is manifested in different communities (because of different frequencies of dominant risk behaviors) and was introduced at different times into different communities and countries. As HIV is still a relatively newly introduced infection in many countries, measurements assessing current HIV prevalence versus governance may as yet be inappropriate. The dynamics of the global HIV epidemic are still fluid and, particularly in Eastern Europe and Asia, future prevalence is uncertain, making an analysis using current prevalence an uneven comparison.

## Case studies

A clearer way to illustrate the relationship between governance and infectious disease is through case studies. This section highlights three diverse examples of infectious diseases, and explains how governance influenced their spread. In the first example, of Guinea worm disease, we will discuss how the neglect of rural and often ethnically marginalized populations led to a disease with a simple lifecycle and relatively easy means of eradication being undercounted and largely ignored despite significant personal, community level, and national economic and health impacts. In the second example, HIV/AIDS, we will discuss how a disease of socially marginalized populations (for example, injection drug users and, in sub Saharan Africa, women) was both initially ignored and subsequently poorly controlled by governments with poor records on human rights and governance. In the third example, we will discuss how the lack of political freedoms in China led to poor recognition and handling of a newly identified infectious disease, severe acute respiratory syndrome (SARS).

### Guinea worm

Guinea worm disease, also known as dracunculiasis, is caused by an infection with the nematode *Dracunculus medinensis*, which lives in the subcutaneous and connective tissues – generally of the legs. Guinea worm disease is characterized by a small, intensely painful blister that is formed by the adult female worm (which, over the course of roughly 12 months' incubation, can grow to a meter in length) as it emerges to release its larvae. Larvae released into stagnant fresh water are ingested by tiny copepods (*Cyclops* spp.), which after two weeks of development are infective when swallowed.

Guinea worm is primarily a rural disease and, despite an enormous economic impact on rural communities, where it is endemic (and an enormous cumulative national economic impact), the disease was little noticed because its burden was primarily felt among the rural poor. Many health ministries in countries where it was endemic – a broad band across sub Saharan Africa, Yemen, Afghanistan, and Pakistan – had little awareness of the disease (Needham and Canning, 2003) and were doing virtually nothing about it.

In reaction to this neglect, in 1991 the Forty-Fourth World Health Assembly (WHA) laid out a strategy for Guinea worm eradication that includes three key governmental steps based on transparency and accountability (Hopkins and Ruiz-Tiben, 1991; WHO, 1991). The widespread availability of information on both epidemic and endemic infectious disease burdens helps to shape the process of the prioritization of response to public health threats. In democratic countries, this information can be used to facilitate the participation of civil society in setting the agenda and goals of public health campaigns and, through civil society involvement, of ensuring their effectiveness.

The first step in the proposed campaign for global Guinea worm eradication was to create a national plan based on a national survey to identify all endemic villages and assess the annual number of cases. In most of the countries affected, prior national surveys had greatly undercounted the number of people infected, and thereby understated the burden of disease. For example, prior to the national eradication program in Nigeria, less than 5000 cases were counted annually. After conducting a true national survey under the newly created Guinea worm program, between 640,000 and 650,000 cases were uncovered (Hopkins and Ruiz-Tiben, 1991).

The second step set out by the WHA strategy was to create safe water supplies using health education messages, "reinforced by religious and traditional/political leaders in the village, schoolteachers, agricultural and other extension workers, community organizations, and by the mass media (radio, posters, etc.) in the local languages" (Hopkins and Ruiz-Tiben, 1991). This kind of effort includes and fosters community participation, but perhaps more importantly encourages and re-emphasizes transparency and openness about infectious disease rather than suppression of information.

The third step, case containment, includes free treatment, which serves to encourage individuals to come forward for Guinea worm screening and health education. The total strategy for Guinea worm eradication illustrates a strategic balancing of concern for human rights and public health. It is an open and participatory approach. Governments demonstrate accountability to individual and community needs, and target the intervention in ways that are inclusive and non-coercive. Based on this strategy, 11 of the 20 nations most impacted by Guinea worm have been successful in eliminating the disease from their countries, and the overall number of people infected has been reduced from an estimated 3.5 million in 1986 to 10,674 in 2005 (Carter Center, 2006).

## HIV and AIDS

When it was first recognized in the early 1980s, AIDS was labeled "gay-related immune deficiency" (GRID). Soon afterwards, it was discovered that HIV was easily transmissible through the sharing of syringes among injecting drug users. The association of a new and poorly understood disease with stigmatizing and criminal behaviors (as well as the widespread blame of "foreigners" for the introduction of the disease) led, quite predictably, to a wide range of ineffective, discriminatory, and stigmatizing public health control strategies. Throughout the history of the HIV epidemic, governments have variously attempted to criminalize HIV transmission, quarantine people living with HIV, and censor factual information about safer sex and drug use. Political leaders have pandered to the stigma surrounding HIV by denying the extent of epidemics in their countries, and, in some cases, suggesting that AIDS is a punishment for perverse and sinful behavior. Such policies and attitudes, largely discredited by experience and research, have pushed already vulnerable groups further to the margins of society and ultimately undermined both HIV prevention and human rights.

One of the most dramatic recent examples of this occurred in Thailand in 2003, when the Thai Government declared a "war on drugs." Prior to this, the Thai Government had received praise for its successful non-punitive approach to AIDS, exemplified by its efforts to promote condom use among sex workers and military conscripts in the 1990s. By contrast, the "war on drugs" was a harsh "zero tolerance" policy that flew in the face of proven HIV prevention strategies, such as provision of sterile syringes or oral methadone to people who inject drugs. Purportedly in response to a rise in methamphetamine use in the country, in January 2003 the then Prime Minister Thaksin Shinawatra called for "ruthless" drug enforcement based on "an eye for an eye" (Human Rights Watch, 2004). There followed a period of mass arrest and incarceration of drug users for even low-level crimes such as possession of narcotics and syringes for personal use. Thaksin instructed local officials to create "blacklists" of suspected drug offenders, and in August 2003 instituted a shoot-to-kill policy against alleged drug traffickers smuggling methamphetamines from neighboring Burma. By the end of the "first phase" of the drug war, an estimated 2275 people had been shot dead in apparent extrajudicial executions (CNN, 2003; AFP, 2003). Thaksin blamed these killings on internecine violence among drug traders, yet at this writing a full investigation of the killings still has not taken place, despite some indication that investigations might occur in the wake of Thaksin's having been overthrown

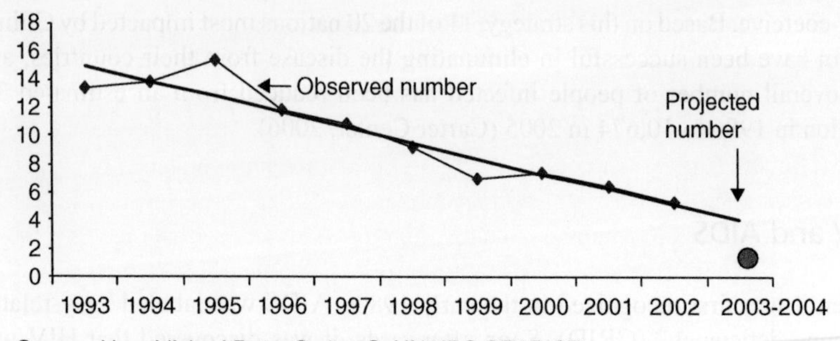


**Figure 15.2** Estimated and observed number of methadone maintenance therapy patients in Thailand, 1992–2004. Source: Sarker, S., Joint United Nations Programme on HIV/AIDS Southeast Asia and Pacific Intercountry Team (unpublished data, 2004).

in late 2006. Throughout the drug war, senior government officials encouraged violence against drug suspects. At one point Thaksin said: "There is nothing under the sun which the Thai police cannot do … If there are deaths among traders, it's normal" (Human Rights Watch, 2004).

In addition to the assault on individual rights, Thailand's "war on drugs" proved to have a negative impact on public health. While pronouncing that drug users were "patients" in need of treatment, Thai police in fact subjected drug users to mass urine testing and detention in military-style boot camps. Many drug users were incarcerated in prisons where syringe-sharing was common and access to HIV prevention information and services was minimal to non-existent. Instead of seeking drug treatment and HIV prevention services, many drug users escaped into hiding (Human Rights Watch, 2004). One study revealed that 37 percent of drug users who had formerly attended drug treatment centers in Chiang Mai were staying away, and that there was an increase in sharing syringes because sterile syringes were more difficult to obtain (Bhatiasevi, 2003). The HIV rate among Thailand's injecting drug users has remained at approximately 40 percent since the 1990s, even as it has declined among sex workers and the general population (see Figure 15.2).

The plight of women in sub Saharan Africa also illustrates the link between human rights abuse, poor governance, and HIV/AIDS. A case in point is the Kingdom of Swaziland – the last remaining absolute monarchy in Africa, and home to the highest estimated rate of HIV infection in the world. As of 2005, an estimated 38 percent of the country's adult population was HIV-positive (Joint United Nations Programme on HIV/AIDS, 2005). Most of those infected are women and girls, due to their increased biological risk of acquiring HIV through unprotected heterosexual sex and also because of the country's pervasive violence, discrimination, and economic marginalization of women. These social forces inhibit the ability of girls and women to make informed decisions about their health in general, and specifically with regard to protecting themselves from HIV (Human Rights Watch, 2003). The country's monarch, King Mswati III, reigns over a system of both absolute monarchy and extreme patriarchy. While a recently ratified constitution contains guarantees of both a balance of powers and gender equality, the mechanisms to enforce these guarantees – independent courts, a robust civil society, and a political opposition – do not exist. The country has been under an official "state of emergency" since 1973, with the King retaining effective control over all branches of government.

In addition to the pervasive authoritarianism and disrespect for the rule of law, the underlying social context of the AIDS epidemic includes the widespread subordination of women to men. Seventy-five percent of all land is considered "Crown land," which is governed by highly patriarchal and gender-biased customary law (Scholz and Gomez, 2004) that prevents women from owning, inheriting, or disposing of property. A Swazi woman enjoys the right to use her father's property, but is expected to marry and depend on the property of her husband. Married women are treated as the legal equivalent of minors, unable to sign contracts or represent themselves in court. If a woman is widowed or separated from her husband, her property typically reverts to (or is grabbed by) her husband's family. Because of these conditions, it is perilous for Swazi women to leave even violent marriages, to refuse sex, to object to polygyny, or to insist on condom use. This helps to explain why in Swaziland, as in numerous African countries, a significant percentage of HIV infections among women occur in marriage. Married women may be unable to seek health care or information because of a lack or resources or dependence on their husband or male relatives. They are also less likely than men to have attended school, where they might have gained access to information about HIV prevention or the skills to become economically independent.

## SARS

*This section stems from discussion with Jennifer Prah Ruger, and draws upon her 2005 article "Democracy and health" in the* Quarterly Journal of Medicine, **98**, 299–304.

The case of Severe Acute Respiratory Syndrome (SARS) in China illustrates how a lack of democratic freedoms can render a country unable to respond promptly to a new health crisis. In 2003, when SARS first emerged in the southern Chinese province of Guangdong, the Chinese Government's immediate response was to cover up, rather than reveal, both the scope and severity of the disease. The government's censorship of news about the spread of SARS ultimately accelerated the spread of the disease (*The Economist*, 2003a; Rosenthal, 2003) by limiting the information available both to citizens (who needed information on precautions and care) and to national and international government health authorities (who needed information to inform decision-making and improve their understanding of a poorly understood disease). Further hindering an effective response, the government threatened citizens with execution and lengthy imprisonment should they become infected with or knowingly spread SARS (Eckholm, 2003).

As news of SARS in China spread through unofficial channels imperfectly monitored and controlled by the Chinese Government, the Chinese Government reversed direction and pledged honest reporting of infections and accountability of public officials (for example, they fired both the Mayor of Beijing and China's Health Minister). While these steps at first brought hope for more effective public health strategies and wider political reform (*The Economist*, 2003b), subsequent efforts fell short of that goal. Far from acting as an independent and free agent, the Communist Party's newspaper, *People's Daily*, instead served as a Party instrument by publicly praising government leadership and strategies and misreporting public opinion. For example, it noted that "the people have become more trusting and supportive of the party and government" (Eckholm, 2003).

China's failure to contain and effectively address SARS ultimately increased pressure on global institutions such as the World Health Organization (WHO) to become more actively involved in "governance" at an international level, resulting in reforms intended to allow it to "fight future international threats" more powerfully (Stein, 2003).

## Limitations

In addition to the methodological limitations to illustrating quantitatively the relationship between governance and infectious disease control, there are several examples that can be cited to suggest that more authoritarian responses lead to more effective disease control, or that governance has little ultimate impact on infectious disease rates. Certainly it is true that, despite the best (or worst) intentions of individuals and governments, infectious pathogens can be stubbornly indifferent or even contrary to theories of health and human rights. HIV transmission can be limited in civil war settings, for example, even when human rights abuses are widespread and governments have collapsed. In Angola, a protracted civil war which reduced cross-border travel and trade is thought to have left the country somewhat protected from the early introduction and spread of disease compared to many of its neighbors; however, the war also impeded the ability of the government to conduct surveillance and education around the disease, and destroyed the health services needed to respond to AIDS. The war also curtailed the formation of a vibrant civil society, such as the development of NGOs and AIDS service organizations, which have been highly effective in both prevention and care elsewhere.

Post-conflict countries often see a sharp increase in HIV prevalence, presumably as measures of governance are improving, indicating that the relationship between HIV and governance may involve a similar time-lag to the relationship between widespread transmission of HIV and the onset and recognition of AIDS. In Mozambique, for example, the HIV rate soared shortly after the cessation of civil conflict in 1992. Another example of the difficulty with the temporal relationship between governance and HIV prevalence comes from the Kingdom of Swaziland, referred to above. The rapid rise in HIV prevalence, from 4 percent in 1992 to 26 percent in 1996, did not occur with a simultaneous deterioration of governance; rather, the precondition and weaknesses in Swaziland allowed for the rapid spread of HIV once it was fully introduced into the country. Conversely, the apparent improvement in the HIV epidemic in Zimbabwe, with HIV prevalence decreasing from 25 percent to 20 percent between 2001 and 2005 despite a worsening human rights and governance environment, may be reflective of earlier actions by the government and, importantly, by international donors, while the impact of current government actions will only be reflected in the years to come (Human Rights Watch, 2006a). Vaccine-preventable diseases may similarly reflect a time-lag between the breakdown of governance (and hence vaccination campaigns) and the accumulation of a large enough susceptible cohort of individuals to sustain disease transmission.

Affluent societies with representative governments are also not immune to leadership failure. For example, during the early years of the AIDS epidemic there was cover-up and scandal over the failure of regulators in Canada, Japan, Ireland, France, and elsewhere to stop promptly the use of tainted blood products that were infecting hemophiliacs with HIV. While broad measures of governance may relate generally to the willingness of governments to adopt effective disease control policies, government policies to control diseases that affect specific (favored or non-favored) populations, or are related to culturally sensitive transmission modalities, may be poorly correlated with overall measures of governance.

## Assessing governance

The diverse chapters in this volume illustrate that factors such as human behavior and socio-economic conditions can have as great or a greater impact on the risk of infectious disease as does the microbiology of disease pathogens. If this is true, it is equally important to examine the factors that structure these socio-ecological factors in the first place. We have argued that government conduct plays an important role in shaping the social context in which individuals live, behave, and make decisions about their health. Not only do governments help to shape the risk environment for infectious disease, but they also determine the manner in which diseases are contained and controlled.

Quantitative analysis of the link between governance and public health provides an interesting starting point for this discussion; however, it has not yet been able to capture the complexity of how governments balance human rights and public health in the context of specific disease threats, and the consequences of the choices and trade-offs made. There are several ways in which the relationship between governance, human rights, and infectious disease may be understood. The degree to which governments are respectful of human rights, and are responsive and accountable to their citizens, will influence the effectiveness with which they respond to disease threats and openly communicate about epidemics. This applies to both epidemic diseases, which governments may conceal in an attempt to maintain social order, as well as to endemic diseases among the poor, which governments may ignore out of political expediency. Individual violations of civil and political rights by governments or non-state actors can constrain the ability of individuals – especially socially marginalized groups – to make informed decisions about their health, thus increasing their vulnerability to infectious disease. These human rights violations may have less to do with health policy *per se*, and more to do with government actions (and inactions) lying outside the health sphere. A government's commitment to human rights will determine the level of coerciveness with which it responds to infectious disease threats and, in turn, how effectively a disease is contained and controlled. Even governments that choose proactively to address infectious diseases may do so in a manner that sacrifices individual rights to a perceived social benefit. A rational and proportionate balancing of individual rights against larger policy objectives may be more likely to reap benefits for public health than ignoring or downplaying human rights.

What is certain is that governance matters to health – that the way in which people are governed, whether their human rights are respected, and the institutions of democracy and civil society can have tangible health consequences at both the individual and population level. Despite the range of work in this area, there have been few attempts to describe the actual mechanisms by which rights and democracy can impact health, and attempts to measure these relationships in quantitative terms have been limited and largely unsuccessful. Even more rare have been efforts to design programmatic interventions that would enact political reforms in countries hard-hit by infectious disease, and evaluate the health impact of these reforms. Operational work in the area of health and human rights has largely been confined to documentation and advocacy work by non-governmental organizations, and has gained little acceptance by the global health community.

If governments are to effectively address the political factors that shape infectious disease epidemics, greater collaboration between health experts, human rights advocates, and legal professionals is likely to be necessary. Human rights law suggests a full range of concrete and justifiable remedies for abuses that fuel infectious disease and impede civil society's response to it (Mann, 1996). Mechanisms of human rights accountability, such as courts, national human rights institutions, United Nations, and other multilateral procedures, and traditional "naming and shaming" techniques by non-governmental organizations have the ability to further human rights goals and thus have an impact on public health. Greater academic research into the precise links between democracy, human rights, and health can further assist policy-makers in implementing rights-based approaches with rigor and integrity. Tools such as the *International Guidelines on HIV/AIDS and Human Rights*, the *Human Rights Impact Assessment* (co-authored by Mann and Gostin), and extensive human rights documentation and advocacy by non-governmental organizations can further operationalize this approach by providing recommendations and policy guidance to governments (Gostin and Mann, 1994; Human Rights Watch, 2006a; Office of the High Commissioner for Human Rights and Joint United Nations Programme on HIV/AIDS, 2006).

Public support for grassroots health activists and community-based organizations, legal protection against political violence, detention, and other human rights abuses, and promotion of a robust and independent civil society should all be recognized as underpinnings for preventing and controlling future infectious epidemic diseases and improving overall health (Navarro, 1978; Committee on Economic, Social and Cultural Rights, 2000; Roth, 2004).

Ultimately, different infectious disease risks present different challenges in balancing human rights and public health goals in the response to epidemics. This balance should be determined by the epidemiologic characteristics of the disease and the methods available and practical for its control, and not by the level of social opprobrium against populations affected, societal discomfort with transmission through illicit or intimate behavior, or fear and ignorance.

## References


AFP (2003). Death toll in Thailand's drug war hits 2,275, say police. *Agence France-Presse* 16 April.

Alvarez-Dardet, C. and Franco-Giraldo, A. (2006). Democratisation and health after the fall of the Wall. *Journal of Epidemiology and Community Health* **60**, 669–671.

Bhatiasevi, A. (2003). 'War on drugs' raises AIDS risk. *Bangkok Post*, 8 July.

Breslow, L. (1999). From disease prevention to health promotion. *Journal of American Medical Association* **281**, 1030–1033.

Carter Center (2006). *Center Receives 2006 Gates Award for Global Health*. Available at: http://www.cartercenter.org/news/documents/doc2354.html.

Committee on Economic, Social and Cultural Rights (CESCR) (2000). *The Right to the Highest Attainable Standard of Health*, General Comment No. 14. CESCR.

CNN (2003). 2,274 dead in Thai drugs crackdown. CNN.com, 7 May.

Dreze, J. and Sen, A. (2002). *India: Development and Participation*. New York: Oxford University Press.

Eckholm E. (2003). China threatens execution in intentional spreading of SARS. *New York Times*, 15 May.

*The Economist* (2003a). China wakes up. *The Economist*, 24 April.

*The Economist* (2003b). China's Chernobyl? *The Economist*, 26 April.

Farmer, P. (1999). *Infections and Inequalities: The Modern Plagues*. Berkeley: University of California Press.

Farmer, P. (2004). Political violence and public health in Haiti. *New England Journal of Medicine* **350**(15), 1483–1486.

Fee, E. and Brown, T.M. (2005). The Public Health Act of 1848. *Bulletin of the World Health Organization* **83**, 866–867.

Franco, A., Alvarez-Dardet, C. and Ruiz, M.-T. (2004). Effect of democracy on health: ecological study. *British Medical Journal* **329**, 1421–1423.

Freedom House (2006). *Freedom in the World, 2006: The Annual Survey of Political Rights and Civil Liberties*. Lanham: Rowman & Littlefield.

Gostin, L. and Mann, J.M. (1994). Towards the development of a human rights impact assessment for the formulation and evaluation of public health policies. *Health and Human Rights* **1**(1), 58–80.

Hopkins, D.R. and Ruiz-Tiben, E. (1991). Strategies for dracunculiasis eradication. *Bulletin of the World Health Organization*, **69**(5), 533–540.

Human Rights Watch (2003). *Policy Paralysis: A Call for Action on HIV/AIDS-Related Human Rights Abuses against Women and Girls in Africa*. Available at: http://www.hrw.org/reports/2003/africa1203/.

Human Rights Watch (2004). Not enough graves: the war on drugs, HIV/AIDS, and violations of human rights in Thailand. *Human Rights Publications* **16**(8c). Available at http://www.hrw.org/reports/2004/thailand0704/thailand0704.pdf.

Human Rights Watch (2006a). *No Bright Future: Government Failures, Human Rights Abuses and Squandered Progress in the Fight Against against AIDS in Zimbabwe*. Available at http://hrw.org/reports/2006/zimbabwe0706 (accessed 16 September 2006).

Human Rights Watch (2006b). *Cumulative Reports, HIV/AIDS Program*. Available at: http://www.hrw.org/doc/?t=hivaids&document_limit=0,2.

International Covenant on Civil and Political Rights (1966). General Assembly Resolution, 2200A (XXI), 21 U.N. GAOR Supp. (No. 16) at 52, U.N. Doc. A/6316.

International Covenant on Economic, Social and Cultural Rights (1966). General Assembly Resolution 2200 (XXI), 21 UN GAOR, 21st Sess., Supp. No. 16, at 49, UN Doc. A/6316.

Joint United Nations Programme on HIV/AIDS (2005). *Report on the Global AIDS Epidemic*. New York: United Nations.

Kaufmann, D., Kraay, A. and Mastruzzi, M. (2003). Governance Matters III: Governance Indicators for 1996–2002. World Bank Policy Research Working Paper, No. 2195.

Kiple, K.F. (ed.) (1995). *The Cambridge World History of Human Disease*. New York: Oxford University Press.

Mann, J.M. (1996). Health and human rights. *British Medical Journal* **312**, 924–925.

Mann, J.M. (1999). Human rights and AIDS: the future of the pandemic. In: J.M. Mann, S. Gruskin, M.A. Grodin and G.J. Annas (eds), *Health and Human Rights: A Reader*. New York: Routledge.

Menon-Johansson, A.S. (2005). Good governance and good health: the role of societal structures in the Human Immunodeficiency Virus pandemic. *BioMed Central International Health and Human Rights*, **5**(4).

Navarro, V. (1978). The economic and political determinants of human (including health) rights. *International Journal of Health Services* **8**(1): 145–168.

Needham, C. and Canning, R. (2003). *The Race for the Last Child.* Washington, DC: ASM Press, p. 134.
Office of the United Nations High Commissioner for Human Rights and the Joint United Nations Programme on HIV/AIDS (2006). *International Guidelines on HIV/AIDS and Human Rights (2006 Consolidated Version).* Geneva: UN Publication No. E.06.XIV.4.
Reidpath, D.D. and Allotey, P. (2006). Structure, (governance) and health: an unsolicited response. *BioMed Central International Health and Human Rights* **6**(12), 1–7.
Rosenthal, E. (2003). AIDS scourge in rural China leaves villages of orphans. *The New York Times*, 25 August.
Roth, K. (2004). Defending economic, social and cultural rights: practical issues faced by an international human rights organization. *Human Rights Quarterly* **26**, 4.
Ruger, J.P. (2005). Democracy and health. *Quarterly Journal of Medicine*, **98**(4), 299–304.
Scholz, B. and Gomez, M. (2004). *Bringing Equality Home: Promoting and Protecting the Inheritance Rights of Women.* Geneva: Center on Housing Rights and Evictions.
Sen, A. (1999). *Development as Freedom.* New York: Oxford University Press.
Stein, R. (2003). SARS prompts WHO to seek more power to fight disease. *Washington Post*, 18 May, A10.
Susser, M. and Susser, E. (1996a). Choosing a future for epidemiology: I. Eras and paradigms. *American Journal of Public Health* **86**, 668–673.
Susser, M. and Susser, E. (1996b). Choosing a future for epidemiology: II. From black box to Chinese boxes and eco-epidemiology. *American Journal of Public Health* **86**, 674–677.
Universal Declaration of Human Rights (1948). General Assembly Resolution 217A (III), U.N. Doc A/810 at 71.
World Health Organization (1986). Ottawa Charter for Health Promotion.
World Health Organization (1991). Elimination of Dracunculiasis: Resolution of the 44th World Health Assembly. Resolution No. WHA 44.5, Geneva.

# International organizational response to infectious disease epidemics

# 16

**Bjorg Palsdottir, Susan H. Baker and André-Jacques Neusy**

The roots of global public health are found in the economics and interface of war, trade, and health, reaching significant proportions during the Age of Exploration, slavery, and colonialism from the early fifteenth century (Kickbush and Buse, 2005). One of the earliest examples of public health policy in action was the routine short-term use of quarantine and isolation of trade ships in order to combat plague, beginning in the fourteenth century (Basch, 1999). Efforts at long-term prevention leading to institutionalizing response began much later.

Today, the social ecology of infectious diseases links the fates of peoples and ecosystems around the globe. Because the world has not dealt with a pandemic caused by a highly contagious, rapidly spreading infectious disease since the 1918 influenza epidemic, assessment of today's true organizational response capacity is speculative at best. Hurricane Katrina, which hit the Gulf Coast of the United States in 2005, illustrates that capacity on paper does not always reflect how people and plans perform during emergencies. The potential for a global outbreak of avian flu in the near future makes the topic politically sensitive, and the existing figures and plans moving targets.

Effective organizational response to the threat of infectious disease involves the integration of two basic models: emergency response, and primary health care and prevention. The emergency response model is a direct response to a specific disease threat. Its main goal is to control, contain, or eliminate an imminent threat. The primary care and prevention model is concerned with prevention and control of a myriad of disease threats through the ongoing supply of basic public health infrastructure and primary health care.

At the time of writing, our research suggests that the bulk of international aid resources for infectious diseases go to short-term technical emergency response measures – this despite the 1978 Alma Ata Declaration, signed by WHO and 134 nation-states, committing to "Health Care for All by the Year 2000," and the acknowledgment that improving primary care services and strengthening weak health systems in the long term is of great importance. The focus tends to be on specific health interventions, such as the development and stockpiling of vaccines; or drug distribution for specific conditions like HIV/AIDS or avian influenza. The overall system is a fragmented and informal collaborative network of a relatively few elite organizations and partnerships that intervene by government invitation and act to bolster weak local and regional public health capacity. As we shall describe, the efficacy of such a system has intrinsic problems based on inadequacies in public health infrastructure; on political, economic and social conditions; and on the inherent difficulties of coordinating the response of numerous diverse organizations operating under different command structures.

A combination of political commitment, infrastructure, and the availability of resources and technical expertise determine a nation's ability to respond to infectious disease threats. The front-lines of the international organizational response to infectious disease threats often are where public health infrastructures are weakest. In the poorest countries, in particular, there is little government planning, few paved roads, no public transportation, limited or no access to electricity, minimal sanitation and clean water supply, no reliable communication and information systems, and a limited supply of goods and services (Ndikuyeze, 2000). Education and veterinary and human health care also may be inaccessible or very much below acceptable standards. Impoverished nations often share borders with countries that have similar problems, thereby creating regions deficient in basic and vital resources. The front-line context of infectious disease control also includes conditions of natural disaster, civil strife, population dislocation, violence, and terror. International health responders thus often function in the face of great insecurity (Connolly, 2000; Kakar, 2000; Shoo, 2000; Human Security Centre, 2005). Even countries with burgeoning economies and rising GDPs, like Brazil, Singapore, South Africa, China, and India, present enormous challenges to health responders because the rural poor are migrating in vast numbers to cities that have *ad hoc* and destitute districts which lack clean water, sanitation, and access to health-care services (Brower and Chalk, 2003; see also Chapter 4).

The front-line responders during epidemics are health workers in public and private agencies, institutions, and non-governmental organizations (NGOs). Controlling global disease threats requires adequate resources, capacity, and action at every decision-making level in every corner of the world – from the international and regional to the nation and its communities. It necessitates close coordination across borders and sectors, and among and within organizations. As we shall describe, organizations like the World Health Organization (WHO) and the United States Centers for Disease Control (CDC) have made great strides to improve coordination in their own efforts and across responders, and to streamline emergency response activities generally. Still, it is fair to observe that the global emergency response infrastructure in both poor and wealthy nations is characterized by weak local capacity, politicization, and fragmentation of funding and effort. The present international focus too often is based on "fire-fighting" through periodic, isolated rapid response to emergencies. We believe this has to change. The effectiveness of international medical SWAT teams is inevitably constrained by the weak or absent health systems where epidemics are most likely to arise. A long-term solution requires improving local primary care and prevention, as well as economic and social infrastructure.

The legal structure for global infectious disease control is provided by the International Health Regulations, administered by the WHO. Adopted in 1951, the IHR require the WHO member states, and other signatories, to report to the WHO the occurrence of infectious diseases in their country, and follow standards and norms to prevent the international spread of these diseases while minimizing impact on trade and travel. In the latest revision, effective in 2007, regulations no longer focus on specific diseases, but require nations to notify the WHO of all events that could potentially constitute an international public health emergency (WHO, 2007). Moving to strengthen the emergency response model, the new IHR establish a unified code of routine procedures and practices at border crossings, ports and airports to prevent the spread of pathogens without unduly disrupting trade and travel. They also provide the legal framework for the WHO and other organizations involved in surveillance, detection, alert and response to global outbreaks and epidemics. They address concerns regarding economically damaging travel and trade restrictions that have created reluctance among some to report disease incidence and outbreaks. The new IHR recommend measures for contending with specific public health threats, and define the basic public health capacities that must be in place at each level within every nation to be able to identify pathogens, and report and take action against public health risks that could spread across borders. While the regulations are binding, they, as many other international agreements, do not include enforcement mechanisms. Hence, compliance is dependent on individual government's willingness and ability to act.

## Today's responders

The world's ability to respond to epidemics rests on the operational capacity of an amalgamation of diverse agencies and organizations. These include several multinational organizations; sundry national, regional, and local government agencies; various non-governmental organizations; and a host of diverse public and private health-care institutions and providers that together form our existing global health infrastructure. Figure 16.1 depicts an example of the vast array of

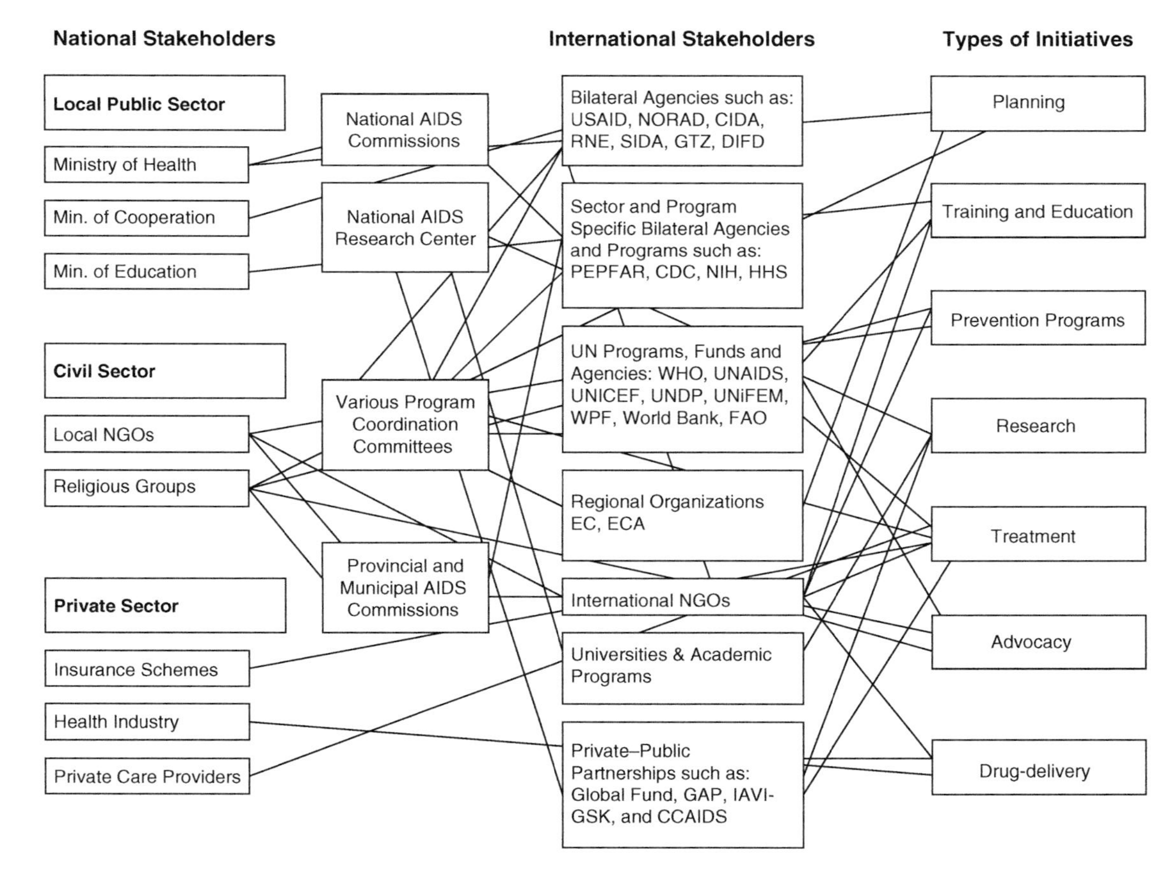


**Figure 16.1** Typical stakeholder relationships in country receiving international aid for HIV/AIDS.

organizations typically involved in programs related to one disease in one developing country. These diverse organizations often compete for funding and have different missions and priorities; sometimes they have overlapping roles; they may also have differing levels of power and resources, and limited (if any) experience in effective inter-institutional, even intra-organizational coordination, communication and collaboration. Without clear protocols, and linked organizational and communication structures, the increasing number of actors involved in infectious disease control and prevention perpetuates fragmentation of services and disparities in policy and program development (Neusy, 2004; Travis *et al.*, 2004; Kickbush and Buse, 2005).

## The World Health Organization

The World Health Organization (WHO) was the first, and today remains the leading, multilateral organization for responding to infectious disease threats. It was established in 1948 as a specialized agency within the United Nations. The WHO is governed by its 192 member states through the annual World Health Assemblies (WHA) held at its headquarters in Geneva. The WHA sets policies and provides the overall governance of the WHO. The WHO also has an Executive Board, composed of 32 members technically qualified in the field of health, that prepares work programs for WHA, provides technical advice, reviews finances, and decides on actions on disasters and epidemics. The WHO is further divided into three levels – headquarters, relatively autonomous regional organizations, and the 142 country offices. This division reflects the WHO's origins as an amalgamation of several organizations dealing with health. Its member nations are grouped into six regions: Africa, the Americas, South-East Asia, Europe, the Eastern Mediterranean and the Western Pacific (Minelli, 2003).

At the time of writing, the WHO operates in three management clusters and seven technical clusters: Communicable Diseases; Non-communicable Diseases and Mental Health; HIV/AIDS, TB and Malaria; Sustainable Development and Healthy Environments; Health Technology and Pharmaceuticals; Family and Community Health; and Evidence and Information for Policy. With the exception of HIV/AIDS, tuberculosis and malaria, and immunization and childhood vaccines, all WHO's epidemic-related activities fall under the auspices of the Communicable Disease Cluster – and, more specifically, under the Department of Epidemic and Pandemic Alert and Response. Initially the WHO's budget was based on member contributions, which were determined by a formula based on population size and gross national product (Walt, 2005). By the 1970s, high-income member states had raised the level of extra-budgetary funding for specific projects. Increasingly seen as donors, their influence grew, challenging the WHO's relative autonomy. By the 1990s, close to 54 percent of the WHO's funding came from extra-budgetary sources, compared with 25 percent in 1971.

Some have raised concern that a few wealthy donor nations are exerting undue influence on priority-setting (Walt, 2005). It could also be argued that creating disease-specific organizational components undermines efforts to integrate disease surveillance and strengthen health systems in low-income countries. For example, while the WHO has a Communicable Diseases Cluster, a separate cluster has been created for HIV/AIDS, TB and malaria that now receives more funds than other infectious diseases.

Early on, the WHO decided against creating its own research institutions, and instead to take advantage of existing research and education organizations in member nations. These partners, called WHO Collaborating Centers, constitute elements of a cross-institutional network to support its programs at every level. The centers also participate in strengthening the national capacity of the health systems by assisting in data collection, research, training, and providing health services (*WHO Collaborating Centers*, at http://whqlily.who.int/). While the WHO's Collaborating Centers on various infectious diseases provide important research data and sentinels around the world, there has been criticism raised that the large number of agencies and organizations on the ground has also created confusion and even chaos (Shoo, 2000; Enserink, 2004). Responding to this criticism, especially to the handling of the Ebola outbreak in Kikwit, the Democratic Republic of the Congo, in 1995, and that of Rift Valley fever in East Africa in 1997, in 2000 the WHO launched the Global Outbreak Alert and Response Network (GOARN) (Enserink, 2004). GOARN's mission is to improve coordination in the field and within the organization by functioning as an operational platform for pooling human and technical resources, and linking this expertise to needs on the ground. The Department of Epidemic and Pandemic Alert and Response manages the network, and links more than 130 laboratory and surveillance networks around the world in order to identify and verify pathogens rapidly, and to coordinate the overall international response to disease outbreaks. GOARN'S partners include government agencies, ministries of health, academic centers, and UN agencies, as well as networks of military laboratories and NGOs located in areas with high risk of epidemic outbreaks. GOARN responds to more than 50 outbreaks a year (Drager and Heymann, 2004; Heymann and Rodier, 2004).

The WHO also works through GOARN to improve global, regional, and national preparedness and response to contain epidemics, emerging diseases, and drug resistance. It helps to establish surveillance standards, create regional or sub-regional preparedness and rapid response networks, improve laboratory capacity, establish laboratory networks, and provide training in field epidemiology and assessment and strengthening of national surveillance systems (*Epidemic and Pandemic Alert and Response*, at http://www.who.int/csr/en/). It also provides manuals, guidelines, and checklists to assist in prevention, detection, and response activities. In addition, the WHO provides collaborative risk assessment, and communication and ongoing advice on infection control.

Realizing it lacked an operational component for managing information and field deployment for rapid response operations, in 2004 the WHO established the Strategic Health Operations Centre (SHOC) to coordinate epidemic response at its headquarters and in the field (Nebehay, 2005). In the event of an outbreak, SHOC becomes a high-tech global command center, and the operational hub of global alert and response activities of GOARN and others. The 2004 Asian tsunami became the first crisis in which SHOC was fully operational, functioning as the virtual and physical place where teams from across various WHO health clusters came together to coordinate an agency-wide response. It served as a focal point for daily briefings, operations and facilities planning, and communications and coordination with the WHO Regional Office for South East Asia, other UN agencies, non-governmental organizations, and member states.

The WHO and its partners have created several tools to support epidemic response activities, including the Global Atlas of Infectious Disease, which provides GOARN members and other networks with an up-to-date interactive single electronic platform to map infectious disease outbreaks. Another example is the Global Public Health Intelligence Network, developed and run by Health Canada, which provides the WHO with a computer-based tool that continuously scans websites, online news services, public health discussion groups, and e-mail services for information that could signal potential disease outbreaks. The WHO has also made efforts to integrate infectious disease programs into health-system development. A good example is the Integrated Disease Surveillance and Response (IDSR) program to help establish nationally owned and maintained disease surveillance systems capable of collecting and reporting data. Implementing such systems worldwide would significantly enhance global capacity to develop plans and public health interventions.

## The US Centers for Disease Control and Prevention

The US Centers for Disease Control and Prevention (CDC) may be considered the core of the world's response SWAT team. Established in 1946 as the Communicable Disease Center to help control malaria in the United States, the CDC has become one of the world's most prestigious organizations for responding to epidemics. Its mission is: "To promote health and quality of life by preventing and controlling disease, injury, and disability." Funded by appropriations from the US Congress, it operates as an independent agency under the United States Department of Health and Human Services (HHS). As such, the bulk of its work is in the United States, but over time it has widened its functions to be worldwide in scope, and so now is an essential cluster group of the WHO Collaborating Centers (*About CDC*, at http://www.cdc.gov/about/default.htm; see also *WHO Collaborating Centers at CDC by CIO*, at http://www2a.cdc.gov/od/gharview/GHARwhocollabs.asp). Like the WHO, the CDC tends to be organized around specific diseases and programs, and is influenced by the politics and funding priorities of the United States Government.

In 2003, the CDC launched a 22-month-long, large-scale strategic planning effort to improve its ability to achieve its mission. The ensuing restructuring effort, approved by Congress in April 2005, sought to streamline management and operations, reduce management levels, and increase coordination. It organized the CDC into coordinating centers and offices – the Coordinating Center for Environmental Health and Injury Prevention; the Coordinating Center for Health Information and Services; the Coordinating Center for Health Promotion; the Coordinating Office for Global Health; the Coordinating Center for Infectious Diseases, which includes the Coordinating Office for Terrorism Preparedness and Emergency Response; the National Center for HIV, STD, and TB Prevention; and the National Institute for Occupational Safety and Health (NIOSH). In addition, the CDC provides administrative support for the Agency for Toxic Substances and Disease Registry (ATSDR), a sister agency of CDC, and the Director of the CDC also serves as the Administrator of the ATSDR.

The Coordinating Center for Infectious Diseases is of particular significance for infectious diseases. It is "responsible for infectious diseases control, HIV/AIDS, STD, and TB prevention, and immunizations in the United States and around the world." The Coordinating Office of Global Health falls under its auspices, and is responsible for "national leadership and support for CDC's global health activities" and collaboration with CDC's global health partners (*National Coordinating Office for Global Health*, at http://www.cdc.gov/ogh/). In addition, the Coordinating Center for Infectious Diseases encompasses the National Center for Infectious Diseases (NCID), the National Immunization Program (NIP), and the National Center for HIV, STD, and TB Prevention (NCHSTP) (see CDC website, at www.cdc.gov).

In the United States, by offer or invitation, the CDC units and centers implement and fund programs on infectious disease control and prevention. They also work with state governments and other national partners to conduct surveillance, epidemic investigations, laboratory and epidemiological research, as well as training and public education programs that develop, evaluate, and promote prevention and control strategies for communicable diseases. With 160 staff based in more than 43 foreign countries, it is also a key player on the global stage, providing technical assistance and support to investigate and respond to epidemics at the request of the WHO and national governments. It is a key contributor of staff and technical resources to WHO's GOARN. CDC experts serve as *ad hoc* technical advisers and trainers at the request of individual US states, nations, and the WHO. The CDC, like the WHO, realized that, to improve response capacity, collaboration and resource sharing are vital. While it remains a vital responder today, it usually does so under the auspices of the WHO, taking advantage of the WHO's logistical resources (Enserink, 2004). Despite the CDC's long-standing history of international outbreak assistance and efforts to help build capacity in developing nations, there is little formal organizational support for these activities (CDC, 2002). While the 2005 restructuring effort makes global health a strategic priority, it remains unclear whether the CDC will be given adequate funding to support the human, epidemiologic, diagnostic, and logistic activities to respond to international outbreaks. If given adequate resources, the mission is also to offer the WHO and host country Ministries of Health support to evaluate public health conditions following the containment of specific epidemic outbreaks. This could include guidance on improving surveillance, disease prevention, future outbreak response and training, and capacity development efforts for local organizations (CDC, 2002).

## Non-governmental organizations

A third key element in the international organization response network is non-governmental organizations (NGOs). They include a wide array of groups involved in funding, advocacy, policy guidance, or service delivery. They can be professional and technical, community-based, faith-based, national, or international. Many play an increasingly significant role in global health. (Walt, 2005) Overseas development assistance from governments channeled through NGOs reached close to US$5 billion in 2004. Private funding for NGOs expanded by 37 percent, from US$6.9 billion in 2000 to US$11.3 billion in 2004, and government funding to NGOs rose to $5 billion. (OECD, 2005).

In some regions of the world where health services are limited, faith-based clinics and hospitals work with local NGOs as the main health providers. During disasters and complex humanitarian emergencies (CHEs), these local and international NGOs are usually the front-line emergency responders for refugees and displaced persons in isolated and insecure areas. The CDC defines CHEs as "situations affecting large civilian populations which usually involve a combination of factors including war or civil strife, food shortages, and population displacement, resulting in significant excess mortality" (Connolly, 2000; Toole *et al.*, 2005). Much like the fragmentation already described on multi-lateral and bi-lateral levels, the services these NGOs provide frequently are not well coordinated either among themselves or with the host government. Particularly during emergencies, NGOs can have diverging priorities and approaches, and adhere to different standards of care. (Toole *et al.*, 2005; Telford *et al.*, 2006). While there are ongoing efforts to set standards and improve coordination across the board, a lack of accountability and transparency causes significant organizational disparities in human and material resources, and this further impacts the ability to provide quality health services (Toole *et al.*, 2005).

As described elsewhere in this text, infectious disease outbreaks are a key concern during natural disasters and complex humanitarian emergencies (Connolly, 2000; Toole *et al.*, 2005; see also Chapters 11 and 13). During July 1994, in the refugee camps of Goma in eastern Zaire, to which one million Rwandan refugees had fled following the genocide, mortality rates increased 30-fold over the rates in Rwanda prior to the conflict. During a period of three weeks, it is estimated that 45,000 people died of cholera in the camps. The main source of water in the refugee camps was Lake Kivu, generally considered to be the source of contamination (Toole *et al.*, 2005). NGOs were the primary responders for people in the camps. While they faced the same challenges and fragmented response issues common at any disaster site, most glaring here was a lack of a single unified disaster response system, which resulted in poor communication and planning across responders. This in turn contributed to inadequate coordination among donors and responders, and led to some duplication of effort, ineffective or slow responses, and inequity of access to health services, culminating in excess morbidity and mortality (Millwood, 1996). Beyond the effects of fragmentation, NGOs face a host of operational challenges that derive from the narrow missions, protocols, and philosophies they must follow. Not only are they often operating in resource-constrained conditions; they also usually have to collaborate closely with (or at least get approval from) sometimes unenthusiastic and ineffective local governments (Lam, 2001). They may go into a setting without clear guidance as to who is in charge (Connolly, 2000; Shoo, 2000), and there are almost always constraints related to funding and supplies (Shoo, 2000), including competition between NGOs for the same funding sources.

Considerable progress is being made by NGOs to develop technical field expertise, and to have their responses driven by performance indicators and codes of ethics. Like other groups working in global health, they must deal with fragmented planning, management, and implementation of programs, and a lack of funding for long-term preventive infrastructure development. Until this paradigm shifts, it is likely their efforts will result in sporadic success and ever-increasing spending on emergency response.

## The private sector

The private sector, including hospitals and clinics, has increasingly become part of the network of key responders during epidemic outbreaks. In the United States, private hospitals and health-care facilities are required to develop emergency response plans. Relying on discretionary preparedness, planning, and responsibility can be disastrous – as the abandonment and death of patients in nursing homes in Louisiana during Hurricane Katrina illustrates. The private sector receives guidance from the government, but is responsible for its own planning. As of yet, there is no common framework, and no common mandates, training, timetables, or chain of command in place. Emergency roles and authority at local, state, and federal levels have not always been clearly defined or established. If this experience has a message to apply going forward, it is that both private and public health facilities will be impacted and possibly overwhelmed by an infectious pandemic (Garrett, 2005). Therefore, defining and strengthening the role of the private sector in relation to the public sector is imperative.

In a global pandemic, a situation might be envisioned in which up to 50 percent of the workforce could become sick, there could be widespread panic, borders could close, and the global economy might shut down. During a global influenza epidemic of 12 to 36 months, there could be significant negative impact on air, ground, and shipping industries essential for the transport of health-care responders, pharmaceuticals, food, water, and other supplies. The ability of other industries and services – such as banks, media outlets, public works and engineering, firefighting, energy suppliers, law enforcement, schools and child-care facilities, producers and distributors of food and essential goods – to deliver services is also vital. According to a Deloitte & Touche survey of US executives, released in 2006, only one-third of companies had adequately prepared for a pandemic such as avian flu (Bradsher and Rosenthal, 2006). Moreover, most of the companies that do have emergency plans focus only on dealing with localized disruptions (Osterholm, 2005). Clearly, more needs to be done to help the private sector prepare to respond to possible upcoming infectious disease emergencies.

## Responding to the challenges ahead

In the scope of human history, the idea of creating a coherent international organizational network regarding infectious disease threats is very new. The first attempts at building public health infrastructure began in the mid-nineteenth century, when the wealthy realized that in order to protect their financial interests and their own health they had to improve conditions among the poor. The problem became particularly acute in large cities with large populations of poor immigrants living in overcrowded, unsanitary conditions (Basch, 1999). As a result, urban trade and industrial centers began to develop basic public health infrastructures, and to use demographic data collection, leading to a greater understanding of how sanitation and hygiene play a major role in spread of disease. By the early twentieth century, megacities like London and New York had put in place the basic pillars of public health infrastructure – public sanitation, clean water, uncontaminated food and drugs, general vaccinations, epidemic-control programs, and basic preventive health measures (Garrett, 2000). The deadly cholera epidemics that devastated Europe in the 1830s also brought forth the first steps towards international collaboration. While the focus of the first international scientific meetings was mostly on sharing epidemic data, they eventually formed the foundation for future international agreements on disease surveillance, control, and treatment (Walt, 2005). Around the turn of the twentieth century, international collaboration became more organized. Eventually, the devastation wrought

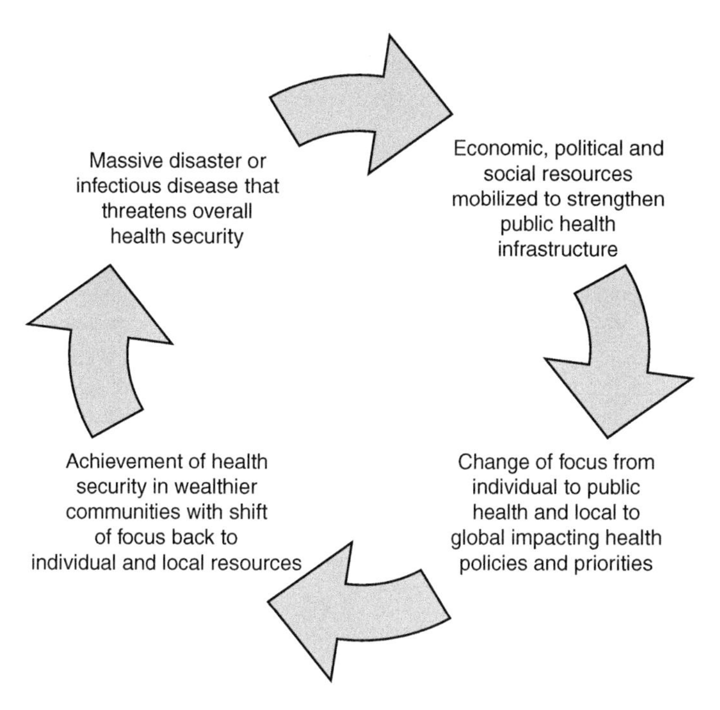


**Figure 16.2** Health policy cycle from local to global.

by two World Wars during the first part of the twentieth century gave birth to today's formal international system, intended to preserve global peace and foster collaboration among nations. Globalization has brought the world back full circle to the same fundamental challenges that public health pioneers in faced in the late nineteenth century, only on a global scale (see Figure 16.2).

There is a strategic framework for improving organizational response that can guide planning in both developed and developing nations (see Table 16.1). The response to an infectious disease outbreak begins at the local level, where, depending on the battleground, the first responder can range from a community health worker in a refugee camp in Africa to an emergency room doctor in a large city hospital. Thus, the global health emergency response infrastructure base is the health workers and resources available at the community level where the first incidents are detected (Kakar, 2000; WHO, 2000). The information and response then has to move outwards, depending on how the health system is organized, to a district level, then perhaps to a Ministry of Health, and then to an international responder, like the WHO. Or, if there is a very weak health system, the next level may be the country headquarters of an NGO, then the WHO, and back to the Ministry of Health. As in a ripple effect, beginning at the local source(s), the

**Table 16.1 Recommendations for improving organizational response**

1. *Develop capacities at the frontlines*:
Strengthen and integrate long-term public health infrastructure in countries and communities where they are absent or weak. Give priority to struggling and middle-income countries, but do not ignore gaps in wealthier countries (Evans *et al.*, 1981; Lee, 2004; Travis *et al.*, 2004; US Government Accountability Office, 2005). Strengthen national leadership, policies and planning within the context of local historical, economic and social circumstances (ActionAid, 2005).

2. *Strengthen capacities at community levels*:
Allocate sufficient resources and mandate local planning, exercises, and adherence to clinical and other critical guidelines; ensure public understanding of emergency preparedness plans; develop clear national incident management structures, policies and roles among primary responders; develop communication structures that educate and inform individuals and communities about what to do in an emergency (Weinberg, 2000; HKSAR, 2003; PHAC, 2003; US Department of Health and Human Services, 2005; *Medical News Today*, 2005; Ballier *et al.*, 2006; Mounier-Jack and Coker, 2006; Tregasis, 2006).

3. *De-politicize health planning and global health governance*:
While politics can neither be ignored nor eliminated, the goal is to move away from short-term response thinking that focuses on protecting those with resources, and towards developing consensus and collective action to protect the health of all (Weinberg, 2000).

4. *Promote a whole system approach*:
Move beyond organizational self-interest to reduce overlap and improve efficiency. It makes no public health or monetary sense to build different or parallel response structures for each disease or crisis response (PHAC, 2003; Chertoff, 2005; Earls and Hearne, 2005; Garrett, 2005). Tightly coordinate national emergency response structures with neighboring and global emergency response structures (Kakar, 2000; Mills *et al.*, 2006).

5. *Coordinate operational protocols*:
In order to respond effectively to an infectious disease emergency, protocols must be in place that cover eight basic areas:

- *Information*: establish surveillance and intelligence systems to warn and inform
- *Communication*: establish vertical and horizontal channels to and from command levels, across responding agencies and organizations, and to and from individuals, affected communities and the public at large
- *Chain of command*: pre-determine and ready clear and organized responsibilities for who gathers what information, who reports to whom, who supports whom, and who supplies what area of expertise or service
- *Key responders*: plan advance identification, definition, and understanding of expertise and roles for all responders, including public and private sector agencies, NGOs, and private sector organizations

(*Continued*)

**Table 16.1 (Continued)**

- *Legal framework of emergency powers:* develop, define and authorize command authority and emergency powers at all relevant levels of responders; determine degrees of flexibility, standards and timetables, penalties for non-compliance
- *Logistics*: identify, establish, and protect essential goods and services and their supply chains
- *Resources*: identify, establish, pre-position, and pre-assign human, financial, material, and transportation resources
- *Testing*: continuously test and readjust, if necessary, these protocols to ensure preparedness.

References: HKSAR, 2003; PHAC, 2003; Earls and Hearne, 2005; *Medical News Today*, 2005; US Department of Health and Human Services, 2005, 2006.

response of each "geo-political ring" to an expanding threat must be meticulous in terms of its surveillance, communication, planning, and implementation.

With an increasing awareness that the first response is local, there is increased realization that in both poor and wealthy countries more emphasis must be placed on building local and national emergency response networks. However, much of the focus continues to be on high-tech surveillance and the development, manufacture, and stockpiling of vaccines and anti-viral pharmaceuticals. Improving local planning and training, although widely encouraged, still often amounts to little more than detailed guidelines which contain few common frameworks, mandates, training, timetables, or chains of command.

The problems faced by low-income nations dwarf the challenges in affluent nations like the United States. However, disparities in preparedness and resource availability at the various response levels plague the system in different but no less serious ways in wealthy nations (Earls and Hearne, 2005; Ballier *et al.*, 2006; Mills *et al.*, 2006; Mounier-Jack and Coker, 2006). For example, in the United States, during a global pandemic the state and local health departments are typically the front-line responders to disease outbreaks (Zwillich, 2005). Yet support for local and state preparedness constitutes only about 10 percent of the $3.3 billion allocated to pandemic preparedness (US Department of Health and Human Services, 2006). While federal and state authorities provide detailed guidance and some financial support to states, the primary responsibility for planning, decision-making, and organizing is the responsibility of individual counties and municipalities, and capacity is uneven (Earls and Hearne, 2005). Thus, for example, New York State's preparedness for a potential influenza pandemic varies widely across localities. Wealthier communities have professional police, fire, and emergency medical service departments, and access to major medical centers, while less affluent ones depend heavily on part-time and volunteer police, fire, and emergency services, and have no access to a major hospital.

Most fall somewhere in between. Meanwhile, they are expected (but not mandated) to develop plans for, among other things, vaccine distribution, vaccination of priority groups, monitoring of adverse events, tracking of vaccine supply and administration, vaccine coverage and effectiveness studies, communications, legal preparedness, and ensure that health-care providers conduct initial screening, assessment, and management of patients. Individual counties are also responsible for conducting appropriate training in infection control and disease containment strategies, to prevent or decrease transmission and a host of related topics (New York State Department of Health, 2006). Left without adequate technical and financial support or strong integration into a common national emergency framework such as the US Department of Homeland Security's National Response Plan's National Incident Management System (NIMS), the US health infrastructure, particularly in poorer communities, might very well not be up to the task of containing or controlling a major influenza epidemic.

SARS erupted in countries with relatively strong health systems, and raised world concern that weak points in any health system can, in addition to undermining patient care, allow emerging infections to intensify and spread beyond local communities and borders. We were fortunate, as containment might have been impossible had the outbreak started in or spread to countries with weak health infrastructure (WHO, 2003). Still, both the Canadian and Hong Kong investigations into the SARS outbreak cited preparation and planning weaknesses, as well as organizational and structural flaws and conflicts in authority, as key reasons for response failures. Both committees recommend better contingency planning in the future, including the development of a clear public health organizational structure under a centralized authority, and a clear chain of command with clear policies, in order to provide effective leadership and action. In addition, both committees recommend that the public health infectious disease response structure form a part of an integrated command and management structure. They recommend pre-clarification of legal and regulatory issues, and pre-establishment of requisite local, regional, national, and international technical and communication networks and links. Both call for the formation of local infectious disease SWAT teams and human resource capacity building (HKSAR, 2003; PHAC, 2003).

Surveillance systems also can be better integrated. Since the early 1990s, there has been widespread recognition that we need better coordination, standardization, and integration of epidemic surveillance activities at national, regional, and global levels (Gouvras, 2004). In 1988 the Institute of Medicine in the United States recommended developing a standardized national data set, but reported only limited progress in its 2003 report (IOM, 2003). It suggests that fragmentation and incompatibility of surveillance efforts at federal, state, and local levels exist because authority rests across local and state entities, and because funding historically tended to be disease-specific (IOM, 2003). Similarly, collaboration among European Union members has often been hampered by differing opinions about priorities, division of roles and responsibilities, and resistance to greater centralization (Gouvras, 2004). Coordinating surveillance systems for the 192 member states of the WHO, therefore, is a daunting task. National sovereignty is a hurdle. The WHO does not have the authority to take control when local governments resist, even when the health of the local and global community is at risk. According to the United States General Accounting Office review of the lessons learned from SARS, the WHO's ability to respond was primarily stymied by the initial lack of cooperation of the Government of China (US Government Accountability Office, 2004a). Without proper authority to intervene to protect global public health, the WHO can only offer guidance and technical support, and must rely on individual nations' willingness to share information. In the United States, the Centers for Disease Control struggles with similar turf issues. These hampered its ability to trace international travelers that might have been exposed to SARS in 2003. Airlines questioned the CDC's authority, and refused or delayed providing passenger information (USGAO, 2004a). Authority to obtain passenger information has still not been resolved.

Even though an unprecedented number of nations, agencies, and organizations are working hard to address the challenges of global infectious disease, there continue to be the common and interrelated systemic problems we already have mentioned: inadequate infrastructure in areas most vulnerable to epidemics, political barriers, too little coordination and funding (ActionAid, 2005; OECD, 2005), and the segregation of long-term health infrastructure from emergency response – especially in low- and middle-income countries (Barnett *et al.*, 2005) There seems to be an ongoing process of dividing and subdividing the organizations that make up the international response network. Planning and funding is accomplished on a program-by-program basis. The WHO's 2006–2007 budget request for "essential health interventions" includes: HIV/AIDS; child and adolescent health; communicable disease prevention and control; making pregnancy safer; malaria; mental health and substance abuse; reproductive health; tuberculosis; emergency preparedness and response; epidemic alert and response; and immunization and vaccine development. Despite their seeming interrelatedness, they are in fact funded as separate programs. Combined, these items constitute 53 percent of the WHO's budget against 11 percent allocated to health policies, systems, and products (WHO Proposed Programme Budget, 2006–2007).

Humanitarian disasters in the 1990s, including the AIDS pandemic and regional wars and civil strife, produced an explosion in the number of non-governmental organizations and public–private partnerships (PPPs) involved in world health (Walt, 2005). With this enormous increase in international funding and effort, there is an even greater need to reduce overlap and improve inefficiency (Worley, 2006). Funding from large philanthropic foundations, like the Rockefeller and the Bill and Melinda Gates Foundations, is changing the landscape of intervention (Cohen, 2006). In addition to funding new initiatives, they are linking the private sector, academic institutions, and non-governmental and multilateral organizations to produce inter-sector groups with unprecedented political and financial influence to tackle specific diseases. While no one knows for certain what the long-term impact of a proliferation of new major initiatives, like the Global Fund to Fight AIDS, TB and Malaria, the Roll Back Malaria campaign, and the Global Alliance for Vaccines and Immunization, will be on international organizational response, the process remains characterized by weak local capacity, politicization, and fragmentation of funding and efforts. To improve global capacity to respond to outbreaks, systemic emergency response capabilities, primary care and prevention activities, and basic infrastructure must be integrated and strengthened locally and globally.

The good news is that progress is being made. Since the late 1990s, the WHO has also helped to reduce competition and pool resources by getting a host of partners, such as the CDC, Health Canada, and the network of Pasteur Institutes, to operate under its auspices. Programs are in place to expand the availability of new technologies, like antibiotics, vaccinations, and insecticides, for developing nations. And in the United States, as of publication, the Department of Homeland Security is making progress moving the National Response Plan from paper to practice, using an all-hazards approach.

Although research and technological advances have a tremendous impact on disease containment and control, our greatest challenge may be to develop the global organizational and operational capacity to provide access to effective emergency response and long-term prevention and care. Once again, wealthy communities need to improve the conditions that cause and spread disease in the world's most vulnerable communities, or risk the loss of their own health and prosperity to the re-emerging threat of infectious disease. It would be shameful if history were to show that the lion's share of today's resources went to emergency response SWAT teams and disease-specific programs, while leaving behind the same weak health system and thereby once more the potential for the reoccurrence and expansion of more epidemics in the future (Neusy, 2005).

## References and further reading

ActionAid International USA (2005). *Changing Course: Alternative Approaches to Achieve the Millennium Development Goals and Fight HIV/AIDS*. Available at: www.actionaidusa.org/pdf/changing %20course %20report.pdf (accessed 11 February 2006).

Ballier, R.D., Omer, S.B., Barnett, D.J. and Everly, G.S. Jr (2006). Local public health workers' perceptions toward responding to an influenza epidemic. *Biomedical Central Journal*, 18 April. Available at: www.biomedcentral.com/1471-2458/6/99 (accessed 26 April 2006).

Barnett, D.J., Balicer, R.D., Lucey, D.R. *et al.* (2005). A systematic analytic approach to pandemic influenza preparedness planning. *Public Library of Science* **2**(12). Available at: http://medicine.plosjournals.org/perlserv/?request=get-document&doi=10.1371/journal (accessed 8 January 2006).

Basch, P.F. (1999). *Textbook of International Health*, 2nd edn. New York: Oxford University Press.
Boshell-Samper, J. (2001). Emerging infections in Columbia: a national perspective. In: J.R. Davis and J. Lederberg (eds), *Emerging Infectious Diseases from the Global to the Local Perspective: A Summary of a Workshop of the Forum on Emerging Infections*. Washington, DC: Institute of Medicine, National Academy Press, pp. 36–39.
Bradsher, K. and Rosenthal, E. (2006). Is business ready for a flu pandemic? *New York Times*, 16 March. Available at: http://www.nytimes.com/2006/03/16/business/16bird.html?ex=1146283200&en=11f40f6084bf6358&ei=5070 (accessed 16 March 2006).
Brower, J. and Chalk, P. (2003). *The Global Threat of New and Emerging Infectious Diseases: Reconciling US National Security and Public Health Policy*. Santa Monica: Rand.
Centers for Disease Control and Prevention (2002). *Protecting the Nation's Health in an Era of Globalization: CDC's Global Infectious Disease Strategy*. Available at: http://www.cdc.gov/globalidplan/global_id_plan.pdf (accessed 10 February 2006).
Centers for Disease Control and Prevention (2006). *Budget Request Summary Fiscal Year 2007*. Available at: http://www.cdc.gov/fmo/PDFs/FY07budgetreqsummary.pdf (accessed 4 March 2006).
Chertoff, M. (2005). Letter to United States Governors, The NIMS Integration Center DHS/FEMA, 4 October. Available at: http://www.fema.gov/nims (accessed 12 February).
Cohen, J. (2006). The new world of global health. *Science* **311**, 162–167.
Connolly, M. (2000). *Outbreak Alert and Response in Emergencies, Global Outbreak Alert and Response*. Report of a WHO meeting, Geneva, Switzerland 26–28 April 2000. World Health Organization, Department of Communicable Diseases Surveillance and Response WHO/CDS/CSR2000.3. Available at: http://www.who.int/emc (accessed 8 January 2006).
Council on Foreign Relations (CFR) (2005). *Summary Report of a Conference on the Global Threat of Pandemic Influenza, 16 November 2005*. Available at: www.cfr.org/content/publications/attachments/pandemic_conference_summary.pdf (accessed 5 January 2006).
Drager, N. and Heymann, D.M. (2004). *Emerging and Epidemic-Prone Infectious Diseases: Threats to Public Health Security*. Commissioned briefing notes for meeting November 12-13. 2004 arranged by Centre for Global Studies (CFGS) and the Centre for International Governance Innovation. Available at: http://www.cigionline.org/publications/docs/g20.sanjose.heymann.pdf (accessed 13 March 2006).
Earls, M.J. and Hearne, S.A. (2005). *Facing the Flu: From the Bird Flu to a Possible Pandemic, Why isn't America Ready?* Trust for America's Health, February 2004. Available at: www.healthyamericans.org/reports/files/AvianFlu.pdf (accessed 10 January 2006).
Enserink, M. (2004). Emerging infectious diseases: a global fire brigade responds to disease outbreaks. *Science* **303**, 1605.
Evans, J.R., Hall, K.L. and Warford, J.J. (1981). Shattuck Lecture – Health Care in the Developing World: Problems of Scarcity and Choice. *New England Journal of Medicine* **305**, 1117–1127.
Fidler, D.P. (1996). Globalization, international law, and emerging infectious diseases. *Emerging Infectious Diseases* **2**(2), 77–84.
Garrett, L. (2000). *Betrayal of Trust: The Collapse of Global Public Health*. New York: Hyperion.
Garrett, L. (2005). *The Next Pandemic?* Foreign Affairs July/August 2005 Council on Foreign Relations. Available at: www.foreignaffairs.org/20050701faessay84401/laurie-garrett/the-next-pandemic.html (accessed 11 January 2006).

Gerberding, J.L. (2005). *Pandemic Planning and Preparedness*. Testimony before House Committee on Energy and Commerce Subcommittee on Health, 26 May 2005. Available at: www.hhs.gov/asl/testify/t050526a.html (accessed 5 January 2006).

Global Forum for Health Research (2004). *The 10/90 Report on Health Research 2003–2004. Executive Summary*. Geneva: Author.

Global Health Facts.Org (2006). Kaiser Foundation. Available at: www.globalhealthfacts.org/index.jsp (accessed 12 February 2006).

Gouvras, G. (2004). The European Centre for Disease Prevention and Control. *Eurosurveillance* **9**(10), 2–5. Available at: http://www.eurosurveillance.org/em/v09n10/0910-221.asp.

Heymann, D. (2001). Introduction. In: J.R. Davis and J. Lederberg (eds), *Emerging Infectious Diseases from the Global to the Local Perspective: A Summary of a Workshop of the Forum on Emerging Infections*. Washington, DC: Institute of Medicine, National Academy Press. Available at: www.nap.edu/books/0309071844/html/ (accessed 8 January 2006).

Heymann, D. and Rodier, G. (2004). Global surveillance, national surveillance, and SARS. *Emerging Infectious Diseases* **10**(2), 173–175.

Hong Kong Special Administrative Region (HKSAR) (2003). *SARS in Hong Kong: From Experience to Action*. SARS Expert Committee Summary Report, 30 September, 2003. Available at: www.sars-expertcom.gov.hk/english/reports/summary (accessed 8 February 2006).

Human Security Centre (2005). *Human Security Report 2005*. New York: Oxford University Press.

Institute of Medicine (2003). *The Future of the Public's Health in the 21st Century: The Governmental Public Health Infrastructure*. Washington, DC: National Academies Press.

Institute of Medicine (2005). *The Threat of Pandemic Influenza: Are We Ready?* Workshop Summary. S.L. Knobler, A. Mack, A. Mahmoud and S.M. Lemon (eds). Washington, DC: National Academies Press.

Kakar, F. (2000). *International Outbreak Response – Afghanistan: A Country Perspective. Global Outbreak Alert and Response*. Report of a WHO meeting, Geneva, Switzerland, 26–28 April 2000. World Health Organization, Department of Communicable Diseases Surveillance and Response WHO/CDS/CSR2000.3. Available at: http://www.who.int/emc (accessed 8 January 2006).

Kickbush, I. and Buse, K. (2005). Global influences and global responses: international health at the turn of the twenty-first century. In: M.H. Merson, R.E. Black and A.J. Mills (eds), *International Public Health: Diseases, Programs, Systems, and Policies*. Sudbury: Jones and Bartlett, pp. 701–737. Originally published by Aspen Publishers, Inc. (2001).

Lam, S.K. (2001). International smart partnership in emerging diseases: sense and sensibility. In: J.R. Davis and J. Lederberg (eds), *Emerging Infectious Diseases from the Global to the Local Perspective: A Summary of a Workshop of the Forum on Emerging Infections*. Washington, DC: Institute of Medicine, National Academy Press, pp. 68–71.

Leavitt, M.O. (2005). *HHS Pandemic Influenza Plan Testimony before House Committee on Energy and Commerce 8 November 2005*. Available at: http://www.hhs.gov/asl/testify/t051108.html (accessed 11 January 2006).

Lederberg, J. (2001). *Summary and Assessment*. In: J.R. Davis and J. Lederberg (eds), *Emerging Infectious Diseases from the Global to the Local Perspective: A Summary of a Workshop of the Forum on Emerging Infections*. Washington, DC: Institute of Medicine, National Academy Press, pp. 1–28.

Lee, J.W. and McKibbin, W.J. (2004). Estimating the global economic cost of SARS. In: S. Knobler, A. Mahmoud, S. Lemon *et al.* (eds), *Learning from SARS: Preparing for the Next Disease Outbreak – Workshop Summary*. Washington, DC: Institute of Medicine, National Academies Press.

Lee, K. (2004). The pit and the pendulum: can globalization take health governance forward? *Development*, **47**(2), 11–17.

MacKellar, L. (2005). Priorities in Global Assistance for Health, AIDS and Population. OECD Development Centre, Working Paper No. 244. Available at: http://www.oecd.org/dataoecd/42/39/34987795.pdf.

Martinez, L.J. (2002). *Global Overview, Global Outbreak Alert and Response.* Report of a WHO meeting, Geneva, Switzerland, 26–28 April 2000. World Health Organization, Department of Communicable Diseases Surveillance and Response WHO/CDS/CSR2000.3. Available at: http://www.who.int/emc (accessed 8 January 2006).

*Medical News Today* (2005). Worrisome gaps in US planning for avian flu outbreak, new analysis. Available at: www.medicalnewstoday.com/newsid=22744 (accessed 8 January 2006).

Mills, C.E., Robins, J.M., Bergstrom, C.T. and Lipsitch, M. (2006). Pandemic influenza: risk of multiple introductions and the need to prepare for them. *Public Library of Science Medicine* **3**(6), e135.

Millwood, D. (ed.) (1996). *The International Response to Conflict and Genocide: Lessons from the Rwanda Experience.* Copenhagen, Denmark: Steering Committee of the Joint Evaluation of Emergency Assistance to Rwanda. Available at: http://www.reliefweb.int/library/nordic/.

Minelli, E. (2003). *World Health Organization: The Mandate of a Specialized Agency of the United Nations.* Geneva: Foundation for Medical Education and Research. Available at: http://www.gfmer.ch/TMCAM/WHO_Minelli/Index.htm/ (accessed 2 December 2005).

Mounier-Jack, S. and Coker, R.J. (2006). *How Prepared is Europe for Pandemic Influenza? Analysis of National Plans.* Available at: www.thelancet.com DOI: 10.1016/S0140-6736(06)68511–5 (accessed 26 April 2006).

Murray, C.J., Lopez, A.D. and Wibulpolprasert, S. (2004). Monitoring global health: time for new solutions. *British Medical Journal* **329**, 1096–1100.

Ndikuyeze, A. (2000). *From Response to Preparedness: How can Better International Coordination lead to Better Epidemic Preparedness at the Regional and National levels? Global Outbreak Alert and Response.* Report of a WHO meeting, Geneva, Switzerland, 26–28 April 2000. World Health Organization, Department of Communicable Diseases Surveillance and Response, WHO/CDS/CSR2000.3. Available at: http://www.who.int/emc (accessed 8 January 2006).

Nebehay, S. (2005). *WHO "War Room" Prepares for Bird Flu Pandemic.* Available at: http://mmrs.fema.gov/news/influenza/2005/oct/nflu2005-11-21a.aspx (accessed 12 January 2006).

Neusy, A.J. (2004). Pandemic politics. *Foreign Policy* **Nov/Dec**, 82–84.

New York State Department of Health (2006). *Pandemic Influenza Plan, February 2006.* Available at: www.health.state.ny.us./disease/communicable/influenza/pandemic/doc.s_inflenza_plan.pdf (accessed 25 February 2006).

Osterholm, M.T. (2003). SARS: *How Effective is the State and Local Response?* Testimony before Senate Committee on Governmental Affairs 21 May, 2003. Available at: www.senate.gov/~govtaff/index.cfm?Fuseaction=Hearings.Testimony&HearingID=72&WitnessID=244 (accessed 12 January 2006).

Osterholm, M.T. (2005). *Avian Flu: Addressing the Global Threat.* Testimony before the House Committee on International Relations, December 7, 2005. Available at: http://wwwa.house.gov/international_relations/109/ost120705.pdf (accessed 17 January 2006).

Public Health Agency of Canada (PHAC) (2003). *A Report on the National Advisory Committee on SARS and Public Health*. Available at: www.phac-aspc.gc.ca/publicat/sars-aras/pdf/sars-e.pdf (accessed 12 January 2006).

Shoo, R. (2000). *Early Warning Response Network, Southern Sudan, Global Outbreak Alert and Response*. Report of a WHO meeting, Geneva, Switzerland, 26–28 April 2000. World Health Organization, Department of Communicable Diseases Surveillance and Response WHO/CDS/CSR2000.3. Available at: http://www.who.int/emc (accessed 8 January 2006).

Telford, J., Cosgrave, J. and Houghton, R. (2006). *Joint Evaluation of the International Response to the Indian Ocean Tsunami: Synthesis Report*. London: Tsunami Evaluation Coalition.

The White House (2006). *The Federal Response to Hurricane Katrina: Lessons Learned*. Available at: http://www.whitehouse.gov/reports/katrina-lessons-learned.pdf (accessed 1 March 2006).

Toole, M.J., Waldman, R.J. and Zwi, A.B. (2005). *Complex Humanitarian Emergencies*. In: M.H. Merson, R.E. Black and A.J. Mills (eds), *International Public Health: Diseases, Programs, Systems, and Policies*. Sudbury: Jones and Bartlett, Inc., pp. 439–513. Originally published by Aspen Publishers, Inc. (2001).

Travis, P., Bennett, S., Haines, A. *et al*. (2004). Overcoming health-systems constraints to achieve the Millennium Development Goals. *Lancet* **364**, 900–906.

Tregasis, S.R. (2006). Ready or Not? *NYU Magazine New York University*, **6**.

United Nations Development Programme (2005). *Human Development Report: International Cooperation at Crossroads: Aid Trade and Security in an Unequal World*. New York: Author.

United States Department of Health and Human Services (2005). *HHS Pandemic Influenza Plan Fact Sheet*. Available at: http://www.hhs.gov/pandemicflu/plan/factsheet.html (accessed 13 March 2006).

United States Department of Health and Human Services (2006). *Pandemic Planning Update: A Report from Secretary Michael O. Leavitt, March 13, 2006*. Available at: http://www.pandemicflu.gov/plan/pdf/panflu20060313.pdf (accessed 2 April 2006).

United States Department of Homeland Security (2004). *National Response Plan*. Available at: http://www.dhs.gov/interweb/assetlibrary/NRPbaseplan.pdf (assessed 17 January 2006).

United States Department of Homeland Security (2006). *National Incident Management System (NIMS) Basic Introduction and Overview*. Available at: http://www.fema.gov/pdf/nims/MIMS_basic_introduction_and_overview.pdf (accessed 20 April 2006).

United States Government Accountability Office (2004a). *Emerging Infectious Diseases: Asian SARS Outbreak Challenged International and National Responses*. Available at: http://www.gao.gov/new.items/d04564.pdf (accessed 13 January 2006).

United States Government Accountability Office (2004b). *Emerging Infectious Diseases: Review of State and Federal Disease Surveillance Efforts*. Available at: http://www.gao.gov/new.items/d04877.pdf (accessed 13 January 2006).

United States Government Accountability Office (2005). *Influenza Pandemic: Challenges in Preparedness and Response*. Available at: http://www.gao.gov/new.items/d05863t.pdf (accessed 13 January 2006).

United States Government Accountability Office (2006a). *Emergency Preparedness and Response: Some Issues and Challenges Associated with Major Emergency Incidents*. Testimony before the Little Hoover Commission, State of California: Statement of

William O. Jenkins, Jr, Director Homeland Security and Justice Issues. Available at: http://www.gao.gov/new.items/d06467t.pdf (accessed 13 March 2006).
United States Government Accountability Office (2006b). *Global Health: Spending Requirement Presents Challenges for Allocating Prevention Funding under the President's Emergency Plan for AIDS Relief*. Available at: http://www.gao.gov/new.items/d06395.pdf (accessed 5 April 2006).
Walt, G. (2005). Global cooperation in international public health. In: M.H. Merson, R.E. Black and A.J. Mills (eds), *International Public Health: Diseases, Programs, Systems, and Policies*. Sudbury: Jones and Bartlett, Inc., pp. 667–699. Originally published by Aspen Publishers, Inc. (2001).
Weinberg, J. (2000). *Outbreak Alert and Specialized Surveillance Networks, Global Outbreak Alert and Response*. Report of a WHO meeting, Geneva, Switzerland, 26–28 April 2000. World Health Organization, Department of Communicable Diseases Surveillance and Response WHO/CDS/CSR2000.3. Available at: http://www.who.int/emc (accessed 8 January 2006).
World Bank Group Operations Evaluation Department (OED) (2005). *Evaluation of the World Bank's Assistance for Fighting the HIV/AIDS Epidemic*. Available at: www.worldbank.org/oed/aids/docs/report/hiv_complete_report.pdf.
World Economic Forum (WEF) (2006a). *Global Health Initiative (GHI) Newsletter February 2006*. Available at: www.weforum.org/documents/ghi/GHI_Newsletter_February_2006.htm.
World Economic Forum (WEF) (2006b). *Global Risks 2006*. Report 28.01.2006. Available at: www.weforum.org/pdf/CSI/Global_Risk_Report.pdf (accessed 19 February 2006).
World Health Organization (2000). *Global Outbreak Alert and Response*. Report of a WHO meeting, Geneva, Switzerland, 26–28 April 2000. World Health Organization, Department of Communicable Diseases Surveillance and Response. Available at: http://www.who.int/emc (accessed 8 January 2006).
World Health Organization (2001). *A Framework for Global Outbreak Alert and Response*. Department of Communicable Disease Surveillance and Response. Available at: http://www.who.int/csr/resources/publications/surveillance/WHO_CDS_CSR_2000_2/en/ (accessed 8 December 2005).
World Health Organization (2003). *The World Health Report 2003: Shaping the Future*. Geneva: Author.
World Health Organization (2004). *Water, Sanitation and Hygiene Links to Health*. Available at: http://www.who.int/water_sanitation_health/publications/facts2004/en/index.html (accessed 13 March 2006).
World Health Organization (2005a). *Report by the Secretariat, Strengthening Pandemic Influenza Preparedness and Response*. Fifty-Eighth World Health Assembly, Provisional Agenda Item 13.9, 7 April 2005.
World Health Organization (2005b). *Responding to the Avian Influenza Pandemic Threat: Recommended Strategic Actions*. Available at: http://www.who.int/csr/resources/publications/influenza/WHO_CDS_CSR_GIP_05_8-EN.pdf (accessed 13 March 2006).
World Health Organization (2006a). *Pandemic Influenza Preparedness and Mitigation in Refugee and Displaced Populations: WHO Guidelines for Humanitarian Agencies*. Available at: http://www.who.int/csr/disease/avian_influenza/guidelines/avian2006-04-9.pdf.
World Health Organization (2006b). Responding to urgent health needs. In: *The World Health Report 2006: Working Together for Health*. Geneva: Author.
Worley, H. (2006). *Intersecting Epidemics Tuberculosis and HIV*. Population Reference Bureau (PRB). Available at: www.prb.org (accessed 26 April 2006).
Yach, D. (2004). Guest Editorial: Politics and health. *Development* **47**(2), 5–10.

Zwillich, T. (2005). President Bush outlines national flu plan. *Medscape Medical News*, 2 November. Available at: http://www.medscape.com/viewarticle/515996?src=mp.

## Useful websites

Centers for Disease Control and Prevention. *About CDC* (http://www.cdc.gov/about/default.htm, accessed 10 February 2006; no longer available).
Centers for Disease Control and Prevention. *WHO Collaborating Centers at CDC by CIO*. Available at http://www2a.cdc.gov/od/gharview/GHARwhocollabs.asp.
Centers for Disease Control and Prevention. *CDC Organization*. Available at: http://www.cdc.gov/about/cio.htm (accessed 10 February 2006).
Centers for Disease Control and Prevention. *CDC – Our Story*. Available at: http://www.cdc.gov/about/ourstory.htm (accessed 10 February 2006).
Centers for Disease Control and Prevention. *National Center for Infectious Disease*. Available at: http://www.cdc.gov/ncidod/ (accessed 10 February 2006).
Centers for Disease Control and Prevention. *Futures Initiatives*. Available at: http://www.cdc.gov/futures/index.htm (accessed 10 February 2006).
Centers for Disease Control and Prevention. *National Coordinating Office for Global Health*. Available at: http://www.cdc.gov/ogh/ (accessed 10 February 2006).
Centers for Disease Control and Prevention. *Emergency Preparedness and Response*. Available at: http://www.bt.cdc.gov/ (accessed 13 March 2006).
Organization for Economic Cooperation and Development. *OECD DAC Development Co-operation Report 2005*. Available at: http://fiordiliji.sourceoecd.org/pdf/dac/stat-analysis.pdf (accessed 12 January 2006).
World Health Organization. *International Health Regulations*. Available at: http://www.who.int/csr/ihr/en/ (accessed 14 December 2005).
World Health Organization. *About WHO*. Available at: http://www.who.int/about/en/ (accessed 22 February 2006).
World Health Organization. *Governance*. Available at: http://www.who.int/governance/en/ (accessed 22 February 2006).
World Health Organization. *WHO Headquarters Structure*. Available at: http://www.who.int/dg/lee/hqstructureenglish_06.pdf (accessed 17 January 2006).
World Health Organization. *WHO Collaborating Centers*. Available at: http://whqlily.who.int/ (accessed 17 January 2006).
World Health Organization. *Epidemic and Pandemic Alert and Response*. Available at: http://www.who.int/csr/en/ (accessed 22 February 2006).
World Health Organization. *Global Health Atlas*. Available at: http://globalatlas.who.int/.
World Health Organization. *Health Systems*. Available at: http://www.who.int/healthsystems/en/ (accessed 23 February 2006).
World Health Organization. *Avian Influenza*. Available at: http://www.who.int/csr/disease/avian_influenza/en/ (accessed 23 February 2006).
World Health Organization. *Health Action in Crisis*. Available at: http://www.who.int/hac/en/ (accessed 3 March 2006).

# 17 Principles of building the global health workforce

Pierce Gardner, Aron Primack, Joshua P. Rosenthal and Kenneth Bridbord

One of the most positive aspects of globalization has been a burgeoning recognition of and interest in global health, and a sharp increase in resources and programs committed to addressing the health disparities that exist between rich and poor nations.

A broadened base of support has been nurtured by a variety of factors:

1. The traditional focus on humanitarian assistance has been strengthened by the power of the media and other forms of communication to focus attention on the enormity of global health problems and stir the moral conscience of the wealthy countries to respond.
2. There is increasing concern about protecting our country against the importation of health threats as exemplified by SARS, influenza, and the recent re-emergence of poliomyelitis.
3. There is an increasing consensus that good health is a driving force in economic development, and that efforts to improve global health will build partners in world trade.
4. Diseases such as HIV/AIDS can be so disruptive as to threaten political, economic, and civil stability, and thus global health is considered a factor in our national security.
5. Health and science have long been recognized as venues for bridge-building among disparate societies, and global health assistance has become an important component of diplomacy in our foreign policy.
6. Expert knowledge of global health is valuable in advising and treating US international travelers as well as immigrants from other countries.

Finally, science conducted in the developing world may have unique benefits for the developed world. For example, because HIV/AIDS vaccine trials are best done in areas of high HIV endemicity, the road to a successful vaccine will be through developing countries. As another example, the pathophysiology of cholera and the development of the oral rehydration mixtures, which are used the world around, were largely accomplished by research in Bangladesh and India.

Thus, the heightened appreciation of the value of investing in global health has broadened the constituency of support well beyond the humanitarian base, to include economists, scientists, diplomats, politicians, the military, national security advisors, and many others.

How has this wellspring of interest been manifest? Recent years (especially since 2000) have seen a dramatic increase in global health expenditures across the entire spectrum of organizations and activities (prevention, therapy, research, and support). Bilateral programs (e.g. the United States Agency for International Development, USAID, and the President's Emergency Plan for AIDS Relief, PEPFAR), multilateral agencies (e.g. the United Nations, the World Health Organization, the Pan American Health Organization, and others), development banks, other multilateral efforts from the European Community and the Global Fund to Fight AIDS, Tuberculosis and Malaria, and private non-profit organizations (e.g. the Bill and Melinda Gates Foundation, the Rockefeller Foundation and many others) in the aggregate provide in excess of $12 billion per year in development assistance for global health. Not included in this figure is the increased funding of global health research by the National Institutes of Health (currently in excess of $500 million per year), the international activities of the Centers for Disease Control and Prevention, and the growing commitment of universities (including medical schools, schools of public health, and other professional schools) to major activities in global health research and training. Similar efforts are taking place in other donor countries.

Significant landmarks of international cooperation are:

- the Group of Eight (G-8) agreements to address global health and to reduce the indebtedness of low income countries
- the United Nation's adoption of eight Millennium Development Goals designating specific targets (three in health) to be achieved in the decade ending 2015.

However, in general the global health assistance activities can be characterized as independent efforts, which have not been well coordinated within an overarching plan or organizational structure (see also Chapter 16).

In the poorest nations, the health disparities and needs are enormous, urgent, and worsening. Many of the social forces discussed in other chapters of this book (e.g. explosive urban growth, social dislocation, and sexual mores) have abetted a deterioration of health-care systems, diminished life expectancy, and also threaten societal unrest in these vulnerable populations. The density of the total health workforce in Africa is less than one-tenth of that in the Americas (see Table 17.1). An estimated global shortfall of more than four million health-care workers exists, with one million of this shortfall in sub Saharan Africa, where the already overburdened workers are further taxed by the added clinical and public health burden of the growing numbers of persons with HIV/AIDS and its complications. Stress, occupational risk of infection, poor working conditions, and low pay have led many health workers to seek improved conditions and opportunities by migrating to other positions or locations, with the result that attrition exceeds the output of new graduates into the workforce in many of the poorest countries.

## Key factors/principles in training the global health workforce

A critical limiting factor in the response to the formidable global health challenge is the paucity of human and institutional resources. This in turn limits the absorptive capacity *vis-à-vis* the ability to use the increasing donor nation funds to build effective programs. The education pipeline is inadequate to meet current needs, and shortages of teachers limit the ability to ramp up the output. Help is needed at all levels, from national leadership and planning, to the most downstream interactions of health workers with individuals and communities.

*Every country should have a national workforce training plan, tailored to its situation and responsive to its short-, medium-, and long-term needs. The plan should include an assessment not only of the number and distribution of health-care workers, but also the skills mix needed, the administrative and support systems required, the rewards/career opportunities, measures to reduce out-migration to other jobs or locations, and a budget. National governments need to identify the resources to implement their plans. These issues have been well addressed in the* 2006 World Health Report: Working Together for Health *(WHO Press, Geneva).*

We believe that certain key factors/principles are core to the success of health training activities, whether the focus is clinical, research, public health, or administration and infrastructure building, and whether the training sponsor is government, non-government, academic, faith-based, or other entity. This section outlines five key factors/principles for success, and gives examples from public–private partnerships as well as from the Fogarty International Center's (FIC) programs that have developed over the past two decades in response to the HIV/AIDS crisis in the developing world. Although the FIC programs are focused on building research capacity, we believe these principles also apply broadly to other global health efforts, including the ramping up of prevention and therapeutic care for the approximately 40 million persons in the world currently living with HIV/AIDS.

**Table 17.1 Global health workforce, by density**

| WHO region | Total health workforce | | Health service providers | | Health management and support workers | |
|---|---|---|---|---|---|---|
| | Number | Density (per 1000 population) | Number | Percentage of total health workforce | Number | Percentage of total health workforce |
| Africa | 1,640,000 | 2.3 | 1,360,000 | 83 | 280,000 | 17 |
| Eastern Mediterranean | 21,000,000 | 4.0 | 1,580,000 | 75 | 520,000 | 25 |
| South-East Asia | 7,040,000 | 4.3 | 4,730,000 | 67 | 23,000,000 | 33 |
| Western Pacific | 10,070,000 | 5.8 | 7,810,000 | 78 | 2,260,000 | 23 |
| Europe | 16,630,000 | 18.9 | 11,540,000 | 68 | 5,090,000 | 31 |
| Americas | 21,740,000 | 24.8 | 12,460,000 | 57 | 9,280,000 | 43 |
| World | 59,220,000 | 9.3 | 39,470,000 | 67 | 19,750,000 | 33 |

All data for latest available year. For countries where data on the number of health management and support workers were not available, estimates have been made based on regional averages for countries with complete data. Source: World Health Organization, *Global Atlas of the Health Workforce* (http://www.who.int/globalatlas/default.asp).

## Key factor/principle #1

*In establishing a training program, the initial planning should encompass and be responsive to local needs and priorities.*

There must be agreement on the goals and objectives, clarity regarding the responsibilities of the trainers and trainees, and a detailed program description written and signed by all parties. In this process, the program leaders at the foreign site must be full and equal partners in identifying the local needs and setting the priorities. The planning should encompass the full scope of training activities, and a timetable with clearly stated and measurable outcome goals should be established.

## Key factor/principle #2

*Workforce capacity building requires long-term commitment and stability together with flexibility to adapt as needs, priorities, and circumstances change.*

Although short-term programs, such as stand alone one- or two-week training courses, often have a profound personal effect on the visitors, they rarely lead to the desired outcome of sustainable programs that will increase in value and productivity with time. The commitment must be for the long run, recognizing that establishing training programs can be particularly challenging in settings where infrastructure and institutional support is weak. Training programs are never static and, despite the best planning efforts, new needs, priorities, and circumstances will arise as programs evolve. This requires flexibility from all parties, and the ability to respond and adapt to new situations.

## Key factor/principle #3

*Establish a partnership and commitment at both the individual and institutional levels, with increasing empowerment of the foreign partner as the collaboration matures.*

In sponsored training programs, a key prerequisite to success is a true partnership and commitment to the program by the individuals and the institutions involved. Strong leadership and support must be sustained. A formal "twinning" arrangement between the institutions is one example that has proved to be successful in solidifying such relationships. Although the overall goal is to build workforce capacity in the developing country, the ways in which this can be done are diverse. Early on, it is common for the predominance of training activities to be at the donor-country institutions, bringing the foreign trainees for long-, medium-, or short-term training. As programs mature and the foreign site becomes better staffed with returning trainees, more of the training activities can take place at the foreign site. This is not only cost-saving but also has the collateral benefits that (i) in-country trainees are less likely to emigrate out of country, and (ii) the trainees become teacher/role models for others and multiply their value by "south–south" teaching. With full maturation of the developing country site, the center of gravity in the collaborative relationship shifts increasingly to the foreign site, which can then function with relative autonomy. Such programs serve not only as national and regional resources, but are also positioned to assume global significance as centers of excellence in training and research.

## Key factor/principle #4

*Human resources must be nurtured by long-term mentoring that offers follow-on opportunities to update and reinforce skills, attention to career development and working conditions, as well as professional and financial rewards.*

The training and equipping of individuals with special skills is just the beginning of building workforce capacity in global health. The task of nurturing throughout a professional career is usually a larger and more complex challenge than the initial training program. Elements of success in this area include:

1. Long-term mentoring with ties to the original training program, together with updates and refresher activities, so that trainees stay connected to resources that will enhance their professional growth. This process has been greatly facilitated by advances in information technology, which allow distance learning and communication that were not possible in an earlier era.
2. Attention to career development and professional opportunities. Having attractive career paths and opportunities is prerequisite to building the global workforce. The low level of health-care funding in the world's poorest countries has severely limited the job opportunities for well-trained individuals at all levels.
3. Attention to professional and financial rewards, and working conditions. These are closely related factors. Poorly equipped and understaffed hospitals and laboratories, the danger of nosocomial infection with pathogens such as HIV or tuberculosis, infrastructure problems, low pay, and increasing demand for services all contribute to stress and dissatisfaction, and result in an acceleration of health-care professionals seeking a better life in other settings. These migrations both within and across borders exacerbate the already critical dearth of health-care workers in the poorest countries.

As the health labor market has become more global, high-income countries increasingly have recruited nurses and doctors to meet clinical demands, and have provided attractive opportunities for foreign scientists and public health workers as well. Efforts to address this "brain drain" include adoption of incentives to foster retention and to promote reverse migration by expatriates, economic and political barriers to out-migration, and creating limited training programs that yield credentials that are only recognized locally and are not valued internationally. Nevertheless, the shortages remain severe and acute. An interesting proposal of the Global Commission on International Migration is the creation of an Education Reinvestment Fund in which the brain-drain beneficiary countries would provide funds to build the educational capacity in the resource-poor countries that are losing substantial numbers of health-care workers to out-migration.

### Key factor/principle #5

*Develop multidisciplinary centers of excellence, which are catalysts for training and research activities in the developing world.*

The magnitude and urgency of the need to build the training capacity for global health will require contributions from many disciplines and backgrounds. In addition to the traditional fields of clinical services, public health, and research, many other fields now have a significant stake in global health. These include the health economists, social scientists, ethicists, ecologists, population demographers, urban planners, and many others. The "silo" approach of individual disciplinary programs is being supplanted by interactive multidisciplinary programs, which approach problems more broadly. For example, addressing the "know–do" gap between ascertaining knowledge and implementation of beneficial action requires that those doing the upstream science connect with those responsible for downstream operations and health services. Basic scientists, clinicians, public health workers, and politicians may all be involved.

Such multidisciplinary activities are most likely to arise in broad-based centers, which are also the logical places to invest in building the information technology and management infrastructure necessary to support centers of excellence. These centers ideally would serve as national and regional hubs for "south–south" training and research, and for "south–north" teaching and research such as HIV/AIDS vaccine trials, and provide training experiences for future generations of global health workers.

## Illustrative successful programs

There are myriad training efforts going on to respond to the immediate and long-term needs of building an expanded global health workforce. Rather than attempt to catalog all of them, we will describe here two successful programs that we believe illustrate the key factors and principles already described. Our goal is to emphasize what we believe works, and discuss pitfalls that can be avoided with proper planning.

*Pierce Gardner, Aron Primack, Joshua P. Rosenthal and Kenneth Bridbord*

## AIDS international training and research program (AITRP), Fogarty International Center, National Institutes of Health

This program began in 1988 as one of a new generation of research training programs designed to help scientists from institutions in low-and middle-income countries build research and public health capacities in those same countries. Grants are awarded to US institutions with strong HIV-related research training experience and with HIV-related research collaboration with institutions in low- and middle-income countries. The grantees, in partnership with their foreign collaborating institutions, identify foreign health scientists, clinicians, and allied health workers from the foreign countries to participate in their joint research training programs. The primary goal of this program is to build multidisciplinary biomedical and social science research capacity to address the HIV/AIDS epidemic in the collaborating country. The training programs provide a variety of short-, medium-, and long-term training opportunities. While academic courses may be taken at either the US or foreign site, to the greatest extent possible the research takes place at the foreign site, working on problems considered to be of high priority for that country. The AITRP supports a broad variety of research training opportunities, including pursuit of MS, MPH, or PhD degrees; postdoctoral experiences; special training in laboratory procedures, data management, administration and other activities; and in-country practical and applied short-term training. In addition, it may support advanced research training for current and/or former trainees. For the US side of the collaboration, support may be provided to the faculty for research training activities at the foreign site and, in some programs, to US health science students to receive overseas health research experiences. Currently, grants to 26 US schools of medicine and public health support research training in nearly 60 low- and middle-income countries where HIV/AIDS is endemic or epidemic. More than 1000 individuals have received long-term training of more than six months in this program – which does not include the secondary training that these individuals have provided to others – and many more have participated in short-term training.

The principles outlined in the previous section have been closely observed throughout the 18 years of the program. From the beginning, the collaboration between the US and foreign institutions and program leaders has been one of full partnership, and the training and research priorities and activities have focused on needs relevant to the foreign site. The long-term stability and mutual trust that has evolved has allowed an unusual degree of flexibility as needs, priorities, and circumstances change. Many returning scientists have established successful centers of clinical research excellence in their home countries to serve national and regional needs, including creating important sites for multicentered clinical trials. These “Fogarty Fellows” have assumed major roles in directing HIV/AIDS activities in government, public health, science, and teaching. They comprise a cadre of committed global health workers who have a major impact on policy and programs, and serve as a network of communication and support for trainees and others in the field.

As the AITRP program has matured it has helped to spawn other global health research training investments, including four second-generation programs designed to strengthen the overall research capacity building effort:

1. *The International Clinical, Operational, and Health Services Research and Training Awards for AIDS and Tuberculosis (ICOHRTA-AIDS/TB).* This program addresses the "know–do" gap by supporting research that spans the spectrum from clinical science to operational and health service measures that will bring tangible benefits to the population. A novel feature of this award is that the foreign site is the initiator in choosing its collaborating partner, and receives direct funding from FIC/NIH.
2. *The Global Health Research Initiative Program for New Foreign Investigators (GRIP).* This initiative fosters career development for NIH-trained foreign investigators by providing salary and research support as they establish themselves upon re-entry to their home countries.
3. *Framework Programs for Global Health.* This program encourages the bringing together of multiple disciplines, such as engineering, business, chemistry, biology, communication, public health, medicine, bioethics, and environmental studies, to form centers of excellence in support of global health activities.
4. *FIC/Ellison Overseas Fellowships in Global Health and Clinical Research.* This program seeks to build both sides of the collaboration bridge by providing early career opportunities for US and developing country graduate students in the health professions to participate in one year of mentored clinical research at an NIH-funded research center in a developing country. It pairs US students with students from the host country, creating partnerships and contributing to building a new international community of global research scholars. The program is designed to give the trainees a substantive experience in global health research at a time when they are making career choices, so that their decisions regarding global health careers will be more fully informed.

## The Academic Alliance for Aids Care and Prevention in Africa (AA)

This program began in 2000 with a strong clinical training focus, and has broadened its activities to include preventive medicine, clinical research, laboratory training, and information technology and other infrastructure building.

It was conceptualized by a small group of academic leaders in the Infectious Diseases Society of America and the leadership at Makerere University School of Medicine in Kampala, Uganda. They formed the Academic Alliance (now a foundation) – a unique public–private partnership between academicians of Africa and North America, the pharmaceutical industry, and other organizations and institutions. The group recognized that "while international resources are needed to combat the ravages of the epidemic, the solutions for longer-term success rest in developing and sustaining African capacity to train, to treat, and to develop the research and care strategies within an African setting." This partnership, in 2001, launched the Academic Alliance for AIDS Care and Prevention in Africa (AA) with the goal of creating "a clinical training and research center where HIV/AIDS patients can receive high quality sustainable care, while clinically relevant scientific research can be used to answer important questions about HIV/AIDS in Africa." A 25,000 square foot Infectious Diseases Institute (IDI) was opened in October 2004, and now cares for more than 8000 patients. Since then, over 900 health-care professionals from 22 different African countries have received training, mostly in short courses in Kampala. Ownership of the building and the program is in Ugandan hands and, although the faculty is very international, Ugandan professionals increasingly are filling the senior positions in the program. An AIDS Treatment Information Center provides a telemedicine referral network, which includes both a service and an educational function and provides free of charge, rapid, high-quality consultation relative to a wide variety of questions regarding HIV/AIDS and other conditions. Currently it is connected to 30 of 56 districts in Uganda. Service is planned for all districts by the end of 2007. The program connects with trainees in the field, and allows for follow-up and continuing medical education of trainees. An important contribution to the infrastructure needs is the establishment of a center of excellence in laboratory training, with a special focus on tests relevant to HIV/AIDS. A Clinical Scholars program has been established to provide five years of support to five young investigators who have completed their clinical training and choose to engage in clinical research under the mentorship of an internationally recognized investigator. Additional training opportunities include a three-year Infectious Diseases Fellowship, an Exchange Program in International Medicine (a six-month opportunity for US or Canadian Infectious Disease fellows to work at the IDI at Makerere University), and a two-year Masters in Medicine program at Makerere University.

Attention to the program's stability and sustainability is evidenced by the endowment of a Chair for the Executive Director of the Infectious Diseases Institute, and by successful applications for training funds from industry and government. For example, financial support from PEPFAR has been obtained to provide HIV/AIDS training to African physicians, nurses, and clinical officers who work in military hospitals and care for African military personnel (considered to be a group with high-risk behavior). The Academic Alliance has an effective public relations effort, which includes a quarterly newsletter to more than 5000 people, and several major fund raising events each year. It has appointed a visible and powerful board of directors (including a former Head of the US Department of Health Care and Human Services) and a distinguished scientific advisory board.

Overall, the Academic Alliance for AIDS Care and Prevention in Africa (AA) has done very well as measured by the five principles/key factors presented in the previous section. However, the enterprise is still very young, and the ability to sustain the same high level of funding and hands-on support will be tested when the founders who currently lead the effort pass responsibility on to their successors.

## New training paradigms

The theme of this book, and a growing consensus in the scientific community, is that the determinants of infectious diseases are much more complex than the traditional host–pathogen interaction, and that social, economic, ecological, and political factors are powerful forces. This is particularly well illustrated by emerging and re-emerging infectious disease problems such as HIV/AIDS, SARS, Nipah virus, malaria, antimicrobial resistant pathogens, and others. Recognition of the complexity of infectious disease determinants has led to a call for transdisciplinary and system-based approaches to understanding and dealing with these issues. Most often, this has taken the form of assembling a collaborative multidisciplinary team to address an identified problem and develop a "common conceptual framework" or action plan. In addition to integration across the different areas of knowledge (horizontal integration), there is greater recognition of the need for vertical integration – i.e. incorporation of individual, community, and other non-academic views that add important practical, ethical, and political perspectives. Each type of integration has its value, but tends to function independently of the other. For example, the stunning success in the recognition and control of the SARS outbreak was largely accomplished by horizontal integration of the efforts of clinicians, epidemiologists, virologists, public health workers, veterinarians, and multinational organizations working in concert. However, the vertical integration of the social, economic, and ecological drivers of the disease, which have great relevance in terms of potential recurrence, received far less attention.

While there is strong consensus for transdiciplinary integration, the implications for how we train and organize the future global health workforce are much less clear. The lead in integration across academic disciplines will almost certainly come from universities (including the schools of public health and the health sciences). As an example, Duke University has made global health a university-wide priority, and is committed to building a program "that truly spans the humanities, social sciences, engineering, environment, law, and divinity as well as the life sciences." The program will include teaching, research, and services, and will be woven into both the undergraduate and graduate curricula, with special courses and certifications. More than a dozen current activities in global health will be brought together within a Duke Global Health Institute, and will build on Duke's core strengths in basic, clinical, and translational research.

On a smaller scale, the need for broadening training and perspectives is recognized by US medical students, one-third of whom overall and up to 70 percent in some medical schools now take one or more years out of the traditional training track to pursue special interests in fields like global health or research, or degrees in business, public health, or policy. Vertical integration of individual, community, political, and other special perspectives is a more difficult organizational task. However, there is increasing emphasis on diversity and lay representation in appointing advisory groups and boards of directors, and attention is given to minorities and under-represented groups in many federal grants and philanthropic activities.

The take-home message for training programs in global health is that trainees must understand the context in which they will work, and that cultural, social, economic, and other relevant areas should be included in the orientation and mentoring activities.

## Current challenges

The magnitude and urgency of identified health disparities, as well as the rapid growth of funding and political will among donor nations, has created a mandate to mount a major response quickly. Laudable goals, such as getting effective antiretroviral therapy to three million HIV/AIDS-infected individuals by the end of 2005, have been set without the operational and health services planning required for such a mammoth undertaking, and as a result the programs have not achieved their goals.

The Peace Corps ideal has spawned suggestions for donor-country basic scientists to build the scientific community in low- and middle-income countries, and for a clinical and preventive medical Peace Corps to provide services and training in areas of greatest need. WHO is addressing workforce issues to respond to emerging diseases, and the World Bank is expanding its health focus to meet the projected needs in surveillance, outbreak investigation, clinical care, and research. Other countries and non-government organizations are planning additional large-scale programs. We believe that the principles outlined in this chapter apply to these efforts, and that great care must be exercised in attempting such ambitious programs.

### Potential pitfalls

Pitfalls to be avoided include the following:

1. Failure to adapt programs to local needs and priorities. Even within countries, different areas have different and unique needs, and careful planning involving the local medical and political community is a prerequisite to success.

2. Medical tourism. The one- to three-week visit by a well-meaning but unprepared health worker takes valuable time and often resources from hosts who are already overburdened. While it may be a profound experience for the visitor, and interesting for the host, the value in terms of sustained benefit to the population served is usually marginal. Some exceptions might be highly technical teams that perform specialized services not available in-country, and which do not require long-term follow-up care after the service is complete (e.g. cataract removal, cleft palate repair, vaginal fistula repair). Another successful model for short-term (four- to six-week) experiences has been the establishment of a mentored teaching service at a particular foreign hospital (usually involving a formal twinning agreement between the US and foreign hospital), so that the residents and students rotate through (as on other services at the home hospital) and work together with the in-country trainees. The US trainees complete preparatory training prior to arrival, and are able to "hit the ground running" and contribute from the start. Although the training benefits both the US and in-country trainees, true reciprocity (bringing the foreign trainees to US clinical facilities for clinical work) is severely limited by hospital regulations and liability considerations.
3. Failure to train in-country staff. In general, all medical and public health activities should include training and transfer of skills that will empower the in-country staff personnel to carry on and sustain such efforts.
4. Siphoning key personnel from other important programs. The critical scarcity of health personnel in low-resource countries is worsened when well-funded donor country programs siphon doctors, nurses, and medical officers away from other vital activities. The rapid expansion of health services demanded by the attempt to improve and broaden HIV/AIDS care thus threatens to detract from other important high-value programs, such as childhood immunization and maternal/child health programs. Now that more donor funds and programs are being introduced, it is important that an assessment of the absorptive capacity of the training system be made at the outset. In order to expand the numbers of available health personnel quickly, new programs to train community-based workers for specific limited health tasks will need to be developed, in the context of a national health manpower plan.
5. Failure to evaluate and adjust. In the eagerness to respond to a crisis, evaluation is often overlooked. Programs such as an international medical Peace Corps should be front-loaded with operations and health services research to ascertain what works best, and to guide program development. A steering committee needs objective information in making necessary adjustments.
6. Failure to recognize the interconnections of research, training, and services. Research guides and validates the science on which training and services, both preventive and clinical, are based. Conversely, services influence the priorities and concerns that drive the science. Each arm strengthens the other. Scientific method is appropriate to virtually all activities of medicine.

Recognition of the importance of "downstream" applied science to study operations and health services has contributed much to current practices. *Ergo*, in the planning phase of any major global health endeavor, the components of service, training, and research should each be discussed.

7. Failure to sustain. Programs must have sustainability built into their planning. Resources, both human and financial, must be arranged at the outset for the future continuation of efforts. Unless the infrastructure issues identified in the 2006 World Health Report are addressed and improved, many of the efforts to train health-care workers will not have a sustainable impact on the dire situation that currently exists in many of the most threatened nations.

The challenges are great, but none of the potential pitfalls is unavoidable, and we are cautiously optimistic that many programs will prove valuable in approaching the targets set for the Millennium Development Goals. However, it is imperative that such programs be part of each country's national manpower training program, and be coordinated with the management, education, and planning activities of the national leadership. Without national leadership and sustained commitment, there is little chance of success.

## Conclusions

The past decade has brought a sea change in the attention focused on global health problems, and a determination to take action to alleviate the distress. There is a broadening of political will and support beyond the traditional humanitarian base to include economic, security, diplomatic, scientific, ecologic, and other self-interests that favor aid growth of global health assistance. The entire spectrum of government and non-government organizations is increasingly focused on global health, as evidenced by the setting of the UN Millennium Development Goals to be achieved over the next decade, and dramatic commitments of additional funds by government (e.g. PEPFAR – $15 billion) and non-government (e.g. the Bill and Melinda Gates Foundation, supplemented by the Warren Buffet fortune) sources.

A critical limiting factor in the response to the formidable global health challenge is the paucity of human (and institutional) resources in most low- and middle-income nations. Help is needed at all levels, from national leadership and planning to the most downstream interactions of health workers with individuals and communities. We believe that certain key factors/principles are core to the success of health-training activities, whether the focus is clinical, research, public health, or administration/infrastructure building. They are as follows:

1. In establishing a training program, the initial planning should encompass and be responsive to local needs and priorities.

2. Workforce capacity building requires long-term commitment and stability together with flexibility to adapt as needs, priorities, and circumstances change.
3. A partnership and commitment must be established at both the individual and institutional levels, with increasing empowerment of the foreign partner as the collaboration matures.
4. Human resources must be nurtured by long-term mentoring, follow-on opportunities to update and reinforce skills, attention to career development and working conditions, as well as professional and financial rewards.
5. Multidisciplinary centers of excellence, which are catalysts for training and research activities in the developing world, must be developed.

Two examples of successful programs are presented in this chapter.

There is a growing consensus in the scientific community that the determinants of infectious diseases are much more complex than simply the traditional host–pathogen interaction, and that social, economic, ecological, and political factors are powerful forces. We discuss the need for new training paradigms that are transdisciplinary and system-based, and cite some efforts now underway at US universities to integrate their global health activities.

Finally, we have addressed, with some caution, the proposals to generate a medical Peace Corps to provide services and training in areas of great need. Pitfalls to be avoided are discussed, and the need to integrate such efforts into each country's national manpower training program is stressed.

## Resources

The principles presented in this chapter are derived from the experience of the Fogarty International Center in developing research and training programs over the past two decades.

The background materials are derived from three comprehensive documents which detail the health challenges in the poor countries in far greater detail, and focus on the geopolitical and economic factors as well as the human resources that are each critical components of health planning. These documents are as follows:


1. World Health Organization (2006). *The World Health Report 2006, Working Together For Health*. Geneva: World Health Organization.
2. Joint Learning Initiative (JLI) (2004). *Human Resources For Health: Overcoming the Crisis*. Cambridge: Harvard University Press.
3. Committee on the Options for Overseas Placement of US Health Professionals, Mullan, F., Panosian, C. and Cuff, P. (eds) (2005). *Healers Abroad: Americans Responding to the Human Resource Crisis in HIV/AIDS*. Washington, DC: National Academies Press.

# Index

Printed and bound by CPI Group (UK) Ltd, Croydon, CR0 4YY
08/05/2025
01864892-0001